Comprehensive Study of Atomic and Molecular Physics

Comprehensive Study of Atomic and Molecular Physics

Edited by **Elisha Franks**

NY RESEARCH PRESS

New York

Published by NY Research Press,
23 West, 55th Street, Suite 816,
New York, NY 10019, USA
www.nyresearchpress.com

Comprehensive Study of Atomic and Molecular Physics
Edited by Elisha Franks

International Standard Book Number: 978-1-63238-087-6 (Hardback)

Printed in China.

Contents

Preface

The field of physics is a vast and detailed one that has many sub-divisions that branch off in diverse directions. There has always been a part of the physical world that escapes our eyes and cannot really be explained through simple terms. The components of matter are minuscule and need especially devoted arenas of research and study to be understood. Atomic and molecular study is an essential part of this world of physics. Atomic physics is an area of study that focuses on the study of atoms in the manner of an isolated system of nucleus and electrons. Primarily focused on the arrangement of electrons around the nucleus, it also studies how these arrangements change. Molecular Physics on the other hand is a field of study that is focused on the physical properties of molecules. It looks at molecules as well as molecular dynamics and bonds. Both fields could be said to be closely related, but they also overlap with physical chemistry, chemical physics and theoretical chemistry. Both atomic and molecular physics are essentially concerned with the electronic structure of atoms and molecules and the dynamic processes through which these structures arrange themselves. The rapidly advancing research techniques bode well for the future of atomic and molecular physics.

This book is an attempt to compile and collate all current and proposed research and data in the field of atomic and molecular physics. I am thankful to all those who's hard work and effort went into these studies. I wish to personally thank all the contributing authors who shared their knowledge in this book and with me throughout the editing process. It was an honour working with you all. I also wish to thank my family who have always been my support system.

Editor

1

Analysis of Water and Hydrogen Bond Dynamics at the Surface of an Antifreeze Protein

Yao Xu, Ramachandran Gnanasekaran, and David M. Leitner

Department of Chemistry and Chemical Physics Program, University of Nevada, Reno, NV 89557, USA

Correspondence should be addressed to David M. Leitner, dml@unr.edu

Academic Editor: Keli Han

We examine dynamics of water molecules and hydrogen bonds at the water-protein interface of the wild-type antifreeze protein from spruce budworm *Choristoneura fumiferana* and a mutant that is not antifreeze active by all-atom molecular dynamics simulations. Water dynamics in the hydration layer around the protein is analyzed by calculation of velocity autocorrelation functions and their power spectra, and hydrogen bond time correlation functions are calculated for hydrogen bonds between water molecules and the protein. Both water and hydrogen bond dynamics from subpicosecond to hundred picosecond time scales are sensitive to location on the protein surface and appear correlated with protein function. In particular, hydrogen bond lifetimes are longest for water molecules hydrogen bonded to the ice-binding plane of the wild type, whereas hydrogen bond lifetimes between water and protein atoms on all three planes are similar for the mutant.

1. Introduction

While the complex dynamics of large biological molecules and the connection to function have fascinated physical scientists for some time, in more recent years researchers have turned their attention to the interface of biomolecules with water. Coupling of protein and water dynamics, for example, has been examined by molecular simulations [1–10] and a growing number of experimental probes [11–14], and a wide variety of dynamical time scales have been found [15, 16] due to the heterogeneity of protein-water interactions. One class of proteins for which protein-water interactions are critical to function is antifreeze proteins (AFPs). AFPs are widely distributed in certain plants, vertebrates, fungi, and bacteria to provide cells protection in cold environments [17–20] but the mechanism for antifreeze activity is still not well understood. In this paper we analyze by all-atom molecular dynamics (MD) simulations the dynamics of water molecules and hydrogen bonds at the protein-water interface of the AFP from the spruce budworm *Choristoneura fumiferana* and a mutant that has little antifreeze activity. We calculate velocity autocorrelation functions and their power spectra for water molecules around the protein and we compute hydrogen bond time correlation functions for bonds between the protein and water. We obtain distinct spectra for the water around different regions of the protein, which are affected by mutation. Moreover, we observe longer hydrogen bonding between water molecules and the ice-binding plane of this AFP compared to other parts of the protein, a difference that nearly disappears with mutation, indicating a correlation between hydrogen bond lifetimes and activity of this AFP.

AFPs were first discovered in several Antarctic fish species [21], AFPs that have since been classified as Type I. The generally accepted mechanism for the Type I AFP is the adsorption-inhibition mechanism [22–24], which proposes that AFPs adsorb onto the preferred growth sites of an ice surface, thereby preventing new ice growth [25]. It was initially thought that ice and AFP interacted through hydrogen bonding [22]. However, when parts of the protein that were thought to facilitate this hydrogen bonding were mutated, the hypothesized decrease in antifreeze activity was not observed, and hydrophobic interactions were suggested instead [26]. MD simulations have been carried out to sort out the possible mechanisms [6], but there is still no consensus on which sites of the protein interact with ice,

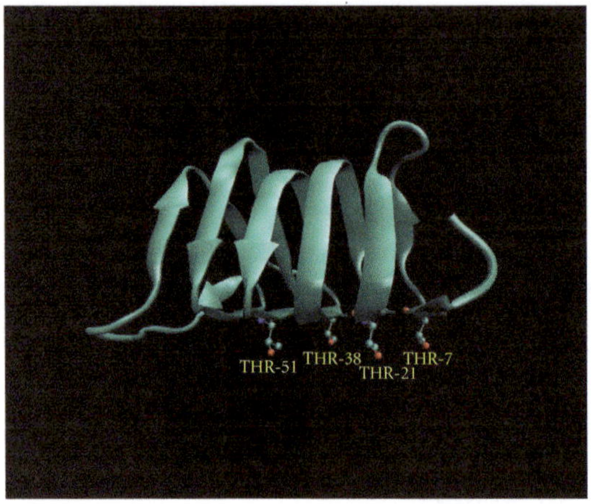

Figure 1: The structure of wild-type AFP from spruce budworm *Choristoneura fumiferana*, indicating the location of the four threonine residues on the ice-binding plane (Plane 1), which in our mutation studies we replace with leucine. Plane 2 is in the front and Plane 3 is in the back.

or whether the protein inhibits growth of ice locally at the protein-water interface or over a larger number of water layers near the protein. Recent THz studies [27] indicate that, at least for AFP in winter flounder, the effect appears to be delocalized.

The antifreeze activity of the AFP from the spruce budworm *Choristoneura fumiferana* [28], shown in Figure 1, can apparently be attributed in part to specific residues located on part of the surface of the protein. This protein is not a member of the Type I family. The protein structure contains three planes and mutation studies demonstrate that threonine-rich Plane 1 is the ice-binding plane. Mutation of just a few of the threonines to leucines (Figure 1) dramatically diminishes antifreeze activity [28]. Nutt and Smith [29] recently carried out MD simulations to examine the water dynamics in the hydration layer around the protein and found distinct dynamics around each of the three planes and noticeably slower dynamics around Plane 1. In this study we observe, like Nutt and Smith, quite distinct differences for the water dynamics around each of the three planes of the protein and in the hydrogen bond lifetimes for hydrogen bonds between the water molecules and the protein. Moreover, we examine a mutant that is antifreeze inactive and find that the mutation affects the hydrogen bond dynamics; that is, hydrogen bond lifetimes around the three planes are much closer to each other than in the wild type.

In the following section, we provide details of the computational methods and analysis. We then report results of our calculation of power spectra for water molecules near the three distinct planes of the protein and of our investigation of hydrogen bond lifetimes for bonds between water molecules and the protein. Concluding remarks are given in the final section.

2. Computational Methods

The initial coordinates of the antifreeze protein from the spruce budworm *Choristoneura fumiferana* were taken from the Protein Data Bank file 1L0S. Missing residues and hydrogen atoms were built into the structure and the iodated tyrosine Y26 required for the structure determination was reverted to a standard tyrosine using Swiss PDB Viewer [30]. For the mutant, four threonine residues on Plane 1 were mutated to leucines (Figure 1), a mutation that significantly reduces the antifreeze properties of the protein [17], to explore the effect of this mutation on the water dynamics and hydrogen bond lifetimes.

Both the wild-type and mutated structure were first minimized for 1000 steps with the steepest descent algorithm using the AMBER03 force field [31], after its solvation in a 70 Å cubic water box of TIP5P water model. Then the systems, each of which contained 10539 water molecules, were equilibrated for 400 ps. For the first 100 ps the positions of the proteins were restrained and in the latter 300 ps they were released. Constraints were applied to all bonds to hydrogen with the SHAKE algorithm and periodic boundary conditions were applied. All the classical MD simulations were performed on the systems in canonical (NVT) ensemble with the GROMACS software package [32]. Following equilibration, trajectories of 2 ns were obtained at 300 K with a Nose-Hoover thermostat [22, 23]. Nonbonded interactions were gradually brought to zero by a shift function for the electrostatics as well as a switch function for van der Waals interactions between 10 and 12 Å [24, 25]. All the simulations were performed by integrating Newton's equations of motion with the Verlet algorithm [26] using 1 fs time steps. The system coordinates and velocities were stored every 5 fs, and the velocity autocorrelation function (VACF) was averaged over 15 ps time segments of the trajectory for the oxygen atoms that survive in the first hydration shell of thickness 5 Å as well as for those that hydrogen bond to the proteins. (Criteria for hydrogen bonds are specified below.) The VACF is defined as

$$C_V(t) = \frac{\langle \mathbf{v}_i(t) \cdot \mathbf{v}_i(0) \rangle}{\langle \mathbf{v}_i(0) \cdot \mathbf{v}_i(0) \rangle}, \quad (1)$$

where $\mathbf{v}_i(t)$ is the velocity vector of the oxygen atom at time, t. The angular brackets denote averaging over all atoms of the particular type present in the hydration shell and over different reference initial times. Power spectra were obtained by Fourier transform of $C_V(t)$. The power spectra correspond to the vibrational density of the water. The vibrational density of protein molecules has been discussed elsewhere [11, 33–36].

Hydrogen bond time correlation functions, $C_{\mathrm{HB}}(t)$, were also computed for bonds between water molecules and the protein at 300 K. $C_{\mathrm{HB}}(t)$ is defined as the probability that, if a hydrogen bond between donor, D, and acceptor, A, exists at $t = 0$, then it still exists at time, t, even if the bond broke at some intermediate time [37]. We adopt a standard criterion for hydrogen bonds, that is, a DA distance of 3.5 Å and a D-H-A angle greater than 150° [1, 38, 39].

3. Results and Discussion

3.1. VACF Power Spectra. A protein molecule perturbs the regular water-water hydrogen bond network in bulk water

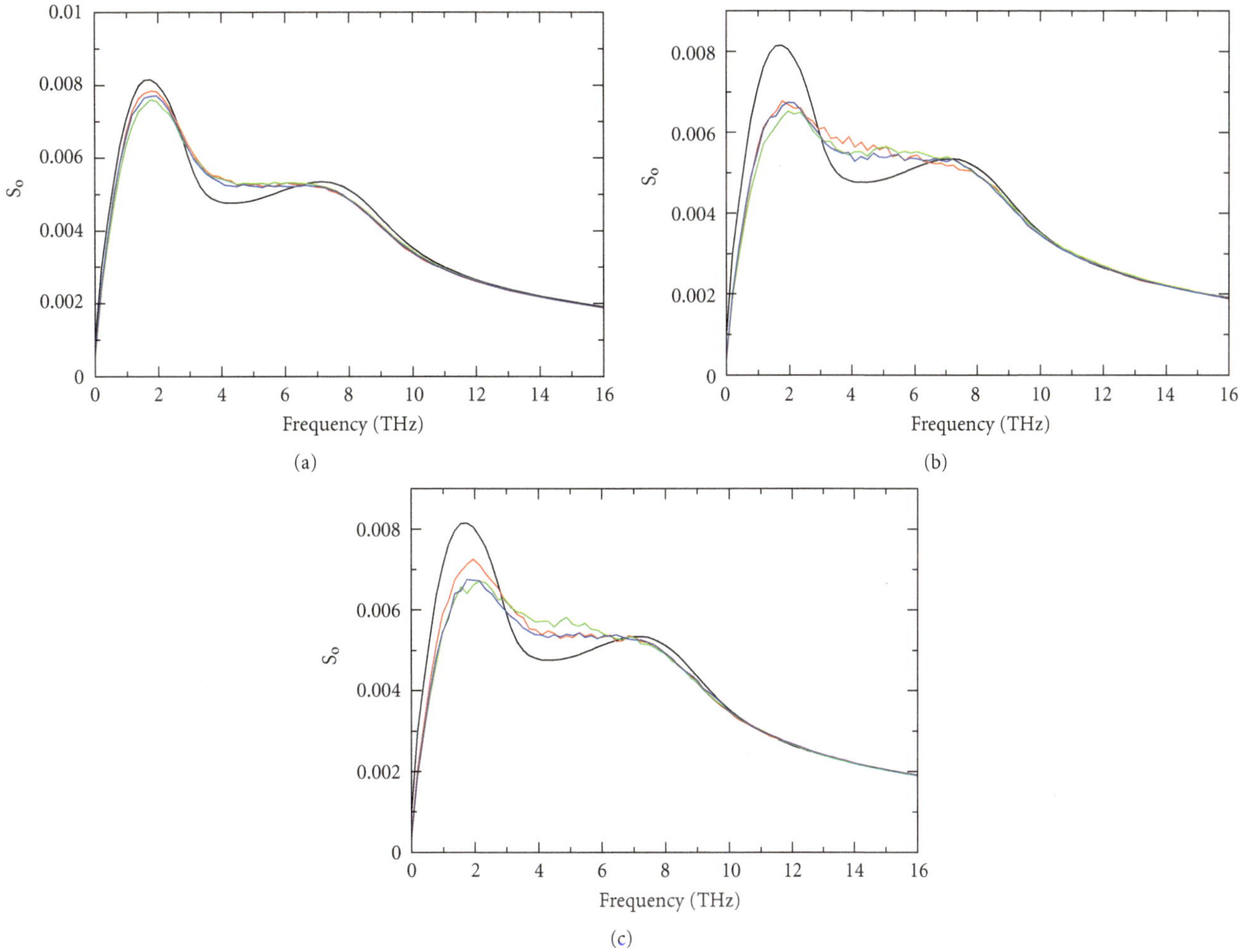

Figure 2: Power spectra of the velocity autocorrelation function of water in hydration layers around Plane 1 (red), Plane 2 (green), and Plane 3 (blue) of the protein, as well as for bulk water (black), at 300 K. (a) Power spectrum of water in the hydration layer, taken to be 5 Å from the surface of the wild-type AFP. (b) Power spectrum of water hydrogen bonding to wild type AFP. (c) Same as (b), but for mutant AFP.

with the formation of protein-water hydrogen bonds and influences the water dynamics in the hydration layer around the protein surface. We have calculated at 300 K the velocity autocorrelation function and its power spectra for the water molecules in the hydration layer around the protein, which can provide insights into THz spectra of solvated proteins [40]. We have carried out this calculation both for the water molecules that form hydrogen bonds with the amino acid residues of the three planes of the protein and for the larger number of water molecules within a layer of thickness 5 Å from the protein [41]. Power spectra are plotted in Figure 2 for the wild-type and mutant at 300 K. The results of a separate MD simulation of pure TIP5P water under the same conditions are also included for comparison.

We consider first the power spectra for bulk water, which appears in each of the panels in Figure 2. We observe two bands in the power spectra of water at about 2 and 8 THz. The lower frequency band has been interpreted [42, 43] as corresponding to the O· · · O· · · O bending mode from triplets of hydrogen-bonded water molecules and the higher frequency band as O· · · O stretching mode between pairs of hydrogen-bonded water molecules. Turning to the hydration water, the results plotted in Figure 2(a) reveal a clear blue shift in $S_O(\omega)$ for the band corresponding to the O· · · O· · · O bending for water. The shift is very similar for the water molecules in the 5 Å hydration layer around each of the three planes. A blue shift in the same spectral region has been observed for water molecules in the hydration layer around helices of the villin headpiece subdomain, HP-36 [44]. Figure 2(b) gives the result for the hydration layer around the wild-type AFP, and we observe similar results for the mutant (not shown). Overall, we find that for the water molecules in the 5 Å hydration layer around the protein there is little difference among the spectra obtained for the water near Planes 1, 2, or 3.

For the water molecules hydrogen bonded to the protein we observe distinct differences in the power spectra of the velocity autocorrelation function for each of the planes. The power spectra for the water hydrogen bonded to the protein exhibit again peaks near 2 and 8 THz, but the intensity of

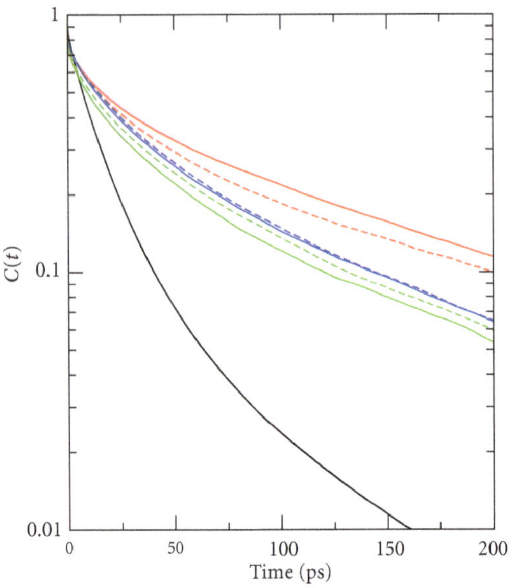

FIGURE 3: The hydrogen-bond time correlation function plotted for water hydrogen bonded to Plane 1 (red), Plane 2 (blue), and Plane 3 (green) of wild-type (solid) and mutant (dotted) AFP at 300 K. The result for hydrogen bonds between water molecules in the bulk is plotted (black) for comparison.

the 2 THz peak is smaller than for the hydration water, and the peak corresponding to the O· · · O· · · O bending appears even further broadened and blue shifted compared to bulk water than the peak for the hydration water in the 5 Å layer around the protein. This could be related to the more restricted dynamics of the water molecules hydrogen bonded to the protein. Indeed, we have computed the power spectrum for bulk water at 250 K and for water molecules in the hydration shell and found the first peak for bulk water at this lower temperature to have a smaller intensity, and similar to that for the hydration water [45]. For the wild type we observe that Plane 1 exhibits a greater intensity on the blue edge of the lower frequency band compared to the other planes, whereas for the mutant the intensity is also greater but on the red side of the band. Because the power spectra for the wild type and the mutant are distinct, we expect that differences in the THz spectra of the wild type and the mutant can be detected.

3.2. Hydrogen Bond Correlation Function. We plot in Figure 3 results for the hydrogen bond correlation function, $C_{\mathrm{HB}}(t)$, defined in Section 2, to times of 200 ps for hydrogen bonds between water molecules and protein atoms on Planes 1, 2, and 3 of the wild type and mutant, as well as between water molecules in the bulk for comparison. All the simulations were carried out at 300 K. Overall, the observed slow rearrangement times for hydrogen bonds between water molecules and the protein compared to hydrogen bonds between water molecules in the bulk are consistent with expectations for water molecules in the hydration layer around a protein [46, 47]. Nutt and Smith [29] computed the hydrogen bond correlation function for bonds between water and the three planes of the wild type, and we focus here mainly on comparison of the wild-type results with the results for the mutant. The hydrogen bond lifetime for bonds between water and the protein survive longer than hydrogen bonds between water molecules in the bulk, as seen in numerous previous simulation studies [1, 9, 38, 48, 49]. However, we also observe differences for hydrogen bonds between water and atoms on different planes of the protein. Considering first the wild type, we find, as did Nutt and Smith [29], that $C_{\mathrm{HB}}(t)$ for hydrogen bonds between water molecules and atoms of Plane 1 decays significantly slower than $C_{\mathrm{HB}}(t)$ for bonds between water molecules and the other two planes.

Interestingly, we find the hydrogen bond correlation functions for hydrogen bonds between water and the three protein planes to be noticeably closer to each other for the mutant than for the wild type out to the 100 picosecond time scale. The antifreeze activity of the protein decreases dramatically when replacing four of the threonines on Plane 1, indicated in Figure 1, with leucines [28] and we observe in our MD simulations that the hydrogen bond lifetimes for bonds between water molecules and atoms of each plane become similar to one another with this mutation. Only four point mutations have a sizable effect on the hydrogen bond dynamics, indicating the effect may not simply be local, but may influence the orientation of many water molecules. Such a nonlocal effect on the orientation of hydration waters by point mutation has been illustrated recently for a simple protein-sized model system [50].

4. Concluding Remarks

In this work, we examined the power spectrum of the velocity autocorrelation function for water molecules near the surface of the antifreeze protein (AFP) from the spruce budworm *Choristoneura fumiferana* and analyzed the hydrogen bond lifetimes for bonds between water molecules and the protein. We explored effects of the heterogeneity of the protein surface, in particular the distinctive properties of the water and protein-water interactions on the three planes of the protein, one of which is vital to the function of this AFP, and how the dynamics is affected by mutation.

For the power spectra of the water in the hydration layer of the AFP and the subset of that water that hydrogen bonds to the protein we find a blue shift of the roughly 2 THz band compared to the same band in bulk water with a more pronounced shift for the water molecules that are hydrogen bonded to the protein residues. Although the power spectra for the water molecules within 5 Å of each of the planes of the protein appear quite similar, power spectra for the water molecules hydrogen bonded to different planes of the protein exhibit distinct spectra in the range 1–4 THz. The differences among the power spectra for the water molecules hydrogen bonding to each of the three planes are influenced by mutation. We expect that THz measurements, which are highly sensitive to the hydration water [48, 51–55], will reveal differences between the wild type and mutant. Recent THz experiments [49] on a λ-repressor fragment indicate that only a few point mutations can give rise to very different THz spectra.

The hydrogen bond time correlation function was computed for hydrogen bonds between water molecules and each of the planes of the protein. For wild type AFP we observe differences in the hydrogen bond lifetimes for bonds between water and the three planes. The longest lifetimes are found for hydrogen bonds between water molecules and Plane 1, the ice-binding plane of the protein, consistent with results of previous simulations [29]. We observe that by introducing only four mutations to Plane 1, mutations that have been observed to substantially diminish the AFP activity of the protein [28], the hydrogen bond correlation function for bonds between water molecules and each of the three planes are similar to one another. Overall, mutation is seen to modify hydrogen bonding over a wide range of time scales observable both in the power spectra and analysis of hydrogen bond lifetimes. These measures of hydrogen bonding at the protein-water interface aid in quantifying the complexity and heterogeneity of the interactions between water and the antifreeze protein and reveal regions of the protein-water interface important for antifreeze activity.

Acknowledgments

Support from the National Science Foundation (NSF CHE-0910669) and from the Volkswagen Foundation (VW Stiftung Az I/84 302) is gratefully acknowledged.

References

[1] D. J. Tobias, N. Sengupta, and M. Tarek, "Molecular dynamics simulation studies of coupled protein and water dynamics," in *Proteins: Energy, Heat and Signal Flow*, D. M. Leitner and J. E. Straub, Eds., pp. 361–386, Taylor & Francis, Boca Raton, Fla, USA, 2009.

[2] M. E. Johnson, C. Malardier-Jugroot, R. K. Murarka, and T. Head-Gordon, "Hydration water dynamics near biological interfaces," *Journal of Physical Chemistry B*, vol. 113, no. 13, pp. 4082–4092, 2009.

[3] A. R. Bizzarri and S. Cannistraro, "Molecular dynamics of water at the protein-solvent interface," *Journal of Physical Chemistry B*, vol. 106, no. 26, pp. 6617–6633, 2002.

[4] P. J. Steinbach and B. R. Brooks, "Protein hydration elucidated by molecular dynamics simulation," *Proceedings of the National Academy of Sciences of the United States of America*, vol. 90, no. 19, pp. 9135–9139, 1993.

[5] D. N. LeBard and D. V. Matyushov, "Ferroelectric hydration shells around proteins: electrostatics of the protein-water interface," *Journal of Physical Chemistry B*, vol. 114, no. 28, pp. 9246–9258, 2010.

[6] X. Yu, J. Park, and D. M. Leitner, "Thermodynamics of protein hydration computed by molecular dynamics and normal modes," *Journal of Physical Chemistry B*, vol. 107, no. 46, pp. 12820–12828, 2003.

[7] F. Despa, A. Fernández, and R. S. Berry, "Publisher's note—dielectric modulation of biological water," *Physical Review Letters*, vol. 93, no. 26, Article ID 228104, 1 pages, 2004.

[8] R. Gnanasekaran, J. K. Agbo, and D. M. Leitner, "Communication maps computed for homodimeric hemoglobin: computational study of water-mediated energy transport in proteins," *Journal of Chemical Physics*, vol. 135, no. 6, Article ID 065103, 10 pages, 2011.

[9] R. Gnanasekaran, Y. Xu, and D. M. Leitner, "Dynamics of water clusters confined in proteins: a molecular dynamics simulation study of interfacial waters in a dimeric hemoglobin," *Journal of Physical Chemistry B*, vol. 114, no. 50, pp. 16989–16996, 2010.

[10] A. Lervik, F. Bresme, S. Kjelstrup, D. Bedeaux, and J. M. Rubi, "Heat transfer in protein-water interfaces," *Physical Chemistry Chemical Physics*, vol. 12, no. 7, pp. 1610–1617, 2010.

[11] D. M. Leitner, M. Havenith, and M. Gruebele, "Biomolecule large-amplitude motion and solvation dynamics: modelling and probes from THz to X-rays," *International Reviews in Physical Chemistry*, vol. 25, no. 4, pp. 553–582, 2006.

[12] L. Mitra, N. Smolin, R. Ravindra, C. Royer, and R. Winter, "Pressure perturbation calorimetric studies of the solvation properties and the thermal unfolding of proteins in solution—experiments and theoretical interpretation," *Physical Chemistry Chemical Physics*, vol. 8, no. 11, pp. 1249–1265, 2006.

[13] S. K. Pal, J. Peon, and A. H. Zewail, "Biological water at the protein surface: dynamical solvation probed directly with femtosecond resolution," *Proceedings of the National Academy of Sciences of the United States of America*, vol. 99, no. 4, pp. 1763–1768, 2002.

[14] W. Doster and M. Settles, "The dynamical transition in proteins: the role of hydrogen bonds," in *Hydration Processes in Biology: Experimental and Theoretical Approaches*, M.-C. Bellissent-Funel, Ed., pp. 177–195, IOS Press, Amsterdam, The Netherlands, 1999.

[15] E. Persson and B. Halle, "Cell water dynamics on multiple time scales," *Proceedings of the National Academy of Sciences of the United States of America*, vol. 105, no. 17, pp. 6266–6271, 2008.

[16] H. Frauenfelder, P. W. Fenimore, G. Chen, and B. H. McMahon, "Protein folding is slaved to solvent motions," *Proceedings of the National Academy of Sciences of the United States of America*, vol. 103, no. 42, pp. 15469–15472, 2006.

[17] S. P. Graether, *Biochemistry and Function of Antifreeze Proteins*, Nova Science, New York, NY, USA, 2011.

[18] J. G. Duman, K. R. Walters, T. Sformo et al., "Antifreeze and ice-nucleator proteins," in *Low Temperature Biology of Insects*, D. L. Delinger and R. E. Lee, Eds., pp. 59–90, Cambridge University Press, New York, NY, USA, 2010.

[19] B. Moffatt, V. Ewart, and A. Eastman, "Cold comfort: plant antifreeze proteins," *Physiologia Plantarum*, vol. 126, no. 1, pp. 5–16, 2006.

[20] L. Pham, R. Dahiya, and B. Rubinsky, "An *in vivo* study of antifreeze protein adjuvant cryosurgery," *Cryobiology*, vol. 38, no. 2, pp. 169–175, 1999.

[21] A. L. DeVries and D. E. Wohlschlag, "Freezing resistance in some antarctic fishes," *Science*, vol. 163, no. 3871, pp. 1073–1075, 1969.

[22] J. A. Raymond and A. L. DeVries, "Adsorption inhibition as a mechanism of freezing resistance in polar fishes," *Proceedings of the National Academy of Sciences of the United States of America*, vol. 74, no. 6, pp. 2589–2593, 1977.

[23] J. A. Raymond, P. W. Wilson, and A. L. DeVries, "Inhibition of growth of nonbasal planes in ice by fish antifreezes," *Proceedings of the National Academy of Sciences of the United States of America*, vol. 86, no. 3, pp. 881–885, 1989.

[24] C. A. Knight, C. C. Cheng, and A. L. DeVries, "Adsorption of α-helical antifreeze peptides on specific ice crystal surface planes," *Biophysical Journal*, vol. 59, no. 2, pp. 409–418, 1991.

[25] J. Duman and A. L. DeVries, "Isolation, characterization, and physical properties of protein antifreezes from the winter flounder, pseudopleuronectes americanus," *Comparative Biochemistry and Physiology*, vol. 54, no. 3, pp. 375–380, 1976.

[26] A. D. Haymet, L. G. Ward, M. M. Harding, and C. A. Knight, "Valine substituted winter flounder "antifreeze": preservation of ice growth hysteresis," *FEBS Letters*, vol. 430, no. 3, pp. 301–306, 1998.

[27] S. Ebbinghaus, K. Meister, B. Born, A. L. Devries, M. Gruebele, and M. Havenith, "Antifreeze glycoprotein activity correlates with long-range protein-water dynamics," *Journal of the American Chemical Society*, vol. 132, no. 35, pp. 12210–12211, 2010.

[28] S. P. Graether, M. J. Kuiper, and S. M. Gagne, "Beta-helix structure and ice-binding properties of a hyperactive antifreeze protein from an insect," *Nature*, pp. 325–328, 2000.

[29] D. R. Nutt and J. C. Smith, "Dual function of the hydration layer around an antifreeze protein revealed by atomistic molecular dynamics simulations," *Journal of the American Chemical Society*, vol. 130, no. 39, pp. 13066–13073, 2008.

[30] N. Guex and M. C. Peitsch, "SWISS-MODEL and the Swiss-PdbViewer: an environment for comparative protein modeling," *Electrophoresis*, vol. 18, no. 15, pp. 2714–2723, 1997.

[31] Y. Duan, C. Wu, S. Chowdhury et al., "A point-charge force field for molecular mechanics simulations of proteins based on condensed-phase quantum mechanical calculations," *Journal of Computational Chemistry*, vol. 24, no. 16, pp. 1999–2012, 2003.

[32] H. J. C. Berendsen, D. Spoel, and R. V. Drunen, "GROMACS: a message-passing parallel molecular dynamics implementation," *Computer Physics Communications*, vol. 91, no. 1–3, pp. 43–56, 1995.

[33] X. Yu and D. M. Leitner, "Vibrational energy transfer and heat conduction in a protein," *Journal of Physical Chemistry B*, vol. 107, no. 7, pp. 1698–1707, 2003.

[34] X. Yu and D. M. Leitner, "Anomalous diffusion of vibrational energy in proteins," *Journal of Chemical Physics*, vol. 119, no. 23, pp. 12673–12679, 2003.

[35] X. Yu and D. M. Leitner, "Heat flow in proteins: computation of thermal transport coefficients," *Journal of Chemical Physics*, vol. 122, no. 5, Article ID 054902, 11 pages, 2005.

[36] D. M. Leitner, "Vibrational energy transfer and heat conduction in a one-dimensional glass," *Physical Review B*, vol. 64, no. 9, Article ID 094201, 9 pages, 2001.

[37] B. Bagchi, "Water dynamics in the hydration layer around proteins and micelles," *Chemical Reviews*, vol. 105, no. 9, pp. 3197–3219, 2005.

[38] M. Tarek and D. J. Tobias, "Role of protein-water hydrogen bond dynamics in the protein dynamical transition," *Physical Review Letters*, vol. 88, no. 13, Article ID 138101, 4 pages, 2002.

[39] A. Luzar and D. Chandler, "Hydrogen-bond kinetics in liquid water," *Nature*, vol. 379, no. 6560, pp. 55–57, 1996.

[40] M. Heyden and M. Havenith, "Combining THz spectroscopy and MD simulations to study protein-hydration coupling," *Methods*, vol. 52, no. 1, pp. 74–83, 2010.

[41] S. Bandyopadhyay, S. Chakraborty, and B. Bagchi, "Secondary structure sensitivity of hydrogen bond lifetime dynamics in the protein hydration layer," *Journal of the American Chemical Society*, vol. 127, no. 47, pp. 16660–16667, 2005.

[42] G. E. Walrafen and Y. C. Chu, "Linearity between structural correlation length and correlated-proton Raman intensity from amorphous ice and supercooled water up to dense supercritical steam," *Journal of Physical Chemistry*, vol. 99, no. 28, pp. 11225–11229, 1995.

[43] G. E. Walrafen, Y. C. Chu, and G. J. Piermarini, "Low-frequency Raman scattering from water at high pressures and high temperatures," *Journal of Physical Chemistry*, vol. 100, no. 24, pp. 10363–10372, 1996.

[44] S. Chakraborty, S. K. Sinha, and S. Bandyopadhyay, "Low-frequency vibrational spectrum of water in the hydration layer of a protein: a molecular dynamics simulation study," *Journal of Physical Chemistry B*, vol. 111, no. 48, pp. 13626–13631, 2007.

[45] Y. Xu, R. Gnanasekaran, and D. M. Leitner, (published results).

[46] N. Nandi and B. Bagchi, "Dielectric relaxation of biological water," *Journal of Physical Chemistry B*, vol. 101, no. 50, pp. 10954–10961, 1997.

[47] S. K. Pal, J. Peon, B. Bagchi, and A. H. Zewail, "Biological water: femtosecond dynamics of macromolecular hydration," *Journal of Physical Chemistry B*, vol. 106, no. 48, pp. 12376–12395, 2002.

[48] S. Ebbinghaus, S. J. Kim, M. Heyden et al., "An extended dynamical hydration shell around proteins," *Proceedings of the National Academy of Sciences of the United States of America*, vol. 104, no. 52, pp. 20749–20752, 2007.

[49] S. Ebbinghaus, S. J. Kim, M. Heyden et al., "Protein sequence- and pH-dependent hydration probed by terahertz spectroscopy," *Journal of the American Chemical Society*, vol. 130, no. 8, pp. 2374–2375, 2008.

[50] A. D. Friesen and D. V. Matyushov, "Non-Gaussian statistics of electrostatic fluctuations of hydration shells," *Journal of Chemical Physics*, vol. 135, no. 10, Article ID 104501, 7 pages, 2011.

[51] B. Born, S. J. Kim, S. Ebbinghaus, M. Gruebele, and M. Havenith, "The terahertz dance of water with the proteins: the effect of protein flexibility on the dynamical hydration shell of ubiquitin," *Faraday Discussions*, vol. 141, pp. 161–173, 2008.

[52] U. Heugen, G. Schwaab, E. Bründermann et al., "Solute-induced retardation of water dynamics probed directly by terahertz spectroscopy," *Proceedings of the National Academy of Sciences of the United States of America*, vol. 103, no. 33, pp. 12301–12306, 2006.

[53] M. Heyden, E. Bründermann, U. Heugen, G. Niehues, D. M. Leitner, and M. Havenith, "Long-range influence of carbohydrates on the solvation dynamics of water—answers from terahertz absorption measurements and molecular modeling simulations," *Journal of the American Chemical Society*, vol. 130, no. 17, pp. 5773–5779, 2008.

[54] J. Knab, J. Y. Chen, and A. G. Markelz, "Hydration dependence of conformational dielectric relaxation of lysozyme," *Biophysical Journal*, vol. 90, no. 7, pp. 2576–2581, 2006.

[55] N. Q. Vinh, S. J. Allen, and K. W. Plaxco, "Dielectric spectroscopy of proteins as a quantitative experimental test of computational models of their low-frequency harmonic motions," *Journal of the American Chemical Society*, vol. 133, no. 23, pp. 8942–8947, 2011.

2

Temperature and H/D Isotopic Effects in the IR Spectra of the Hydrogen Bond in Solid-State *2-Furanacetic Acid* and *2-Furanacrylic Acid*

Henryk T. Flakus and Anna Jarczyk-Jędryka

Institute of Chemistry, University of Silesia, 9 Szkolna Street, 40-006 Katowice, Poland

Correspondence should be addressed to Henryk T. Flakus, flakus@ich.us.edu.pl

Academic Editor: Joanna Sadlej

Polarized IR spectra of *2-furanacetic acid* and of *2-furanacrylic acid* crystals were measured at 293 K and 77 K in the ν_{O-H} and ν_{O-H} band frequency ranges. The corresponding spectra of the two individual systems strongly differ, one from the other, by the corresponding band shapes as well as by the temperature effect characterizing the bands. The crystal spectral properties remain in a close relation with the electronic structure of the two different molecular systems. We show that a vibronic coupling mechanism involving the hydrogen bond protons and the electrons on the π-electronic systems in the molecules determines the way in which the vibrational exciton coupling between the hydrogen bonds in the carboxylic acid dimers occurs. A strong coupling in *2-furanacrylic acid* dimers prefers a "*tail-to-head-*" type *Davydov coupling* widespread by the π-electrons. A weak *through-space* coupling in *2-furanacetic acid* dimers is responsible for a "*side-to-side-*" type coupling. The relative contribution of each exciton coupling mechanism in the dimer spectra generation is temperature and the molecular electronic structure dependent. This explains the observed difference in the temperature-induced evolution of the compared spectra.

1. Introduction

Infrared spectroscopy still constitutes a basic tool in the research of the hydrogen bond dynamics. The ν_{X-H} bands measured in the highest frequency range of the mid-infrared attributed to the proton stretching vibrations in X–H$\cdots$Y hydrogen bonds are the source of wealth data system in this matter. Complex fine structure patterns of these bands are considered as the result of anharmonical coupling mechanisms involving the proton stretching vibrations and other normal vibrations occurring in associated molecular systems, mainly the low-frequency X$\cdots$Y hydrogen bridge stretching vibrational motions [1–5]. The band contour shapes are extremely susceptible on the influences exerted by diverse physical factors, such as changes of temperature, changes in the matter state of condensation, pressure, and solvents [1–5].

Among the contemporary theories of the IR spectra of the hydrogen bond, formed in molecular systems, quantitative theoretical models elaborated for the description of the ν_{X-H} band generation mechanisms are of the particular importance. There are two most advanced quantitative theoretical models, namely, the "*strong-coupling*" theory [6–8] (the elder theory) and the "*relaxation*" (*linear response*) theory, the novel model [9, 10]. Both models are of a purely vibrational nature. Over the last four decades, by using of these theories, IR spectra of diverse hydrogen bond systems have been reproduced satisfactorily. The model calculations concerned quantitative interpretation of spectra of single, isolated hydrogen bonds [7, 11], spectra of cyclic dimeric hydrogen bond systems [7, 12–14], and the IR spectra of hydrogen-bonded molecular crystals [15]. Simultaneously, the H/D isotopic effects observed in the spectra of the deuterium-bonded corresponding systems have been interpreted [7–15].

Nevertheless, despite the doubtless successes achieved in this area, when interpreting the hydrogen bond system spectra, it seems that a number of basic theoretical problems still remain unsolved. It also seems that the main source in the understanding of many spectral phenomena characterizing systems consisting with a number of mutually coupled hydrogen bonds, in terms of the two different quantitative approaches, is in the early history of these studies. In practice, up to the beginning of the 90s of the 20th century, these studies were restricted to the interpretation of spectra of a number of very simple hydrogen bond systems, mainly to the spectra of cyclic acetic acid dimers formed in the gaseous phase [7, 12–14]. The extension of this research over other, more diversified and complex hydrogen bond aggregates allowed us to recognize numerous puzzling spectral effects attributed to these systems. Interpretation of these effects seemed to be beyond the contemporary quantitative theoretical models of the hydrogen bond IR spectra without assuming that some not revealed yet mechanisms codecide in the spectra generation.

For the last decade, spectroscopy in polarized light of hydrogen-bonded molecular crystals has provided key experimental data in this area. By measuring of polarized IR spectra of spatially oriented molecular crystals, characterized by a rich diversity of hydrogen bond arrangements met in their lattices, the most complete information has been be obtained about the coupling mechanisms involving hydrogen bonds in these systems. It appeared that the investigation of spectra of even so simple mutually interacting hydrogen bond aggregates like cyclic dimers (e.g., carboxylic acid dimers) allowed to reveal new H/D isotopic effects, namely, the H/D isotopic *self-organization* effects. They depend on a nonrandom distribution of protons and deuterons in the crystal lattices of isotopically diluted hydrogen bond systems. These spectral effects may be considered as the manifestation of a new kind of *cooperative interactions* involving hydrogen bonds, that is, the so-called *dynamical cooperative interactions* [16–18]. This revealing has emphasized the role of the vibronic coupling between the electronic and the proton vibrational motions taking place in hydrogen bond aggregates, in the generation of the very nature of the hydrogen bond as the natural phenomenon and in the interhydrogen bond interaction mechanisms [17, 18].

In the lattices of carboxylic acid crystals, centrosymmetric hydrogen bond dimers, present in the $(COOH)_2$ cycles, are frequently met [19, 20]. These dimers are the bearers of the main crystal spectral properties in the frequency ranges of the ν_{O-H} bands attributed to the proton stretching vibrations. One might expect that regardless of the molecular structure of carboxylic acids in their fragments placed outside the carboxyl groups, the ν_{O-H} band contour shapes should be fairly similar one to the other. This presumption is based on the considerations of the classic vibrational analysis, which predicted that the proton stretching vibrations in these molecules practically do not mix with vibrations of other atomic groups [21]. The experiment learns, however, that spectra of diverse carboxylic acid crystals considerably differ, one from the other, with regard to their ν_{O-H} band contour shapes as well as with regard to the temperature effects measured in the spectra. Qualitatively similar conclusion is valid for the ν_{O-D} bands in the spectra of the deuterium-bonded species [22–27]. Our hitherto estimations, resulting from the comparison of the IR crystalline spectra of diverse carboxylic acid molecular systems, ascribe the differences between the compared spectra in relation to the differences in the electronic structure of carboxylic acid molecules. For instance, π-electronic systems of aromatic rings or other larger conjugated π-electronic systems, linked directly to carboxyl groups, strongly change the basic spectral properties of carboxylic acid dimers in comparison with the analogous properties of aliphatic carboxylic acids [22–27]. The generation mechanism of these effects still remains unknown.

This paper deals with IR spectra of the hydrogen bond in crystals of two different carboxylic acids, namely, of *2-furanacetic acid* and *2-furanacrylic acid*. In these crystalline systems, associated molecules form hydrogen-bonded cyclic, centrosymmetric dimers (Complete crystallographic data for *2-furanacetic acid* and (excluding structure factors) have been deposited at the Cambridge Crystallographic Data Centre under the number CCDC-885823. Copies can be obtained free of charge from CCDC, 12 Union Road, Cambridge CB2 1EZ, U.K. (Fax: Int.+1223-336-033; e-mail: deposit@ccdc.cam.ac.uk)). The crystallographic data for 2-furanacrylic acid can be found in [28, 29]. Molecules of these two individual molecular systems differ, one from the other, by their electronic structures. In the latter case, the carboxyl groups are directly linked to the large π-electronic systems. In the *2-furanacetic acid* crystal case, *methylene* groups separate the hydrogen bonds, formed by the associated carboxyl groups, from the π-electronic system of *furan* rings.

The aim of the study reported in this paper was to provide new arguments of experimental nature about the role of the electronic structures of carboxylic acid molecules in the generation of IR spectra of cyclic hydrogen bond dimers. The investigation results presented constitute a part of results obtained in the frames of a wider project, which also assumed measuring of crystalline spectra of other *carboxylic acids*, mainly of *furan* and *thiophene* derivatives. Our choice of these model molecular systems was strongly supported by advantageous well-developed ν_{O-H} and ν_{O-D} band contour shapes in the IR spectra of these systems. We expected that the quantitative analysis of the polarized IR spectra of *2-furanacetic acid* and *2-furanacrylic acid* crystals and also of the spectra of relative carboxylic acid crystals should provide new arguments for the formulation of a new theoretical approach for the description of the hydrogen bond dimer spectra. The understanding of the temperature effects and the generation mechanism of the intensity distribution patterns in the ν_{O-H} and ν_{O-D} bands in the spectra of diverse carboxylic acid crystals are of the particular interest and importance in this project.

2. X-Ray Structures of *2-Furanacetic Acid* and *2-Furanacrylic Acid*

Crystals of *2-furanacetic acid* are monoclinic and the space-symmetry group is $P2_1/c$, $Z = 4$. The lattice constants at 100 K: $a = 13.0525(4)$ Å; $b = 4.85360(10)$ Å; $c = 9.4107(3)$ Å, $\beta = 103.832(3)°$. In a unit cell four translationally nonequivalent molecules form two plain centrosymmetric cyclic hydrogen-bonded dimers (Complete crystallographic data for *2-furanacetic acid* (excluding structure factors) have been

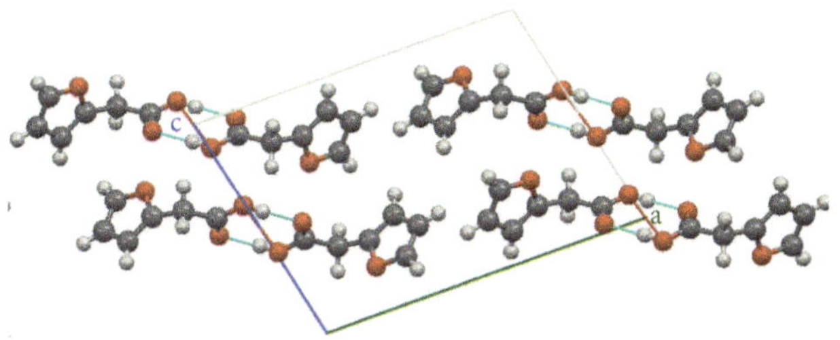

FIGURE 1: The X-ray structure of *2-furanacetic acid* crystal. Projection of the lattice onto the "*ac*" plane.

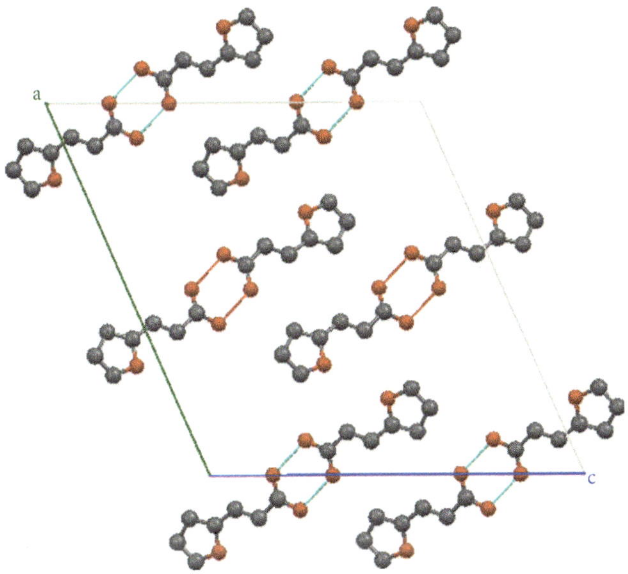

FIGURE 2: The X-ray structure of *2-furanacrylic acid* crystal. Projection onto the "*ac*" plane.

deposited at the Cambridge Crystallographic Data Centre under the number CCDC-885823). The molecules of *2-furanacetic acid* in the lattice are linked together by the O–H$\cdots$O hydrogen bonds, forming centrosymmetric dimers. A view of the crystal lattice of *2-furanacetic acid* is shown in Figure 1.

Crystals of *2-furanacrylic acid* are also monoclinic, the space-symmetry group is C2/c and Z = 8. The unit cell parameters are a= 18.975 Å; b = 3.843 Å; c = 20.132 Å, β = 113.9°. The molecules of *2-furanacrylic acid* in the lattice are linked together by the O–H$\cdots$O hydrogen bonds, forming cyclic approximately centrosymmetric dimers [28, 29]. The X-ray structure of *2-furanacrylic acid* crystals is shown in Figure 2.

3. Experimental

2-Furanacetic acid (C_4H_3O–CH_2–COOH) and *2-furanacrylic acid* (C_4H_3O–CH=CH–COOH) used for our studies were the commercial substance (Sigma-Aldrich). *2-furanacetic acid* was employed without further purification, while *2-furanacetic acid* was purified by crystallization from its acetone solution. The d_1 deuterium derivatives of the compounds (C_4H_3O–CH_2–COOD and C_4H_3O–CH=CH–COOD) were obtained by evaporation of D_2O solution of each compound at room temperature and under reduced pressure. It was found that the deuterium exchange rate for the COOH groups varied from 60 to 90% and from 70 to 90% for different samples, respectively.

Crystals suitable for further spectral studies were obtained by melting solid samples between two closely compressed spaces CaF_2 windows, followed by a very slow cooling of the liquid film. By that means, reasonably thin crystals could be received, characterized by their maximum absorbance at the ν_{O-H} band frequency range near to 0.5 at room temperature. From the crystalline mosaic, adequate monocrystalline fragments, having dimensions of at least 2×2 mm, were selected and then spatially oriented with the help of a polarization microscope. It was found that in each system case the crystals most frequently developed the "*ac*" crystalline face. These crystals were selected to the experiment by use of a thin, tin plate diaphragm with a 1.5 mm diameter hole, and then IR spectra of these crystalline fragments were measured by a transmission method. Spectral experiments were accomplished at room temperature and also at the temperature of liquid nitrogen, using polarized IR radiation. In each measurement, two different, mutually perpendicular orientations of the incident beam electric field vector "*E*" were applied, with respect to the developed face of the crystal lattice. The solid-state polarized spectra were measured with a resolution of 2 cm^{-1}, for the normal incidence of the IR radiation beam with respect to the crystalline face. The IR spectra were measured with the Nicolet Magna 560 FT-IR spectrometer. Measurements of the spectra were repeated for ca. 8 crystals of each isotopomer of an individual compound. Spectra were recorded in a similar manner for the deuterium derivatives.

The Raman spectra of polycrystalline samples of *2-furanacetic acid* and 2-furanacrylic acid were measured at room temperature with the use of the Bio-Rad FTS-175C FT-IR spectrometer at the 1 cm^{-1} resolution.

4. Results

The preliminary experimental studies of spectral properties of *2-furanacetic acid* and *2-furanacrylic acid* based on the measurements in CCl_4 solution in the frequency range of the ν_{O-H} proton stretching vibration bands. The results are shown in Figure 3.

In Figure 4 are shown the ν_{O-H} bands from the IR spectra of the polycrystalline acid samples in KBr pellets, measured at 298 K and 77 K, and in Figure 5 the ν_{O-D} bands spectra of the deuterium derivatives samples in the same conditions. The comparatively wealth spectrum of ν_{O-H} and ν_{O-D} bands for *2-furanacrylic acid* molecules may be predictable, based on earlier results for *cinnamic acid* crystals [24], while the ν_{O-H} and ν_{O-D} bands for *2-furanacetic acid* crystals are relatively poorer, similarly as in the *phenylacetic acid* crystal case [25].

Polarized IR spectra of the two crystalline systems measured at the room temperature in the ν_{O-H} band frequency

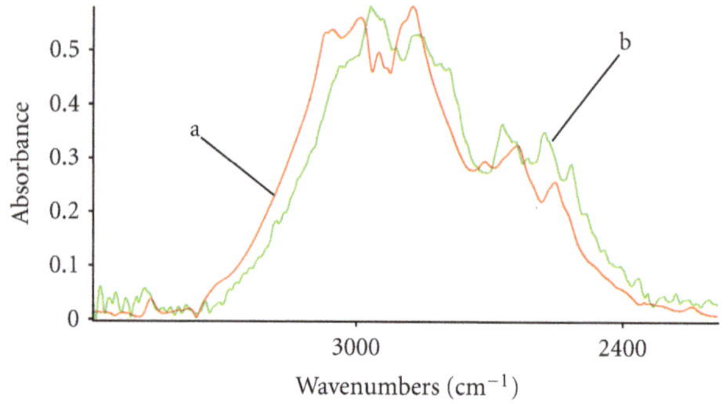

FIGURE 3: The ν_{O-H} band in the IR spectra of (a) *2-furanacetic acid* and (b) 2-furanacrylic acid in CCl_4 solution.

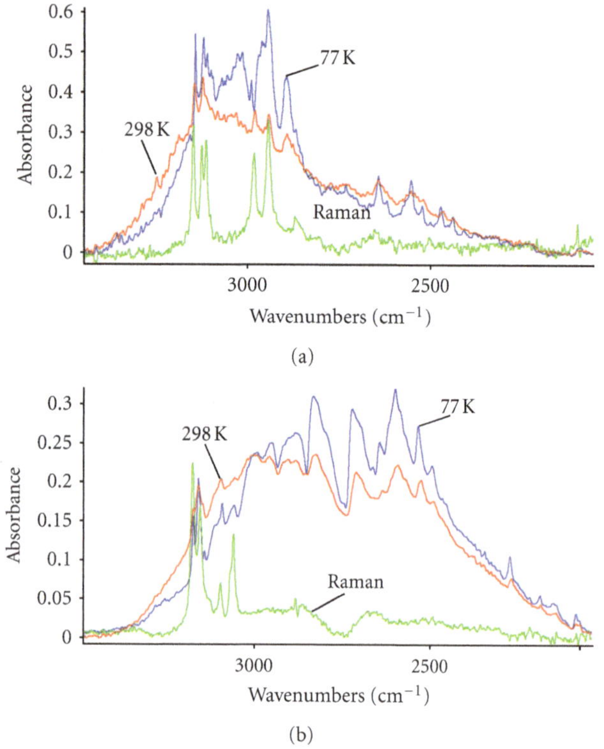

FIGURE 4: The ν_{O-H} bands in the IR spectra of polycrystalline samples of (a) *2-furanacetic acid* and (b) 2-furanacrylic acid, dispersed in KBr pellets. Temperature effect in the spectra. The Raman spectra measured for polycrystalline samples of the compounds at room temperature are also shown.

range are presented in Figure 6, whereas the corresponding low-temperature spectra are shown in Figure 7.

The corresponding spectra of isotopically diluted crystals recorded in the ν_{O-D} band range are shown in Figures 8 and 9.

The temperature effect in the crystalline spectra in the most intense polarized components of the ν_{O-H} bands is shown in Figure 10 and in the ν_{O-D} bands is given in Figure 11.

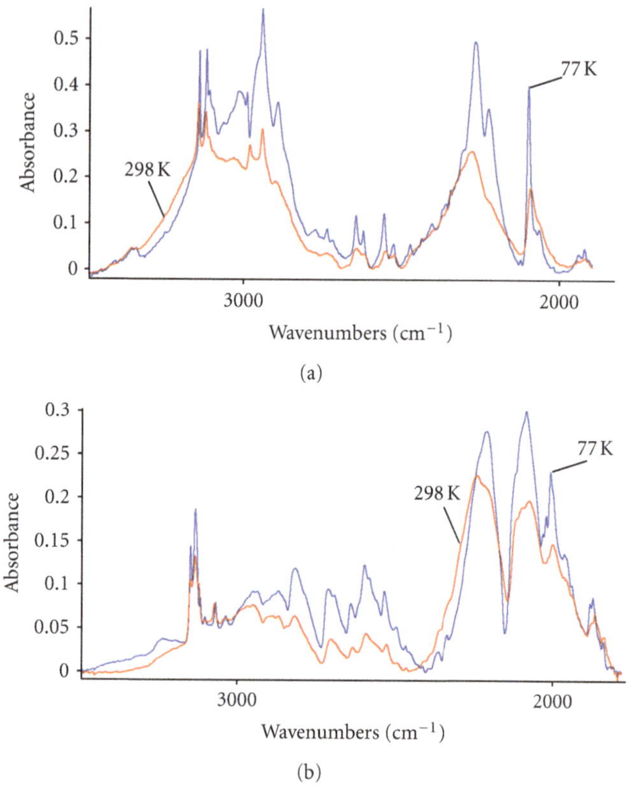

FIGURE 5: The ν_{O-D} bands in the IR spectra of polycrystalline samples of (a) d_1*-2-furanacetic acid* (ca. 45% H and 55% D) and (b) d_1*-2-furanacrylic acid* (ca. 20% H and 80% D) dispersed in KBr pellets. Temperature effect in the spectra.

5. Isotopic Dilution Effects in the Crystalline IR Spectra

On comparing the spectra in Figures 3 and 6–9, it can be noticed that the replacement of the major part of the hydrogen bond protons by deuterons changed the dichroic properties in the "*residual*" ν_{O-H} band substantially. The band shapes no longer depended on the crystal orientation investigated and resembled the spectrum measured for the CCl_4 solution of the compounds. Regardless of the increase in the rates of deuterium substitution in the samples, the "*residual*" ν_{O-H} band still retained its "*dimeric*" character. This is due to the fact that the hydrogen-bonded dimeric spectrum measured in the "*residual*" ν_{O-H} band range is still under the influence of the interhydrogen bond vibrational exciton interactions occurring within each individual carboxylic acid dimer [22–27].

The unusual properties of the "*residual*" ν_{O-H} bands have proved that the distribution of protons and deuterons between the hydrogen bonds of the isotopically diluted crystalline samples is nonrandom and in an individual dimer the coexistence of two identical hydrogen isotope atoms, proton or deuterons, is preferred. As a result, the interhydrogen bond exciton interactions still occur in each dimeric system and consequently the "*residual*" ν_{O-H} bands retain their "*dimeric*" properties. These spectral effects, that is, the so-called H/D isotopic "*self-organization*" effects, are

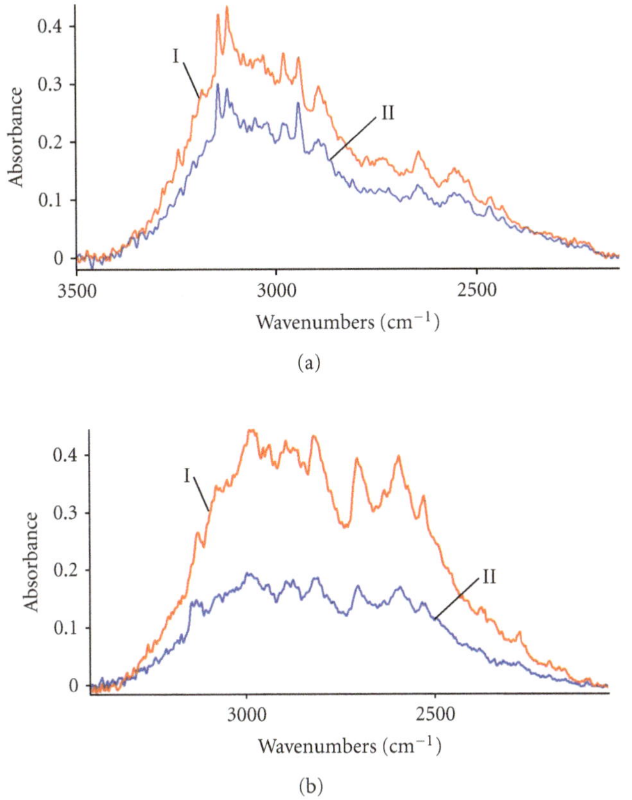

FIGURE 6: Polarized IR spectra of *2-furanacetic acid* and *2-furanacrylic acid* crystals measured at room temperature in the ν_{O-H} band frequency range for the IR radiation of the normal incidence with respect to the "*ac*" crystal faces. (a) *2-furanacetic acid* crystal. (I) The electric field vector *E* of the incident beam of IR radiation parallel to the *a*-axis (II) The E vector parallel to the c^*-axis (the c^*-symbol denotes the vector in the reciprocal lattice). (b) *2-furanacrylic acid* crystal. (I) The electric field vector E parallel to the *c*-axis. (II) The E vector parallel to the a^*-axis.

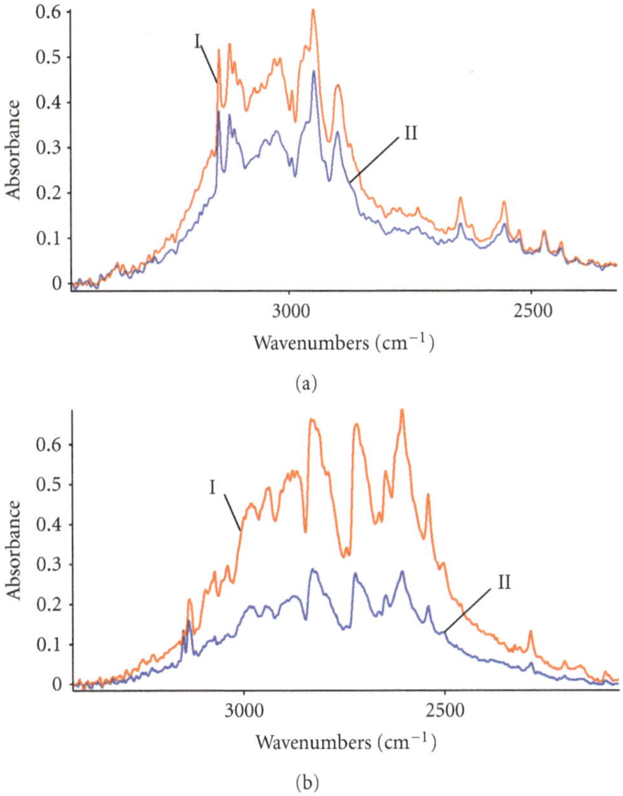

FIGURE 7: Polarized IR spectra of (a) *2-furanacetic acid* and (b) *2-furanacrylic acid* crystals measured at 77 K in the ν_{O-H} band frequency range. (a) *2-furanacetic acid* crystal. (I) The electric field vector E parallel to the *a*-axis. (II) The *E* vector parallel to the $c*$-axis. (b) 2-furanacrylic acid crystal. (I) The electric field vector *E* parallel to the *c*-axis. (II) The E vector parallel to the a^*-axis.

the attribute of the "*dynamical cooperative interactions*" involving hydrogen bonds in the dimers [16–18].

In the case of high excess of protons in the crystals qualitatively similar spectral effects can be identified in the "*residual*" ν_{O-D} bands, located in the range of 1900–2300 cm^{-1}, as those observed in the "*residual*" ν_{O-H} bands. In the low concentration of deuterons, the "*residual*" ν_{O-D} bands still retain the characteristic linear dichroic effects accompanying them (see Figures 4–9). For the two compared "*residual*" bands, ν_{O-H} and ν_{O-D}, not only the linear dichroic but also the temperature effects appear to be similar to the corresponding effects measured in the spectra of isotopically neat crystals.

This property results from the "*dynamical cooperative interactions*" in the hydrogen-bonded systems which lead to the appearance of the so-called H/D isotopic *self-organization* effects in the hydrogen bond IR spectra [17, 18]. The source of these nonconventional interactions in the hydrogen bond dimers is a vibronic coupling mechanism involving the totally symmetric proton stretching vibrations and the electronic motions in the systems [17, 18]. According to the theory of the "*dynamical cooperative interactions*," the symmetric hydrogen bond dimers of the HH or DD-type, with identical hydrogen isotope atoms, are thermodynamically more stable than the non-symmetric dimers of the HD type. The distribution of the HH- or DD-type dimers in the lattice sites is random. The energy difference between the two forms of dimers, the HH and the HD types was estimated as approximately equal to 1.5 kcal/mole of the dimers. Therefore, the relative concentration of the HD-type dimers is negligibly low and practically nondetectable with the use of the IR spectroscopic methods [16–18].

From the experimental studies presented in Figures 3–11 it also results that hydrogen-bonded cyclic centrosymmetric dimers are the bearers of the crystal spectral properties, since the inter-dimer vibrational exciton interactions are negligibly small.

6. Model

6.1. Carboxylic Acid Dimers the Basic Idea. The problem of the quantitative theoretical treatment of the spectral properties of systems composed with mutually interacting hydrogen bonds still constitute a real challenge in the area of the hydrogen bond research. There are still many problems to solve in this matter, since even the most advanced theories,

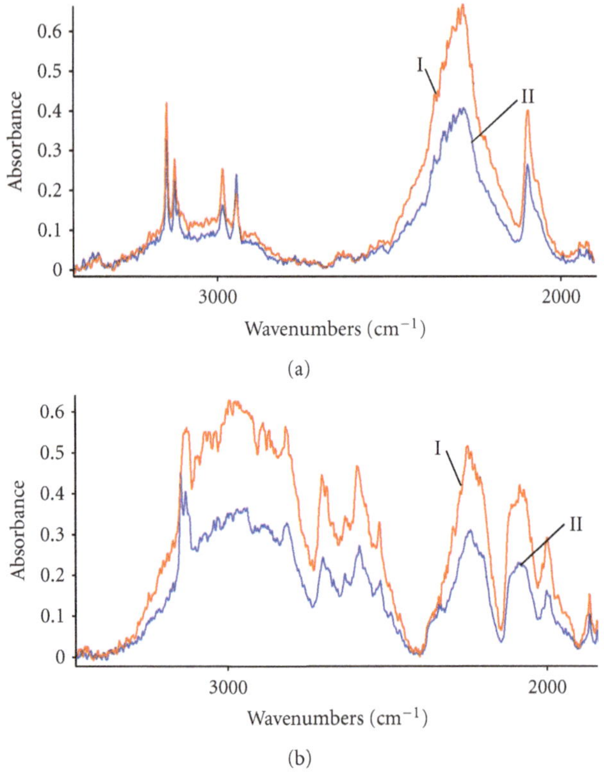

FIGURE 8: Polarized IR spectra of (a) d_1-*2-furanacetic acid* and (b) d_1-2-furanacrylic acidcrystals measured at room temperature in the ν_{O-D} band frequency range. (a) *2-furanacetic acid* crystal (ca. 10% H and 90% D). (I) The electric field vector E parallel to the a-axis, (II) The E vector parallel to the c^*-axis (b) 2-furanacrylic acid crystal (ca. 65% H and 35% D). (I) The electric field vector E parallel to the c-axis, (II) The E vector parallel to the a^*-axis.

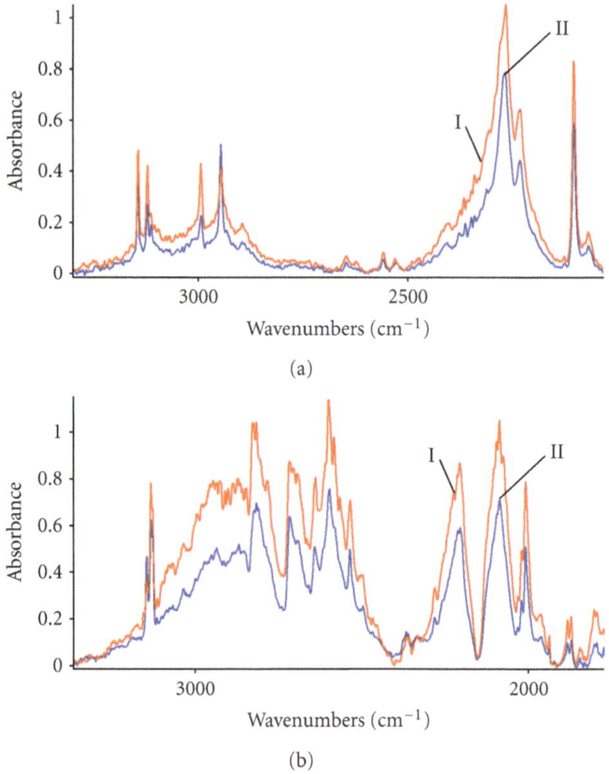

FIGURE 9: Polarized IR spectra of (a) d_1-*2-furanacetic acid* and (b) d_1-2-furanacrylic acid crystals measured at 77 K in the ν_{O-D} band frequency range. (a) *2-Furanacetic acid* crystal (ca. 10% H and 90% D). (I) The electric field vector E parallel to the a-axis, (II) The E vector parallel to the c^*-axis. (b) 2-Furanacrylic acid crystal (ca 65% H and 35% D). (I) The electric field vector E parallel to the c-axis, (II) The E vector parallel to the a^*-axis.

elaborated for the description of the IR spectra of hydrogen bond systems, are unable to reliably explain a number of effects observed in the dimeric spectra. Despite of spectacular achievements in the quantitative description of the intensity distribution in the ν_{X-H} bands, which are the attribute of the proton stretching vibrations in the X–H$\cdots$Y bridges and in the description of the H/D isotopic effects, the understanding of temperature effects in the spectra seems to be totally incomplete.

Cyclic hydrogen bond dimers, formed by associated carboxyl groups of diverse carboxylic acid molecules, are the most frequently studied model systems investigated in this research area. They exhibit some unusual spectral properties in IR connected with the highly abnormal thermal evolution of the ν_{O-H} and ν_{O-D} band contour shapes. One could expect that the hydrogen bond spectra of diverse carboxylic acid dimers, measured in the ν_{O-H} and ν_{O-D} band frequency ranges, should be fairly similar one to another due to the identical structural units of the molecular dimers, namely, the $(COOH)_2$ rings, in which two hydrogen bonds exist forming hydrogen bond dimers. However, on comparison of the crystalline spectra of diverse carboxylic acids, a considerable variation degree of the analyzed band contour shapes can be found. This fact undoubtedly remains in a close connection with differences in the electronic structures of diverse carboxylic acid molecules. Simultaneously, these spectra strongly differ, one from the other, by temperature effects characterizing them. Also these effects undoubtedly remain in a close relation with the electronic structures of the associating molecules. The basic experimental facts supporting the hypothesis given above are presented in the following.

6.2. Electronic Structure of Carboxylic Acid Molecules versus the Temperature Effects in Their Crystalline IR Spectra. Based on our previous studies, at this point, let us summarize the basic properties of the ν_{O-H} bands in the IR spectra of the hydrogen bond cyclic dimers formed by diverse carboxylic acid molecules, in relation to their electronic structures.

(a) In the case of carboxylic acid molecules in which the aliphatic fragments are connected directly with carboxyl groups (e.g., *aliphatic monocarboxylic acids* [11–13, 30, 31] and *dicarboxylic acids* [22]), the ν_{O-H} bands are characterized by different intensity distribution patterns, when compared with the corresponding band properties in the IR spectra of *arylcarboxylic acids* [23, 26]. In the first case,

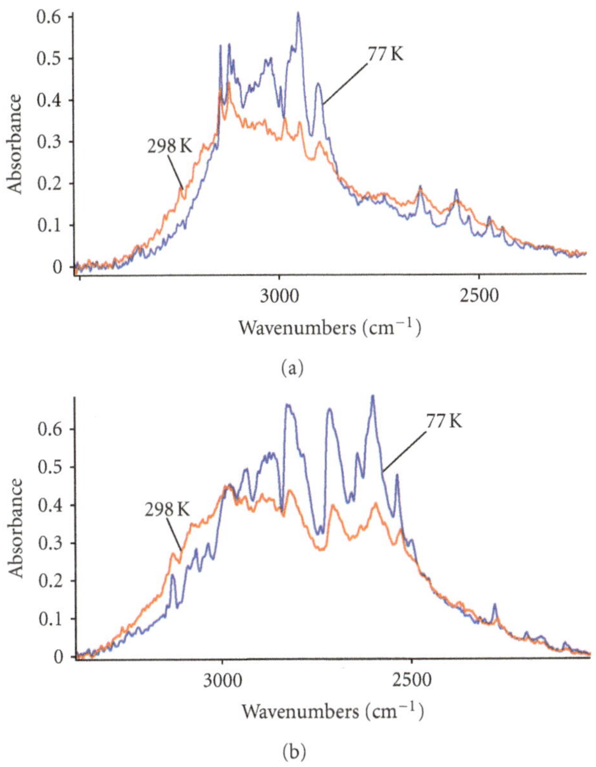

FIGURE 10: The ν_{O-H} bands in the IR spectra of monocrystalline samples of (a) *2-furanacetic acid* and (b) 2-furanacrylic acid. Temperature effect in the spectra.

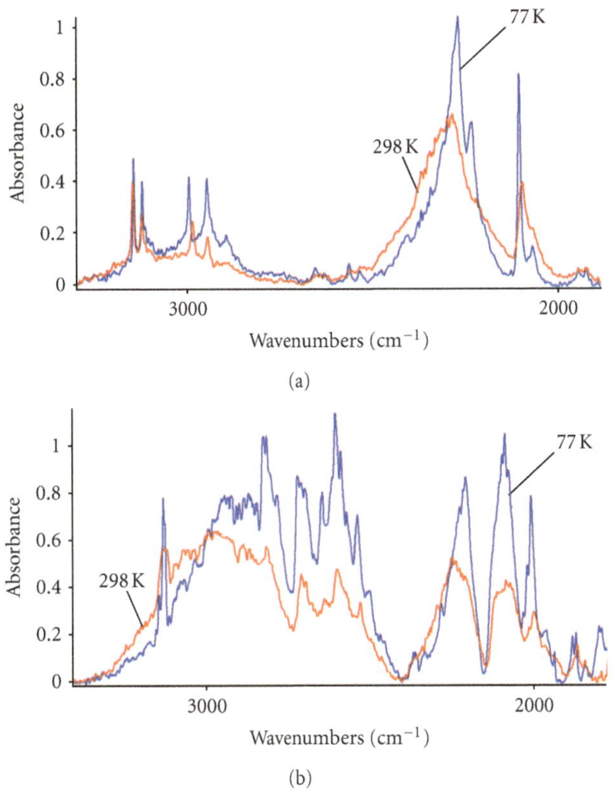

FIGURE 11: The ν_{O-D} bands in the IR spectra of monocrystalline samples of (a) d_1*-2-furanacetic acid* (ca 10% H and 90% D). and (b) d_1-2-furanacrylic acid (ca 65% H and 35% D). Temperature effect in the spectra.

the higher-frequency branch of the ν_{O-H} band is more intense in relation to the intensity of the lower-frequency band branch.

(b) In the case of hydrogen-bonded molecular systems, in which carboxyl groups are directly linked to π-electronic systems (e.g., *arylcarboxylic* [23, 26] and *arylacrylic acids* [24]), the ν_{O-H} band contours are a "*mirror reflection*" of the band shapes of systems from the point "a." In this case, the lower-frequency branch of the band is the most intense one. Similar property characterizes spectra of carboxylic acids with other large π-electronic systems in their molecules, for example, *cinnamic acid* [24], *2-naphthoic acid* [26], and *1-naphthylacrylic acid* [32].

(c) For other carboxylic acids, in which aromatic radicals are separated from carboxyl groups by fragments of aliphatic hydrocarbon chains (e.g., *arylacetic acid* [25, 27] and *styrylacetic acid* [33]), the ν_{O-H} band contour shapes are fairly similar to the corresponding band characteristics from the point "a," that is, to the corresponding spectra of *aliphatic monocarboxylic acids* [30, 31] and *dicarboxylic acids* [22]).

The ν_{O-H} bands in the spectra of the hydrogen bond of carboxylic acid crystals from the "a" and "c" groups, measured at room temperature, are characterized by relatively low intensity of the lower-frequency branch of he band in comparison with the higher-frequency band branch intensity. On the decrease of temperature to 77 K, only a relatively small growth of the relative intensity of the lower-frequency branch of each band can be observed. This band branch still remains of the lower intensity in the low-temperature spectra.

In the case "b," even at room temperature spectra, the ν_{O-H} bands exhibit relatively high intensity of their lower-frequency branch in relation to the higher-frequency branch. On the temperature decrease up to 77 K, a considerable growth of the relative intensity of the lower-frequency branch of each analyzed band can be observed. As the result of the band contour thermal evolution, in the low-temperature spectra of carboxylic acid crystals of this group the lower-frequency branch is of the dominant intensity in the bands.

According to the "*state-of-art*" in our contemporary knowledge about the quantitative description of the IR spectra of the hydrogen bond in carboxylic acid dimers, the following interpretation of the ν_{O-H} band generation mechanisms seemed to be valid: the lower-frequency branch of the ν_{O-H} band is generated by the transition occurring to the Ag-symmetry excited state of the totally symmetric proton stretching vibrations in the dimers. This transition, forbidden by the symmetry rules, becomes allowed via a vibronic mechanism, which is a kind of reverse of the familiar *Herzberg-Teller* mechanism, originally responsible for the promotion of forbidden electronic transitions in UV spectra of aromatic hydrocarbons [34]. Within this approach of the reverse *Herzberg-Teller* vibronic coupling mechanism, electronic properties of single hydrogen bonds themselves, as well as electronic properties of the whole associated molecules and the proton vibration anharmonicity, are responsible for the magnitude of the forbidden transition promotion effects in the dimeric spectra [35]. The promotion mechanism determines a unique property of centrosymmetric hydrogen bond dimeric system. This effect found no counterpart in the vibrational spectroscopy of single centrosymmetric molecules.

On the other hand, the higher-frequency spectral branch of the band corresponds with the symmetry-allowed transition to the Au-state of the nontotally symmetric proton vibrations in the centrosymmetric hydrogen bond dimers. One should expect that the higher-frequency branch of the ν_{O-H} band, attributed to the allowed transition, should be more intense than the other band branch related with the forbidden transition. Therefore, based on these intuitive predictions, the spectral properties of the carboxylic acid dimers from the "b" group seem to be highly surprising, contradicting the interpretation of the spectra of systems belonging to the "a" and "c" groups. The particular electronic properties of the carboxylic acid molecules from the "c" group can anyway explain the extremely high integral intensity of the forbidden lower-frequency branch of the band and its strong temperature dependence.

In order to propose a reliable explanation of this paradox in our analysis, one should also recall the hydrogen bond IR spectra of other hydrogen bond dimeric systems, including spectra of hydrogen-bonded heterocycles. On comparison of the IR spectra of diverse crystalline systems containing cyclic hydrogen bond dimers as the structural units of their lattices, the following general conclusions can be made: most of centrosymmetric hydrogen bond dimers exhibit regular enough spectral properties characterizing their hydrogen bond spectra. Usually, the ν_{X-H} bands have the lower-frequency (i.e., the "*forbidden*") branch of a lower intensity, even in their low-temperature spectra. However, in some rare cases, for example, *3-hydroxy-4-methyl-2(3H)-thiazolethione* [36], *2-tiopyridone* [37], and *2-pyridone* [38], the ν_{O-H} and ν_{N-H} bands are characterized by an abnormal, that is, by a "*reverse*" intensity distribution patterns in their contours. In the latest cases, the lower-frequency branch of each band is more intense when compared with the higher-frequency band intensities. It fairly resembles the properties of the spectral properties at 77 K of carboxylic acid crystals of the "b" group. In the case of the dimeric spectra of the reverse intensity distribution patterns in the bands, for example, *3-hydroxy-4-methyl-2(3H)-thiazolethione* [36] and *2-tiopirydone* [37], this effect was ascribed previously to the influence of the extreme lengths of the O–H· · · S and N–H· · · S hydrogen bonds in the dimeric systems.

The recent considerations, aiming to explain these phenomena, were performed in terms of the *dipole-dipole* model of the vibrational exciton interactions involving the hydrogen bonds in the dimers. In the case of the interpretation of the spectra of *3-hydroxy-4-methyl-2(3H)-thiazolethione* [36] and *2-tiopirydone* [37], the hydrogen bond geometry

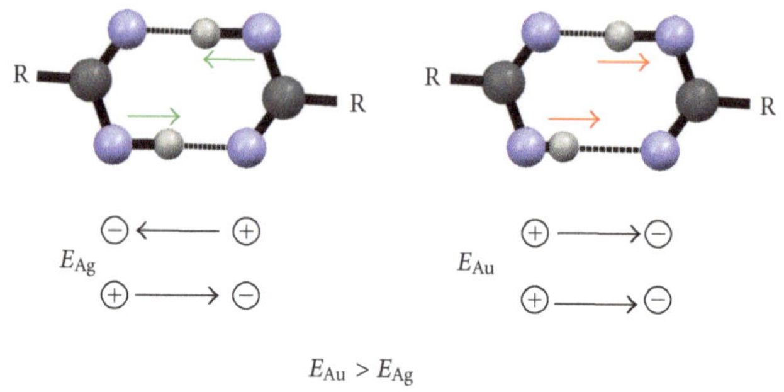

FIGURE 12: The "*side-to-*side" (SS) exciton coupling involving the proton stretching vibrations in a cyclic centrosymmetric hydrogen bond dimer.

was considered to be responsible for the unusual spectral property of these dimers. However, this approach fails in the interpretation of the spectra of *2-pyridone* cyclic dimers [38], in which the N–H· · · O hydrogen bonds are considerably shorter when compared with the N–H· · · S bond lengths in *2-thiopyridone* cyclic dimers [37] and their spectra qualitatively fairly resemble the corresponding spectra of *2-pyridone* [38]. On the other hand, even among the hydrogen bond dimers of diverse molecular systems with the N–H· · · S hydrogen bonds, for which the extreme spectral properties were found, a substantial diversification in the analyzed spectral properties has been found, despite the extremely long hydrogen bonds in these cases. The IR spectra of *2-mercaptobenzothiazole* cyclic dimers [39] exhibit regular properties of the intensity distribution pattern in their ν_{N-H} band contours, similarly as the carboxylic acid dimers in the crystals of the groups "a" an "b," regardless of the extreme N–H· · · S bond lengths, like these found in *2-thiopyridone* dimers [37].

6.3. Spectra of Cyclic Dimers versus Spectra of Chain Hydrogen Bond Systems. It is surprising that spectra of cyclic hydrogen bond dimers in *3-hydroxy-4-methyl-2(3H)-thiazolethione* [36], *2-thiopyridone* [37], *and 2-pyridone* [38] crystals fairly resemble by their intensity distribution patterns of the ν_{N-H} bands the spectra of chain hydrogen bond systems in a particular group of molecular crystals. In the hydrogen bond spectra of *pyrazole* [40] and *4-thiopyridone* [41] crystals, with hydrogen-bonded molecules forming infinite chains in their lattices, strong linear dichroic effects can be observed, which prove a considerable influence of the exciton interactions involving the adjacent hydrogen bonds in each chain. Figures 12 and 13 explain the source of the differences in the hydrogen bond dimers, the cyclic and the chain ones.

The analysis of this inter-hydrogen-bond coupling, in case of cyclic centrosymmetric dimers and in linear dimers, requires taking into consideration two situations of the vibrational transition moment directions for hydrogen bonds in the dimers. For cyclic dimers, the parallel mutual orientation of the dipole transition moments, the exciton interaction energy E_{Au} in the limits of the *dipole-dipole* model is of the positive sign. The vibrational transition corresponding to such arrangement of the vibration dipole moments is

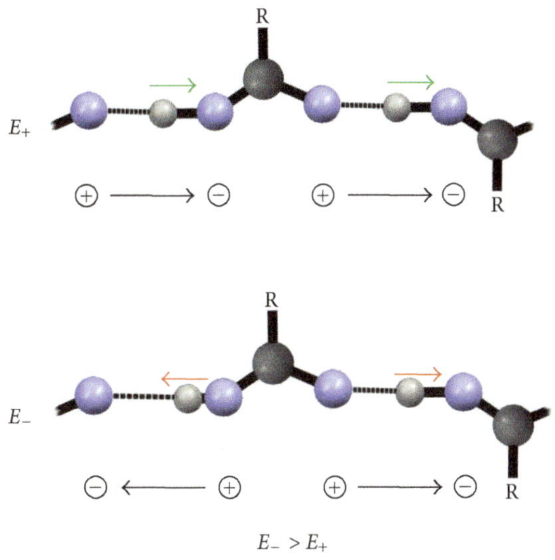

FIGURE 13: The "*head-to-tail*" (TH) exciton coupling involving the proton stretching vibrations in an infinite chain of associated hydrogen bonds.

responsible for generation of the intense, symmetry-allowed, shorter-wave branch of the dimeric spectra. In contrast, when the dipole transition moments are of the antiparallel arrangement (see Figure 12), the energy exciton interaction energy value E_{Ag} is negative, so the band generated by this situation is placed at the lower frequency and it corresponds to the symmetry-forbidden excitation of the totally symmetric proton vibrations. Such sequence of the spectral branches in the hydrogen bond stretching bands is typical for cyclic, centrosymmetric hydrogen bond dimers.

When the vibrating transition moment dipoles in a linear dimer, in the case of the totally symmetric proton vibrations, are oriented axially as "*tail-to-head*" (Figure 5), the sign of the exciton interaction energy value E_+ is negative, so the intense branch corresponding to the symmetry-allowed transition is placed at the lower-frequency range. On the contrary, the forbidden by the symmetry rules spectral branch, situated at the higher frequency, is generated by the antiparallel orientation of the vibrating dipoles (see Figure 13). In this case, the exciton coupling energy E_- is of the positive sign. The sequence and the properties of the branches in the proton stretching vibration bands in the discussed case are reverse to those observed in the IR spectra of hydrogen bond cyclic dimeric systems.

Therefore, the following problem demands explanation: why do some individual cyclic hydrogen bond dimeric systems exhibit similar spectral properties to the corresponding properties of a particular group of crystals with chain structures of hydrogen-bonded associates (*formic acid* [31], *pyrazole* [40], and *4-thiopyridone* [41] crystals). Undoubtedly, this property remains in a close connection with the π-electronic properties of the associating molecules. In the associated molecular systems, vibrational exciton couplings are of the "*tail-to-head*" (TH) type. They involve the adjacent hydrogen bonds within each individual chain in the lattice. The electronic structure of molecules of this group is most probably the key factor governing these interhydrogen bond interactions.

Nevertheless, the majority of crystals with hydrogen-bonded molecular chains in their lattices surprisingly exhibit the spectral properties similar to the analogous properties of cyclic hydrogen bond dimer spectra from the "a" and "c" groups (e.g., *acetic acid* [30], *N-methylthioacetamide* [42], or *acetanilide* [18] crystals). In the latest case the exciton interactions of the "*side-to-side*" (SS) type involve the closely spaced hydrogen bonds where each moiety belongs to a different chain. In molecules of this group, large π-electronic systems are absent. Only *carbonyl* or *thiocarbonyl* groups, each with a small π-electronic system, are present in these molecules.

From the above-presented data, it results that the way of realization of the vibrational exciton interactions in various hydrogen bond aggregates (cyclic dimers, infinite chains), affecting the ν_{X-H} and ν_{X-H} band fine structures, does not directly depend on the hydrogen bond system geometry. It is rather determined by the electronic structure of the associating molecules.

7. Theoretical Approach Proposed

The *dipole-dipole* interaction model, widely used for a simplified description of the exciton interactions between hydrogen bonds, seems to be nonadequate in the explanation of the wide diversity of the spectra of cyclic hydrogen bond dimers. There is some experimental data indicating that these couplings do not always occur as "*through-space*" and they are also widespread by the hydrogen bond electrons as well as by electrons of the molecular skeletons. Therefore, in terms of the theory of molecular vibrational excitons [43, 44], the exciton interaction integrals in some cases may also considerably strongly depend on the electronic coordinates. In advantageous circumstances, resulting from a proper electronic structure of the associating molecules, the proton stretching vibrations can induce electric current oscillating around a cyclic hydrogen bond dimer, or in the other case, oscillating along a hydrogen bond chain. However, only the totally symmetric proton vibrations are able to effectively induce the electric current in the ring or in the chain, while the nontotally symmetric vibrations are inactive in this mechanism, since currents induced in each individual hydrogen bond are annihilated in a dimer. The formalism of the model of the electric current generated by oscillating protons in cyclic hydrogen bond dimers was proposed by *Nafie* three decades ago [45].

In the scope of the considerations given above, it seems justified to treat formally a cyclic hydrogen bond dimer by the following two ways, taking into account the exciton interactions in the system.

(1) As a closed chain in which the adjacent hydrogen bonds are strongly exciton-coupled, similarly as in the chain associates in *pyrazole* [40] and *4-thiopyridone* [41] crystals. This is the coupling of the TH type occurring around the molecular cycle. This

way the coupling occurs via the easy-polarizable electrons on the π-orbitals. Therefore, the cyclic dimer spectrum is fairly similar to the spectrum of a chain system, with a low intensity of the higher-frequency band branch.

(2) As a pair of partially independent hydrogen bonds, which remains only "*through-space*" exciton coupled. It can be considered as a coupling of the SS type, without the generation of the ring electric current in the dimer. This behavior characterizes the associated molecular systems with no large π-electronic systems in their structures, where only small π-electronic systems are present in *carbonyl* and *thiocarbonyl* groups. In these circumstances, the dimeric spectra are of the standard form, with a low intensity of the lower-frequency ν_{X-H} band branch. For the quantitative description of the exciton interactions involving hydrogen bonds, influencing the dimer spectra, the *dipole-dipole* model is sufficiently adequate.

The ν_{X-H} band shapes in the two types of the dimer spectra are related one with the other by the approximate mirror reflection symmetry. In the case 1, the lower-intensity spectral branch appears in the higher-frequency range and is generated by the quasiforbidden vibrational transition in a dimer, occurring to the excited state of the totally symmetric proton stretching vibrations. In case 2 the lower intensity spectral branch appears in the lower-frequency range. It corresponds with the quasi-forbidden vibrational transition in a dimer. The above-presented spectral properties of diverse hydrogen bond cyclic dimers may allow explaining the thermal evolution effects in the hydrogen bond IR spectra of carboxylic acid crystals.

It seems that in order to explain the temperature effects in the IR spectra of cyclic hydrogen bond dimers the following hypothesis concerning the mechanisms of the spectra generation should be accepted: let us assume that two competing mechanisms of vibrational exciton interactions involving hydrogen bonds in cyclic dimers are simultaneously responsible for the formation of the ν_{X-H} band contour shapes. The contribution of each individual mechanism depends on the electronic structure of the associating molecules, on the electronic properties of the heavy atoms forming the hydrogen bridges as well as on temperature.

(A) The first mechanism depends on the "*side-to-side*" (SS)-type vibrational exciton coupling between the hydrogen bonds in cyclic dimers. In this case, the dimer hydrogen bonds interact one with the other as *through-space* via the *van der Waals* forces.

(B) The other mechanism assumes a "*tail-to-head*" (TH)-type exciton coupling involving the hydrogen bonds in the dimers. These interactions occur around the cycles via electrons.

The "B" mechanism seems to be privileged in the case of the particular kind of associated molecules, in which hydrogen bonds couple with large π-electronic systems, for example, for aromatic carboxylic acid molecules. The "A" mechanism seems to dominate in the case of molecular systems with small π-electronic systems, for example, for aliphatic carboxylic acid molecules.

It seems obvious that for an individual hydrogen-bonded dimeric system the contribution of each mechanism is temperature dependent. For molecules with large π-electronic systems directly coupled with the hydrogen bonds, the "B" mechanism should be privileged at very low temperatures. Temperature growth, influencing the increase of atomic vibration amplitudes, should annihilate the electric current induced by the totally symmetric proton vibrations in the cycles. In these circumstances, the role of the "A" mechanism increases, namely, of the "*through-space*" vibrational exciton coupling between the hydrogen bonds in a dimer. This should, therefore, result in a particularly strong temperature-induced evolution of the ν_{X-H} bands, especially in the case of the spectra of *2-thiopyridone* [37] and *2-pyridone* [38] type dimers. Even when the lower-frequency branch of the band is less intense when compared with the higher-frequency one, the temperature decrease till 77 K causes its considerable intensity growth, and in these circumstances the lower-frequency branch becomes more intense than the higher-frequency band branch.

In the spectra of cyclic dimers, with only small π-electronic systems in the associating molecules, the temperature decrease usually does not cause a considerable intensity growth of the lower-frequency band branch. It still remains less intense when compared with the higher-frequency branch of the band. It means that, due to the molecular electronic properties of this group of dimers, the "B" mechanism cannot be activated effectively enough even at very low temperatures.

8. Spectral Consequences of the Model for Carboxylic Acids

From the above assumptions, it results the choice of the proper way of the model calculations of the ν_{X-H} and ν_{X-D} band contours in IR spectra of hydrogen bond dimers. In the limits of the proposed approach, a theoretical spectrum of the model system can be derived, formally treated as a superposition of two component spectra, where each individual spectrum corresponds with a different mechanism of the exciton interactions, SS (A) and TH (B), involving the dimer hydrogen bonds. In terms of the "*strong-coupling*" theory [6–8], in each exciton interaction mechanism case, the ν_{X-H} band in the dimeric spectrum is a superposition of two component bands, "*Plus*" and "*Minus*," each of a different origin.

The "*Plus*" band is generated by the dipole allowed transition to the excited state of the nontotally symmetric proton stretching vibrations in a centrosymmetric dimer, belonging to the Au representation. On the other hand, the "*Minus*" band is connected with the symmetry forbidden transition to the Ag-symmetry state of the totally symmetric proton vibrations in the dimers, activated by a vibronic mechanism [35]. In the case when the mechanism "A" exclusively decides about the dimer spectra generation mechanism, the "*Minus*" band appears in the lower "B" mechanism frequency range in relation to the "*Plus*" band location. In the other case, when the "B" mechanism governs the dimer spectra generation, the two component bands appear in the reverse sequence than in the case "A." It means that in the case of the "B" mechanism governing the spectra generation, the "*Minus*" band representing the forbidden transition appears in the higher-frequency range than the "*Plus*" band connected with the allowed transition.

9. Model Calculations of the Band Contours

In the two cases, A and B, model calculations, aiming at reconstituting of the "*residual*" ν_{O-H} and ν_{O-D} band shapes, were performed within the limits of the "*strong-coupling*" theory, for a model centrosymmetric O–H$\cdots$O hydrogen bond dimeric system. [6–8, 46]. We assumed that the main ν_{O-H} and ν_{O-D} band shaping mechanism involved strongly anharmonically coupled the high-frequency proton (or deuteron) stretching vibrations and the low-frequency O$\cdots$O hydrogen bridge stretching vibrational motions. Calculation of the hydrogen bond system IR spectra in terms of the "*strong-coupling*" model allows to obtain results fairly comparable with the results of the spectra calculation performed using the "*relaxation*" theory [9, 10, 47–49].

According to the formalism of the "*strong-coupling*" theory [6–8, 46], the ν_{O-H} band shape of a dimer depends on the following system of dimensionless coupling parameters: (i) on the distortion parameter, "b_H", and (ii) on the resonance interaction parameters, "C_O" and "C_1". The "b_H" parameter describes the change in the equilibrium geometry for the low-energy hydrogen bond stretching vibrations, accompanying the excitation of the high-frequency proton stretching vibrations ν_{O-H}. The "C_O" and "C_1" parameters are responsible for the exciton interactions between the hydrogen bonds in a dimer. They denote the subsequent expansion coefficients in the series on developing the resonance interaction integral "C" with respect to the normal coordinates of the $\nu_{O\cdots O}$ low-frequency stretching vibrations of the hydrogen bond. This is in accordance with the formula:

$$C = C_O + C_1 Q_1, \tag{1}$$

where Q_1 represents the totally symmetric normal coordinate for the low-frequency hydrogen bridge stretching vibrations in the dimer. This parameter system is closely related to the intensity distribution in the dimeric ν_{N-H} band. The "b_H" and "C_1" parameters are directly related to the dimeric ν_{N-H} component bandwidth. The "C_O" parameter defines the splitting of the component bands of the dimeric spectrum corresponding to the excitation of the proton vibrational motions of different symmetries, Ag and Au. In its simplest, original version, the "*strong-coupling*" model predicts reduction of the distortion parameter value for the deuterium bond systems according to the relation:

$$b_H = \sqrt{2} b_D. \tag{2}$$

For the " C_O" and "C_1" resonance interaction parameters, the theory predicts the isotopic effect expressed by the 1.0 to $\sqrt{2}$-fold reduction of the parameter values for *D*-bonded dimeric systems.

As the consequence of the "*strong-coupling*" model, the ν_{O-H} and ν_{O-D} band contour fine structures were treated as a superposition of two component bands. They correspond to the excitation of the two kinds of proton stretching vibrations, each exhibiting a different symmetry. In the case of the A exciton coupling mechanism and for the C_i point symmetry group of the model dimer, the excitation of the Ag vibrations in the dimer generates the lower-frequency transition branch of the ν_{O-H} band when the Au vibrations are responsible for the higher-frequency band branch. In the case of the B mechanism, the component subbands appear in reverse sequence.

Here, we consider an identical anharmonic coupling parameter system for the two individual mechanism cases A and B although diversification of the coupling parameter value systems seems to be better justified. We assume the contribution of each mechanism as governed by a *Boltzmann*-type relation. In addition, for the statistical weight parameters of each individual mechanism, $P_A(T)$ and $P_B(T)$, one must distinguish which state is dominant, that is, when the SS (A) state is of the lower energy and the TH (B) state is of a higher energy value and vice versa. In order to reproduce the temperature dependence of experimental spectra particularly for its width and the position of its first moment, we used for the $P_A^{AB}(T)$ exponential temperature dependence according to

$$P_A^{AB}(T) = 1 - \exp\left(-\frac{\alpha^{AB}}{k_B T}\right), \tag{3}$$

where is α^{AB} the activation energy parameter when the SS state is dominant and k_B is the constant of Boltzmann. In such circumstance, $P_B^{AB}(T)$ takes the following expression:

$$P_B^{AB}(T) = \exp\left(-\frac{\alpha^{AB}}{k_B T}\right), \tag{4}$$

It is interesting to note that, in the case of A, for very low temperatures, the statistical weight $P_A^{AB}(T)$ parameter is close to 1.0 and $P_B^{AB}(T)$ is almost equal 0.0. In these circumstances, the SS-type interaction is the basic type of the exciton coupling involving the dimer hydrogen bonds. For high temperatures, the $P_B^{AB}(T)$ parameter values are different from 0.0 and they are intermediate between 0.0 and 1.0 (rather closer to 0.5) and $P_A^{AB}(T)$ approaches 0.5. When the temperature increases, $P_B^{AB}(T)$ also increases. It means that, the TH coupling, occurring via the electric current in the ring is activated in higher temperatures in a magnitude depending of the energy gap between these two states of the vibrationally excited dimer. From our experimental estimations, the energy gap for some dimeric system cases is relatively large and in another cases it may be relatively low.

In the case B, where the TH state is of a lower energy value, we assume the same formula but the energy barrier α^{BA} height is relatively low. In such a circumstance, the statistical weight parameters, $P_A(T)$ and $P_B(T)$, may be written as follows:

$$P_A^{BA}(T) = \exp\left(-\frac{\alpha^{BA}}{k_B T}\right),$$
$$P_B^{BA}(T) = 1 - \exp\left(-\frac{\alpha^{BA}}{k_B T}\right). \qquad (5)$$

As we can see, for very low temperatures, $P_A^{AB}(T)$ may be practically equal to 1.0. For this kind of dimeric systems, the TH-type exciton coupling is the basic natural way in which the inter-hydrogen bond interactions occur. The growth in temperature annihilates this way of the coupling, due to the vanishing of the electronic current induced in the cycles, accompanied by large-amplitude thermal motions of atoms in the dimers. For high temperatures, $P_A^{AB}(T)$ decreases and becomes of an intermediate value between 0.0 and 1.0 (rather closer to 0.5), while the statistical weight $P_A^{AB}(T)$ grows declining from 0.0 up to 0.5. The energy gap between the two states in some molecular cases is usually relatively large, and in other cases it may be relatively small. It depends of the electronic properties of the associating molecules forming the dimers. From our experimental data, it can be concluded that the cases A and B represent the extreme cases of the interhydrogen bond coupling in cyclic hydrogen bond dimers. There are also many systems exhibiting an intermediate behavior. For a relatively small magnitude of the absolute values of the energy barrier height, the two cases A and B are practically nondistinguishable.

The theoretical spectra reconstituting the ν_{O-H} band contours measured at the two different temperatures, 293 K and 77 K, were calculated in terms of the two different individual coupling mechanisms, SS and TH, which generate the two component bands, "*plus*" and "*minus*" in a different sequence. The following coupling parameter values, identical in both molecular system cases, were used.

For the *2-furanacetic acid* crystal spectra: $b_H = 1.6$, $C_0 = 1.5$, $C_1 = -0.2$, $F^+ = 1.0$, $F^- = 0.2$, $\Omega_{O\cdots O} = 100\,\text{cm}^{-1}$, and we used the same parameter system for calculation of the *2-furanacrylic acid* crystal spectra: $b_H = 1.6$, $C_0 = 1.5$, $C_1 = -0.2$, $F^+ = 1.0$, $F^- = 0.2$, $\Omega_{O\cdots O} = 100\,\text{cm}^{-1}$.

The F^+ and F^- symbols denote the statistical weight parameters for the "*plus*" and "*minus*" theoretically derived subspectra contributing at the band formation.

The coupling parameter values used for calculation of the ν_{O-D} band contour shapes were as follows.

For *2-furanacetic acid* crystal, spectrum $b_D = 0.7$, $C_0 = 0.7$, $C_1 = -0.1$, $F^+ = 1.0$, $F^- = 0.2$, $\Omega_{O....O} = 100\,\text{cm}^{-1}$ and for *2-furanacrylic acid* crystal spectrum: $b_H = 0.7$, $C_0 = 0.7$, $C_1 = -0.1$, $F^+ = 1.0$, $F^- = 0.2$, $\Omega_{O\cdots O} = 100\,\text{cm}^{-1}$.

For the *2-furanacetic acid* crystal spectra the statistical weight parameter ratio, $P_A(T)$: $P_B(T)$, for the SS and TH mechanisms was estimated as equal to 1.0 : 0.0 in the case of the room temperature spectrum reconstitution. For the low-temperature spectrum case, this parameter ratio value is very similar and equal to 1.0 : 0.0. Among various parameter ratio values for the SS and TH mechanisms contributing in the band generation, this parameter ratio value allowed for the most adequate reproduction of the temperature effect in the crystal spectra.

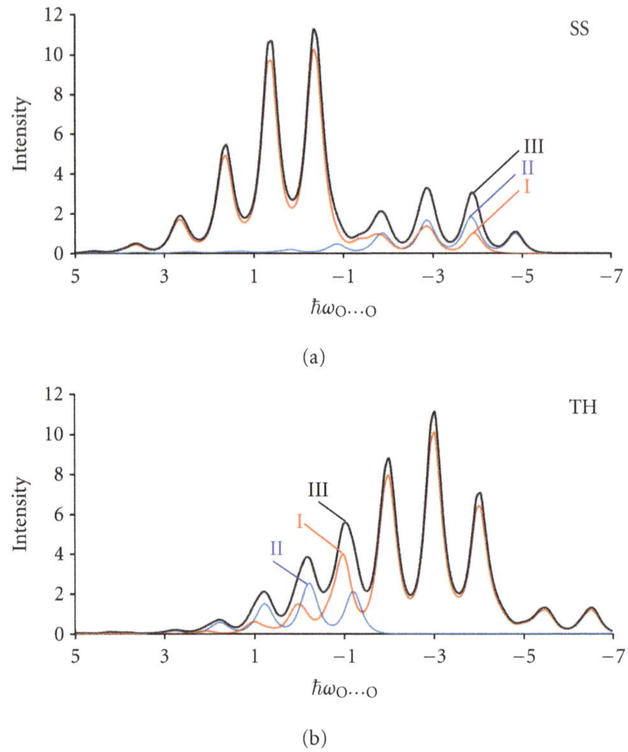

FIGURE 14: The theoretically derived ν_{O-H} band contours calculated in terms of the "*strong-coupling*" theory in the limits of the two different vibrational exciton coupling mechanisms involving the cyclic dimer hydrogen bonds, that is, "*side-to-side*" (SS) and "*tail-to-head*" (TH). (a) The SS coupling mechanism. (b) The TH coupling mechanism. (I) The "*minus*" band. (II) The "*plus*" band (III) Superposition of the I and II spectra, each taken with its appropriate individual statistical weight parameter, F^- and F^+. In both mechanism cases, the same coupling parameter value system was used for calculations: $b_H = 1.4$, $C_0 = 1.5$, $C_1 = -0.2$, $F^+ = 1.0$, $F^- = 0.2$, $\Omega_{O\cdots O} = 100\,\text{cm}^{-1}$. The transition frequencies are in the $\omega_{O\cdots O}$ vibrational quantum units, and the transition frequencies are expressed with respect to the gravity center of the hypothetical spectrum of a monomeric hydrogen bond in the cyclic hydrogen bond dimer. Transition intensities are in arbitrary units.

For the *2-furanacrylic acid* crystal spectra, the statistical weight parameter ratio, $P_A(T) : P_B(T)$, for the SS and TH mechanisms were estimated as equal to 0.35 : 0.65 in the case of the room temperature spectrum reconstitution. For the low-temperature spectrum case, this parameter ratio value is equal to 0.55 : 0.45.

In Figures 14 and 15, we present the theoretical ν_{O-H} and ν_{O-D} band contours calculated in terms of the two individual mechanisms of the vibrational exciton interactions involving the dimer hydrogen bonds, SS and TH.

In Figures 16 and 17, the evolution of the ν_{O-H} and ν_{O-D} band contour shapes accompanying the variation in the relative contribution of the SS and TH coupling mechanisms in generation of a dimeric spectra is shown. Similar band

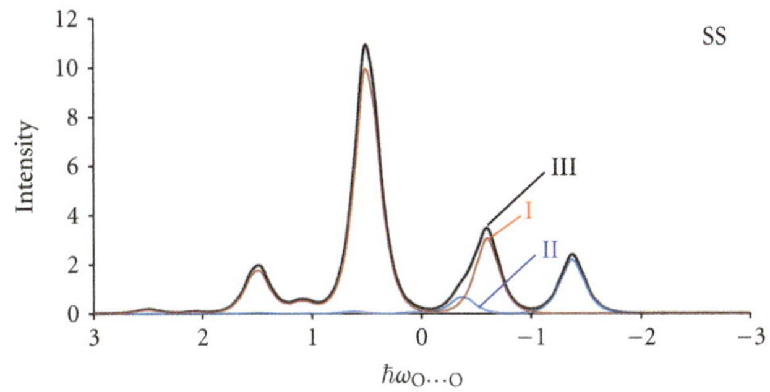

(a)

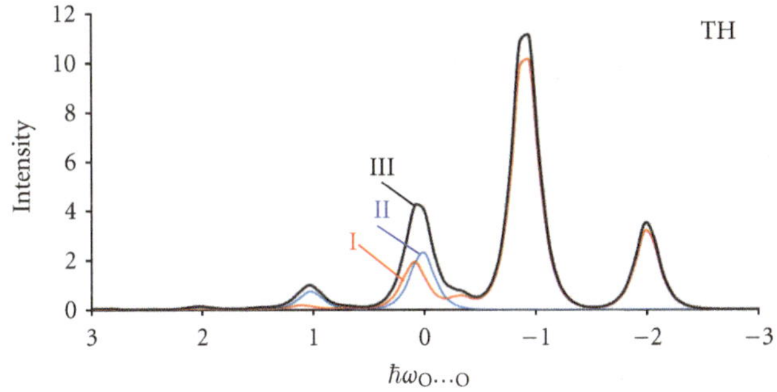

(b)

FIGURE 15: The theoretically derived ν_{O-D} band contours calculated in terms of the "*strong-coupling*" theory in the limits of the two different vibrational exciton coupling mechanisms involving the cyclic dimer hydrogen bonds, that is, "*side-to-side*" (SS) and "*tail-to-head*" (TH). (a) The SS coupling mechanism. (b) The TH coupling mechanism. (I) The "*minus*" band. (II) The "*plus*" band. (III) Superposition of the spectra I and II, each taken with its appropriate individual statistical weight parameter, F^- and F^+. In both mechanism cases, the same coupling parameter value system was used for calculations: $b_H = 0.7$, $C_0 = 0.7$, $C_1 = -0.2$, $F^+ = 1.0$, $F^- = 0.2$, $\Omega_{O\cdots O} = 100\,\text{cm}^{-1}$. The transition frequencies are in the $\omega_{O\cdots O}$ vibrational quantum units, and the transition frequencies are expressed with respect to the gravity center of the hypothetical spectrum of a monomeric hydrogen bond in the cyclic deuterium bond dimer. Transition intensities are in arbitrary units.

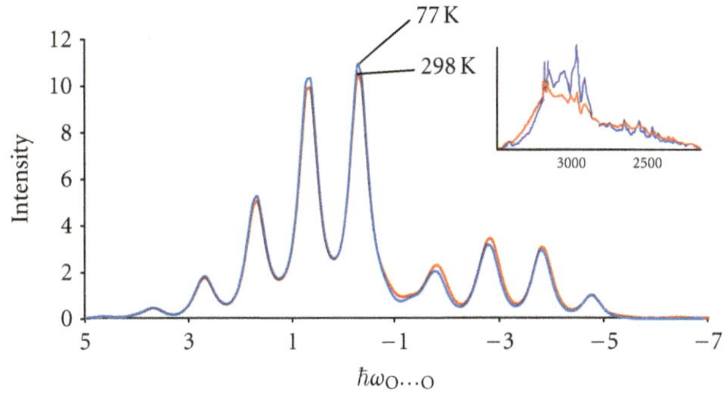

(a)

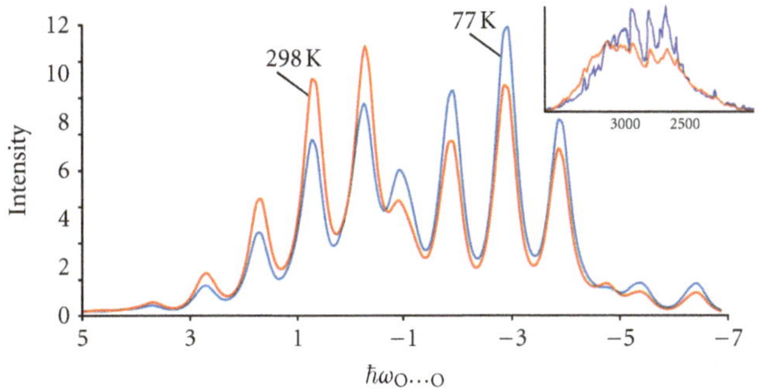

(b)

FIGURE 16: Temperature-induced evolution of the ν_{O-H} band contour shapes accompanying the variation in the contribution rate of the two different exciton coupling mechanisms, that is, "*side-to-side*" (SS) and "*tail-to-head*" (TH). Numerical reproduction of the temperature effect in the spectra of hydrogen-bonded (a) *2-furanacetic acid* crystal (b) *2-furanacrylic acid* crystal. The relative contribution ratio of the SS and TH mechanisms in the ν_{O-H} band generation is, for *2-furanacetic acid* crystal: 0.95 : 0.05 at 293 K and 0.95 : 0.05 at 77 K and for *2-furanacrylic acid* crystal: 0.65 : 0.35 at 293 K and 0.40 : 0.60 at 77 K. The experimental spectra are shown in inset.

shape evolution accompanies temperature changes during the spectral experiments.

From the comparison of the corresponding calculated and experimental spectra, it results that the intensity distribution patterns and the temperature effects in the spectra of the two different crystalline systems have been at least semiquantitatively reproduced via the model calculations.

10. Spectra of *2-Furanacetic* and *2-Furanacrylic Acid* Crystals

On comparing the IR spectra of the hydrogen bond for the two crystalline systems, essential differences analyzed crystalline spectra othe ν_{O-H} and ν_{O-D} bands. In the case of *2-furanacetic acid* spectra, the fine structure pattern of each band, ν_{O-H} and ν_{O-D}, is relatively simple. Each band consists of a low number of well-separated spectral lines. In the spectra of *2-furanacrylic acid,* each considered band is composed of a noticeably larger number of lines (ca. 2 times larger). It seems to prove a more complex mechanism of the spectra generation in the case of *2-furanacrylic acid* in relation to the mechanism governing the spectra generation of *2-furanacetic acid.*

The analyzed crystalline spectra of *2-furanacetic acid* seem to fully belong to the case A. On the other hand, the crystalline spectra of *2-furanacrylic acid* seem to satisfy the demands of the case B. The analyzed difference in the spectral properties of *arylacetic acid* dimers and the *arylacrylic acid* dimers most probably results from the influences exerted on to the hydrogen bond dimers, present in the $(COOH)_2$ cycles, by the aromatic rings. The direct contact between the furan rings with carboxyl groups (*arylcarboxylic, furanacrylic,* and *thiopheneacrylic acids*) most likely influences the electric charge density in the $(COOH)_2$ cycles. This in turn strengthens the vibronic mechanism of the electronic current generation in the hydrogen bond cycles [45]. Separation of the carboxyl groups from aromatic rings by *methylene*

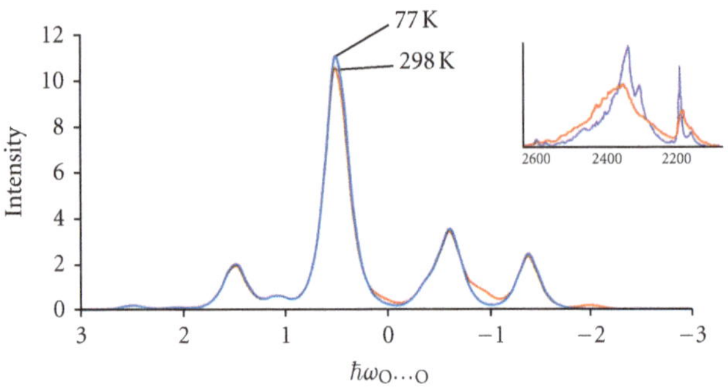

(a)

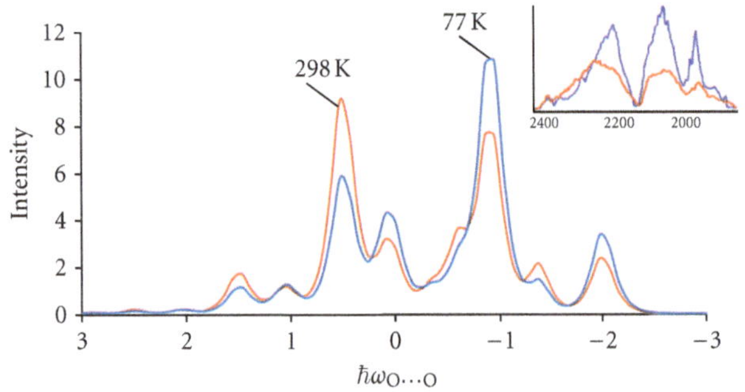

(b)

FIGURE 17: Temperature-induced evolution of the ν_{O-D} band-contour shapes accompanying the variation in the contribution rate of the two different exciton coupling mechanisms, that is, "*side-to-side*" (SS) and "*tail-to-head*" (TH). Numerical reproduction of the temperature effect in the spectra of deuterium-bonded: (a) *2-furanacetic acid* crystal (b) *2-furanacrylic acid* crystal.The relative contribution ratio of the SS and TH mechanism in the ν_{O-D} band generation is, for *2-furanacetic acid* crystal: 0.95 : 0.05 at 293 K and 0.95 : 0.05 at 77 K and for *2-furanacrylic acid* crystal: 0.65 : 0.35 at 293 K and 0.40 : 0.60 at 77 K. The experimental spectra are shown in inset.

groups (*arylacetic acids, furanacetic acids,* and *thiopheneacetic acids*) effectively weakens the vibronic coupling mechanism. Therefore, these latter systems belong to the A case.

The analyzed spectral properties of the two different crystalline systems, *2-furanacetic acid* and *2-furanacrylic acid,* are in a good agreement with the described above vibrational exciton interaction mechanisms of the spectra generation for cyclic hydrogen bond dimer. This remains in a close relation to the electronic properties of the two carboxylic acid molecules. For *2-furanacetic acid* dimers, the exciton interactions involving the dimer hydrogen bonds of a SS-type is only weakly temperature dependent. In the case of *2-furanacrylic acid* dimers, due to their electronic structure, the interhydrogen bond exciton coupling mechanism changes its character along with the changes in temperature. At very low temperatures, the TH-type interactions, transferred in the $(COOH)_2$ cycles via electrons are dominating. When temperature increases, this mechanism becomes less privileged as being annihilated by the hydrogen-bond atom thermal vibrational motions. It is replaced by the other mechanism depending of the SS-type interactions. Each individual mechanism generates its own spectrum characterized by its unique intensity distribution pattern. Therefore, the ν_{O-H} and ν_{O-D} bands in the spectra of *2-furanacrylic acid* crystals exhibit more complex fine structure patterns, since they are superposition of two different spectra, where each component spectrum is of a different origin. Each component spectrum contributing to the ν_{O-H} and ν_{O-D} band formation, with its statistical weight parameter depended of temperature, corresponds with another exciton interaction mechanism in the cyclic hydrogen bond dimers in the lattice.

Spectra of *2-thiopheneacrylic acid* crystals [50] exhibit qualitatively fairly similar properties as the spectra of *2-furanacrylic acid* crystals. Their ν_{O-H} and ν_{O-D} bands also demonstrate complex and dense fine structure patterns. They also show very similar temperature effects when compared with the corresponding spectra of *2-furanacrylic acid* crystals.

In turn, the spectra of *2-thiopheneacetic acid* crystals [50] exhibit qualitatively very similar properties as the spectra of *2-furanacetic acid* crystals. Their ν_{O-H} and ν_{O-D} bands also exhibit relatively simple fine structure patterns. They also demonstrate fairly similar temperature effects when compared with the corresponding spectra of *2-furanacetic acid* crystals.

From the comparison of the spectra of the two different groups of carboxylic acid crystals, it results that the electronic structure of the associating molecules is the main factor determining the crystal spectral properties in IR, differentiating the spectral properties of the two groups of hydrogen-bonded systems. Namely, the temperature effects registered in IR spectra of the hydrogen bond in carboxylic acid crystals remain in a close connection with the electronic spectra of the associating molecules forming cyclic hydrogen-bonded dimers in the lattices.

11. The Problem of the Vibrational Selection Breaking in IR Spectra of Centrosymmetric Hydrogen Bond Dimers

The mechanism proposed in this paper for understanding the sources of temperature effects in the IR spectra of cyclic centrosymmetric hydrogen bond dimers explains the generation of the lower-frequency ν_{O-H} and ν_{O-D} band branches of extremely high intensities in IR spectra of carboxylic acid crystals. However, at this stage, the relation with the formerly published vibronic mechanism of the vibrational rule selection breaking in the IR spectra of centrosymmetric hydrogen bond dimers [35] ought to be discussed since both mechanisms can generate and also explain qualitatively fairly similar spectral effects.

The vibronic mechanism was originally elaborated in the past for the understanding of the fine structure patterns of the published earlier IR spectra of the cyclic, centrosymmetric N–H··· S bond dimers formed by 2-thiopyridone and 2-mercaptobenzothiazole molecules as well as extremely nonregular H/D isotopic effects in the spectra [37, 39, 51]. The isotopic effects were expressed by the unusually narrow ν_{N-D} bands in correspondence to the very wide ν_{N-H} bands characterized by complex fine structure patterns. In terms of the vibronic model, these effects were explained by the disappearance of the intensity of the lower-frequency branch of the ν_{N-D} bands attributed to the N–D bond totally symmetric stretching vibrations in the dimers, due to the weakening of the forbidden transition promotion mechanism [35]. In the case of the ν_{N-H} bands, the promotion mechanism was effective enough generating the forbidden transition spectral branch of noticeably high intensity. Nevertheless, this branch appeared to be less intense when compared with the allowed transition, higher-frequency branch of the ν_{N-H} band. The vibronic model ascribed these effects to the difference in the proton and deuteron vibration anharmonicity and to the extremely high polarizability of the N–H$\cdots$S hydrogen bonds in 2-thiopyridone and 2-mercaptobenzothiazole dimers. These factors were considered as responsible for the magnitude of the vibrational selection rule breaking effects in the dimeric IR spectra [35].

The IR spectra of carboxylic acid crystals with cyclic dimers in their lattices considerably differ by the analogous H/D isotopic effects from the spectra of the N–H$\cdots$S bonded dimers [22–27, 37, 39]. In the case of carboxylic acid crystals practically no impact of the isotopic substitution onto the relative intensity of the lower-frequency band branch intensities of the ν_{O-H} and ν_{O-D} bands in relation to the corresponding higher-frequency band branch intensities. can be noticed. Also the incidentally observed very high intensities of the forbidden transition bands distinguish these IR spectra of carboxylic acid crystals. This proves that the spectra generation mechanism for the carboxylic acid dimers in the crystals essentially differs from the vibronic selection rule breaking mechanism [35].

The following question arises in the scope of our latest estimations: should the vibronic mechanism be definitively rejected as inadequate in the description of the IR spectral properties of centrosymmetric hydrogen bond dimers, especially carboxylic acid dimers in the solid state?

From our hitherto studies of IR spectra of hydrogen-bonded molecular crystals, it results that the two different mechanisms forming the band structures act parallel, each with its individual statistical weight, depending of the electronic properties of the molecular systems forming the dimers. In the case of cyclic dimeric N–H$\cdots$S bonded molecular systems, the vibronic mechanism appeared to be relatively very sufficient, leading to the appearance of intense forbidden transition ν_{N-H} band branches. On the other hand, the ν_{N-D} bands are extremely narrow as practically devoid of the forbidden band branch [37, 39]. The vibronic mechanism is also effective in the generation of IR spectra of crystals with infinite open chains of hydrogen bonded molecules, for example, N-methylthioacetamide [42] or N-phenylacrylamide [52] crystals. Also the H/D isotopic effects in their spectra are fairly similar to the analogous isotopic effects in the corresponding spectra of the N–H$\cdots$S bonded cyclic dimers. In these chain structures, centrosymmetric hydrogen bond dimeric systems are composed of hydrogen bonds, where each moiety belongs to another chain of associated molecules penetrating a unit cell. Most probably, the chain structure of the molecular associates, which excludes the possibility of the induction to circulating electric currents in such dimers, as well as the polarization properties of these hydrogen bonds, is responsible for the existence of the vibronic mechanism [35] in the pure form, influencing the band contour formation.

For the carboxylic acid dimer spectra, the mechanism proposed in this work is dominant regardless of the electronic structure of the substituent atomic groups linked to the carboxyl groups in the molecules. On the basis of the "*state-of-art*" in the spectral studies of the hydrogen bond systems in molecular crystals, the H/D isotopic effects in the spectra seem to be the main criterion for distinguishing these two individual mechanisms. However, this problem demands further intensive studies in the future.

12. Conclusions

In this paper, we report experimental and theoretical study of IR spectra of *2-furanacetic* acid and of *2-furanacrylic acid* crystals measured at 293 K and 77 K in the ν_{O-H} and ν_{O-D} band frequency ranges. The corresponding spectra of the two individual systems strongly differ. Indeed, in the case of *2-furanacetic acid* spectra, the fine structure pattern of each band, ν_{O-H} and ν_{O-D}, is relatively simple. Each band consists of a low number of well-separated spectral lines. In the spectra of *2-furanacrylic acid,* each considered band is composed of a noticeably larger number of lines. In addition, the temperature effect characterizing the bands is not the same for the two compounds. The results presented in this paper for *2-furanacetic acid* and *2-furanacrylic acid* allow for the following observations and conclusions.

(1) The crystal IR spectral properties remain in a close relation with the electronic structure of the two different molecular systems. The vibronic coupling mechanism involving the hydrogen bond protons and the electrons on the π-electronic systems in the molecules determines the way in which the vibrational exciton coupling between the hydrogen bonds in the carboxylic acid dimers occurs.

(2) The analyzed spectral properties of the two different crystalline systems, *2-furanacetic acid* and 2-furanacrylic acid, are in a good agreement with the vibrational exciton interaction mechanisms of the spectra generation for cyclic hydrogen bond dimer.

(3) For *2-furanacetic acid* dimers, the exciton interactions involving the dimer hydrogen bonds of the SS type are only weakly temperature dependent. A weak "through-space" coupling in *2-furanacetic acid* dimers of a van der Waals type is responsible for the SS-type coupling.

(4) In the case of *2-furanacrylic acid* dimers, due to their electronic molecular structure, the interhydrogen bond exciton coupling mechanism strongly changes its character along with the changes in temperature. Strong coupling in *2-furanacrylic acid* dimers prefers a TH-type Davydov coupling widespread by the π-electrons. At very low temperatures, the TH-type interactions, transferred in the $(COOH)_2$ cycles via

electrons are dominating. This mechanism becomes less privileged at higher temperature as annihilated by the hydrogen-bond atom thermal vibrational motions.

(5) Each individual mechanism, that is, the TH and SS, generates its own spectrum characterized by its unique individual intensity distribution pattern. As we can see, the ν_{O-H} and ν_{O-D} bands in the spectra of *2-furanacrylic acid* crystals exhibit more complex fine structure patterns, since they are superposition of two different spectra, where each component spectrum is of a different origin. Each component spectrum contributing to the ν_{O-H} and ν_{O-D} bands formation, with its temperature-dependent statistical weight, corresponds with the different exciton interaction mechanism, TH or SS, acting in the cyclic hydrogen bond dimers in the lattice. This explains the observed difference in the temperature-induced evolution of the compared spectra.

References

[1] C. Pimentel and A. L. McClellan, *The Hydrogen Bond*, W. H. Freeman and Co, San Francisco, Calif, USA, 1960.

[2] P. Schuster, G. Zundel, and C. Sandorfy, *The Hydrogen Bond*, vol. 1–3, North-Holland, Amsterdam, The Netherlands, 1976.

[3] G. L. Hofacker, Y. Marechal, and M. A. Ratner, "The dynamical aspects of hydrogen bonds," in *In The Hydrogen Bond, Recent Developments in Theory and Experiment*, W. P. Schuster, G. Zundel, and C. Sandorfy, Eds., vol. 1, p. 295, North-Holland, Amsterdam, The Netherlands, 1976.

[4] P. Schuster and W. Mikenda, *Hydrogen Bond Research, Monatshefte für Chemie, Chemical Monthly*, vol. 130, Springer, New York, NY, USA, 8th edition, 1999.

[5] D. Hadzi, Ed., *Theoretical Treatments of Hydrogen Bonding*, Wiley, New York, NY, USA, 1997.

[6] A. Witkowski, "Infrared spectra of the hydrogen-bonded carboxylic acids," *The Journal of Chemical Physics*, vol. 47, no. 9, pp. 3679–3680, 1967.

[7] Y. Marechal and A. Witkowski, "Infrared spectra of H-bonded systems," *The Journal of Chemical Physics*, vol. 48, no. 8, pp. 3697–3705, 1968.

[8] S. F. Fischer, G. L. Hofacker, and M. A. Ratner, "Spectral behavior of hydrogen-bonded systems: quasiparticle model," *The Journal of Chemical Physics*, vol. 52, no. 4, pp. 1934–1947, 1970.

[9] O. Henri-Rousseau and P. Blaise, "The infrared spectral density of weak hydrogen bonds within the linear response theory," *Advances in Chemical Physics*, vol. 103, pp. 1–137, 1998.

[10] O. Henri-Rousseau and P. Blaise, "The V_{X-H} line shapes of centrosymmetric cyclic dimers involving weak hydrogen bonds," *Advances in Chemical Physics*, vol. 139, pp. 245–496, 2008.

[11] M. J. Wójcik, "Theoretical interpretation of infrared spectra of the Cl–H stretching vibration in the gaseous (Ch3)2O· · · HCl complex," *International Journal of Quantum Chemistry*, vol. 29, no. 4, pp. 855–865, 1986.

[12] J. L. Leviel and Y. Marechal, "Infrared spectra of H-bonded systems: anharmonicity of the H-bond vibrations in cyclic dimers," *The Journal of Chemical Physics*, vol. 54, no. 3, pp. 1104–1107, 1971.

[13] J. Bournay and Y. Maréchal, "Dynamics of protons in hydrogen-bonded systems: propynoic and acrylic acid dimers," *The Journal of Chemical Physics*, vol. 55, no. 3, pp. 1230–1235, 1971.

[14] P. Excoffon and Y. Marechal, "Infrared spectra of H-bonded systems: saturated carboxylic acid dimers," *Spectrochimica Acta A*, vol. 28, no. 2, pp. 269–283, 1972.

[15] M. J. Wójcik, "Theory of the infrared spectra of the hydrogen bond in molecular crystals," *International Journal of Quantum Chemistry*, vol. 10, no. 4, pp. 747–760, 1976.

[16] H.T. Flakus and A. Bańczyk, "Abnormal distribution of protons and deuterons between the hydrogen bonds in cyclic centrosymmetric dimers in partially deuterated samples," *Journal of Molecular Structure*, vol. 476, no. 1–3, pp. 57–68, 1999.

[17] H. T. Flakus, "Vibronic model for H/D isotopic self-organization effects in centrosymmetric dimers of hydrogen bonds," *Journal of Molecular Structure*, vol. 646, no. 1–3, pp. 15–23, 2003.

[18] H. T. Flakus and A. Michta, "Investigations of interhydrogen bond dynamical coupling effects in the polarized IR spectra of acetanilide crystals," *Journal of Physical Chemistry A*, vol. 114, no. 4, pp. 1688–1698, 2010.

[19] R. W. G. Wyckoff, *Crystal Structures*, vol. 5, Wiley, New York, NY, USA, 1972.

[20] Z. Berkovitch-Yellin and L. Leiserowitz, "Atom-atom potential analysis of the packing characteristics of carboxylic acids. A study based on experimental electron density distributions," *Journal of the American Chemical Society*, vol. 104, no. 15, pp. 4052–4064, 1982.

[21] E. B. Wilson, J. C. Decius, and P. C. Cross, *Molecular Vibrations; The Theory of Infrared and Raman Vibrational Spectra*, McGraw- Hill, New York, NY, USA, 1955.

[22] H. T. Flakus and A. Miros, "Infrared spectra of the hydrogen bonded glutaric acid crystals: polarization and temperature effects," *Journal of Molecular Structure*, vol. 484, no. 1–3, pp. 103–115, 1999.

[23] H. T. Flakus and M. Chelmecki, "Infrared spectra of the hydrogen bond in benzoic acid crystals: temperature and polarization effects," *Spectrochimica Acta A*, vol. 58, no. 1, pp. 179–196, 2002.

[24] H. T. Flakus and M. Jabłońska, "Study of hydrogen bond polarized IR spectra of cinnamic acid crystals," *Journal of Molecular Structure*, vol. 707, no. 1–3, pp. 97–108, 2004.

[25] H. T. Flakus and M. Chełmecki, "Polarization IR spectra of the hydrogen bond in phenylacetic acid crystals: H/D isotopic effects-temperature and polarization effects," *Spectrochimica Acta Part A*, vol. 58, no. 9, pp. 1867–1880, 2002.

[26] H. T. Flakus and M. Chełmecki, "Polarization IR spectra of hydrogen bonded 1-naphthoic acid and 2-naphthoic acid crystals: electronic effects in the spectra," *Journal of Molecular Structure*, vol. 659, no. 1–3, pp. 103–117, 2003.

[27] H. T. Flakus and M. Chełmecki, "Polarization IR spectra of the hydrogen bond in 1-naphthylacetic and 2-naphthylacetic acid crystals: H/D isotopic effects. Temperature and polarization effects," *Journal of Molecular Structure*, vol. 705, no. 1–3, pp. 81–89, 2004.

[28] S. E. Filippakis and G. M. J. Schmidt, "Topochemistry. Part XVI. The crystal structure of trans-β-2-furylacrylic acid," *Journal of the Chemical Society B*, pp. 229–232, 1967.

[29] M. Danish, S. Ali, M. Mazhar, A. Badshah, and E. R. T. Tieking, "Crystal structure of 3-(2-Furyl)acrylic Acid, $C_7H_6O_3$," *Zeitschrift für Kristallographie*, vol. 210, no. 9, p. 703, 1995.

[30] H. T. Flakus and A. Tyl, "Polarized IR spectra of the hydrogen bond in acetic acid crystals," *Chemical Physics*, vol. 336, no. 1, pp. 36–50, 2007.
[31] H. T. Flakus and B. Stachowska, "A systematic study of polarized IR spectra of the hydrogen bond in formic acid crystals," *Chemical Physics*, vol. 330, no. 1-2, pp. 231–244, 2006.
[32] A. Tyl, E. Chełmecka, M. Jabłońska et al., "X-ray analysis at 150 K, synthesis and theoretical calculations of 1-naphthalene-acrylic acid," *Structural Chemistry*, vol. 23, no. 2, pp. 325–323, 2012.
[33] H. T. Flakus, M. Jabłońska, and P.G. Jones, "Study of polarized IR spectra of the hydrogen bond system in crystals of styrylacetic acid," *Spectrochimica Acta A*, vol. 65, no. 2, pp. 481–489, 2006.
[34] G. Fisher, *Vibronic Coupling*, Acadamic Press, London, UK, 1984.
[35] H. T. Flakus, "On the vibrational transition selection rules for the centrosymmetric hydrogen-bonded dimeric systems," *Journal of Molecular Structure C*, vol. 187, pp. 35–53, 1989.
[36] H. T. Flakus, A. Pyzik, A. Michta, and J. Kusz, "'Reversal' exciton coupling effect in the IR spectra of the hydrogen bond cyclic dimers; polarized IR spectra of 3-hydroxy-4-methyl-2(3H)-thiazolethione crystals," *Vibrational Spectroscopy*, vol. 44, no. 1, pp. 108–120, 2007.
[37] H. T. Flakus and A. Tyl, "Strong vibrational exciton coupling effects in polarized IR spectra of the hydrogen bond in 2-thiopyridone crystals," *Vibrational Spectroscopy*, vol. 47, no. 2, pp. 129–138, 2008.
[38] H. T. Flakus, A. Tyl, and A. Maślankiewicz, "Electron-induced phase transition in hydrogen-bonded solid-state 2-pyridone," *Journal of Physical Chemistry A*, vol. 115, no. 6, pp. 1027–1039, 2011.
[39] H. T. Flakus, A. Miros, and P. G. Jones, "Influence of molecular electronic properties on the IR spectra of dimeric hydrogen bond systems: polarized spectra of 2-hydroxybenzothiazole and 2-mercaptobenzothiazole crystals," *Journal of Molecular Structure*, vol. 604, no. 1, pp. 29–44, 2002.
[40] H. T. Flakus and A. Machelska, "Polarization IR spectra of hydrogen bonded pyrazole crystals: self-organization effects in proton and deuteron mixture systems. Long-range H/D isotopic effects," *Spectrochimica Acta Part A*, vol. 58, no. 314, pp. 553–566, 2002.
[41] H. T. Flakus, A. Tyl, and P. G. Jones, "'Self-organization' processes in proton and deuteron mixtures in open-chain hydrogen bond systems: Polarization IR spectra of 4-mercaptopyridine crystals," *Spectrochimica Acta A*, vol. 58, no. 2, pp. 299–310, 2002.
[42] H. T. Flakus, W. Śmiszek-Lindert, and K. Stadnicka, "Strong vibronic coupling effects in polarized IR spectra of the hydrogen bond in N-methylthioacetamide crystals," *Chemical Physics*, vol. 335, no. 2-3, pp. 221–232, 2007.
[43] C. A. Davydov, *Teorya Molekularnykh Ekscitonov*, Nauka, Moscow, Russia, 1968.
[44] R. L. Hochstrasser, *Molecular Aspects of Symmetry*, W. A. Benjamin Inc., New York, NY, USA, 1966.
[45] L. A. Nafie, "Adiabatic molecular properties beyond the Born-Oppenheimer approximation. Complete adiabatic wave functions and vibrationally induced electronic current density," *The Journal of Chemical Physics*, vol. 79, no. 10, pp. 4950–4957, 1983.
[46] H.T. Flakus, "The effect of strong coupling between vibrations in hydrogen bonds on the polarized spectra of the mercaptobenzothiazole crystal: an "anomalous" isotopic effect," *Chemical Physics*, vol. 62, no. 1-2, pp. 103–114, 1981.
[47] P. Blaise, M. J. Wojcik, and O. Henri-Rousseau, "Theoretical Interpretation of the Lineshape of the Gaseous Acetic Acid Dimer.," *Journal of Chemical Physics*, vol. 122, Article ID 064306, 2005.
[48] N. Rekik, H. Ghalla, H. T. Flakus, M. Jabłońska, P. Blaise, and B. Oujia, "Polarized infrared spectra of the H(D) bond in 2-thiophenic acid crystals: a spectroscopic and computational study," *ChemPhysChem*, vol. 10, no. 17, pp. 3021–3033, 2009.
[49] R. Najeh, G. Houcine, H. T. Flakus, M. Jablonska, and O. Brahim, "Experimental and theoretical study of the polarized infrared spectra of the hydrogen bond in 3-thiophenic acid crystal," *Journal of Computational Chemistry*, vol. 31, no. 3, pp. 463–475, 2010.
[50] H. T. Flakus, N. Rekik, and A. Jarczyk, "Polarized IR spectra of the hydrogen bond in 2-thiopheneacetic acid and 2-thiopheneacrylic acid crystals: H/D isotopic and temperature effects," *The Journal of Physical Chemistry A*, vol. 116, no. 9, pp. 2117–2130, 2012.
[51] L. J. Bellamy and P. E. Rogasch, "Proton transfer in hydrogen bonded systems," *Proceedings of the Royal Society A*, vol. 257, pp. 98–108, 1960.
[52] H. T. Flakus, A. Michta, M. Nowak, and J. Kusz, "Effects of dynamical couplings in IR spectra of the hydrogen bond in N-phenylacrylamide crystals," *Journal of Physical Chemistry A*, vol. 115, no. 17, pp. 4202–4213, 2011.

3

An Analytic Analysis of the Diffusive-Heat-Flow Equation for Different Magnetic Field Profiles for a Single Magnetic Nanoparticle

Brenda Dana[1] and Israel Gannot[2]

[1] *Department of Electrical Engineering, Faculty of Engineering, Tel-Aviv University, 69978 Tel-Aviv, Israel*
[2] *Department of Biomedical Engineering, Faculty of Engineering, Tel-Aviv University, 69978 Tel-Aviv, Israel*

Correspondence should be addressed to Israel Gannot, gannot@eng.tau.ac.il

Academic Editor: Yuval Garini

This study analytically analyzes the changes in the temperature profile of a homogenous and isotropic medium having the same thermal parameters as a muscular tissue, due to the heat released by a single magnetic nanoparticle (MNP) to its surroundings when subject to different magnetic field profiles. Exploring the temperature behavior of a heated MNP can be very useful predicting the temperature increment of it immediate surroundings. Therefore, selecting the most effective magnetic field profile (MFP) in order to reach the necessary temperature for cancer therapy is crucial in hyperthermia treatments. In order to find the temperature profile caused by the heated MNP immobilized inside a homogenous medium, the 3D diffusive-heat-flow equation (DHFE) was solved for three different types of boundary conditions (BCs). The change in the BC is caused by the different MF profiles (MFP), which are analyzed in this article. The analytic expressions are suitable for describing the transient temperature response of the medium for each case. The analysis showed that the maximum temperature increment surrounding the MNP can be achieved by radiating periodic magnetic pulses (PMPs) on it, making this MFP more effective than the conventional cosine profile.

1. Introduction

Magnetic Hyperthermia (MH) is one of many approaches currently being tested for cancer therapy [1–3]. The goal of this approach is to specifically heat the regions containing the cancerous cells by means of the magnetic losses caused by the physical properties of the magnetic nanoparticles (MNPs) when being exposed to an external magnetic field (MF).

The MNPs that are often used in MH, are usually made of ferromagnetic or ferrimagnetic materials which strongly react to the externally applied MF [4]. This magnetic reaction is converted by the two dominant relaxation mechanisms, the Néel mechanism and Brownian mechanism, into power dissipation or heat [1, 4].

The eddy currents losses contribution may be neglected due to the low electrical conductivity that characterizes ferro- or ferrimagnetic materials and due to the small particle radius [5–9].

Fannin et al. [10] pointed out that for small enough particles, the anisotropy energy barrier, E_a, may become so small, that thermal energy fluctuations can overcome it and spontaneous reverse the magnetization of a particle from one easy direction to the other, even in the absence of an applied field. The time it takes for a spontaneously fluctuation to reverse the magnetization after overcoming the energy barrier is characterized by a time constant, referred to as the Néel relaxation time, or τ_N. The probability of such a transition is proportional to $\exp(\sigma)$, where σ is the ratio of anisotropy energy to thermal energy or (E_a/k_BT) [11].

The other distinct mechanism, by which the magnetization of MNPs may relax after an applied field has been removed, is the physical rotational Brownian motion of the particle immobilized inside a medium. When a magnetic field is applied to MNPs, they rotate and progressively align with the magnetic field due to the torque generated by the interaction of the magnetic field with the magnetization [12].

The time associated for an MNP to align with a small external magnetic field is given by the Brownian relaxation time, τ_B [13].

Because these relaxation mechanisms happen simultaneously, they both contribute to the total magnetization and the heat losses and their total influences can be express by an effective relaxation time, τ_e, which is a combination of τ_N and τ_B [14–16].

The two relevant mechanisms to change the magnetization of magnetic particles in an external field are given in Figure 1.

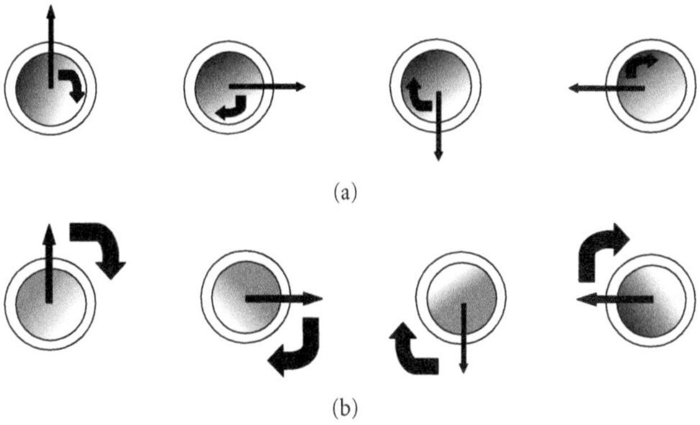

FIGURE 1: (a) Néel rotation of magnetization inside a fixed magnetic particle due to the spontaneously reversing the magnetization from one easy direction to the other (the particle does not rotate); (b) Brownian rotation of an MNP due to the rotation of the torque generated by the interaction of the magnetic field with the magnetization (the particle rotates as a whole) [17].

Moreover, our interests in MNPs as heat sources derive from the fact that they are vastly used as MRI agents [17] and their controllable size, ranging from few nanometers to tens of nanometers [18]. This means that the MNP size is smaller than or as same as that of a protein (5–50 nm), a virus (20–450 nm) or a gene (2 nm wide and 10–100 nm long) [11], which enables them to penetrate through the leaky pathological vasculature into the tumor interstitial easily reaching any cell of interest in the body including cancerous cells [19].

In addition, the MNPs can be attached to a specific type of cancerous cells causing a controllable elevation of temperature in them with almost no effects on healthy cells [20, 21]. By selectively heating the cancerous cells to a temperature ranging from 42 to 46°C, one can damage the tumors without causing vast harm to the healthy surrounding tissue [17, 19, 22].

Furthermore, in order to ensure that the treatment is biologically noninvasive and thermally tolerated for extended period of time, an experimentally measured tolerable limit of induced heating by an alternative MF was conducted defining a limit to the product of the MF strength (H) and the frequency (f) of the MF (e.g., $H \cdot f \leq 4.85 \cdot 10^8$ A/m·s [23] or a less rigid criterion $H \cdot f \leq 5 \cdot 10^9$ A/m·s [24]).

Due to the MNPs submicron length size, the conventional approach to heat conduction problems using macroscopic empirical laws such as Fourier's law or Joule's law of heat generation requires justification and even breakdown when the length scale of the system is comparable to the carrier mean free path or when the time scale of the physical process is smaller than the relaxation time of the heat carriers [25]. In this case, transport of heat carriers must be treated using the Boltzmann transport equation as Chen et al. pointed out [26].

Chen [27] suggested that heat is transported by carriers comprising of electrons, phonons, and photons. In dielectric materials, the heat conduction is dominated by phonons, in pure metals predominantly by electrons, and in impure metals or alloys by both phonons and electrons [27, 28]. Therefore, the mean free path of the heat carriers, for an MNP with a Fe core is approximately 0.8 nm [29, 30] and for a biological tissue 0.5 nm [26, 31, 32] allowing the conventional approach to be used for particles having a radius bigger than 10 nm.

Consequently, the temperature gradient caused by the release of the magnetic energy which an MNP absorbs to its immediate surroundings can be found analytically when applying Fourier transforms (FTs) to the DHFE [33] as Shih et al. [34] and Yuan et al. [35] suggested. Using this technique, Liu and Xu [36] analyzed the influences that a sinusoidal heat flux source placed on the skin surface have on the temperature inside it, and Tjahjono [37] analytically analyzed the heating temperature of a slab embedded with gold NPs due to a constant magnetic flux.

By analytically solving the DHFE for different boundary conditions, one can easily describe the dependence of the solution on each parameter composing it, such as the radius of the MNP, the frequency of the MF, and the material properties [38]. This allows us to optimize the solution for better performances, reaching the highest temperature elevation under specific constrains, for example the radius of the MNP. Moreover, when exploring the solution analytically, other parameters and their influence may be observed more clearly, which are usually neglected, or not explored (e.g., the MFP and its effects on the temperature gradient).

Until recently, the MFP was poorly analyzed in context of hyperthermia treatments and how it influences the temperature distribution concerning biological materials and tissues surrounding MNPs. Previous work focused on exploring the influences of different magnetic profiles on biological tissues. These studies were mostly experimental and did not focus on the MNPs contribution to the temperature change when exposed to different types of magnetic field profiles [39–43].

Recently, a numerical simulation model based on the Landau-Lifshitz-Gilbert equation was created for simulating MNPs ensembles when exposed to an incident square wave [44], as opposed to the usual sine wave. This work showed an increase in the normalized heat released by MNPs by at least 50%, as well as a more constant heating efficiency over the spectrum of particle anisotropies due to the infinite number of harmonics contained in an ideal square wave with the possibility of much greater improvement depending on the magnetic anisotropies, volumes, and angles to the incident radiation. However, Morgan and Victora [44] did not elaborate on the temperature dependencies on space and time near the MNP surface, but mostly focused on the dependencies between the angle of the incident wave relative to the anisotropy axis of the MNPs and the magnitude of the normalized output power released from them.

Therefore, the primary aim of this paper is to explore the transient analysis of the changes in temperature (from the steady state temperature $T_b = 310.15°K$), as a function of the external MFP applied to a single MNP. By doing so, one can select the most efficient MFP that may improve the efficiency of MH treatments allowing the MF strength and the frequency reductions in order to meet the requirement $H \cdot f \leq 4.85 \cdot 10^8$ A/m·s [23].

The aims of this paper are as follows:

(i) To construct a theoretical model of the magnetic losses for the three different MF profiles studied in this article as follows:

(1) Case 1—a cosine profile [18, 37],
(2) Case 2—a PMP [45, 46],
(3) Case 3—a discontinuous cosine profile.

(ii) To explore the maximal temperature elevation and the rate of the temperature change near the MNP's surface and into the tissue-like surrounding it for each of the above cases.

(iii) To investigate who the core radius influences the maximal rate of the temperature change and the maximal temperature value, in order to find the optimal core radius that should be used for each of the above cases [5, 47].

(iv) To study the effective confining heat depth (ECHD), symbolized as δ (see Figure 2), where the temperature elevation has a significant influence for each of the above cases.

2. Methods

In this study, we analytically model the transient temperature field (TTF) produced by a single MNP inside a homogenous and uniform medium having the same thermal parameters as a cancerous muscle cell. The analysis for each of the three MFPs mentioned in Section 1 is presented, after solving the DHFE inside the medium surrounding the MNP with the proper BC corresponding to its specific MFP.

In order to simplify the solution of DHFE that gave us the TTF and the temperature rate change due to the magnetic losses, some assumptions were made:

(a) The properties of the surrounding medium are constant and homogeneous having the same thermal properties as the macroscopic-scale muscular tissue [48].

(b) The temperature on the surface of the MNP is uniform.

(c) There is a negligible emission and evaporation.

(d) There are no "thermally significant" blood vessels near the zone of interest therefore the perfusion is negligible.

(e) The metabolic-heat generation is neglected.

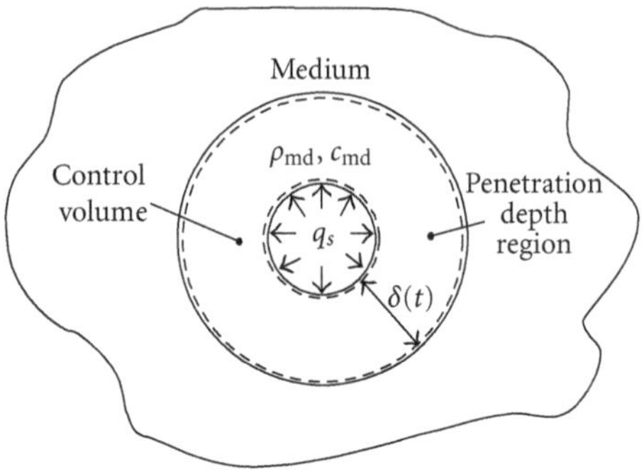

FIGURE 2: The control volume where the conductive analysis is preformed. q_s'' is the constant heat flux released from the MNP after absorbing the magnetic energy and δ is the thickness of the penetration region [37].

2.1. The Thermodynamic Analysis. The TTF originating from the surface of a single MNP can be found after solving the 3D DHFE in the homogenous medium surrounding it [39]. The general DHFE can be written in spherical coordinates (due to the problem's symmetry) as follows:

$$k_m \nabla^2 T_m(r,t) = \rho_m c_m \frac{\partial T_m(r,t)}{\partial t}, \tag{1}$$

where ρ_m (kg m^{-3}) is the mass density, c_m (J kg^{-1}°C^{-1}) the specific heat, and k_m (Wm^{-1}°C^{-1}) the thermal conductivity of the phantom tissue.

This equation was also used by Keblinski et al. [38] and Govorov et al. [49] for solving nanoscale heat problems.

The general BC for this heat problem is given as follows:

$$-k_m \cdot \nabla T_m(r,t)|_{r=a} = q_s''(t), \tag{2}$$

where a: is the radius of the MNP in meters and $q_s''(t)$ (Wm^{-2}) is the heat flux.

The DHFE is valid if the mean free path of the heat carrier phonon or electron is smaller than the characteristic feature size as mentioned in Section 1. For amorphous solids and liquids, due to lack of crystalline, the mean free path is very short and of the order of atomic distances. Consequently, the heat flow in cells can be well described by the diffusive heat equation, even when nanoscopic length scales are involved [31].

Based on the above considerations, we evaluate the temperature field arising from continuous heating of a single particle by solving the heat equation (1) in the region outside the solid sphere surrounding the MNP where there are no heat sources, using a constant heat-flux-boundary condition at the particle surface, caused by the magnetic losses inside the MNP. The constant heat flux from the MNP's surface becomes heat input to the medium which is then stored within the volumetric penetration depth region as shown in Figure 2.

After solving (1) and (2) (see the detailed formulations in Appendix A (A.1)–(A.12)), the temperature elevation inside the medium surrounding the MNP can be expressed using (A.12) as follows:

$$\Delta T_m(r,t) = \frac{\theta(R,t)}{k_m r}\sqrt{\alpha_m} = \frac{\phi(R,t)^* q_s''(t)}{k_m r}\sqrt{\alpha_m}, \quad (3)$$

where $*$ symbolizes the convolution between two functions, $R = \sqrt{1/\alpha_m} r$ and $\alpha_m = k_m/\rho_m c_m$.

In order to analytically calculate (3), the general expression of $q_s''(t)$ must be found for each case, which depends on $H(t)$ (Am^{-1}), the magnetic field, and on $M(t)$ (Am^{-1}), the magnetization.

When a linear and isotropy material is assumed, the relation between $M(t)$ and $H(t)$ in the frequency domain (using FTs) may be described by the magnetic susceptibility [5]

$$\widetilde{M}(\omega) = \int dt' \chi(t') e^{-i\omega t'} \int dt H(t) e^{-i\omega t} = \widetilde{\chi}(\omega)\widetilde{H}(\omega). \quad (4)$$

The magnetic susceptibility $\widetilde{\chi}(\omega)$ in the frequency domain can be expressed as [4]

$$\widetilde{\chi}(\omega) = \frac{\chi_0}{1+i\omega\tau_e} = \frac{\phi\mu_0 M_s{}^2 v_p}{3k_B T}\frac{1}{1+i\omega\tau_e}, \quad (5)$$

where χ_0 is the static susceptibility, $\tau_e = \tau_{\text{Neel}} || \tau_{\text{Brown}}$ is the effective relaxation time given by Fannin [14], ϕ is the volume fraction solid [4], v_p the particle volume, μ_0 the vacuum permeability, k_B is the Boltzmann constant, and M_s is the magnetic saturation.

Moreover, in order to calculate the total heat generated by a single MNP caused by the conversion of the absorbed magnetic energy to heat inside a linear ferromagnetic medium, we must introduce Poynting's theorem for small electric fields and neglecting ohmic losses [5, 6] as follows:

$$\begin{aligned} \nabla \cdot S''_{\text{out}}(t) &= -\int d\omega \int d\omega' \overline{H}(\omega) \cdot H(\omega)\omega\mu_0 \operatorname{Im}(\mu(\omega)) e^{i(\omega-\omega')t} - \frac{\partial U(t)}{\partial t} \\ &= -P_{\text{Loss}}(t) - \frac{\partial U(t)}{\partial t}, \end{aligned} \quad (6)$$

where $S''_{\text{out}}(t)$ represents the energy flowing out through the boundary surfaces of the volume per unit time, $\overline{H}(\omega)$ is the conjugate of $H(\omega)$, $\mu_0 = 4\pi 10^{-7}$ (Vs/Am) is the vacuum permeability, $\mu(\omega) = \mu_r(\omega) - i\mu_{\text{im}}(\omega) = \mu_0(1+\widetilde{\chi}(\omega))$ is the complex magnetic permeability [5, 6], Im( ) is the imaginary part of $\mu(\omega)$, $\partial U(t)/\partial t$ is the time rate change of the effective electromagnetic energy density given by (7), and $P_{\text{Loss}}(t)$ represents the conversion of the magnetic energy into heat not counting conductive losses. It is worth mentioning that only the imaginary part of the complex permeability is causing heat losses, and $\partial U(t)/\partial t$ can be found using [5, 6] as follows:

$$\begin{aligned} H \cdot \frac{\partial B}{\partial t} &= \int d\omega \int d\omega' \overline{H}(\omega) \cdot H(\omega)\omega\mu_0 \operatorname{Im}\mu(\omega) e^{i(\omega-\omega')t} \\ &\quad + \frac{\mu_0 \partial}{2\partial t}\int d\omega \int d\omega' \overline{H}(\omega) \cdot H(\omega)\frac{d}{d\omega}[\omega\overline{\mu}(\omega)] \\ &\quad \times e^{i(\omega-\omega')t} \\ &= P_{\text{Loss}}(t) + \frac{\partial U(t)}{\partial t}, \end{aligned} \quad (7)$$

where $\overline{\mu}(\omega)$ is the conjugate of $\mu(\omega)$.

Next, the explicit analytic expressions for the temperature gradient profile from the equilibrium tempertaure, $\Delta T_m(r,t)$ and $\partial T_m(r,t)/\partial t$ are deduced for three different BC derived from the three MFPs mentioned earlier in Section 1.

2.2. *The Analytic Expressions of the TTF for Three Different MFPs*

Case 1 (a cosine profile). The magnetic field has a cosine profile so

$$H(t) = A\cos(\omega_0 t). \quad (8)$$

Taking the inverse FT of (A.19) deduced in Appendix A using (A.13)–(A.19), one can find that $\theta(R,t)$ can be written in this case as

$$\begin{aligned} \theta(R,t) &= \frac{a\mu_0 A^2\omega_0}{6}\frac{\omega_0\chi_0\tau}{1+(\omega_0\tau)^2} \\ &\quad \cdot \left(\frac{aR_0 e^{i2\omega_0 t}}{2(R_0\sqrt{i2\omega_0}+1)}\exp\left(-\sqrt{i2\omega_0}R\right)\right. \\ &\quad \left.+\frac{aR_0 e^{-i2\omega_0 t}}{2(R_0\sqrt{-2\omega_0 i}+1)}\exp\left(-\sqrt{-i2\omega_0}R\right)+1\right), \end{aligned} \quad (9)$$

where $\theta(R,t)$ is a function in the complex domain, therefore the temperature profile has a magnitude and phase as often occurs in many problems of physics or engineering such as theory of heat conduction, particularly when nonsteady heat conduction is concerned [50, 51]. Moreover, (9) is related to the TTF by (3).

Sometimes, the derivative of the temperature profile, or the rate of the change in the temperature surrounding the MNP is taken in consideration in order to verify that the treatment is safe for inducing controlled MH [47, 52]. For Case 1, this equals

$$\begin{aligned} \frac{\partial\theta(R,t)}{\partial t} &= i\omega_0\frac{a\mu_0 A^2\omega_0}{6}\frac{\omega_0\chi_0\tau}{1+(\omega_0\tau)^2} \\ &\quad \cdot \left(\frac{aR_0 e^{i2\omega_0 t}}{(R_0\sqrt{i2\omega_0}+1)}\exp\left(-\sqrt{i2\omega_0}R\right)\right. \\ &\quad \left.-\frac{aR_0 e^{-i2\omega_0 t}}{(R_0\sqrt{-2\omega_0 i}+1)}\exp\left(-\sqrt{-i2\omega_0}R\right)\right). \end{aligned} \quad (10)$$

Case 2 (a PMP). The magnetic field has a rectangular pulse shape profile with a period of $T_s = 2\pi/\omega_0$ and a pulse width of Δ, so

$$H(t) = 2A \quad 0 \le t \le \Delta, \qquad H(t) = 0 \quad \Delta \le t \le T_s. \tag{11}$$

The amplitude of the pulse wave was chosen to be twice the amplitude of the cosine single in order to maintain the same peak-to-peak value, for this case and the previous one. For this case, the temperature elevation as a function of time can be expressed using the inverse FT of (A.25) found in Appendix A that was deduced using (A.20)–(A.25) to receive the following:

$$\begin{aligned}
\theta(R,t) &= \frac{a}{3}\frac{4\mu_0 A^2\omega_0}{\pi^2}\sum\sum\frac{\sin(m\pi\Delta/T_s)}{m}\sin\left(\frac{n\pi\Delta}{T_s}\right)\frac{n\omega_0\chi_0\tau}{1+(n\omega_0)^2\tau^2} \\
&\cdot\left(\frac{aR_0 e^{i(n+m)\omega_0 t}}{R_0\sqrt{i(n+m)\omega_0}+1}\exp\left(-\sqrt{i(n+m)\omega_0}R\right)\right. \\
&\quad+\frac{aR_0 e^{-i(n+m)\omega_0 t}}{R_0\sqrt{-i(n+m)\omega_0}+1}\exp\left(-\sqrt{-i(n+m)\omega_0}R\right) \\
&\quad+\frac{aR_0 e^{i(m-n)\omega_0 t}}{R_0\sqrt{i(m-n)\omega_0}+1}\exp\left(-\sqrt{i(m-n)\omega_0}R\right) \\
&\quad\left.+\frac{aR_0 e^{-i(m-n)\omega_0 t}}{R_0\sqrt{-i(m-n)\omega_0}+1}\exp\left(-\sqrt{-i(m-n)\omega_0}R\right)\right) \\
&+\frac{a}{3}4\omega_0\mu_0 A^2\cdot\frac{\Delta}{\pi T_s}\sum\sin\left(\frac{n\pi\Delta}{T_s}\right)\frac{n\omega_0\chi_0\tau}{1+(n\omega_0)^2\tau^2} \\
&\times\left(\frac{aR_0 e^{in\omega_0 t}}{R_0\sqrt{in\omega_0}+1}\exp\left(-\sqrt{in\omega_0}R\right)\right. \\
&\quad\left.+\frac{aR_0 e^{-in\omega_0 t}}{R_0\sqrt{-in\omega_0}+1}\exp\left(-\sqrt{-in\omega_0}R\right)\right).
\end{aligned}\tag{12}$$

Again, (12) is related to the TTF by (3).

As for Case 1, we can calculate the rate of the change in the temperature surrounding the MNP and receive the following:

$$\begin{aligned}
\frac{\partial\theta(R,t)}{\partial t} &= i\omega_0^2\frac{a}{3}\frac{4\mu_0 A^2}{\pi^2}\sum\sum\frac{\sin(m\pi\Delta/T_s)}{m} \\
&\times\sin\left(\frac{n\pi\Delta}{T_s}\right)\frac{n\omega_0\chi_0\tau}{1+(n\omega_0)^2\tau^2} \\
&\cdot\left((n+m)\frac{aR_0 e^{i(n+m)\omega_0 t}}{R_0\sqrt{i(n+m)\omega_0}+1}\right. \\
&\times\exp\left(-\sqrt{i(n+m)\omega_0}R\right) \\
&-(n+m)\frac{aR_0 e^{-i(n+m)\omega_0 t}}{R_0\sqrt{-i(n+m)\omega_0}+1} \\
&\times\exp\left(-\sqrt{-i(n+m)\omega_0}R\right) \\
&+(m-n)\frac{aR_0 e^{i(m-n)\omega_0 t}}{R_0\sqrt{i(m-n)\omega_0}+1} \\
&\times\exp\left(-\sqrt{i(m-n)\omega_0}R\right) \\
&-(m-n)\frac{aR_0 e^{-i(m-n)\omega_0 t}}{R_0\sqrt{-i(m-n)\omega_0}+1} \\
&\left.\times\exp\left(-\sqrt{-i(m-n)\omega_0}R\right)\right) \\
&+i\omega_0^2\frac{a}{3}4\mu_0 A^2\frac{\Delta}{\pi T_s}\sum\sin\left(\frac{n\pi\Delta}{T_s}\right) \\
&\cdot\frac{n^2\omega_0\chi_0\tau}{1+(n\omega_0)^2\tau^2} \\
&\times\left(\frac{aR_0 e^{in\omega_0 t}}{R_0\sqrt{in\omega_0}+1}\exp\left(-\sqrt{in\omega_0}R\right)\right. \\
&\quad\left.-\frac{aR_0 e^{-in\omega_0 t}}{R_0\sqrt{-in\omega_0}+1}\exp\left(-\sqrt{-in\omega_0}R\right)\right).
\end{aligned}\tag{13}$$

Case 3 (a discontinuous cosine profile). The magnetic field has a periodic discontinuous cosine profile with a time constant of $T_s = 2\pi/\omega_1$ and a pulse width of Δ, so

$$\begin{aligned}
H(t) &= A\cos(\omega_0 t) \quad 0 \le t \le \Delta, \\
H(t) &= 0 \quad \Delta \le t \le T_s \quad \omega_0 \ne \omega_1.
\end{aligned}\tag{14}$$

For this third case, the temperature elevation as a function of time can be expressed using the inverse FT of (A.35) found in Appendix A that was deduced using (A.30)–(A.35) to receive:

$$\begin{aligned}
\theta(R,t) &= \frac{a}{3}\mu_0\left(\frac{A}{T_s}\right)^2 aR_0 \\
&\times\sum\sum\left[\left(\frac{\sin((\Delta/2)(\omega_0+m\omega_1))}{\omega_0+m\omega_1}\right.\right. \\
&\quad\left.+\frac{\sin((\Delta/2)(\omega_0-m\omega_1))}{\omega_0-m\omega_1}\right) \\
&\times\left(\frac{\sin((\Delta/2)(\omega_0+n\omega_1))}{\omega_0+n\omega_1}\right. \\
&\quad\left.+\frac{\sin((\Delta/2)(\omega_0-n\omega_1))}{\omega_0-n\omega_1}\right)
\end{aligned}$$

$$\cdot \frac{(n\omega_1)^2 \chi_0 \tau}{1+(n\omega_1)^2\tau^2} \times \left(\frac{e^{i(n+m)\omega_1 t}}{R_0\sqrt{i(n+m)\omega_1}+1} \times \exp\left(-\sqrt{i(n+m)\omega_1}R\right) + \frac{e^{-i(n+m)\omega_1 t}}{R_0\sqrt{-i(n+m)\omega_1}+1} \times \exp\left(-\sqrt{-i(n+m)\omega_1}R\right) + \frac{e^{i(m-n)\omega_1 t}}{R_0\sqrt{i(m-n)\omega_1}+1} \times \exp\left(-\sqrt{i(m-n)\omega_1}R\right) + \frac{e^{-i(m-n)\omega_1 t}}{R_0\sqrt{-i(m-n)\omega_1}+1} \times \exp\left(-\sqrt{-i(m-n)\omega_1}R\right)\right)\Bigg]. \tag{15}$$

Again, (15) is related to the TTF by (3).

As for Cases 1 and 2, we can calculate the rate of the change in the temperature surrounding the MNP to receive the following:

$$\frac{\partial\theta(R,t)}{\partial t} = i\omega_1 \frac{a}{3}\mu_0 \left(\frac{A}{T_s}\right)^2 \times \sum\sum\Bigg[\left(\frac{\sin((\Delta/2)(\omega_0+m\omega_1))}{\omega_0+m\omega_1} + \frac{\sin((\Delta/2)(\omega_0-m\omega_1))}{\omega_0-m\omega_1}\right) \times \left(\frac{\sin((\Delta/2)(\omega_0+n\omega_1))}{\omega_0+n\omega_1} + \frac{\sin((\Delta/2)(\omega_0-n\omega_1))}{\omega_0-n\omega_1}\right) \cdot \frac{(n\omega_1)^2 \chi_0 \tau}{1+(n\omega_1)^2\tau^2} \times \left((n+m)\frac{aR_0 e^{i(n+m)\omega_1 t}}{R_0\sqrt{i(n+m)\omega_1}+1} \times \exp\left(-\sqrt{i(n+m)\omega_1}R\right) - (n+m)\frac{aR_0 e^{-i(n+m)\omega_1 t}}{R_0\sqrt{-i(n+m)\omega_1}+1} \times \exp\left(-\sqrt{-i(n+m)\omega_1}R\right) + (m-n)\frac{aR_0 e^{i(m-n)\omega_1 t}}{R_0\sqrt{i(m-n)\omega_1}+1} \times \exp\left(-\sqrt{i(m-n)\omega_1}R\right) - (m-n)\frac{aR_0 e^{-i(m-n)\omega_1 t}}{R_0\sqrt{-i(m-n)\omega_1}+1} \times \exp\left(-\sqrt{-i(m-n)\omega_1}R\right)\right)\Bigg]. \tag{16}$$

For Cases 2 and 3, there is a limitation regarding the MFS and the frequency in order for the MH treatment to be safe (see (A.30) and (A.39)). Moreover, for frequencies lower than 10 MHz, there is essentially no attenuation of the MFS within muscle-equivalent materials, limiting the maximal harmonic frequency to 10 MHz [16].

In conclusion, (9)–(16) can be used to predict the TTP and the special temperature profile for a single-MNP subject to three different magnetic field profiles and using the same equations we can also explore the influence that the core radius has on the temperature profile estimating the ECHD for each case.

2.3. The Simulations Parameters. The mathematical expressions of the TTP were simulated using MATLAB and COMSOL (COMSOL results can be seen in Appendix B) for a single MNP immobilized inside a uniform and isotropic phantom medium having the same biological thermal properties as a muscular tissue [48] and are summarized in Table 1. These assumptions were made in order to simplify the theoretical calculations.

The thermal parameters are considered to be constant with temperature and space as will be latter proven. Moreover, the magnetic parameters of the MNP were measured at $T_b = 310.15°\text{K}$ based on the findings of Fannin [14] and are summed up in Table 2.

The external magnetic field strength (MFS) for all three cases, was chosen as $8.8\ \text{kAm}^{-1}$ and the MF frequency as $f_0 = 400\ \text{KHz}$. These values are based on previous works made by Kettering et al. [52], Hergt et al. [53], and Hilger et al. [54].

For all three profiles mentioned in this section, the simulations were plotted for $0 \le r - a \le 10\,\text{nm}$ and $0 \le t \le 5\,\mu\text{s}$, where r is the distance from the center of the MNP and a its radius. The upper value for distance simulation was chosen accordantly to the thickness of the cell membrane thatis about 5–10 nm [55–57] and damaging it can cause the destruction of the cell [58]. The upper time value was chosen so several cycles of the magnetic field could be simulated and plotted.

For all the simulations the volume fraction solid was defined as $\phi = 0.032$. This value is been justified in Section 4.

In Section 3, as already mentioned in Section 1, the maximal temperature elevation and the temperature change rate near the MNP's surface and into the tissue surrounding it, are simulated. Moreover, the influence the core radius has on the maximal temperature change rate and on the maximal temperature elevation was also explored, in order to find the

TABLE 1: Thermal tissue properties for the phantom muscle cell at a temperature of 310.15 K [48].

c_m (J kg^{-1}°C^{-1})	ρ_m (kg m^{-3})	k_m (W m^{-1}°C^{-1})
3500	1047	0.518

TABLE 2: The magnetic parameters of the MNP [14].

a (nm)	V_{hyd} (m^3)	K_u (kJm^{-3})	M_s (kAm^{-1})	γ (kμs^{-1}A^{-1}m)	α
10	$3.34 \cdot 10^{-26}$	9.6	300	202	0.1

optimal value that must be used for each case as suggested by Rosensweig [4] and Kappiyoor et al. [47]. Furthermore, the ECHD that is defined by the point the temperature reaches e^{-1} of the maximal value, has also been explored for each case, defining the confining heat region and can be compared with the thickness of a cell membrane which varies between 3–10 nm [53–56].

3. Results

The mathematical expressions of the TTP were simulated in this section, using MATLAB, for a single MNP immobilized inside a uniform and isotropic phantom medium having the same biological thermal properties as a muscular tissue (Table 1). Moreover, the MNP's magnetic parameters are summarized in Table 2.

For Case 1, the mathematical expression of the temporal and spatial temperature increment, (9), is presented in Figure 3 for $T_s = 2.5\,\mu$s.

It can be seen from Figure 3 that the temperature changes periodically with a time period of 1.25 μs that is equivalent to a frequency oscillation of 800 kHz, which is twice the frequency of the external applied MF, as predicted by (9). This can be explained by the multiplication of the magnetic field and the magnetic induction, both being a function of f_0 or ω_0.

Moreover, the temperature increment reached its maximum value after 0 μs, reaching $\Delta T_{\max} = 2.1$ n°K on the surface of the MNP. As expected, the hottest spots are on the surface of the MNP, and as the point of view gets further from the surface the temperature declines as (9) predicted. This value causes only a low-temperature gradient in the thermal properties of the surrounding medium therefore, the thermal parameters of the phantom cell can be considered constants, as assumed.

According to Figure 3, the temperature profile has a "DC" level that can be found from calculating the first term of (9) making the temperature increment to be always higher than the initial temperature as expected, because the magnetic losses inside the MNP irradiate heat to the medium surrounding it, at all times [59, 60].

Furthermore, at a distance of 12 nm apart from the MNP's surface the temperature maximal value equals 0.8 n°K that is equivalent to e^{-1} of the absolute maximal value, defining the ECHD or $\delta = 12$ nm.

In order to have a unique quantity to be compared with each case and does not depend on time, we averaged the TTP

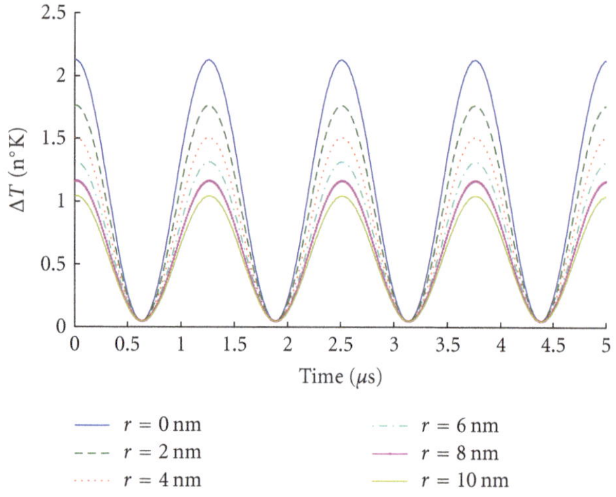

FIGURE 3: The temperature profile for a cosine MFP plotted as a function of time and as a function of the observation point located at a distance, r, from the surface of the MNP; r ranging from 0 nm to 10 nm for a core radius of 10 nm.

over one cycle. In Case 1, the averaged value over one time period equals 0.55 n°K.

Next, we explored the maximal temperature rise rate as a function of the core radius using (9) and (10) receiving the data in Figure 4.

As can be seen from Figure 4, the maximal temperature rate change equals 0.011°Ks^{-1} and the maximal temperature rise equals 4.7 n°K, both received for a core radius of 9.3 nm. The temperature rate rise and the maximal temperature are considerably small, due to the relaxation time that depends on the volume of the particle making this MFP to be safe to use for MH treatments [47]. For radii larger or smaller than 9.3 nm, the magnetic heat dissipation start to decrease as the magnetic relaxation time gets bigger or smaller, respectively, reducing the denominator or numerator in (9) and (10).

Equations (9) and (10) enable us to understand that the changes in the temperature depend on many parameters such as, the magnetic field strength, the magnetic field frequency, the magnetic properties of the material, and the core radius. Consequently, in order to optimize the heat losses we must select the most effective radius for a specific type of MNP.

For Case 2, the mathematical expression of the temporal and spatial temperature increment are plotted in Figure 5 for the summation of 25 indexes (not to exceed 10 MHz [16]) and $\Delta = 0.2\,T_s$, where $T_s = 2.5\,\mu$s.

For convenience, Figure 5 describes the temperature profile for the first two cycles as given by (12). This equation shows that the characteristic behavior of the temperature repeats itself every $T_s = 2.5\,\mu$s, that is, the cycle of the magnetic field, therefore one can limit the study to only a final number of cycles.

As can be seen from Figure 5, the temperature builds up very fast due to the steep elevation of the magnetic field caused by the Heaviside-shaped MP and reaches a maximal value of $\Delta T_{\max} = 8.8$ n°K on the MNP's surface after 0.45 μs. Then the temperature begins to drop after 0.1 μs from the

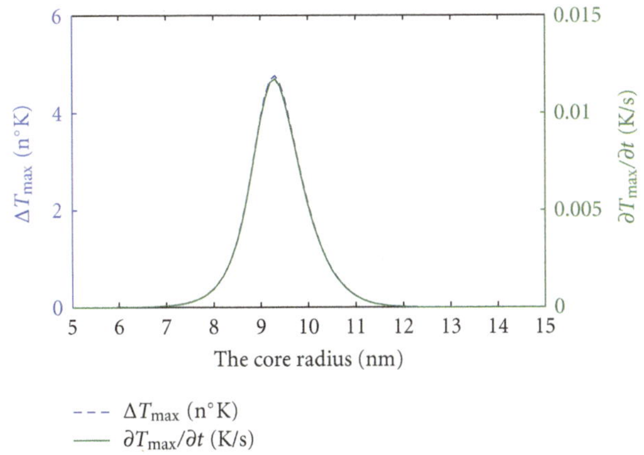

FIGURE 4: The maximal temperature rise rate and the maximal temperature for a cosine MFP plotted as a function of the core radius; a ranging from 5 nm to 15 nm, where the observation points are on the surface of the MNP.

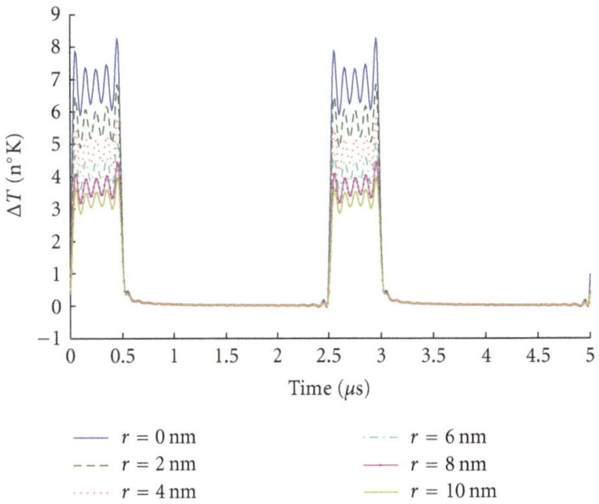

FIGURE 5: The temperature rise for a periodic pulse-shaped MFP having a pulse width of $0.2\,T_s$ plotted as a function of the observation point located at a distance, r from the surface of the MNP, r ranging from 0 nm to 10 nm for a core radius of 10 nm, and the number of indexed summed is 25.

time the MF was turned on reaching a minimal value of 6 n°K near the surface of the MNP. From that point on, the temperature profile temporal behavior is defined by the summation of the total numbers of harmonics composing the MF until the MF is turned off again as can be found from (12). Furthermore, the temperature reaches its maximum value close to the surface of the MNP and decreases with distance reaching a maximal value of 3 n°K, 12 nm apart from the MNP's surface.

For this case, the maximal value is 4 times higher than the one received in Case 1, making it a preferable MFP to be used in MH as Morgan and Victora suggested [44].

Again, the thermal parameters can be considered constant and not dependent on temperature near the MNP's surface because the temperature rise is less than 1°K.

From Figure 5, the ECHD can be found as $\delta = 12$ nm, which is the same as the value received in Case 1, meaning that the temperature decreases as fast as the cosine case and is confined to a specific area near the MNP's surface.

Moreover, in order to have a unique quantity to be compared in each case that does not depend on time, we averaged the total temperature rise over one cycle. In this case the averaged temperature elevation was 1.3 n°K after been normalized to the time period. This value is about 2.4 times higher than the value received in Case 1, making this MFP a better candidate for MH treatments.

Next, we explored the maximal temperature rise rate as a function of the core radius. For Case 2, we can use (11) and (12) receiving the data in Figure 6.

As seen from Figure 6, the absolute maximal temperature elevation equals $0.032\,\mu$°K received for a core radius of 8.3 nm, and the maximal temperature derivative 1.01 °Ks^{-1} is received for a core radius of 8.2 nm. Because this MFP produces temperature changes that are too rapid to be safe for inducing MH [47], the radius that we chosen for a safer treatment is in consistence with Case 1 and equals 10 nm. Consequently, the NP size plays an important role in

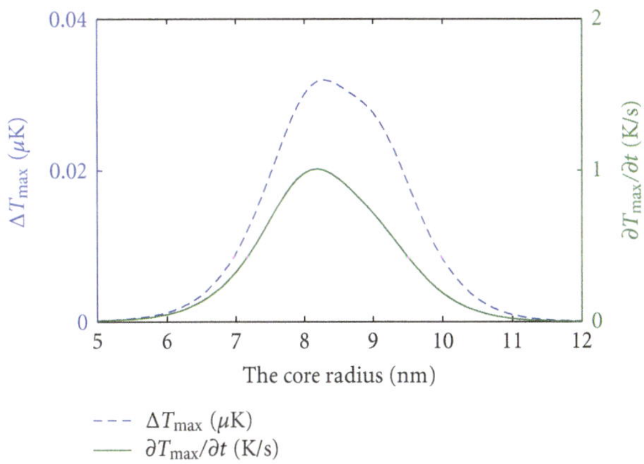

FIGURE 6: The absolute maximal temperature rise rate and the maximal temperature for a periodic pulse-shaped MFP having a pulsed width of $0.2T_s$ plotted as a function of the core radius; a ranging from 5 nm to 15 nm, the observation point are on the surface of the MNP, and the number of indexed summed is 25.

determining the amount of heating that an MFH treatment can provide as Kappiyoor et al. [47] already mentioned.

Again, the maximal temperature rate rise and the maximal temperature are considerably small due to the relaxation time that depends on the volume of the particle. For radii larger or smaller than 8.4 nm, the magnetic heat dissipation starts to decrease as the magnetic relaxation time gets bigger or smaller, respectively, due to its affect on the relaxation time, reducing the denominator or numerator in (11) and (12).

By comparing Case 2 to Case 1 we see that for the same MNP radius (10 nm) having the same magnetic material properties (given by Table 2), the maximal temperature rise received is about 4 times higher for Case 2 than in Case 1 and

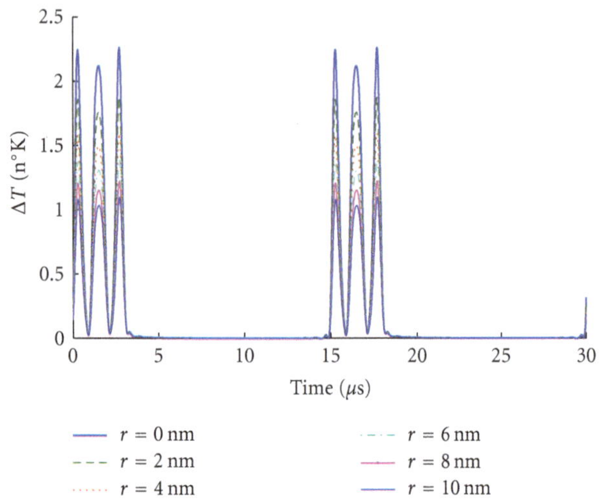

FIGURE 7: The temperature profile for a periodic discontinuous cosine MFP having a pulse width of $0.2T_s$ plotted as a function of time and as a function of the observation point located at a distance, r, from the surface of the MNP, r ranging from 0 nm to 10 nm for a core radius of 10 nm, and the number of indexed summed is 25.

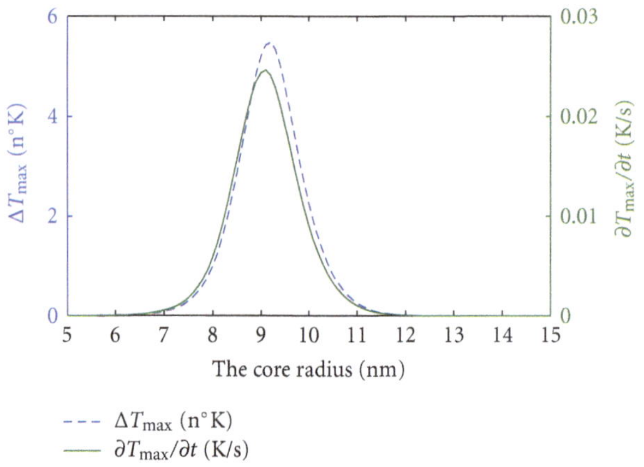

FIGURE 8: The absolute maximal temperature rise rate and the maximal temperature rise for a periodic discontinuous cosine MFP having a pulse width a pulsed width of $0.2T_s$ plotted as a function of the core radius, a ranging from 5 nm to 15 nm, the observation point are on the surface of the MNP, and the number of indexed summed is 25.

the maximal temperature derivative for this case is 40 times higher than Case 1, making the periodic pulse-shaped MFP a better magnetic field source for MH treatments.

For Case 3 the analytic expression for the TTP can be plotted for $\Delta = 0.2\,T_s$ and $T_s = 15\,\mu$s and are shown in Figure 7 for the summation of 25 indexes (not to exceed 10 MHz [16]). The cosine MF time period that multiplies the Heaviside function equals 2.5 μs, and is equivalent to a frequency of 400 KHz.

As Figure 7 shows, the changes in the temperature profile result from the MNP reaction to two different MFPs, the cosine profile and the periodic rectangular pulse profile. The last is responsible for switching on and off the MF.

The influence that the periodic rectangular-pulse-shaped MF has on the temperature gradient can be seen by the steep temperature elevation at the beginning and at the end of every cycle caused by the derivative of Heaviside function composing the magnetic flux density, $B(t)$, and the influence that the cosine MFP has on the temperature gradient can be seen as the cosine "ripple" that is added. This "ripple" has 3 peaks that are separated 1.25 μs apart, which is twice the cosine MF frequency received in Case 1. On the MNP surface, the maximal temperature gradient reaches $\Delta T_{\max} = 2.3$ n°K after 2 ns from the time the MF was turned on, and repeats itself every 15 μs, which is equivalent to the time period of the signal. This value is higher than the value received for the cosine-shaped MF, but lower than the one received in Case 2. However, after the highest peak the maximal value of the cosine "ripple" reaches the same one as in the cosine case or 2.1 n°K as expected.

For this case, the ECHD equals $\delta = 12$ nm that is the same as for the other two cases where the temperature change reaches a value of 0.8 n°K. After $0.2\,T_s$, the temperature elevation becomes insignificant as the MF is turned off.

Again, in order to have a unique quantity to be compared in each case that does not depend on time, we can average the total temperature rise over one cycle. In this case the average temperature elevation equals 0.235 n°K after we normalize it to the time period. This value is about 2.5 times lower than the value received in Case 1, making this MFP a less preferable candidate for MH treatments.

Next we explored the maximal temperature rise rate as a function of the core radius. For Case 3, we used (15) and (16) receiving the data in Figure 8.

As seen from Figure 8, the absolute maximal temperature rate elevation equals 5.4 n°K received for a core radius of 9.2 nm. The maximal temperature derivative $0.024°\mathrm{Ks}^{-1}$ is received for a core radius of 9.1 nm. Again, the maximal temperature rate rise and the maximal temperature are considerably small due to the relaxation time that depends on the volume of the particle. For radii larger or smaller than 9.2 nm, the magnetic heat dissipation starts to decrease as the magnetic relaxation time gets bigger or smaller, respectively, reducing the denominator or the numerator in (15) and (16).

By comparing Case 3 to Case 1 for a core radius of 10 nm and the same magnetic material properties (given by Table 2), the maximal temperature rise received for Case 3 is about two times higher than Case 1, and the maximal temperature derivative for this case is 2.2 times higher than Case 1. However, due to the long period for which the MF is turned off, and consequently the lower heat released from the MNP over one cycle, this MFP is less preferable than the cosine MFP, for MH treatments.

In order to make it easier to understand the differences between the three cases analyzed in this paper, Table 3 is added that summarizes the most significant parameters.

Moreover, a summarizing figure, Figure 9, describing the temperature rise as a function of time is also added for a particle with a core radius of 10 nm when the observation

TABLE 3: Summary of the most significant parameters received from all the simulations.

MATLAB simulation for a single MNP						
	The maximal temperature for a core radius of 10 nm (n°K)	ECHD (nm)	The temperature 12 nm apart from the surface of the MNP (n°K)	The most effective radius for maximal temp. rise (nm)	The maximal temperature derivative at the optimal radius (K/s)	The average temperature over one cycle for a core radius of 10 nm (n°K)
Case 1	2.1	12	0.8	9.3	0.011	0.55
Case 2	8.8	12	3	8.3	1.1	1.36
Case 3	2.3	12	0.8	9.2	0.024	0.235

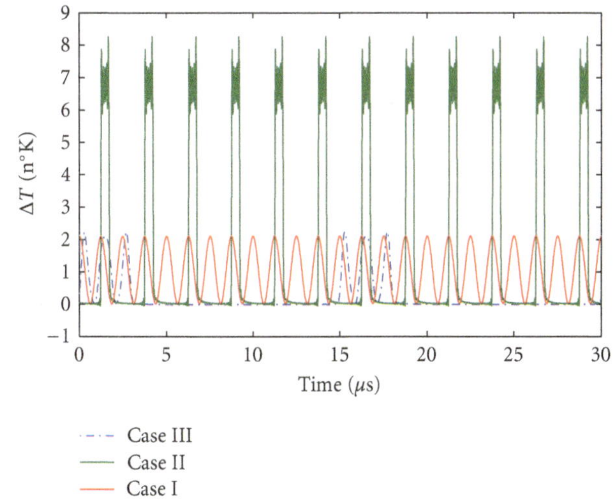

FIGURE 9: The temperature rise profiles plotted as a function of time for a core radius of 10 nm and an observation point located on surface of the MNP, for all the cases explored in this paper Case 1 (red) the cosine MFP, Case 2 (green) the periodic pulse-shaped MFP with a duty cycle of 0.2 and $N = 25$, Case 3 (blue) the periodic discontinuous cosine MFP with a duty cycle of 0.2 and $N = 25$.

point is on the MNP surface, in order to easily evaluate the differences in the three cases studied in this article.

4. Discussion

An analytical analysis of the TTP was preformed for three MFPs. The mathematical models were received by solving the DHFE for different BC matching each MFP, using the FTs.

Major work have been done in the past to solve the DHFE equation for a cosine-MF source as can be found in [36, 37, 49, 59]. Keblinski et al. [38] found that a laser source having a constant power of $1.4 \cdot 10^{-8}$ W heating a single MNP with a radius of 65 nm can cause a temperature change of 0.06 K at the particle surface. Moreover, for a cosine-MF heat source the local temperature was found to be even lower, causing a maximum change in temperature of 0.1 m°K for a particle having a radius of 50 nm at a frequency of 2 MHZ [48]. Both results are negligible from the point of view of biological applications as expected.

However, Keblinski et al. [38] and others [4, 20] solved the DHFE equation only for a constant heat flux having the average power of a cosine-MF, without exploring the temperature temporal behavior. In addition, until now there has not been an explicit mathematical formulation that solves the DHFE equation for other periodic MFPs that can be used as radiation sources for MH treatments. Morgan and Victora [44] showed that the use of an incident square wave, as opposed to the usual sine wave, increases the normalized power heat by at least 50%, however this conclusion was based on calculating only the Poynting vector and not based the solving the DHFE in order to find the explicit temperature change.

In consequence to the above, we should explore the influences that different magnetic irradiation profiles have on the induced temperature gradients inside tumor cells, for the same physical and thermal MNP's parameters in order to verify what Morgan and Victora [44] suggested. Furthermore, optimizing the heat power is of great importance from biological point of view. A typical cell having a diameter ranging from 2–10 μm [61] can uptake a maximal quantity of anionic MNPs, that varies between $2.8 \cdot 10^5$ and $7.2 \cdot 10^6$ per cell, consequently limiting the total amount of magnetic material per cell. Moreover, high concentration of MNPs with different types of coatings can cause a toxic reaction to the central nervous system [62] or may cause cellular perturbations [63] therefore, it is important to reduce the MNP's concentration. Nevertheless, reaching these quantities *in vivo* proves to be a very difficult task all types of cancerous cells [64, 65]. Hence, one must optimize other parameters such as, the profile of the MF in order to use lower magnetic concentration in order to reach the same temperature gradient values.

Consequently, this paper focuses on the influences that three different MFPs have on the temperature surrounding a single MNP, as mentioned in Section 1, when being exposed to it, analytically proving to be the most effective one in causing the highest temperature rise, using the same magnetic and thermal parameters.

For all three cases, the MATLAB (in Section 3) and COMSOL (in Appendix B) simulations results showed, that the maximum temperature rise for a given core radius of 10 nm ranges between 2.1 n°K and 8.8 n°K depending on the MFP.

Similar results were received by Keblinski et al. [38] and Rabin [31] for a constant heat flux and an MNP's having approximately the same physical and magnetic properties. The very low absolute change in temperature caused by a single MNP can be explained by its low magnetic susceptibility χ_0, and by the effective relaxation time that changes drastically with the MNP's volume [15]. Therefore, a single MNP can release only a small amount of heat causing a very small change in the temperature surrounding it.

However, for larger magnetic concentration occupying a single cell such as 1 ng of Fe_3O_4 per human cell, that is, equivalent to $5 \cdot 10^7$ MNPs per cell (for a particle radius 10 nm and cell radius 5 μm), Linh et al. [66] and Balivada et al. [67] showed that a local temperature elevation of several degrees can be reached making MH treatments effective. In addition, these quantities were also proposed by Vera and Bayazitoglu [59], Chan et al. [61], Huang [64], and Melancon et al. [65] who proved their efficiency in inducing MH. Consequently, one can produce a significant global temperature increment inside a cell even if the local temperature increment of each particle is negligible as long as we heat many particles in the same volume of interest.

For the multiparticle case, Rabin [31], and Keblinski et al. [38] calculated the temperature rise inside a spherical region with radius R (m) consisting of many randomly dispersed heat sources using $\Delta T_{\text{global}}(t) = (R^2 \Delta T_{\text{nano}}(t)/r_p^2)(4\pi/3)r_p^3 \rho_N$, where ρ_N (m^{-3}) is the number of MNPs per unit volume, k ($\text{Wm}^{-1\circ}\text{C}^{-1}$) is the thermal conductivity of the medium, r_p (m) is the radius of MNPs, and $\Delta T_{\text{nano}}(t)$ is the temperature gradient caused by a single MNP.

For $\rho_N = 5 \cdot 10^{21}$ (m^{-3}) and an average tumor cell radius of $r_{\text{cell}} = 7\,\mu\text{m}$ [52], the number of MNPs inside a single cell can be calculated to equal $8 \cdot 10^6$ that fits the concentrations found by Linh et al. [66] and Balivada et al. [67]. From ρ_N, we calculated the distances between two neighboring particles that is approximately 58 nm. This means that the volume fraction of the MNPs inside the cell is about 0.02.

By choosing a solid volume fraction of $\phi = 0.032$, the calculated distance between two neighboring particles is about 50 nm fitting a concentration of $\rho_N = 8 \cdot 10^{21}\,\text{m}^{-3}$, that is, in the toxicity safety range, for a tumor cell having an average radius of 7 μm [52, 66].

Due to the large distances between the particles, we assumed that the interparticle interactions are negligible, so the relaxation time and magnetic susceptibility can be calculated using the same expression as (5) [68].

The total temperature increment for the three cases analyzed in this paper can be found by substituting the received values for the single-MNP case, (9), (12), and (15) into $\Delta T_{\text{global}}(t)$ when average tumor radius of $R = 4$ mm was assumed in consistence with magnitudes of cancer tumors [31, 38, 67].

For the cosine MFP an average value of $\Delta T_{\text{global_cos}}(t) = 2.9^\circ$K over one cycle is received near the MNP's surface. This means that the MNP's concentration is not sufficient to give increment to a dramatic temperature gradient under the parameters summarized in Tables 1 and 2.

In order to receive a 6° increment, that is needed for MH, in the temperature near the MNP's surface, a larger amount than the proposed of particles is required.

For the PMP, an average value of $\Delta T_{\text{global_pulsed}}(t) = 7.2^\circ$K over one cycle is received for the same parameters summarized in Tables 1 and 2 that is sufficient to induce MH.

For the discontinuous pulse-shaped MF, a maximum peak of $\Delta T_{\text{global_pulsed_cos}}(t) = 1.23^\circ$K over one cycle is received, meaning that the MNP's concentration in this case as in Case 1 is not sufficient to ensure that MH can occur.

The comparison between the three temperature gradients received for each case, shows that the preferable MFP for MH is the PMP one, compared to Case 1 and Case 3. For Case 2, the temperature gradient at the surface of the MNP is sufficient to cause damage to biologic cells [58, 69, 70]. Therefore, using a periodic pulse MFP can reduce the necessary amount of MNPs by a factor or even more, allowing a wider range of markers to be used for hyperthermia treatments, and simplifying the biological processes to conjugate them to a cell.

In addition, we also explored the influence that the MNP's radius has on the maximal temperature gradient an on its rate rise. As seen from Figures 4, 6, and 8, the NP size has a great influence on determining the amount of heat released from the MFP's surface effecting both the temperature gradient as well as the temperature rise rate, as previous works showed [50, 51, 70].

For the first Case 1 studied, the optimal core radius was found as 9.3 nm where the maximal temperature reaches 4.7 n°K and the temperature change rate equals $0.011^\circ\text{Ks}^{-1}$ (Figure 4). This optimal radius was also received by Kappiyoor et al. [47] for almost the same MF properties and magnetic material properties. However, because the equation solved by Kappiyoor et al. [47] is different than (1), the maximal value is slightly lower that the values received by Rosensweig [4] and Kappiyoor et al. [47]. Moreover, the maximal value is also affected by the parameters chosen to describe the magnetic properties of the MNP, as demonstrated by Kappiyoor et al. [47]. Our magnetic parameters are slightly different than the ones used by Rosensweig [4] and Kappiyoor et al. [47], which may account for the differences in the maximal values in this study as Kappiyoor et al. [47] showed.

For Case 2 studied, the optimal core radius was found as 8.3 nm where the maximal temperature gradient reaches 32 n°K, and the temperature change rate equals 1.1°Ks^{-1} for a summation of 25 indexes (Figure 6). As can be seen by comparison, there is a benefit in using a PMP rather than a cosine MF, due to the higher temperature gradient received in the MNP's surrounding and the sufficient average temperature gradient received per cycle that is about 2.5 times higher in Case 2 than in Case 1.

Although for a total summation of 25 indexes, the temperature change rate is approximately 1°Ks^{-1}, (suggested to be less safe [47]), one can reduce the received value by limiting the number of the summation indexes composing the MF to a lower number such as $N = 10$ instead of $N = 25$, making the treatment safer but also maintaining higher temperature values that in Case 1 (Figure 14). Furthermore, when looking at the results of multiplying each coefficient's amplitude with its matched harmonic, the limitation for the treatment to be biologically noninvasive remains valid as long as $A_{\text{eff}} f_0 \le 5 \cdot 10^9$ A/m·s as mathematically justified in (A.29) and (A.30) limiting the total summation index to a value lower than $N = 25$.

For the Case 3 studied, the optimal core radius was found as 9 nm where the maximal temperature reaches $0.64\,\mu^\circ$K and the temperature change rate equals $27.3^\circ \mathrm{Ks}^{-1}$ (Figure 8). Although the values are higher than the ones received for the first case, the average temperature elevation was found to be lower after normalizing it to the time period, making this type of MFP less preferable.

As we can see for all the three cases analyzed in this paper, the optimal radius depends very much on the magnetic material properties [47] and on the profile of the magnetic field as we have proven in Section 2. Therefore, for each case studied and for each magnetic material the equations, developed for Cases 1–3 must be solved separately in order to optimize the MH treatment.

Another interesting finding driven from the mathematical equations, is the confinement of the temperature to an area having an average radius of 12 nm from the MNP's surface for all three cases. This means that most of the heat dissipation occurs in the vicinity of the heat sources confining the temperature increment in the proximity of tumor cells alone unaffecting the healthy cells.

The importance of this paper lies in the fact that, until now there was no explicit mathematical formulation that solves the DHFE equation for other types of periodic MFPs used as excitation sources for MH treatments. As we found out, changing the profile of the MF radiation can induce higher temperature gradients in tumor cells, for the same physical and thermal parameters enabling reduction of the MNPs concentration per cell. This is of great importance because, a typical cell has a maximal quantity of MNPs that it can uptake, and because high concentration of MNPs with different types of coatings can cause a toxic reaction to the central nervous system [62]. Therefore, lowering the magnetic concentration per cell, but still receiving the same temperature gradients may be of great use.

With the outcome of this paper, we are moving forward to in vitro studies in order to verify the theoretical results received in this paper experimentally.

5. Conclusions

This study investigates the effects of different heat-flux profiles on a single MNP immobilized inside a phantom having the same thermal properties as a muscle tissue. The exact solution of DHFE was solved for different boundaries conditions using FTs. According to the analytic solutions, the PMP profile was found to be the more effective in rising the temperature of the medium surrounding the MNP than the cosine profile, making it a better candidate for hyperthermia treatments rather than the conventional cosine MP.

Moreover, in order to reach a significant temperature gradient for all cases studied (a) a cosine profile, (b) a PMP profile, and (c) a discontinuous cosine profile, there is a need for a larger number of MNPs to be immobilized inside the cell medium as Rabin [31] and Keblinski et al. [38] previously suggested. Using their techniques, a significant temperature rise was achieved for the periodic pulse-shaped MF in comparison to the other two cases studies.

In order to understand the influences that a denser cluster has on the temperature gradient, other studies should be done investigating the interparticle interactions affecting the temperature increment and its derivative.

Appendices

A. Methods

In this appendix, we are deducing the equations for the temperature profiles introduced in Section 2, step by step. For simplicity, new variables are used to solve (1), where:

$$R = \sqrt{\frac{1}{\alpha_m}}r, \qquad \alpha_m = \frac{k_m}{\rho_m c_m},$$
$$\theta(R,t) = k_m(T_m(R,t) - T_b)R, \qquad R_0 = \sqrt{\frac{1}{\alpha_m}}a. \tag{A.1}$$

Therefore, by substituting the new variables from (A.1) into the left part of (1) we receive the following:

$$k_m \nabla^2 T_m(r,t) = \frac{k_m}{r}\frac{\partial^2 r T_m(r,t)}{\partial^2 r}\bigg|_{\substack{T_m(r,t)\to\theta(R,t)/k_m R\\ r\to R\sqrt{\alpha_m}}} = \frac{1}{\alpha_m}\frac{\partial^2\theta(R,t)}{R\partial^2 R}. \tag{A.2}$$

And by substituting the new variables from (A.1) into the right part of (1) we receive that:

$$\rho_m c_m \frac{\partial T_m(r,t)}{\partial t} = \frac{\rho_m c_m}{k_m}\frac{\partial\theta(R,t)}{R\partial t}\bigg|_{\alpha_m\equiv k_m/(\rho_m c_m)} = \frac{1}{\alpha_m}\frac{\partial\theta(R,t)}{R\partial t}. \tag{A.3}$$

So (1) can be rewritten as follows:

$$\frac{\partial^2\theta(R,t)}{\partial^2 R} = \frac{\partial\theta(R,t)}{\partial t}. \tag{A.4}$$

The same procedure can be done to the BC, substituting the new variables from (A.1) into the left part of (2) to receive the following:

$$-k_m\nabla\left(\frac{T_m(r,t)}{r}\right)\bigg|_{r=a} = -k_m\frac{\partial T_m(r,t)}{\partial r}\bigg|_{\substack{T_m(r,t)\to\theta(R,t)/k_m R\\ r\to R\sqrt{\alpha_m}}} = -\frac{1}{\sqrt{\alpha_m}}\frac{\partial}{\partial R}\left(\frac{\theta(R,t)}{R}\right) = -\frac{1}{\sqrt{\alpha_m}}\nabla_R\left(\frac{\theta(R,t)}{R}\right)\bigg|_{R=R_0}, \tag{A.5}$$

And (2) can be rewritten as follows:

$$-\nabla\left(\frac{\theta(R,t)}{R}\right)\bigg|_{R=R_0} = q_s''(t)\sqrt{\alpha_m}. \tag{A.6}$$

By taking the FT of (A.4) (defined as in (4)) one receives the transformation in the frequency domain, so

$$0 = i\omega\widetilde{\theta}(R,\omega) - \frac{\partial^2\widetilde{\theta}(R,\omega)}{\partial^2 R}. \tag{A.7}$$

The general solution of (A.7) can be found as follows:

$$\widetilde{\theta}(R,\omega) = c_2(\omega)e^{-\sqrt{i\omega}R}. \quad \text{(A.8)}$$

Substituting (A.8) into the LT of (A.6), the BC can be written as follows:

$$c_2\frac{a\sqrt{i\omega}+\sqrt{\alpha_m}}{a^2} = \widetilde{q}_s''(\omega)e^{\sqrt{i\omega}R_0}. \quad \text{(A.9)}$$

So

$$\begin{aligned}\widetilde{\theta}(R,\omega) &= \frac{a^2}{\sqrt{\alpha_m}\left((a/\sqrt{\alpha_m})\sqrt{i\omega}+1\right)}\widetilde{q}_s''(\omega)e^{-\sqrt{i\omega}(R-R_0)}\\ &= \frac{aR_0}{R_0\sqrt{i\omega}+1}\widetilde{q}_s''(\omega)e^{-\sqrt{i\omega}(R-R_0)}.\end{aligned} \quad \text{(A.10)}$$

Using technical computing software (Maple or/and Wolfram Mathematica) the inverse FT of $\widetilde{\theta}(R,\omega)/\widetilde{q}_s''(\omega) = \widetilde{\phi}(R,\omega)$ for $t > 0$, can be found by substituting $i\omega \to s$ in (A.10) and taking the inverse Laplace transform of the received equation so that

$$\phi(R,t) = a\left(\frac{e^{-(R-R_0)^2/4t}}{\sqrt{\pi t}} - \frac{\operatorname{erfc}\left((R-R_0)/2\sqrt{t}+\sqrt{t}/R_0\right)e^{(R-R_0)/R_0+t/R_0^2}}{R_0}\right). \quad \text{(A.11)}$$

This function converges to 0 for $t \to \infty$ or/and for $R - R_0 \to \infty$.

So the changes in the temperature can be found using (A.1) and (A.10) as follows.

$$\Delta T_m(r,t) = \frac{\theta(R,t)}{k_m r}\sqrt{\alpha_m} = \frac{\phi(R,t)^* q_s''(t)}{k_m r}\sqrt{\alpha_m}. \quad \text{(A.12)}$$

Equation (A.12) slightly differs than the one received by Keblinski et al. [38] due to the BC that define the heat flux coming from the surface of the MNP defining the heat created by the magnetic losses inside it, whereas Keblinski et al. [38] suggested that the heat sources are inside the medium of interest and that the heat-power density is constant in time. In order to analytically calculate (3) or (A.12), the general expression of $q_s''(t)$ (Wm^{-2}) must be found for each case.

Case 2 (a cosine MFP). For Case 1, the magnetization, $M(t)$, can be found in the time domain after substituting (5) and the MF in (4) and taking the inverse FT of it, that results in

$$\begin{aligned}M(t) &= \chi(t)^* H(t)\\ &= \frac{A\chi_0}{\tau}\left(\frac{\cos(\omega_0 t)}{\tau} + \omega_0\sin(\omega_0 t)\right)\frac{1}{(1/\tau)^2+\omega_0{}^2}.\end{aligned} \quad \text{(A.13)}$$

Substituting (A.13) into the magnetic induction [5] results in

$$B(t) = \mu_0 H(t) + \mu_0 M(t). \quad \text{(A.14)}$$

Further substituting the received magnetic induction described in (A.14) into (7), one can calculate the conversion of the magnetic energy into heat losses, resulting in

$$\begin{aligned}&\mathbf{H(t)}\cdot\frac{\partial\mathbf{B(t)}}{\partial t}\\ &= \iint d\omega\, d\omega' \overline{H}(r,\omega')(i\omega\mu_0\mu(\omega))\cdot H(r,\omega)e^{i(\omega-\omega')t}\\ &= \mu_0 A^2\cos(\omega_0 t)\\ &\quad\cdot\int d\omega\left(i\omega\left(1+\frac{\chi_0}{1+i\omega\tau}\right)\right)\\ &\quad\times\frac{\delta(\omega-\omega_0)+\delta(\omega+\omega_0)}{2}e^{i\omega t}\\ &= \mu_0 A^2\cos(\omega_0 t)\omega_0\\ &\quad\times\left(\frac{\omega_0\chi_0\tau}{1+\omega_0{}^2\tau^2}\cos(\omega_0 t) - \sin(\omega_0 t)\left[1+\frac{\chi_0}{1+\omega_0{}^2\tau^2}\right]\right)\\ &= P_{\mathrm{Loss}}(t) + \frac{\partial U}{\partial t}.\end{aligned} \quad \text{(A.15)}$$

Or

$$P_{\mathrm{Loss}}(t) = \mu_0 A^2\cos(\omega_0 t)\omega_0\frac{\omega_0\chi_0\tau}{1+\omega_0{}^2\tau^2}\cos(\omega_0 t). \quad \text{(A.16)}$$

Because $P_{\mathrm{Loss}}(t)$ is only a function of time between $0 < r < a$ (isotropic and homogeneous material), then the outward heat flux at $r = a$ can be calculated as follows:

$$q_s''(r=a,t)4\pi a^2 = \frac{4\pi a^3}{3}P_{\mathrm{Loss}}(t). \quad \text{(A.17)}$$

Or

$$q_s''(r=a,t) = \frac{a\mu_0 A^2\omega_0}{6}\left(\frac{\omega_0\chi_0\tau}{1+(\omega_0\tau)^2}(\cos(2\omega_0 t)+1)\right). \quad \text{(A.18)}$$

Taking the FT of (A.18) and substituting it in (A.10) one can calculate the FT of $\theta(R,t)$ to receive the following:

$$\begin{aligned}\widetilde{\theta}(R,\omega) &= \frac{a\mu_0 A^2\omega_0}{6}\\ &\quad\times\left(\frac{\omega_0\chi_0\tau}{1+(\omega_0\tau)^2}\left(\frac{\delta(\omega-2\omega_0)+\delta(\omega+2\omega_0)}{2}+\delta(\omega)\right)\right)\\ &\quad\cdot\frac{aR_0}{R_0\sqrt{i\omega}+1}\exp\left(-\sqrt{i\omega}R\right).\end{aligned} \quad \text{(A.19)}$$

Case 3 (a PMP profile). The PMP, (11), can be decomposed using the theory of Fourier's series into its harmonics to receive [71, 72] the following:

$$H(t) = 2A\cdot\frac{\Delta}{T_s} + \sum_{n=1}^{\infty}\frac{4A}{\pi n}\sin\left(\frac{n\pi\Delta}{T_s}\right)\cos(n\omega_0 t). \quad \text{(A.20)}$$

By substituting (A.20) into (A.14) and then using them in (7) we can calculate the total heat dissipation for this case as follows:

$$\begin{aligned}
&\mathbf{H(t)} \cdot \frac{\partial \mathbf{B(t)}}{\partial t} \\
&= \iint d\omega\, d\omega' \overline{H}(r,\omega')(i\omega\mu_0\mu(\omega)) \cdot H(r,\omega)e^{i(\omega-\omega')t} \\
&= \mu_0\left(2A \cdot \frac{\Delta}{T_s} + \sum \frac{4A}{\pi m}\sin\left(\frac{m\pi\Delta}{T_s}\right)\cos(m\omega_0 t)\right) \\
&\quad \cdot \sum \frac{4A}{\pi n}\sin\left(\frac{n\pi\Delta}{T_s}\right)\int d\omega\left(i\omega\left(1+\frac{\chi_0}{1+i\omega\tau}\right)\right) \\
&\quad \times \frac{\delta(\omega - n\omega_0) + \delta(\omega + n\omega_0)}{2}e^{i\omega t} \\
&= \mu_0\left(2A \cdot \frac{\Delta}{T_s} + \sum \frac{4A}{\pi m}\sin\left(\frac{m\pi\Delta}{T_s}\right)\cos(m\omega_0 t)\right) \\
&\quad \cdot \sum \frac{4A}{\pi}\sin\left(\frac{n\pi\Delta}{T_s}\right)\omega_0 \\
&\quad \times \left(\frac{n\omega_0\chi_0\tau}{1+(n\omega_0)^2\tau^2}\cos(n\omega_0 t)\right. \\
&\quad \left. - \sin(n\omega_0 t)\left[1 + \frac{\chi_0}{1+(n\omega_0)^2\tau^2}\right]\right) \\
&= P_{\text{Loss}}(t) + \frac{\partial U}{\partial t}.
\end{aligned} \tag{A.21}$$

Therefore, we can find that the heat losses equal to

$$\begin{aligned}
P_{\text{Loss}}(t) &= \mu_0\left(2A \cdot \frac{\Delta}{T_s} + \sum \frac{4A}{\pi m}\sin\left(\frac{m\pi\Delta}{T_s}\right)\cos(m\omega_0 t)\right) \\
&\quad \cdot \sum \frac{4A}{\pi}\sin\left(\frac{n\pi\Delta}{T_s}\right)\omega_0\left(\frac{n\omega_0\chi_0\tau}{1+(n\omega_0)^2\tau^2}\cos(n\omega_0 t)\right).
\end{aligned} \tag{A.22}$$

And for $\Delta < t < T_s$:

$$\mathbf{H(t)} \cdot \frac{\partial \mathbf{B(t)}}{\partial t} = 0. \tag{A.23}$$

Using (A.17) and (A.22) we can calculate the heat flux at the surface of the MNP.

$$\begin{aligned}
&q_s''(r = a, t) \\
&= P_{\text{Loss}}(t)\frac{a}{3} \\
&= \frac{a}{3}\mu_0\left(2A \cdot \frac{\Delta}{T_s} + \sum \frac{4A}{\pi m}\sin\left(\frac{m\pi\Delta}{T_s}\right)\cos(m\omega_0 t)\right) \\
&\quad \cdot \sum \frac{4A}{\pi}\sin\left(\frac{n\pi\Delta}{T_s}\right)\omega_0\left(\frac{n\omega_0\chi_0\tau}{1+(n\omega_0)^2\tau^2}\cos(n\omega_0 t)\right).
\end{aligned} \tag{A.24}$$

By taking the FT of the resulted heat flux and substituting it in (A.10) one can receive

$$\begin{aligned}
&\widetilde{\theta}(R,\omega) \\
&= \frac{a}{3}8\omega_0\mu_0 A^2 \cdot \frac{\Delta}{\pi T_s}\sum \sin\left(\frac{n\pi\Delta}{T_s}\right)\frac{n\omega_0\chi_0\tau}{1+(n\omega_0)^2\tau^2} \\
&\quad \times \left(\frac{\delta(\omega - n\omega_0) + \delta(\omega + n\omega_0)}{2}\right)\frac{aR_0}{R_0\sqrt{i\omega}+1}\exp\left(-\sqrt{i\omega}R\right) \\
&\quad + \frac{a}{3}\frac{8\mu_0 A^2\omega_0}{\pi^2}\sum\sum \frac{\sin(m\pi\Delta/T_s)}{m}\sin\left(\frac{n\pi\Delta}{T_s}\right) \\
&\quad \times \frac{n\omega_0\chi_0\tau}{1+(n\omega_0)^2\tau^2} \\
&\quad \cdot \left(\frac{\delta(\omega - (n+m)\omega_0) + \delta(\omega + (n+m)\omega_0)}{2}\right. \\
&\quad \left. + \frac{\delta(\omega - (m-n)\omega_0) + \delta(\omega + (m-n)\omega_0)}{2}\right) \\
&\quad \times \frac{aR_0}{R_0\sqrt{i\omega}+1}\exp\left(-\sqrt{i\omega}R\right).
\end{aligned} \tag{A.25}$$

When looking at (A.25), the multiplication of each coefficient's amplitude with its matched harmonic must meet the biologically noninvasive limitation $A_m \cdot f_m \leq 5 \cdot 10^9$ A/m·s. The mathematical justification is deduced next.

Looking at the eddy currents that evolve in the body [73]

$$\begin{aligned}
E(\omega) &= \frac{-i\omega r}{2}B_z, \\
J(\omega) &= -\frac{i\omega r\sigma}{2}B_z \longrightarrow E(t) = -\frac{r}{2}\frac{\partial B_z(t)}{\partial t} = -\mu_0\frac{r}{2}\frac{\partial H(t)}{\partial t}, \\
J(t) &= -\frac{r\sigma}{2}\frac{\partial B_z(t)}{\partial t} = -\frac{\mu_0 r\sigma}{2}\frac{\partial H(t)}{\partial t}.
\end{aligned} \tag{A.26}$$

They can be written using (A.20) as follows:

$$\begin{aligned}
E(t) &= -\mu_0\frac{r}{2}\frac{\partial H(t)}{\partial t} = \mu_0\frac{r}{2}\omega_0\sum\frac{4A}{\pi}\sin\left(\frac{m\pi\Delta}{T_s}\right)\sin(m\omega_0 t), \\
J(t) &= -\frac{\mu_0 r\sigma}{2}\frac{\partial H(t)}{\partial t} = \frac{\mu_0 r\sigma}{2}\omega_0\sum\frac{4A}{\pi}\sin\left(\frac{n\pi\Delta}{T_s}\right)\sin(n\omega_0 t).
\end{aligned} \tag{A.27}$$

So the eddy losses inside the body can be found using Poynting theory [5] as follows:

$$\begin{aligned} P_{\text{average}} &= \frac{1}{T_s}\int P(t)dt = \frac{1}{T_s}\int E(t)J(t)dt \\ &= \frac{\sigma}{T_s}\left(\frac{2r\omega_0\mu_0 A}{\pi}\right)^2 \int \sum \sin\left(\frac{m\pi\Delta}{T_s}\right)\sin(m\omega_0 t) \\ &\quad \times \sum \sin\left(\frac{n\pi\Delta}{T_s}\right)\sin(n\omega_0 t)dt \\ &= \frac{\sigma}{T_s}(\pi r f_0\mu_0)^2\left(\frac{4A}{\pi}\right)^2 \sum \sin^2\left(\frac{m\pi\Delta}{T_s}\right) \\ &\quad \times \int \sin^2(m\omega_0 t)dt \\ &= \sigma(\pi r f_0\mu_0)^2\left(\frac{4A}{\pi}\right)^2 \frac{1}{2}\sum \sin^2\left(\frac{m\pi\Delta}{T_s}\right) \\ &= A_{\text{eff}}^2\sigma(\pi r f_0\mu_0)^2. \end{aligned} \tag{A.28}$$

The last expression is the same as the one received by Atkinson et al. [16]. Therefore the limitation on the MFS and the frequency can be summarized as follows [11, 16, 24]:

$$A_{\text{eff}} f_0 = f_0\left(\frac{4A}{\pi\sqrt{2}}\right)\sqrt{\sum \sin^2\left(\frac{m\pi\Delta}{T_s}\right)} \le 5\cdot 10^9\ \text{A/m}\cdot\text{s}. \tag{A.29}$$

For $N = 25$ and a duty cycle of $d = \Delta/T_s = 0.2$, the treatment is safe as long as

$$\begin{aligned} A_{\text{eff}} f_0 &= f_0\left(\frac{4A}{\pi\sqrt{2}}\right)\sqrt{\sum \sin^2\left(\frac{m\pi\Delta}{T_s}\right)} \\ &= A f_0 3.1 \le 5\cdot 10^9\ \text{A/m}\cdot\text{s}. \end{aligned} \tag{A.30}$$

Consequently, as long as (A.30) is valid the treatment is safe. Choosing other maximal summation index values such as $N = 20$ will result in a new constraint over the frequency and the MFS that must fulfill $A_{\text{eff}} f_0 = f_0(4A/\pi\sqrt{2})\sqrt{\sum \sin^2(m\pi\Delta/T_s)} \approx A f_0 2.8 \le 5\cdot 10^9$ A/m·s and so on.

Moreover, for frequencies lower than 10 MHz there is essentially no attenuation of the MFS within cylinders of muscle-equivalent material, therefore the maximal harmonic frequency should not exceed 10 MHz [16].

Case 4 (a discontinuous cosine MFP). As for the previous case, we decompose the MF using the theory of Fourier's series into its harmonics for $0 \le t \le \Delta$ to receive [71, 72]

$$\begin{aligned} H(t) &= \sum_{n=1}^{\infty} \frac{2A}{T_s}\left[\frac{\sin((\Delta/2)(\omega_0+n\omega_1))}{\omega_0+n\omega_1} + \frac{\sin((\Delta/2)(\omega_0-n\omega_1))}{\omega_0-n\omega_1}\right] \\ &\quad \times \cos(n\omega_1 t). \end{aligned} \tag{A.31}$$

For $\Delta \le t \le T_s$, the magnetic power losses are zero, because the MF dose not exists.

By substituting (A.31) into (A.14) and then using them in (7) we can calculate the total heat dissipation for this case as follows:

$$\begin{aligned} &\mathbf{H(t)}\cdot\frac{\partial \mathbf{B(t)}}{\partial t} \\ &= \iint d\omega\, d\omega' \overline{H}(r,\omega')(i\omega\mu_0\mu(\omega))\cdot H(r,\omega)e^{i(\omega-\omega')t} \\ &= \mu_0\left(\frac{2A}{T_s}\right)^2 \\ &\quad \cdot\Bigg(\sum\Bigg[\frac{\sin((\Delta/2)(\omega_0+m\omega_1))}{\omega_0+m\omega_1} + \frac{\sin((\Delta/2)(\omega_0-m\omega_1))}{\omega_0-m\omega_1}\Bigg] \\ &\quad \times\cos(m\omega_0 t)\Bigg) \\ &\quad \cdot\sum\left[\frac{\sin((\Delta/2)(\omega_0+n\omega_1))}{\omega_0+n\omega_1} + \frac{\sin((\Delta/2)(\omega_0-n\omega_1))}{\omega_0-n\omega_1}\right] \\ &\quad \times\int d\omega\left(i\omega\left(1+\frac{\chi_0}{1+i\omega\tau}\right)\right) \\ &\quad \times\frac{\delta(\omega-n\omega_1)+\delta(\omega+n\omega_1)}{2}e^{i\omega t} \\ &= \mu_0\Bigg(\left(\frac{2A}{T_s}\right)^2 \\ &\quad \times\Bigg[\frac{\sin((\Delta/2)(\omega_0+m\omega_1))}{\omega_0+m\omega_1} + \frac{\sin((\Delta/2)(\omega_0-m\omega_1))}{\omega_0-m\omega_1}\Bigg] \\ &\quad \times\cos(m\omega_1 t)\Bigg) \\ &\quad \cdot\sum\left(\frac{\sin((\Delta/2)(\omega_0+n\omega_1))}{\omega_0+n\omega_1} + \frac{\sin((\Delta/2)(\omega_0-n\omega_1))}{\omega_0-n\omega_1}\right) \\ &\quad \times\Bigg(\frac{(n\omega_1)^2\chi_0\tau}{1+(n\omega_1)^2\tau^2}\cos(n\omega_1 t) \\ &\quad - \sin(n\omega_1 t)\left[1+\frac{\chi_0}{1+(n\omega_1)^2\tau^2}\right]\Bigg) \\ &= P_{\text{Loss}}(t) + \frac{\partial U}{\partial t}. \end{aligned} \tag{A.32}$$

Therefore, we can find that the heat losses equal to

$$\begin{aligned} &P_{\text{Loss}}(t) \\ &= \mu_0\left(\left(\frac{2A}{T_s}\right)^2\right. \end{aligned}$$

$$\times \left[\frac{\sin((\Delta/2)(\omega_0+m\omega_1))}{\omega_0+m\omega_1} + \frac{\sin((\Delta/2)(\omega_0-m\omega_1))}{\omega_0-m\omega_1} \right] \times \cos(m\omega_1 t) \Big) \cdot \sum \left(\frac{\sin((\Delta/2)(\omega_0+n\omega_1))}{\omega_0+n\omega_1} + \frac{\sin((\Delta/2)(\omega_0-n\omega_1))}{\omega_0-n\omega_1} \right) \times \left(\frac{(n\omega_1)^2 \chi_0 \tau}{1+(n\omega_1)^2\tau^2} \cos(n\omega_1 t) \right). \quad (A.33)$$

Using (A.17) and (A.33) we can calculate the heat flux at the surface of the MNP.

$$\begin{aligned} q_s''(r=a,t) &= P_{\text{Loss}}(t)\frac{a}{3} \\ &= \frac{a}{3}\mu_0 \left(\left(\frac{2A}{T_s}\right)^2 \sum \left[\frac{\sin((\Delta/2)(\omega_0+m\omega_1))}{\omega_0+m\omega_1} + \frac{\sin((\Delta/2)(\omega_0-m\omega_1))}{\omega_0-m\omega_1} \right] \cos(m\omega_1 t) \right) \\ &\quad \cdot \sum \left(\frac{\sin((\Delta/2)(\omega_0+n\omega_1))}{\omega_0+n\omega_1} + \frac{\sin((\Delta/2)(\omega_0-n\omega_1))}{\omega_0-n\omega_1} \right) \\ &\quad \times \left(\frac{(n\omega_1)^2 \chi_0 \tau}{1+(n\omega_1)^2\tau^2} \cos(n\omega_1 t) \right). \end{aligned} \quad (A.34)$$

By taking the FT of the resulted heat flux and substituting it in (A.10) one can receive

$$\begin{aligned} \tilde{\theta}(R,\omega) &= \frac{2a}{3}\mu_0 \left(\frac{A}{T_s}\right)^2 \frac{aR_0}{R_0\sqrt{i\omega}+1} \exp\left(-\sqrt{i\omega}R\right) \\ &\quad \times \sum\sum \Bigg[\left(\frac{\sin((\Delta/2)(\omega_0+m\omega_1))}{\omega_0+m\omega_1} + \frac{\sin((\Delta/2)(\omega_0-m\omega_1))}{\omega_0-m\omega_1} \right) \\ &\quad \times \left(\frac{\sin((\Delta/2)(\omega_0+n\omega_1))}{\omega_0+n\omega_1} + \frac{\sin((\Delta/2)(\omega_0-n\omega_1))}{\omega_0-n\omega_1} \right) \cdot \frac{(n\omega_1)^2 \chi_0 \tau}{1+(n\omega_1)^2\tau^2} \\ &\quad \times \left(\frac{\delta(\omega-(n+m)\omega_1)+\delta(\omega+(n+m)\omega_1)}{2} + \frac{\delta(\omega-(m-n)\omega_1)+\delta(\omega+(m-n)\omega_1)}{2} \right) \Bigg]. \end{aligned} \quad (A.35)$$

When looking at (A.35) the multiplication of each coefficient's amplitude with its matched harmonic must meet the biologically noninvasive limitation $A_m \cdot f_m \leq 5 \cdot 10^9$ A/m·s. The mathematical justification is deduced next.

For $f_0 > f_1$ and $T_s = 1/f_1$, we find that (A.26) becomes

$$\begin{aligned} E(t) &= -\mu_0 \frac{r}{2} \frac{\partial H(t)}{\partial t} \\ &= \mu_0 \frac{r}{2} \omega_1 \sum \frac{n2A}{T_s} \left[\frac{\sin((\Delta/2)(\omega_0+n\omega_1))}{\omega_0+n\omega_1} + \frac{\sin((\Delta/2)(\omega_0-n\omega_1))}{\omega_0-n\omega_1} \right] \sin(n\omega_1 t), \\ J(t) &= -\frac{\mu_0 r\sigma}{2} \frac{\partial H(t)}{\partial t} \\ &= \frac{\mu_0 r\sigma}{2}\omega_1 \times \sum \frac{m2A}{T_s} \left[\frac{\sin((\Delta/2)(\omega_0+m\omega_1))}{\omega_0+m\omega_1} + \frac{\sin((\Delta/2)(\omega_0-m\omega_1))}{\omega_0-m\omega_1} \right] \sin(m\omega_1 t). \end{aligned} \quad (A.36)$$

Consequently, (A.28) becomes

$$\begin{aligned} P_{\text{avr}} &= \frac{1}{T_s}\int P(t)dt = \frac{1}{T_s}\int E(t)J(t)dt \\ &= \frac{\sigma}{T_s}\left(\frac{2r\omega_1\mu_0 A}{T_s}\right)^2 \cdot \int \sum n \left[\frac{\sin((\Delta/2)(\omega_0+n\omega_1))}{\omega_0+nn\omega_1} + \frac{\sin((\Delta/2)(\omega_0-n\omega_1))}{\omega_0-n\omega_1} \right] \sin(n\omega_1 t) \\ &\quad \times \sum m \left[\frac{\sin((\Delta/2)(\omega_0+n\omega_1))}{\omega_0+n\omega_1} + \frac{\sin((\Delta/2)(\omega_0-n\omega_1))}{\omega_0-n\omega_1} \right] \sin(n\omega_1 t)dt \\ &= \frac{\sigma}{T_s}\left(\frac{2r\omega_1\mu_0 A}{T_s}\right)^2 \times \sum n^2 \left[\frac{\sin((\Delta/2)(\omega_0+n\omega_1))}{\omega_0+n\omega_1} + \frac{\sin((\Delta/2)(\omega_0-n\omega_1))}{\omega_0-n\omega_1} \right]^2 \times \int \sin^2(n\omega_1 t)dt \\ &= \sigma(\pi r f_1 \mu_0)^2 \left(\frac{4A}{T_s}\right)^2 \end{aligned}$$

$$\times \frac{1}{2}\sum n^2\left[\frac{\sin((\Delta/2)(\omega_0+n\omega_1))}{\omega_0+n\omega_1} + \frac{\sin((\Delta/2)(\omega_0-n\omega_1))}{\omega_0-n\omega_1}\right]^2$$
$$= {A_{\text{eff}}}^2\sigma(\pi r f_1\mu_0)^2. \tag{A.37}$$

The last expression is the same as the one received by Atkinson et al. [16]. Therefore, the limitation on the MFS and the frequency can be summarized as follows [11, 16, 24]:

$$A_{\text{eff}}f_1 = f_1\left(\frac{4A}{T_s\sqrt{2}}\right) \times \sqrt{\sum n^2\left[\frac{\sin((\Delta/2)(\omega_0+n\omega_1))}{\omega_0+n\omega_1}+\frac{\sin((\Delta/2)(\omega_0-n\omega_1))}{\omega_0-n\omega_1}\right]^2} \le 5\cdot 10^9\ \text{A/m}\cdot\text{s}. \tag{A.38}$$

For $N = 25$, a duty cycle of $d = \Delta/T_s = 0.2$ and $\omega_0 = 6\omega_1$, the treatment is safe as long as

$$A_{\text{eff}}f_1 = f_1\left(\frac{4A}{T_s\sqrt{2}}\right) \times \sqrt{\sum n^2\left[\frac{\sin((\Delta/2)(\omega_0+n\omega_1))}{\omega_0+n\omega_1}+\frac{\sin((\Delta/2)(\omega_0-n\omega_1))}{\omega_0-n\omega_1}\right]^2} = Af_1 4.6 < Af_0 \le 5\cdot 10^9\ \text{A/m}\cdot\text{s}. \tag{A.39}$$

Consequently, as long as (A.39), the treatment will be safe. Moreover, for frequencies lower than 10 MHz, there is essentially no attenuation of the MFS within cylinders of muscle-equivalent material, therefore the maximal harmonic frequency should not exceed 10 MHz [16].

B. COMSOL: Results

In order to validate the analytic solutions and the MATLAB simulations, a numerical simulation was performed using COMSOL, for the same thermal and magnetic properties given in Tables 1 and 2. The simulation results can be seen for each case studied in Methods and Results parts in this Appendix.

For Case 1, the mathematical expression of the temperature increment, (9), was plotted as a function of time and space, where the results are given in Figure 10 for $T_s = 2.5\,\mu$s.

The maximal temperature elevation in Figure 10 reached a value of 2.25 n°K on the surface of the MNP which is 0.15 n°K higher than the one received for the analytic simulation, Figure 3.

At 2 nm apart from the surface of the MNP surface, the temperature elevation reached a value of 2.05 n°K that is 0.2 n°K higher than the one receive in Figure 3. Again, there is a small difference between both simulations results. As the observation point gets further from the surface of the MNP, the temperature differences get bigger; reaching a value of 0.4 n°K at an observation point located 10 nm apart from the surface.

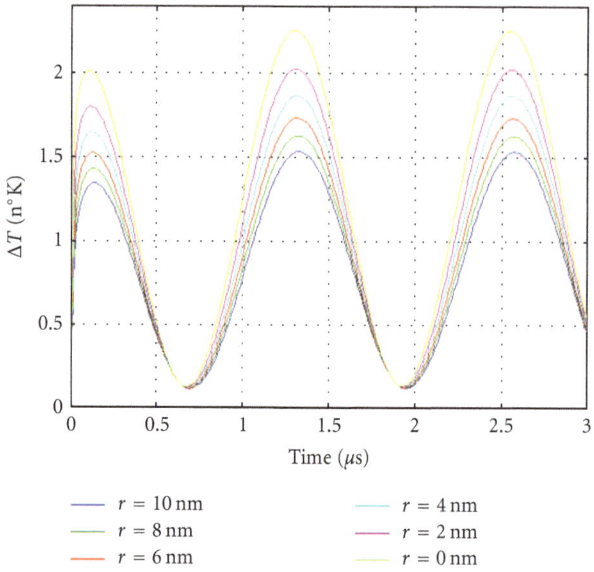

Figure 10: The temperature profile for a cosine MFP plotted as a function of time and as a function of the observation point located at a distance, r, from the surface of the MNP; r ranging from 0 nm to 10 nm for a core radius of 10 nm.

This may be caused by the triangles constructing the COMSOL's numeric mesh, which are used to solve numerically the heat problem, that get larger and bigger, as the observation point gets further from the MNP surface contributing to the error.

Comparing between Figures 3 and 10 we conclude that the numerical simulation fits the analytic solution.

For Case 2, as in Case 1, in order to validate the analytic solution a numerical simulation was also performed using COMSOL for the same thermal and magnetic properties (Tables 1 and 2). The simulation result can be seen in Figure 11.

The maximal temperature elevation in Figure 11 reaches a value of 8.5 n°K on the surface of the MNP, which is 0.4 n°K higher than the one receive in Figure 5.

At 2 nm apart from the surface of the MNP surface, the temperature elevation reached a value of 7.5 n°K, that is 0.2 n°K higher than the one receive in Figure 5. Again, it seems that there exists a small difference between the simulations results. As the observation point gets further from the surface of the MNP, the differences gets bigger reaching a value of 0.8 n°K at an observation point located 10 nm a part from the surface.

This may be caused by the bigger triangles in the mesh that are formed in the COMSOL software as the observation point gets further from the MNP surface contributing to the error.

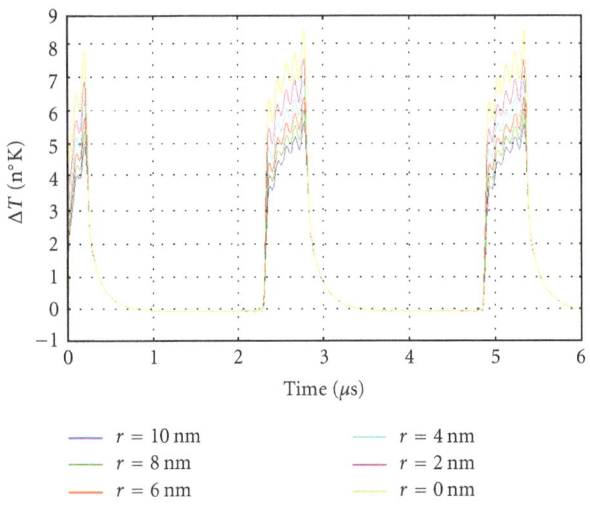

FIGURE 11: The temperature profile for a PMP-shaped MFP plotted as a function of time and as a function of the observation point located at a distance, r, from the surface of the MNP; r ranging from 0 nm to 10 nm for a core radius of 10 nm for $N = 25$.

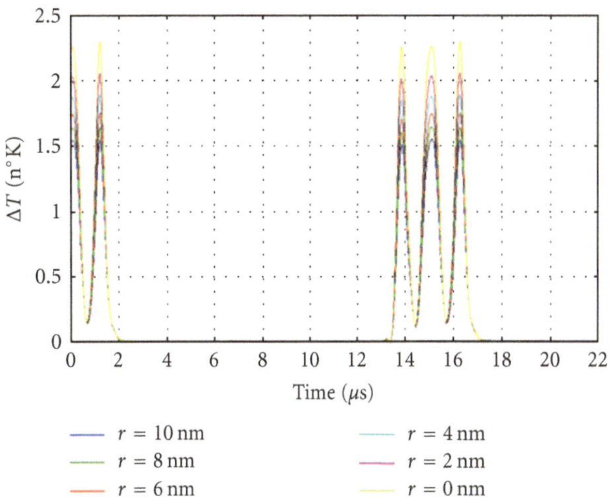

FIGURE 12: The temperature profile for a discontinuous cosine MFP plotted as a function of time and as a function of the observation point located at a distance, r, from the surface of the MNP; r ranging from 0 nm to 10 nm, for a core radius of 10 nm for $N = 25$.

As can be seen from Figure 11, there are 5 peaks during the time that the MF is tuned on that fit the number of peaks in Figure 5, these peaks evolve due to the final number of harmonics that form the PMP MF, as given by (11). However, there is a slightly difference in the temperature profiles between Figures 11 and 5; in Figure 11 the first peak is lower than the others in comparison to Figure 5 where the first peak is about the same high as the last peak.

Again, there are some small changes between both software simulations as expected; however the results for both simulations conclude that there is a benefit in using the PMPs instead of the cosine MFP due to the higher temperature rise values received for the same magnetic parameters.

For Case 3, we used again the numerical simulation COMSOL in order to validate the analytic solution for the same thermal and magnetic properties. The simulation result can be seen in Figure 12.

The maximal temperature elevation in Figure 12 reached a value of 2.3 n°K on the surface of the MNP which is the same as the one receive in Figure 7.

At 2 nm apart from the surface of the MNP surface the temperature elevation reached a value of 2 n°K, that is, 0.2 n°K higher than the one received in Figure 7. Again, there is a small difference between the simulations results. As the observation point get further from the surface of the MNP, the differences gets bigger reaching a value of 0.3 n°K at an observation point located 10 nm a part from the surface.

As explain before, this may be caused by the bigger triangles in the mesh that are formed in the COMSOL software as the observation point gets further from the MNP surface contributing to the error. Although, there are some small changes between both simulations as expected, the maximal temperature rise is almost the same as the cosine MFP.

C. The Effects the Maximal Number of Indexes Has on Cases 2 and 3: Results

In Appendix C, we examined the influences that the maximal numbers of indexes composing the MF signal have on the temperature rise and on the temperature rate rise for Case 2 and Case 3.

The maximal index numbers for summation were chosen as $N = 100$, $N = 25$, $N = 15$, $N = 10$, and $N = 1$. Above $N = 25$ the MF is practically absorbed in the tissue [14], but this fact was not taken in consideration in the simulations results.

Case 3. The temperature rise for Case 2, as a function of the maximal summation indexes can be seen in Figure 13.

From Figure 13, we concluded that the maximal temperature rise depends on the number of harmonics composing the MF signal. For $N = 100$, the maximal temperature rise reaches a value of 50 n°K for a core radius of 7.7 nm, that is, 1.6 times higher than the maximal value received for $N = 25$. The summation of 100 indexes can be seen as the ideal PMPs shaped MFP.

For $N = 15$, we receive a temperature rate of 28 n°K for a core radius of 8.8 nm and for $N = 10$ we received a value of 26 n°K for a core radius of 9 nm. Furthermore, we can see that the number of indexes composing the MF changes the optimal radius as it gets smaller as the index number gets bigger.

Now, we examined the influences that the number of maximal summation indexes composing the MF signal has on the temperature rate rise. The chosen numbers were $N = 100$, $N = 25$, $N = 15$, $N = 10$, and $N = 1$.

From Figure 14, we concluded that the maximal temperature rate rise depends on the number of harmonics composing the MF signal. For $N = 100$, the maximal

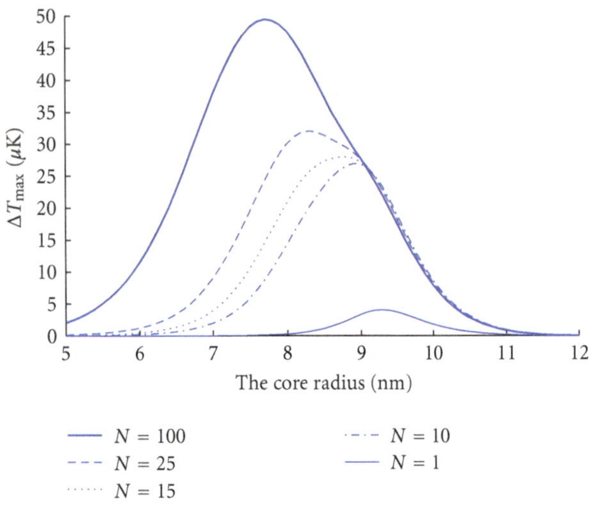

FIGURE 13: The absolute maximal temperature rise was plotted as a function of the number of harmonics summed, N, and as a function of the core radius for a periodic pulse-shaped MFP having a pulsed width of $0.2T_s$; the maximal summation value N ranges from $N = 1$ to $N = 100$.

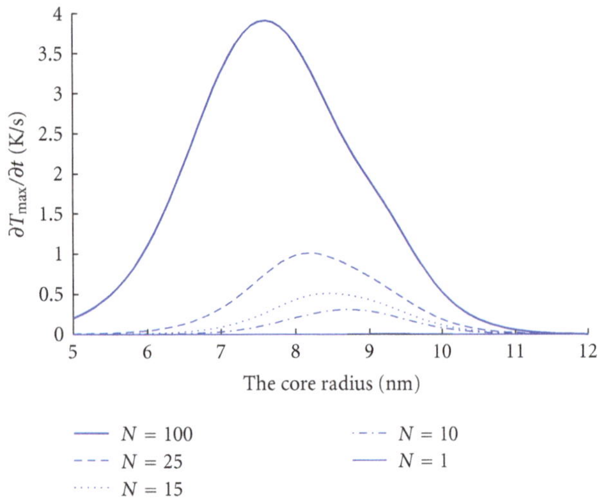

FIGURE 14: The absolute maximal temperature rise rate was plotted as a function of the number of harmonics summed, N, and as a function of the core radius for a periodic pulse-shaped MFP having a pulsed width of $0.2\,T_s$ plotted as a function of the core radius; a: ranging from 5 nm to 15 nm, the observation points are on the surface of the MNP.

temperature rate reaches a value of $3.9°\mathrm{Ks}^{-1}$ is received for a core radius 7.6 nm, and is 3.9 times higher than the value received for $N = 25$. For $N = 15$ we receive a temperature rate of $0.5°\mathrm{Ks}^{-1}$ for a core radius of 8.2 nm, that is, half the value received for $N = 25$, and for $N = 10$, we received a value of $0.3°\mathrm{Ks}^{-1}$ for a core radius of 8.5 nm.

Case 4. Now, we examined the influences that the maximal number of summation indexes composing the MF signal has

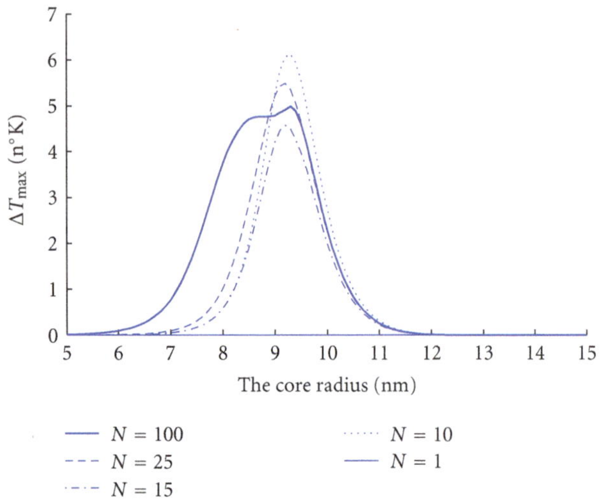

FIGURE 15: The absolute maximal temperature rise was plotted as a function of the number of harmonics summed, N, and as a function of the maximal summation value for a periodic discontinuous cosine MFP having a pulsed width of $0.2\,T_s$. N ranging from $N = 1$ to $N = 100$ for a core radius of 10 nm, the observation point are on the surface of the MNP.

on the temperature rise. The chosen numbers were $N = 100$, $N = 25$, $N = 15$, $N = 10$, and $N = 1$.

From Figure 15, we concluded that the maximal temperature rise depends on the number of harmonics composing the MF signal. For $N = 100$, the maximal temperature rise reaches a value of 5 n°K for a core radius of 9.3 nm, that is, 1.1 times higher than the maximal value received for $N = 25$. For $N = 15$, we receive a temperature rate of 5.5 n°K for a core radius of 9.2 nm, and for $N = 10$, we received a value of 6.1°K for a core radius of 9.3 nm. As already mentioned, there is a limitation to the highest frequency that can be used for MH and should not exceed 10 MHz [16], in our case this limits the summation to 25 indexes that compose the MF signal. Moreover, we can see that the number of indexes composing the MF changes the optimal radius; it gets smaller as the index number gets higher. Furthermore, we can see that the number of indexes composing the MF changes the optimal radius by getting smaller as the index number gets bigger.

Now, we examined the influences that the number of indexes composing the MF signal has on the temperature rate rise. The chosen numbers were $N = 100$, $N = 25$, $N = 15$, $N = 10$, and $N = 1$.

From Figure 16, we concluded that the maximal temperature rise rate depends on the number of harmonics composing the MF signal. For $N = 100$, the maximal temperature rate reaches a value of $0.09°\mathrm{Ks}^{-1}$, is received for a core radius 8.4 nm, and is 4.5 times higher than the value received for $N = 25$. For $N = 15$, we receive a temperature rate of $0.016°\mathrm{Ks}^{-1}$ for a core radius of 9.3 nm, that is, half the value received for $N = 25$, and for $N = 10$ we received a value of $0.015°\mathrm{Ks}^{-1}$ for a core radius of 9.3 nm.

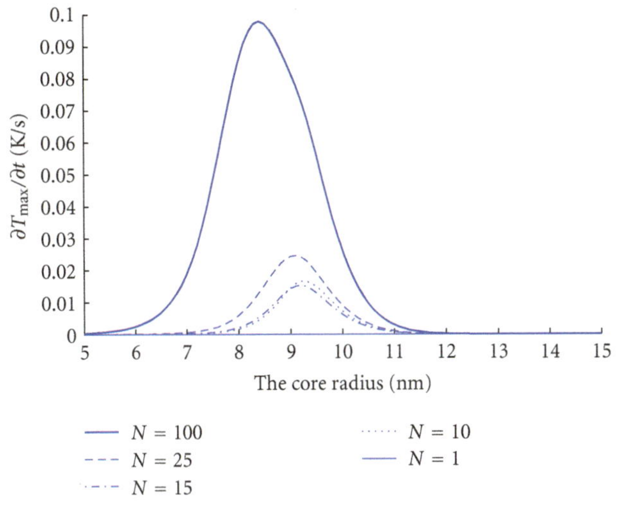

FIGURE 16: The absolute maximal temperature rise rate was plotted as a function of the number of harmonics summed, N, and as a function of the core radius for a periodic discontinuous cosine MFP having a pulsed width of $0.2T_s$; a—ranging from 5 nm to 15 nm, the observation point are on the surface of the MNP.

As already mentioned there is a limitation to the highest frequency that can be used for MH and should not exceed 10 MHz [16], in our case this limits the summation to 25 indexes that compose the MF signal. Moreover, we can see that the number of indexes composing the MF changes the optimal radius and it gets smaller as the index number gets higher.

References

[1] I. M. Gescheit, M. Ben-David, and I. Gannot, "A proposed method for thermal specific bioimaging and therapy technique for diagnosis and treatment of malignant tumors by using magnetic nanoparticles," *Advances in Optical Technologies*, vol. 2008, Article ID 275080, 7 pages, 2008.

[2] H. G. Bagaria and D. T. Johnson, "Transient solution to the bioheat equation and optimization for magnetic fluid hyperthermia treatment," *International Journal of Hyperthermia*, vol. 21, no. 1, pp. 57–75, 2005.

[3] M. A. Giordano, G. Gutierrez, and C. Rinaldi, "Fundamental solutions to the bioheat equation and their application to magnetic fluid hyperthermia," *International Journal of Hyperthermia*, vol. 26, no. 5, pp. 475–484, 2010.

[4] R. E. Rosensweig, "Heating magnetic fluid with alternating magnetic field," *Journal of Magnetism and Magnetic Materials*, vol. 252, no. 1–3, pp. 370–374, 2002.

[5] J. D. Jackson, *Classical Electrodynamics*, John Wily & Sons, 1998.

[6] L. D. Landau, L. P. Pitaevskii, and E. M. Lifshitz, *Electrodynamics of Continuous Media*, vol. 8, 2nd edition, 1984.

[7] N. Guskos, E. A. Anagnostakis, V. Likodimos et al., "Ferromagnetic resonance and ac conductivity of a polymer composite of Fe_3O_4 and Fe_3C nanoparticles dispersed in a graphite matrix," *Journal of Applied Physics*, vol. 97, no. 2, Article ID 024304, 6 pages, 2005.

[8] X. Zhang, B. Q. Li, and S. S. Pang, "A perturbational approach to magneto-thermal problems of a deformed sphere levitated in a magnetic field," *Journal of Engineering Mathematics*, vol. 2-3, no. 4, pp. 337–355, 1997.

[9] A. Jordan, P. Wust, H. Fahlin, W. John, A. Hinz, and R. Felix, "Inductive heating of ferrimagnetic particles and magnetic fluids: physical evaluation of their potential for hyperthermia," *International Journal of Hyperthermia*, vol. 9, no. 1, pp. 51–68, 1993.

[10] P. C. Fannin, Y. L. Raikher, A. T. Giannitsis, and S. W. Charles, "Investigation of the influence which material parameters have on the signal-to-noise ratio of nanoparticles," *Journal of Magnetism and Magnetic Materials*, vol. 252, no. 1–3, pp. 114–116, 2002.

[11] Q. A. Pankhurst, J. Connolly, S. K. Jonesand, and J. Dobson, "Applications of magnetic nanoparticles in biomedicine," *Journal of Physics D*, vol. 36, no. 13, pp. 167–181, 2003.

[12] J. Weizenecker, B. Gleich, J. Rahmer, and J. Borgert, "Particle dynamics of mono-domain particles in magnetic particle imaging," in *Proceedings of the 1st International Workshop on Magnetic Particle Imaging*, Magnetic Nanoparticles, pp. 3–15, World Scientific, 2010.

[13] P. C. Fannin, "Magnetic spectroscopy as an aide in understanding magnetic fluids," *Journal of Magnetism and Magnetic Materials*, vol. 252, no. 1–3, pp. 59–64, 2002.

[14] P. C. Fannin, "Characterisation of magnetic fluids," *Journal of Alloys and Compounds*, vol. 369, no. 1-2, pp. 43–51, 2004.

[15] P. C. Fannin and S. W. Charles, "On the calculation of the Neel relaxation time in uniaxial single-domain ferromagnetic particles," *Journal of Physics D*, vol. 27, no. 2, pp. 185–188, 1994.

[16] W. J. Atkinson, I. A. Brezovich, and D. P. Chakraborty, "Usable frequencies in hyperthermia with thermal seeds," *IEEE Transactions on Biomedical Engineering*, vol. 31, no. 1, pp. 70–75, 1984.

[17] S. Mornet, S. Vasseur, F. Grasset, and E. Duguet, "Magnetic nanoparticle design for medical diagnosis and therapy," *Journal of Materials Chemistry*, vol. 14, no. 14, pp. 2161–2175, 2004.

[18] C. H. Moran, S. M. Wainerdi, T. K. Cherukuri et al., "Size-dependent joule heating of gold nanoparticles using capacitively coupled radiofrequency fields," *Nano Research*, vol. 2, no. 5, pp. 400–405, 2009.

[19] V. P. Torchilin, "Targeted pharmaceutical nanocarriers for cancer therapy and imaging," *The AAPS Journal*, vol. 9, no. 2, pp. E128–E147, 2007.

[20] T. R. Sathe, *Integrated magnetic and optical nanotechnology for early cancer detection and monitoring [Ph.D. thesis]*, Georgia Institute of Technology, 2007.

[21] N. Gigel, "Magnetic nanoparticles impact on tumoral cells in the treatment by magnetic fluid hyperthermia," *Digest Journal of Nanomaterials and Biostructures*, vol. 3, no. 3, pp. 103–107, 2008.

[22] F. Matsuoka, M. Shinkai, H. Honda, T. Kubo, T. Sugita, and T. Kobayashi, "Hyperthermia using magnetite cationic liposomes for hamster osteosarcoma," *BioMagnetic Research and Technology*, vol. 2, no. 3, pp. 1–6, 2004.

[23] Q. A. Pankhurst, "Nanomagnetic medical sensors and treatment methodologies," *BT Technology Journal*, vol. 24, no. 3, pp. 33–38, 2006.

[24] E. Kita, T. Oda, T. Kayano et al., "Ferromagnetic nanoparticles for magnetic hyperthermia and thermoablation therapy,"

Journal of Physics D, vol. 43, no. 47, Article ID 474011, 9 pages, 2010.

[25] L. Pilon and K. M. Katika, "Modified method of characteristics for simulating microscale energy transport," *Journal of Heat Transfer*, vol. 126, no. 5, pp. 735–743, 2004.

[26] G. Chen, R. Yang, and X. Chen, "Nanoscale heat transfer and thermal-electric energy conversion," *Journal de Physique IV*, vol. 125, no. 1, pp. 499–504, 2005.

[27] G. Chen, "Non local and nonequilibrium heat conduction in the vicinity of nanoparticles," *Journal of Heat Transfer*, vol. 118, no. 3, pp. 539–546, 1996.

[28] C. Kittel, *Introduction to Solid-State Physics*, John Wiley & Sons, New York, NY, USA, 1996.

[29] R. Röhlsberger, W. Sturhahn, T. S. Toellner et al., "Phonon damping in thin films of Fe," *Journal of Applied Physics*, vol. 86, no. 1, pp. 584–592, 1999.

[30] K. E. Goodson and M. I. Flik, "Electron and phonon thermal conduction in epitaxial high-Tc superconducting films," *Journal of Heat Transfer*, vol. 115, no. 1, pp. 17–25, 1993.

[31] Y. Rabin, "Is intracellular hyperthermia superior to extracellular hyperthermia in the thermal sense?" *International Journal of Hyperthermia*, vol. 18, no. 3, pp. 194–202, 2002.

[32] G. Chen, "Ballistic-diffusive heat-conduction equations," *Physical Review Letters*, vol. 86, no. 11, pp. 1197–2300, 2000.

[33] E. H. Wissler, "Pennes' 1948 paper revisited," *Journal of Applied Physiology*, vol. 85, no. 1, pp. 35–41, 1998.

[34] T. C. Shih, P. Yuan, W. L. Lin, and H. S. Kou, "Analytical analysis of the Pennes bioheat transfer equation with sinusoidal heat flux condition on skin surface," *Medical Engineering & Physics*, vol. 29, no. 9, pp. 946–953, 2007.

[35] P. Yuan, H. E. Liu, C. W. Chen, and H. S. Kou, "Temperature response in biological tissue by alternating heating and cooling modalities with sinusoidal temperature oscillation on the skin," *International Communications in Heat and Mass Transfer*, vol. 35, no. 9, pp. 1091–1096, 2008.

[36] J. Liu and L. X. Xu, "Estimation of blood perfusion using phase shift in temperature response to sinusoidal heating at the skin surface," *IEEE Transactions on Biomedical Engineering*, vol. 46, no. 9, pp. 1037–1043, 1999.

[37] I. K. Tjahjono, *An analytical model for near-infrared light heating of a slab by embedded gold nanoshells [Ph.D. thesis]*, Rice University, 2006.

[38] P. Keblinski, D. G. Cahill, A. Bodapati, C. R. Sullivan, and T. A. Taton, "Limits of localized heating by electromagnetically excited nanoparticles," *Journal of Applied Physics*, vol. 100, no. 5, Article ID 054305, 5 pages, 2006.

[39] E. Gescheidtova, R. Kubásek, and K. Bartušek, "Quality of gradient magnetic fields estimation," *Journal of EE*, vol. 57, no. 8, pp. 54–57, 2006.

[40] M. Squibb, "A guide to experimental exposure of biological tissue to pulsed magnetic fields," PEMF Usage Guide, 2007.

[41] G. C. Goats, "Pulsed electromagnetic (short-wave) energy therapy," *British Journal of Sports Medicine*, vol. 23, no. 4, pp. 213–216, 1989.

[42] T. Niwa, Y. Takemura, N. Aida, H. Kurihara, and T. Hisa, "Implant hyperthermia resonant circuit produces heat in response to MRI unit radiofrequency pulses," *The British Journal of Radiology*, vol. 81, no. 961, pp. 69–72, 2008.

[43] P. Cantillon-Murphy, L. L. Wald, M. Zahn, and E. Adalsteinsson, "Proposing magnetic nanoparticle hyperthermia in low-field MRI," *Concepts in Magnetic Resonance A*, vol. 36, no. 1, pp. 36–47, 2010.

[44] S. M. Morgan and R. H. Victora, "Use of square waves incident on magnetic nanoparticles to induce magnetic hyperthermia for therapeutic cancer treatment," *Applied Physics Letters*, vol. 97, no. 9, Article ID 093705, 3 pages, 2010.

[45] K. E. Oughstun, *Electromagnetic and Optical Pulse Propagation*, Springer, 2006.

[46] L. R. Squire and J. A. Zouzounis, "ECT and memory: brief pulse versus sine wave," *The American Journal of Psychiatry*, vol. 143, no. 5, pp. 596–601, 1986.

[47] R. Kappiyoor, M. Liangruksa, R. Ganguly, and I. K. Puri, "The effects of magnetic nanoparticle properties on magnetic fluid hyperthermia," *Journal of Applied Physics*, vol. 108, no. 9, Article ID 094702, 8 pages, 2010.

[48] A. Trakic, F. Liu, and S. Crozier, "Transient temperature rise in a mouse due to low-frequency regional hyperthermia," *Physics in Medicine and Biology*, vol. 51, no. 7, pp. 1673–1691, 2006.

[49] A. O. Govorov, W. Zhang, T. Skeini, H. Richardson, E. J. Lee, and N. A. Kotov, "Gold nanoparticle ensembles as heaters and actuators: melting and collective plasmon resonances," *Nanoscale Research Letters*, vol. 1, no. 1, pp. 84–90, 2006.

[50] O. N. Strand, "A method for the computation of the error function of a complex variable," *Mathematics of Computation*, vol. 19, pp. 127–129, 1965.

[51] J. Kestin and L. N. Persen, "On the error function of a complex argument," *Zeitschrift für Angewandte Mathematik und Physik*, vol. 7, no. 1, pp. 33–40, 1956.

[52] M. Kettering, J. Winter, M. Zeisberger et al., "Magnetic nanoparticles as bimodal tools in magnetically induced labelling and magnetic heating of tumour cells: an in vitro study," *Nanotechnology*, vol. 18, no. 17, Article ID 175101, 9 pages, 2007.

[53] R. Hergt, S. Dutz, R. Müller, and M. Zeisberger, "Magnetic particle hyperthermia: nanoparticle magnetism and materials development for cancer therapy," *Journal of Physics*, vol. 18, no. 38, pp. S2919–S2934, 2006.

[54] I. Hilger, R. Hergt, and W. A. Kaiser, "Use of magnetic nanoparticle heating in the treatment of breast cancer," *IEE Proceedings-Nanobiotechnology*, vol. 152, no. 1, pp. 33–39, 2005.

[55] A. A. Velayati, P. Farnia, and T. A. Ibrahim, "Differences in cell wall thickness between resistant and nonresistant strains of Mycobacterium tuberculosis: using transmission electron microscopy," *Chemotherapy*, vol. 55, no. 5, pp. 303–307, 2009.

[56] V. Dupres, Y. F. Dufreȇne, and J. J. Heinisch, "Measuring cell wall thickness in living yeast cells using single molecular rulers," *American Chemical Society Nano*, vol. 4, no. 9, pp. 5498–5504, 2010.

[57] "What is the thickness of the cell membrane?" http://www.weizmann.ac.il/plants/Milo/images/membraneThickness110109RM.pdf.

[58] M. Kaiser, J. Heintz, I. Kandela, and R. Albrecht, "Tumor cell death induced by membrane melting via immunotargeted, inductively heated core/shell nanoparticles," *Microscopy and Microanalysis*, vol. 13, supplement 2, pp. 18–19, 2007.

[59] J. Vera and Y. Bayazitoglu, "Gold nanoshell density variation with laser power for induced hyperthermia," *International Journal of Heat and Mass Transfer*, vol. 52, no. 3-4, pp. 564–573, 2009.

[60] E. I. Gabrielle, *Biology The Easy Way*, Barron's Educational Series, New York, NY, USA, 1990.

[61] B. Chan, B. D. Chithrani, A. A. Ghazani, and C. W. Warren, "Determining the size and shape dependence of gold nanoparticle uptake into mammalian cells," *Nano Letters*, vol. 6, no. 4, pp. 662–668, 2006.

[62] Z. Yang, Z. W. Liu, R. P. Allaker et al., "A review of nanoparticle functionality and toxicity on the central nervous system," *Journal of the Royal Society Interface*, vol. 7, no. 4, pp. S411–S422, 2010.

[63] N. Singha, G. J. S. Jenkinsa, R. Asadib, and S. H. Doaka, "Potential toxicity of superparamagnetic iron oxide nanoparticles," *Nano Reviews*, vol. 1, pp. 1–15, 2010.

[64] H. Huang, *Magnetic nanoparticles based magnetophresis for efficient separation of foodborne pathogenes [M.S. thesis]*, University of Arkansa, 2009.

[65] M. P. Melancon, W. Lu, Z. Yang et al., "In vitro and in vivo targeting of hollow gold nanoshells directed at epidermal growth factor receptor for photothermal ablation therapy," *Molecular Cancer Therapeutics*, vol. 7, no. 6, pp. 1730–1739, 2008.

[66] P. H. Linh, N. C. Thuan, N. A. Tuan et al., "Invitro toxicity test and searching the possibility of cancer cell line extermination by magnetic heating with using Fe_3O_4 magnetic fluid," *Journal of Physics*, vol. 187, no. 1, Article ID 012008, 9 pages, 2009.

[67] S. Balivada, R. S. Rachakatla, H. Wang et al., "A/C magnetic hyperthermia of melanoma mediated by iron(0)/iron oxide core/shell magnetic nanoparticles: a mouse study," *Bio Med Centeral Cancer*, vol. 10, article 119, 9 pages, 2010.

[68] S. Bedanta and W. Kleemann, "Supermagnetism," *Journal of Physics D*, vol. 42, no. 1, Article ID 013001, 28 pages, 2009.

[69] M. Lewin, N. Carlesso, C. H. Tung et al., "Tat peptide-derivatized magnetic nanoparticles allow in vivo tracking and recovery of progenitor cells," *Nature Biotechnology*, vol. 18, no. 4, pp. 410–414, 2000.

[70] S. Purushotham and R. V. Ramanujan, "Modeling the performance of magnetic nanoparticles in multimodal cancer therapy," *Journal of Applied Physics*, vol. 107, no. 11, Article ID 114701, 9 pages, 2010.

[71] J. S. Walker, *Encyclopedia of Physical Science and Technology*, Elsevier Science, 3th edition, 2003.

[72] S. W. Smith, *The Scientist and Engineer's Guide to Digital Signal Processing*, chapter 13, California Technical Publishing, 1997.

[73] J. Pellicer-Porres, R. Lacomba-Perales, J. Ruiz-Fuertes, D. Martínez-García, and M. V. Andrés, "Force characterization of eddy currents," *American Journal of Physics*, vol. 74, no. 4, pp. 267–271, 2006.

4

Helmholtz Bright Spatial Solitons and Surface Waves at Power-Law Optical Interfaces

J. M. Christian,[1,2] E. A. McCoy,[1] G. S. McDonald,[1] J. Sánchez-Curto,[2] and P. Chamorro-Posada[2]

[1] *Joule Physics Laboratory, School of Computing, Science and Engineering, Materials & Physics Research Centre, University of Salford, Greater Manchester M5 4WT, UK*

[2] *Departamento de Teoría de la Señal y Comunicaciones e Ingeniería Telemática, Universidad de Valladolid, ETSI Telecomunicación, Campus Miguel Delibes Paseo Belén 15, E-47011 Valladolid, Spain*

Correspondence should be addressed to J. M. Christian, j.christian@salford.ac.uk

Academic Editor: Alan Migdall

We consider arbitrary angle interactions between spatial solitons and the planar boundary between two optical materials with a single power-law nonlinear refractive index. Extensive analysis has uncovered a wide range of new qualitative phenomena in non-Kerr regimes. A universal Helmholtz-Snell law describing soliton refraction is derived using exact solutions to the governing equation as a nonlinear basis. New predictions are tested through exhaustive computations, which have uncovered substantially enhanced Goos-Hänchen shifts at some non-Kerr interfaces. Helmholtz nonlinear surface waves are analyzed theoretically, and their stability properties are investigated numerically for the first time. Interactions between surface waves and obliquely incident solitons are also considered. Novel solution behaviours have been uncovered, which depend upon a complex interplay between incidence angle, medium mismatch parameters, and the power-law nonlinearity exponent.

1. Introduction

A light beam impinging on the interface between two dissimilar dielectric materials is a fundamental optical geometry [1–12]. After all, the single-interface configuration is an elemental structure that facilitates more sophisticated device designs and architectures for a diverse range of photonic applications. The seminal work of Aceves et al. [6, 7] some two decades ago considered perhaps the simplest scenario, where a spatial soliton (i.e., a self-trapped and self-stabilizing optical beam) is incident on the boundary between two different Kerr-type materials. Their intuitive approach reduced the full complexity of the electromagnetic interface problem to something far more tractable, namely, the solution a scalar equation of the inhomogeneous nonlinear Schrödinger (NLS) type. The development of an equivalent-particle theory [3–6] provided an enormous level of insight into the behaviour of scalar solitons at material boundaries. The adiabatic perturbation technique developed by Aliev et al. [13, 14] provides another toolbox for analyzing interface phenomena (e.g., light incident on the boundary between a linear and a nonlinear medium). Photorefractive [15] and quadratic [16] materials have also been considered.

A recurrent feature of the *waves at interfaces* literature is the appearance of the paraxial approximation, which combines the assumptions of broad (predominantly transverse-polarized) beams and slowly varying envelopes [1–16]. The adoption of this ubiquitous mathematical device can impose some strong physical constraints that should be borne in mind when modelling precisely these types of angular geometries. Indeed, the class of problem at hand is inherently nonparaxial since impinging beams may be arbitrarily oblique with respect to the interface. External refraction (where the refracted beam deviates *away from* the interface) provides a specific context where beam refraction cannot be described using conventional approaches. Paraxial wave optics must be applied with care since, in potentially off-axis regimes, it holds true only where angles (in the laboratory frame) of incidence, reflection, and refraction with respect to the reference direction are negligibly (or near-negligibly) small.

Recently, we proposed the first scalar model of spatial solitons at interfaces that is valid across the entire angular range [17, 18]. By respecting the essential role played by Helmholtz diffraction [19–24], the angular restriction was lifted while retaining an intuitive and manageable envelope equation. Preliminary analyses considered bright [17, 18] and dark [25, 26] spatial solitons incident on the boundary between dissimilar Kerr-type materials. They focused on establishing and developing the *propagation* aspects of our Helmholtz interfaces approach. By enforcing appropriate continuity conditions at the interface, a Snell's law for Kerr spatial solitons was derived whose validity was tested and confirmed by extensive numerical computations. Here, we take the first steps in a systematic study of the *materials* aspects of nonlinear beam-interface interactions. The simplest non-Kerr system one might consider is a class of host media whose refractive index $n_{\mathrm{NL}}(E)$ has a generic power-law dependence on the (complex) electric field amplitude E [27–29]:

$$n_{\mathrm{NL}}(E) = \frac{\alpha}{2n_0}|E|^q, \qquad (1)$$

where α is a positive coefficient, n_0 is the linear index (at the optical frequency), and the exponent q lies within the range $0 < q < 4$. Typically, the nonlinear response of the medium is assumed to be weak so that $\alpha E_0^q/n_0 \ll O(1)$, where E_0 is the peak field amplitude.

Power-law models have played a key role in the theory of nonlinear waves for the past three decades [30, 31]. Indeed, [32] provides an excellent review of the fundamental importance of model (1) in photonics contexts. Materials that fall into this broad category include some semiconductors (e.g., InSb [33] and GaAs/GaAlAs [34]), doped filter glasses (e.g., CsS_xSe_{x-2} [35, 36]), and liquid crystals [32]. One expects non-Kerr regimes (where q deviates from the value of 2) to give rise to a diverse range of new quantitative and qualitative phenomena. The physical basis for this expectation lies in the idealized nature of the Kerr response. In a range of materials, one can often find higher-order nonlinear effects coming into play. Perhaps the most obvious example of the breakdown of Kerr-type behaviour is optical saturation, where the refractive index change becomes bleached in the presence of sufficiently high-intensity illumination. In such cases, model (1) with $q < 1$ can be used to describe generic leading-order corrections from a saturable (dispersive) nonlinearity [35, 36].

In this paper, a detailed account is presented of arbitrary-angle refraction of spatial solitons at the interface between dissimilar power-law materials. Also of intrinsic interest are nonlinear surface waves (i.e., localized modes that travel along the boundary). This fundamental class of excitation has been the subject of previous power-law studies involving a single interface [35–39] and guided waves in multilayer structures (e.g., slab waveguides) [40–43]. Stability characteristics have been inferred from inspection of *power-propagation constant* solution branches. However, to the best of our knowledge, direct verification of such predictions [37–43] (e.g., through numerical solution of the underlying nonlinear Helmholtz equation) has been absent from the literature to date. Rather, computational studies of surface waves tend to have been in the limit of slowly varying envelopes and nonlinear Schrödinger-type models, typically of the diffusive-Kerr [44, 45], thermal [46], or saturable [47] type. Here, we investigate the stability of exact analytical Helmholtz surface waves through direct numerical calculations. As a fairly stringent test of solution robustness, we also report on some key findings concerning arbitrary-angle interactions between surface waves and solitons. In beam-refraction and surface-wave contexts, simulations have uncovered strikingly distinct behaviours as the exponent q is varied and across a range of quasi-paraxial and fully nonparaxial angular regimes.

The layout of this paper is as follows. In Section 2, we propose a governing equation for scalar optical fields in two adjoining power-law materials with dissimilar medium coefficients. Exact analytical bright solitons are presented for both media, and these solutions are used as a nonlinear basis to derive a generalized *Helmholtz-Snell law*. In Section 3, extensive computations test predictions of the new refraction law over a range of system parameters. We also extend our first calculations of the Goos-Hänchen (GH) shifts [48] in the Helmholtz angular regime [49] with power-law nonlinearities. Nonlinear surface waves are derived in Section 4, and simulations provide what appears to be the first full investigation of the stability properties of this new class of Helmholtz solution. We conclude, in Section 5, with some comments about the impact of our results.

2. Helmholtz Model for Scalar Soliton Refraction

The formalism of Helmholtz soliton theory [23, 24] is now deployed to develop a framework for describing refraction phenomena in wider classes of nonlinear optical materials. This type of modelling approach is valid when the beam waist w_0 is much broader than its free-space carrier wavelength λ, such that $\varepsilon \equiv \lambda/w_0 \ll O(1)$. Ultranarrow beam corrections to the governing equation, typically obtained from single-parameter (i.e., ε-based) order-of-magnitude analyses of fully-nonlinear Maxwell equations [50–55], are unnecessary in such regimes.

2.1. Governing Equation. Within the scalar approximation [19–24], we consider an electric field of the form

$$E(x,z,t) = E(x,z)\exp(-\mathrm{i}\omega t) + E^*(x,z)\exp(+\mathrm{i}\omega t), \qquad (2)$$

which is time harmonic with angular frequency ω. The laboratory space and time coordinates are (x,z) and t, respectively. In medium j (where $j = 1$ and 2), it is well known that the complex spatial field $E(x,z)$ satisfies the Helmholtz equation

$$\frac{\partial^2 E}{\partial z^2}+\frac{\partial^2 E}{\partial x^2}+\frac{\omega^2}{c^2}n_j^2(E)E=0, \tag{3}$$

where c is the vacuum speed of light. The refractive index distribution $n_j(E)$ on either side of the boundary is introduced through $n_j^2(E) \equiv n_{0j}^2+\alpha_j|E|^q$, where n_{0j} is the linear index at frequency ω and α_j is a nonlinearity coefficient. To facilitate comparison with our earlier work [17, 18, 25, 26], we look for travelling-wave solutions to (3) of the form $E(x,z) = E_0u(x,z)\exp(ik_1z)$. Here, E_0 is a (real) scale factor determining electric-field units, $u(x,z)$ is the dimensionless envelope, and $\exp(ik_1z)$ biases the solution in the forward longitudinal direction (taken to be z), where $k_1 \equiv n_{01}\omega/c$ is the (linear) propagation constant of the carrier wave in medium 1. It then follows that u satisfies the inhomogeneous equation

$$\frac{\partial^2 u}{\partial z^2}+\mathrm{i}2k_1\frac{\partial u}{\partial z}+\frac{\partial^2 u}{\partial x^2}+\frac{\omega^2}{c^2}\alpha_1E_0^q|u|^qu = \left[k_1^2\left(1-\frac{n_{02}^2}{n_{01}^2}\right)+\frac{\omega^2}{c^2}\alpha_1E_0^q\left(1-\frac{\alpha_2}{\alpha_1}\right)|u|^q\right]h(x,z)u, \tag{4}$$

where $h(x,z)$ is a Heaviside function that is equal to zero (unity) in the domain of medium 1 (medium 2). Equation (4) may be normalized with respect to the parameters in medium 1, in which case the following governing equation may be derived without further approximation [17, 18, 56, 57]:

$$\kappa\frac{\partial^2 u}{\partial \zeta^2}+\mathrm{i}\frac{\partial u}{\partial \zeta}+\frac{1}{2}\frac{\partial^2 u}{\partial \xi^2}+|u|^qu=\left[\frac{\Delta}{4\kappa}+(1-\alpha)|u|^q\right]h(\xi,\zeta)u. \tag{5}$$

The dimensionless longitudinal and transverse coordinates are $\zeta = z/L_D$ and $\xi = 2^{1/2}x/w_0$, respectively, where $L_D = k_1w_0^2/2$ is the diffraction length of a reference (paraxial) Gaussian beam. The inverse beam width is quantified by $\kappa = 1/(k_1w_0)^2 = \varepsilon^2/4\pi^2n_{01}^2 \ll O(1)$, where $\varepsilon \equiv \lambda/w_0$, and the field amplitude scales with $E_0 = (2n_{01}^2/\alpha_1k_1L_D)^{1/q}$. Model (5) is supplemented by the mismatch parameters [17, 18, 25, 26]

$$\Delta \equiv 1-\frac{n_{02}^2}{n_{01}^2}, \tag{6a}$$

$$\alpha \equiv \frac{\alpha_2}{\alpha_1}, \tag{6b}$$

which determine how the linear and nonlinear refractive properties of the system change as one traverses the boundary.

Equation (5) allows one access to material scenarios where $\Delta < 0$ (i.e., configurations with $n_{02} > n_{01}$) [17, 18, 25, 26]. By contrast, the scalings deployed in classic paraxial theory [8, 9] restrict those models to consideration of regimes with $\Delta > 0$. It is also apparent that setting $\kappa \approx 0$ in an attempt to recover the paraxial model is going to lead to complications when handling the linear mismatch term $\Delta/4\kappa$. The physical and mathematical difficulties of interpreting the paraxial approximation as the single-parameter limit $\kappa \approx 0$ have been discussed at length elsewhere [23, 24]; it is particularly well illustrated by interface geometries.

2.2. Solitons as a Nonlinear Basis. When investigating the refraction of nonlinear light beams at material boundaries, it is essential to have an appropriate set of basis functions with which to formulate the problem. Such a basis is provided by the underlying exact analytical Helmholtz solitons [56]. In the following analysis, we align the interface along the z axis so that it is located at transverse position $x = 0$. Medium 1 (the domain of the incident beam, where $h = 0$) is taken to occupy the half-plane $-\infty \le x < 0$, while medium 2 (the domain of the refracted beam, where $h = +1$), occupies $0 \le x \le +\infty$.

In medium 1, the governing equation (5) becomes

$$\kappa\frac{\partial^2 u}{\partial \zeta^2}+\mathrm{i}\frac{\partial u}{\partial \zeta}+\frac{1}{2}\frac{\partial^2 u}{\partial \xi^2}+|u|^qu=0. \tag{7}$$

Sufficiently far from the interface, (7) admits exact analytical solitons of the form [56]

$$u(\xi,\zeta)=\eta_0\operatorname{sech}^{2/q}\left(a\frac{\xi-V_{\mathrm{inc}}\zeta}{\sqrt{1+2\kappa V_{\mathrm{inc}}^2}}\right) \times\exp\left[\pm\mathrm{i}\sqrt{\frac{1+4\kappa\beta_0}{1+2\kappa V_{\mathrm{inc}}^2}}\left(V_{\mathrm{inc}}\xi+\frac{\zeta}{2\kappa}\right)\right] \times\exp\left(-\mathrm{i}\frac{\zeta}{2\kappa}\right), \tag{8a}$$

where η_0 is the peak amplitude of the beam, $a = q[\eta_0^q/(2+q)]^{1/2}$ determines the (inverse) solution width, and

$$\beta_0 = 2\frac{\eta_0^q}{2+q} \tag{8b}$$

quantifies nonlinear phase shift through the (typically small) quantity $4\kappa\beta_0$. The $\pm$ sign flags evolution in the forward/backward longitudinal direction. The propagation angle of the beam in the laboratory (i.e., the (x,z)) frame, denoted by θ_{inc} and measured with respect to the z axis, is related to the transverse velocity parameter V_{inc} through $\tan\theta_{\mathrm{inc}} = (2\kappa)^{1/2}V_{\mathrm{inc}}$ [23, 24]. In medium 2, u satisfies

$$\kappa\frac{\partial^2 u}{\partial \zeta^2}+\mathrm{i}\frac{\partial u}{\partial \zeta}+\frac{1}{2}\frac{\partial^2 u}{\partial \xi^2}-\frac{\Delta}{4\kappa}u+\alpha|u|^qu=0, \tag{9}$$

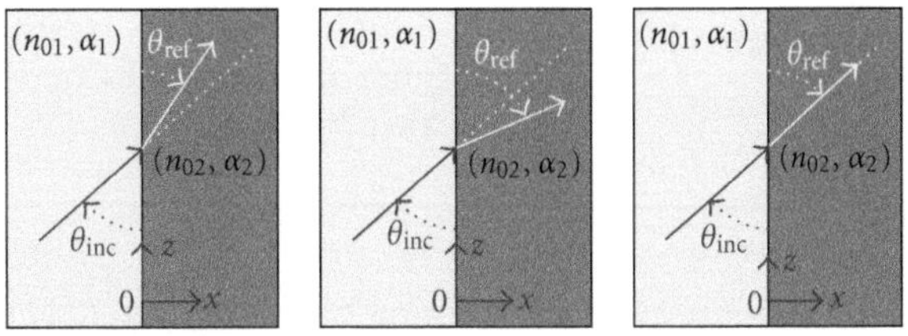

(a) (b) (c)

FIGURE 1: Schematic diagram illustrating (a) internal ($\theta_{ref} < \theta_{inc}$) and (b) external ($\theta_{ref} > \theta_{inc}$) refraction in the laboratory frame. The transparency condition ($\theta_{ref} = \theta_{inc}$) is shown in part (c). External refraction regimes tend to be highly angular and cannot be adequately described by the paraxial approximation.

and one may derive similar families of solitons,

$$u(\xi,\zeta) = \eta_0 \operatorname{sech}^{2/q}\left(a\sqrt{\alpha}\frac{\xi - V_{ref}\zeta}{\sqrt{1+2\kappa V_{ref}^2}}\right) \times \exp\left[\pm i\sqrt{\frac{1-\Delta+4\kappa\beta_0\alpha}{1+2\kappa V_{ref}^2}}\left(V_{ref}\xi + \frac{\zeta}{2\kappa}\right)\right] \times \exp\left(-i\frac{\zeta}{2\kappa}\right). \tag{10}$$

Note that the connection between transverse velocity V_{ref} and propagation angle θ_{ref}, that is, $\tan\theta_{ref} = (2\kappa)^{1/2}V_{ref}$, is unaffected by the (additional, linear) term $\Delta/4\kappa$ in (9) or by the nonlinear coefficient α. The geometry of these solitons, and their inherent stability against perturbations to the local beam shape, was explored in detail in [56].

2.3. Phase Continuity and Refraction. In recent analyses, we have shown that arbitrary-angle refraction is well described by anticipating that the phase distribution of the light be continuous across the interface [17, 18, 25, 26]. Matching the phases of solutions (8a) and (10) at $x = 0$ leads to the requirement that

$$\pm\sqrt{\frac{1+4\kappa\beta_0}{1+2\kappa V_{inc}^2}} = \pm\sqrt{\frac{1-\Delta+4\kappa\beta_0\alpha}{1+2\kappa V_{ref}^2}}. \tag{11}$$

Hence, continuity is possible if and only if the incident and refracted solitons share a common longitudinal sense (i.e., both must be in either the forward or backward directions). By rearranging (11), one can show that V_{ref} is related to V_{inc} through

$$V_{ref}^2 = V_{inc}^2 - \frac{1}{2\kappa}\left(\frac{1+2\kappa V_{inc}^2}{1+4\kappa\beta_0}\right)[\Delta + 4\kappa\beta_0(1-\alpha)]. \tag{12}$$

Expressed in this way, (12) provides a helpful form "$V_{ref}^2 = V_{inc}^2$ + *deviation*," where the sign of the *deviation* can be analysed separately. It is then instructive to define a net mismatch parameter δ as [17, 18]

$$\delta \equiv \Delta + 4\kappa\beta_0(1-\alpha). \tag{13}$$

This parameter can be interpreted as the sum of linear and nonlinear mismatches in material parameters. Its sign fully characterizes beam refraction. When $\delta > 0$, one has that $V_{ref}^2 < V_{inc}^2$, which is equivalent to $\theta_{ref} < \theta_{inc}$. This regime is referred to as *internal refraction*, and it corresponds to the situation where the beam in medium 2 is deviated toward the interface (see Figure 1(a)). Conversely, $\delta < 0$ implies that $V_{ref}^2 > V_{inc}^2$ or, equivalently, $\theta_{ref} > \theta_{inc}$. This *external refraction* regime corresponds to the beam in medium 2 being bent away from the interface (see Figure 1(b)). The special case of $\delta = 0$ is the transparency condition, where linear and nonlinear index mismatches oppose each other exactly so that $V_{ref}^2 = V_{inc}^2$ (or $\theta_{ref} = \theta_{inc}$). The interface is thus essentially transparent to the incident beam (see Figure 1(c)), which experiences no net change in dielectric properties as it crosses the boundary. It is interesting to note that the absence of an interface provides a parameter subset (i.e., $\Delta = 0$ and $\alpha = 1$) that satisfies the transparency condition identically.

2.4. The Helmholtz-Snell Law for Spatial Solitons. By recognizing the rotational symmetry inherent to Helmholtz spatial solitons [23, 24, 56], it becomes clear that "forward" and "backward" designations are arbitrary. The only physical distinction between the two families is the propagation direction relative to the observer. By deploying trigonometric identities to eliminate velocities V_{inc} and V_{ref}, the forward and backward solutions in each medium may be written as

$$u(\xi,\zeta) = \eta_0 \operatorname{sech}^{2/q}\left[a\left(\xi\cos\theta_{inc} - \frac{\zeta}{\sqrt{2\kappa}}\sin\theta_{inc}\right)\right] \times \exp\left[i\sqrt{\frac{1+4\kappa\beta_0}{2\kappa}}\left(\xi\sin\theta_{inc} + \frac{\zeta}{\sqrt{2\kappa}}\cos\theta_{inc}\right)\right] \times \exp\left(-i\frac{\zeta}{2\kappa}\right), \tag{14a}$$

and

$$u(\xi,\zeta) = \eta_0 \operatorname{sech}^{2/q}\left[a\sqrt{\alpha}\left(\xi\cos\theta_{ref} - \frac{\zeta}{\sqrt{2\kappa}}\sin\theta_{ref}\right)\right] \times \exp\left[i\sqrt{\frac{1-\Delta+4\kappa\beta_0\alpha}{2\kappa}}\right.$$

$$\times\left(\xi\sin\theta_{\text{ref}}+\frac{\zeta}{\sqrt{2\kappa}}\cos\theta_{\text{ref}}\right)\Bigg]$$
$$\times\exp\left(-\mathrm{i}\frac{\zeta}{2\kappa}\right). \tag{14b}$$

In this representation, the angles are bounded by $-180^\circ < \theta_{\text{inc, ref}} \leq +180^\circ$ with respect to the z-axis. By matching the solution phase at $\xi = 0$, one can obtain a compact Helmholtz-Snell refraction law involving laboratory-frame angles:

$$\gamma n_{01}\cos\theta_{\text{inc}} = n_{02}\cos\theta_{\text{ref}}, \tag{15a}$$

where

$$\gamma \equiv \left[\frac{1+4\kappa\beta_0}{1+4\kappa\beta_0\alpha(1-\Delta)^{-1}}\right]^{1/2}. \tag{15b}$$

It is worthwhile noting that (15a) has a form which is almost exactly identical to that encountered when studying the classic electromagnetic problem of *plane wave* refraction at the boundary between different *linear* dielectrics. Thus, the single correction factor γ captures the interplay between finite-waist beams (through the appearance of κ) and discontinuities in both the linear *and nonlinear* properties of the adjoining media. The exponent q appears implicitly through β_0.

When a beam encounters the boundary with a rarer medium, there is little penetration of light across that boundary until the incidence angle exceeds a critical value, denoted by θ_{crit}. At criticality, where $\theta_{\text{inc}} = \theta_{\text{crit}}$, the trajectory of the incident beam is deviated so that, in principle, the outgoing beam travels along the interface (i.e., $\theta_{\text{ref}} = 0$). Applying this condition to law (15a) and (15b) leads to an analytical prediction for θ_{crit} in terms of the mismatch parameters Δ and α and also the solution parameter $4\kappa\beta_0$:

$$\tan\theta_{\text{crit}} = \left[\frac{\Delta+4\kappa\beta_0(1-\alpha)}{1-\Delta+4\kappa\beta_0\alpha}\right]^{1/2}. \tag{16}$$

In practice, one rarely finds the refracted soliton travelling along the interface boundary since other effects tend to appear (we will return to this point later).

2.5. Universal versus Specific Representations. There is clearly a universal flavour about (12), (13), (15a), (15b) and (16). For instance, there is no explicit mention of the system nonlinearity so that refraction is fully described by the mismatch parameters Δ and α and the beam parameter $4\kappa\beta_0$. These equations are, in fact, more general than they first appear; for instance, laws of exactly the same structure govern the refraction of plane waves in power-law materials: a wave with real amplitude u_0 has $\beta_0 \equiv u_0^q$ (it is noteworthy that the refraction analysis for plane waves does not capture the modulational instability of such solutions in the single power-law context [58]).

The power-law nature of the problem becomes apparent after one substitutes for β_0 from (8b). The γ factor (c.f. (15b)) then becomes

$$\gamma = \left[\frac{1+8\kappa\eta_0^q(2+q)^{-1}}{1+8\kappa\eta_0^q\alpha(2+q)^{-1}(1-\Delta)^{-1}}\right]^{1/2}, \tag{17a}$$

while the relation for the critical angle (c.f. (16)) is given by

$$\tan\theta_{\text{crit}} = \left[\frac{\Delta+8\kappa\eta_0^q(2+q)^{-1}(1-\alpha)}{1-\Delta+8\kappa\eta_0^q\alpha(2+q)^{-1}}\right]^{1/2}, \tag{17b}$$

and the net mismatch parameter (c.f. (13)) is $\delta = \Delta + 8\kappa\eta_0^q(1-\alpha)/(2+q)$.

3. Simulations of Solitons at Power-Law Interfaces

The Helmholtz type of off-axis nonparaxiality demands that the inequalities $\kappa \ll O(1)$ and $4\kappa\beta_0 \ll O(1)$ are always met, which is equivalent to the *simultaneous* requirements of broad beams with moderate intensities, respectively [23, 24, 56]. Here, attention is restricted to configurations where the mismatch parameters are relatively small, typically $\alpha = O(1)$ and $|\Delta| \ll O(1)$. We now proceed with a three-stage analysis. The simplest case to consider first is that of linear interfaces. We then move on to investigate nonlinear interfaces and conclude by noting the dependence of GH shifts [48, 49] on the nonlinearity exponent q. Stable solitons of the homogeneous power-law Helmholtz model tend to exist in the continuous interval $0 < q < 4$ [27, 56]. For definiteness, we consider here only three discrete values: $q = 1$ (sub-Kerr), 2 (Kerr), and 3 (super-Kerr).

3.1. Solitons at Linear Interfaces. From (13), linear interfaces are defined by the inequality $4\kappa\beta_0|1-\alpha| \ll |\Delta|$. To isolate the effects of a linear-index change alone, we set $\alpha = 1.0$ so that $\delta = \Delta$. One therefore finds the existence of a critical angle in regimes where $\Delta > 0$ (since $n_{02} < n_{01}$). The following simulations consider $q = 1$. Figure 2 shows generally good agreement between theoretical predictions and full numerical calculations when $\kappa = 2.5\times10^{-3}$; the level of agreement is improved even further when $\kappa = 1.0\times10^{-4}$.

The fact that smaller values of κ give rise to better theory-numerics agreement, despite the increased magnitude of the linear-interface perturbation term at $\Delta/4\kappa$, invites comment. We suspect that one possible explanation may lie in the origin of the Helmholtz-Snell law, whereby one matches solution phase (but not amplitude) at the boundary: the matching condition thus takes no account of amplitude curvature. In the laboratory frame, broader beams (i.e., those characterized by smaller κ values) tend to have lower amplitude curvature, and the corresponding spatial solitons (which play the role of nonlinear basis functions) thus map much more consistently onto the inherent assumptions of the analytical approach.

Upon crossing the interface, the refracted soliton may suffer small oscillations (in its amplitude, width, and area)

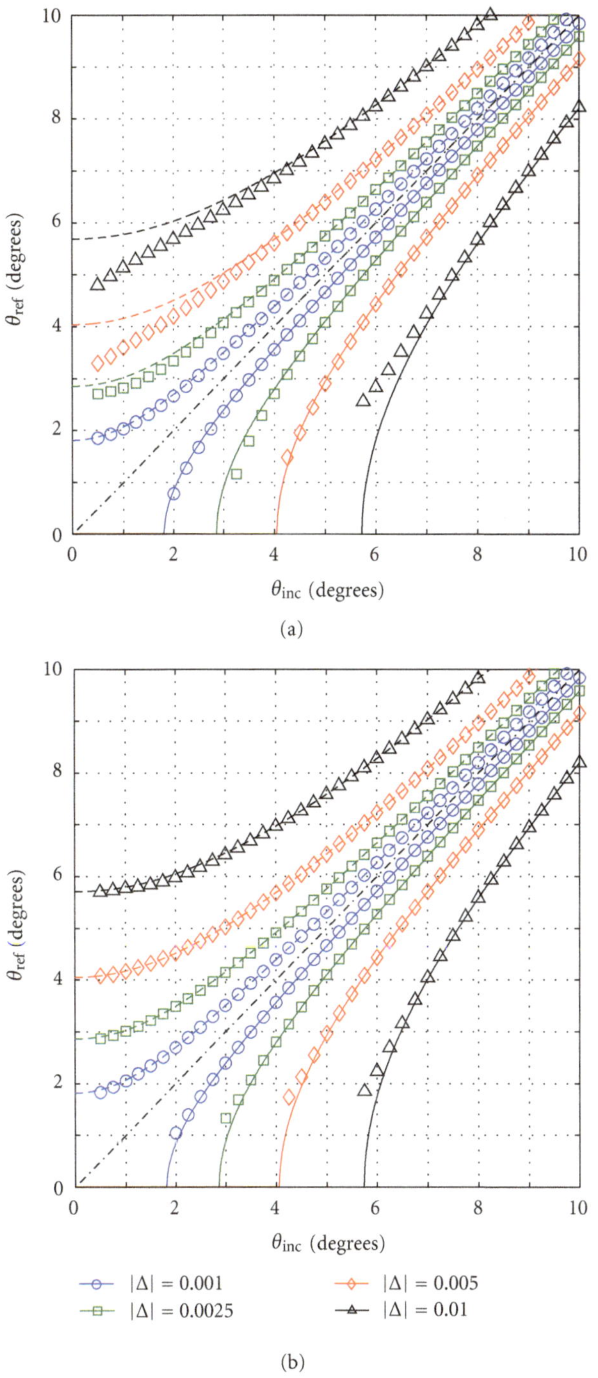

Figure 2: Comparison of the theoretical Snell's law given by (15a) and (15b) (lines) against full numerical computations (points) for a unit-amplitude ($\eta_0 = 1.0$) spatial soliton at a linear interface ($\alpha = 1.0$) with $q = 1$ and when (a) $\kappa = 2.5 \times 10^{-3}$ and (b) $\kappa = 1.0 \times 10^{-4}$. Curves below (above) the $\theta_{\text{ref}} = \theta_{\text{inc}}$ line have $\Delta > 0$ ($\Delta < 0$), so that the refraction is internal (external).

reminiscent of those reported in previous studies [56], and be accompanied by a radiation pattern. Computations [59] have verified the effective independence of the refraction angle θ_{ref} with respect to the incident amplitude η_0. Accordingly, the curves in Figure 2 are essentially insensitive to q; they are quantitatively very similar to those obtained for $q = 2$ [10] and (when θ_{inc} is sufficiently above θ_{crit} in internal-refraction regimes) for $q = 3$.

Any interaction between a spatial soliton and an interface generally involves three distinct components: a reflected beam, a refracted beam (sometimes more than one), and radiation. The way in which the incident energy is distributed amongst these components depends on a complicated interplay between the interface and beam parameters, and also the incidence angle. At very small angles (e.g., $\theta_{\text{inc}} < 1°$), the interaction can be highly inelastic and nonadiabatic (especially in external refraction regimes). Crucially, the single refracted soliton (as predicted in Section 2) dominates as θ_{inc} approaches even modest nonparaxial angles, with reflected and radiation components hardly excited at all. The Helmholtz-Snell law embodied by (15a) and (15b) is, of course, most valid in such regimes.

3.2. Solitons at Nonlinear Interfaces. Nonlinear interface effects dominate beam refraction when $4\kappa\beta_0|1 - \alpha| \gg |\Delta|$. Without loss of generality, we isolate such effects by setting $\Delta = 0$ so that the net mismatch parameter is given by $\delta = 4\kappa\beta_0(1 - \alpha) = 8\kappa\eta_0^q(1 - \alpha)/(2 + q)$. Refraction thus becomes far more sensitive to κ in nonlinear regimes (compare this to linear regimes, where $\delta = \Delta$ is independent of κ). Theoretical predictions are shown in Figure 3. While there is generally good agreement with numerics for both $\kappa = 2.5 \times 10^{-3}$ and $\kappa = 1.0 \times 10^{-4}$ when $\alpha \approx 1.0$, the fit becomes less reliable for $\alpha = 2.0$ and $\alpha = 0.3$. For such parameters, the nonlinear refractive index change across the boundary is no longer small: one cannot expect to find such a close match because of strong nonlinear effects (e.g., beam splitting and radiation phenomena). While the fit is clearly better for broader beams ($\kappa = 1.0 \times 10^{-4}$), the Helmholtz-Snell predictions for narrower beams ($\kappa = 2.5 \times 10^{-3}$) are still in good qualitative agreement.

Detailed attention is first paid to regimes with $\alpha > 1$ (external refraction, since $\delta < 0$), where the nonlinearity is stronger in the second medium. Since the width of the refracted soliton is proportional to $\alpha^{-1/2}$, it follows that the beam must become narrower as it crosses the interface. In this type of material regime, the incident soliton always has sufficient energy flow to excite a self-trapped soliton-like state in medium 2.

Simulations have shown that nonlinear external refraction tends to induce stronger oscillations in the parameters (amplitude, width, and area = amplitude × width) of the outgoing beam than in the linear case. Such oscillations are not captured by the adiabatic analysis in Section 2 (which anticipates a stationary state), but one expects their appearance intuitively. Qualitatively different effects can appear at quasi-paraxial incidence angles as the exponent q is varied; an illustrative example is shown in Figure 4 for $\theta_{\text{inc}} = 3°$ when $\alpha = 2.0$. A unit-amplitude soliton exhibits a pronounced splitting phenomenon when $q = 1$ (see Figure 4(b)), whereby the field distribution in the second medium is shared between a dominant externally refracted beam (as predicted by analysis) and a weaker internally refracted component (there is also a low-amplitude reflected

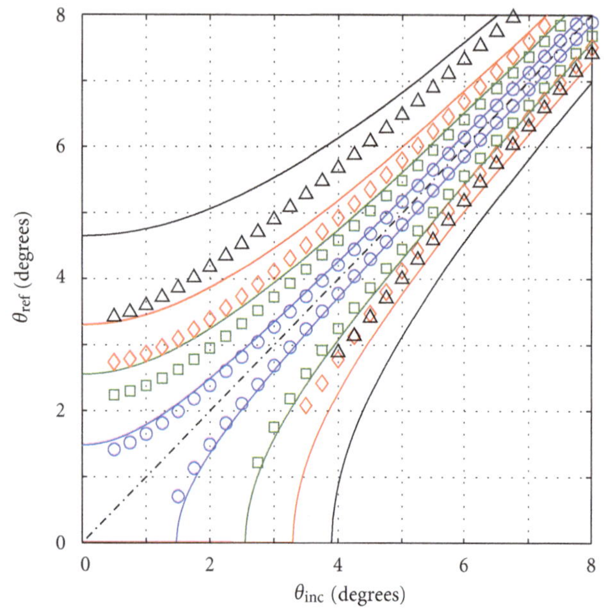

(a)

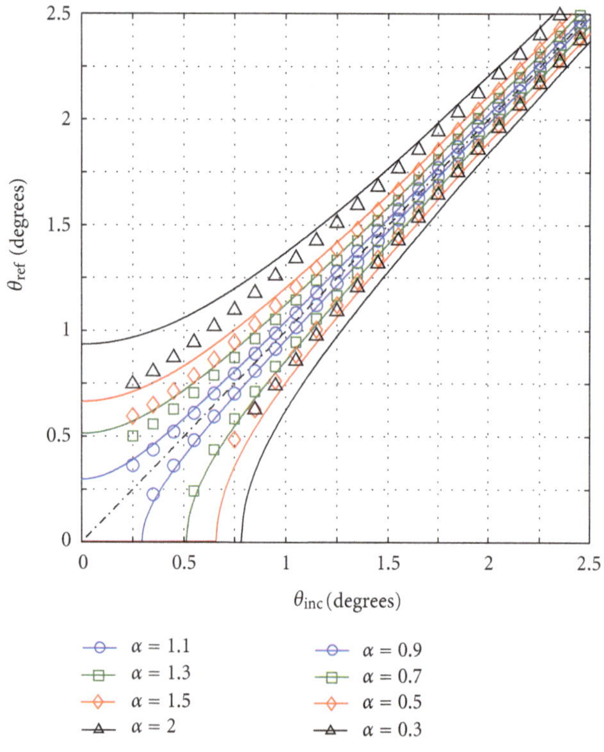

(b)

FIGURE 3: Comparison of the theoretical Snell's law given by (15a) and (15b) (lines) against full numerical computations (points) for a unit-amplitude ($\eta_0 = 1.0$) spatial soliton at a nonlinear interface ($\Delta = 0$) with $q = 1$ and when (a) $\kappa = 2.5 \times 10^{-3}$ and (b) $\kappa = 1.0 \times 10^{-4}$. Curves below (above) the $\theta_{ref} = \theta_{inc}$ line are labelled by the right-hand (left-hand) legend and have $\alpha < 1$ ($\alpha > 1$) so that refraction is internal (external). Note that the numerical datapoints for $\alpha = 0.3$ and $\alpha = 0.5$ are very close together in both panes.

component in the form of radiation modes). Since the internally refracted beam carries away some of the momentum of the incident beam, it follows that the dominant refracted beam travels at a smaller angle than that predicted by (15a) and (15b). This type of splitting is not present for unit-amplitude solitons with $q = 2$ (see Figure 4(b)), though it may appear for incident solitons with higher peak intensities [60]. In such cases, the properties of the daughter solitons may be quantified with recourse to inverse scattering techniques. Splitting is also absent at $q = 3$ (see Figure 4(c)), though one finds quite a complicated radiation ripple pattern in the second medium.

Refraction in nonparaxial regimes tends to be a much cleaner process, with little radiation generated by the beam-interface interaction in comparison with quasi-paraxial regimes. Even at modest angles (e.g., $\theta_{inc} = 30°$), where the interface perturbation is distributed over a relatively short interaction length, the quantitative characteristics of the outgoing beam depend crucially on the power-law exponent. Both the depth of modulation and (longitudinal spatial) frequency of the oscillations tend to increase with q, as shown in Figure 5(a). When $q = 2$, the oscillations tend to vanish in ζ; for $q = 1$ and 3, they survive in the long-term evolution (this is also true for the oscillations shown in Figure 4(a)). A more detailed comparison of how the q affects beam refraction is presented in Figures 5(b)–5(d).

For material combinations with $\alpha < 1$ (internal refraction, since $\delta > 0$), the nonlinearity is weaker in the second medium. In that case, one should expect a critical angle to exist (in accordance with (17b)). If the incident soliton survives the interaction with the interface, then the refracted beam may be expected to undergo self-reshaping oscillations in its parameters, with the overall trend being toward an increase in solution width. Simulations have confirmed this to be the case, with diffractive broadening generally accompanied by a reduction in peak amplitude (see Figure 6(a))—these oscillations are reminiscent of those uncovered previously for perturbed initial-value problems [56].

Computations have uncovered a range of q-dependent effects, an illustrative sample of which is shown in Figure 6 for beams with $\kappa = 2.5 \times 10^{-3}$, a nonparaxial incidence angle $\theta_{inc} = 30°$, and a nonlinear mismatch of $\alpha = 0.5$. The (longitudinal spatial) frequency of the reshaping oscillations tends to decrease with increasing q (c.f., the *increase* with q when $\alpha > 1$). Also at higher q values (e.g., for $q = 3$), a threshold phenomenon can appear whereby the energy-flow [56] of the incident soliton may not be great enough to excite a refracted beam (if the energy flows associated with solutions (8a) and (10) are denoted by W_{inc} and W_{ref}, respectively, then it can be shown that $W_{ref} \approx W_{inc}/\alpha^{1/2}$). This instability is shown in Figure 6(d): upon colliding with the interface, the beam breaks up into radiation (this scenario is also present at quasi-paraxial incidence angles above the critical angle θ_{crit}).

3.3. Snaking at Nonparaxial Angles. Equations (15a) and (15b) show that, at nonlinear interfaces, the refraction

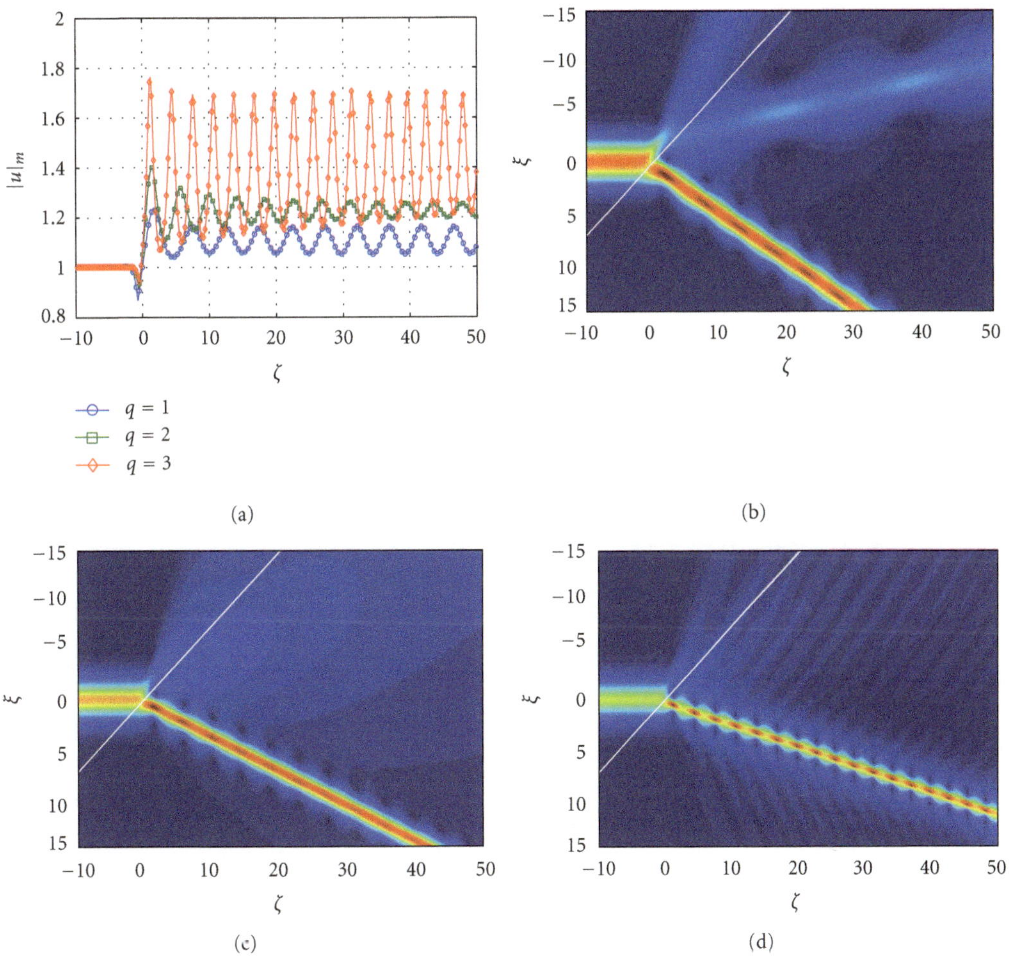

FIGURE 4: External refraction of a unit-amplitude ($\eta_0 = 1.0$) spatial soliton at a nonlinear interface with $\alpha = 2.0$ and a quasi-paraxial incidence angle $\theta_{\text{inc}} = 3^\circ$ when $\kappa = 2.5 \times 10^{-3}$. (a) Evolution in ζ of the peak amplitude $|u|_m$ of the beam. (b), (c), and (d) show the full numerical solution $|u(\xi, \zeta)|$ of (5) when the nonlinearity exponent is $q = 1$, 2, and 3, respectively.

angle must depend on q (a prediction supported by simple inspection of Figures 4, 5, and 6). At this point, it also becomes instructive to consider the trajectory of refracted beams more carefully. Detailed numerical calculations reveal that at quasi-paraxial incidence angles, the beam in the second medium tends to follow a straightline path. Such a simple notion of refraction, founded upon intuition from plane wave theory, is illustrated in Figure 7(a) for a nonlinear interface with $\alpha = 2.0$ and a beam with $\theta_{\text{inc}} = 3^\circ$ and $\kappa = 2.5 \times 10^{-3}$. However, if the incidence angle is increased into the nonparaxial domain (e.g., $\theta_{\text{inc}} = 30^\circ$), a qualitatively different picture emerges. Now, the straightline path $\xi - V_{\text{ref}}\zeta = 0$ predicted by solution (10) defines an average trajectory about which the refracted beam "snakes." Figure 7(b) quantifies this snaking effect for the external refraction simulations shown in Figures 5(b)–5(d). Snaking is more apparent with sub-Kerr nonlinearities (i.e., where $q < 2$), and it increases for narrower beams (i.e., larger values of κ) at a fixed amplitude (see Figure 8(a), where $\eta_0 = 1.0$). Beams with larger amplitudes also exhibit snaking, but oscillations tend to be more rapid in the longitudinal direction (see Figure 8(b)).

The snaking phenomenon is most pronounced in regimes with $\alpha > 1$, where the nonlinearity is stronger in the second medium. There is also an intrinsic dependence on θ_{inc} that can be seen in Figure 7. For small angles of incidence, the incoming soliton experiences an interface perturbation that is distributed over a relatively long distance. The refracting beam is able to accommodate the inhomogeneity in the system since changes in focusing properties of the host medium occur gradually in the longitudinal direction. For larger-incidence angles, the effective beam-interface interaction length may be much shorter. Solitons impinging on the boundary then exhibit a sharp (rather than a gradual) perturbation whose action is to induce sustained oscillations.

3.4. Goos-Hänchen Shifts at Power-Law Interfaces. Recently, GH shifts [48] have been investigated within the context of Helmholtz spatial solitons at Kerr-type material interfaces [49]. These shifts describe the translation in the trajectory

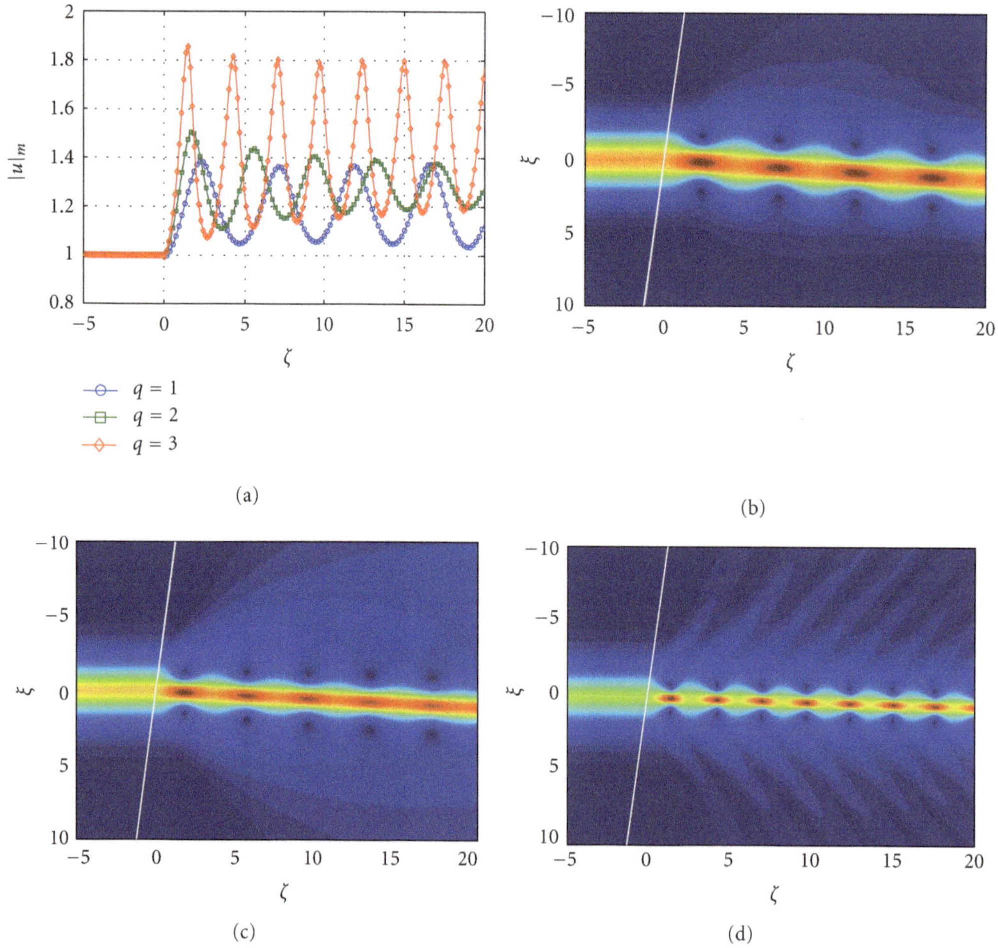

FIGURE 5: External refraction of a unit-amplitude ($\eta_0 = 1.0$) spatial soliton at a nonlinear interface with $\alpha = 2.0$ and a nonparaxial incidence angle $\theta_{\text{inc}} = 30^\circ$ when $\kappa = 2.5 \times 10^{-3}$. (a) Evolution in ζ of the peak amplitude $|u|_m$ of the beam. (b), (c), and (d) show the full numerical solution $|u(\xi, \zeta)|$ of (5) when the nonlinearity exponent is $q = 1, 2$, and 3, respectively.

of a reflected beam relative to its position as predicted by geometrical optics. Extensive numerical investigations considered the interplay between incidence angle θ_{inc}, material mismatches (Δ, α), and the nonparaxial parameter κ. Radiation-induced trapping was found to play a key role in determining the magnitude of the shift. Also uncovered were giant *external* GH shifts (in regimes with $\delta > 0$ but where the second medium has a *weaker* nonlinearity (i.e., $\alpha < 1$)). While a similar investigation of GH shifts in the power-law context is certainly outside our current scope, a small selection of results will now be presented to illustrate how they depend upon the nonlinearity exponent q.

We begin by considering linear interfaces and unit-amplitude incident solitons with $\kappa = 2.5 \times 10^{-3}$. According to (16), interfaces with $\Delta = 0.0025$ have a theoretical critical angle of $\theta_{\text{crit}} \approx 2.86^\circ$ (this value depends only very weakly on q). Figure 9(a) gives a representative set of results. Inspection shows that, for any θ_{inc}, the magnitude of the shift is generally greater for systems with $q = 1$ than for $q = 2$ or $q = 3$. The true critical angle (which can only be found through full simulations) is also slightly greater than that predicted by theory (for $q = 1$ and $q = 2$, $\theta_{\text{crit}} \approx 3.016^\circ$ and $\theta_{\text{crit}} \approx 3.030^\circ$; both angles exceed their theoretical values of $\theta_{\text{crit}} \approx 2.857^\circ$ and $\theta_{\text{crit}} \approx 2.859^\circ$, respectively). While the qualitative behaviour of systems with $q = 1$ and $q = 2$ is largely very similar, strong qualitative differences have been uncovered in the case of $q = 3$. As θ_{inc} approaches the theoretical critical angle, the incident soliton often becomes unstable against the interface perturbation. Large amounts of radiation tend to be generated by the interaction (c.f. Figure 9(d)), so that there is essentially no reflected or refracted beam and a GH shift is thus not easily quantifiable (or even meaningful). However, when θ_{inc} is sufficiently above θ_{crit}, the refraction angle is still well described by theory.

GH shifts at nonlinear interfaces have also been analyzed; results are presented in Figure 10 for $\alpha = 0.7$ and where system nonlinearity has been augmented by considering incident solitons with $\eta_0 = 2.0$. Regimes with $\Delta = -0.001$ and $\Delta = -0.0025$ are associated with linear *external* refraction, while (13) shows that $\delta > 0$ (i.e., for these parameter sets, net refraction is *internal* so that a critical angle should still exist). One general trend to emerge is that the true critical angle is slightly less than the theoretical value (c.f. linear interfaces of Figure 9, where

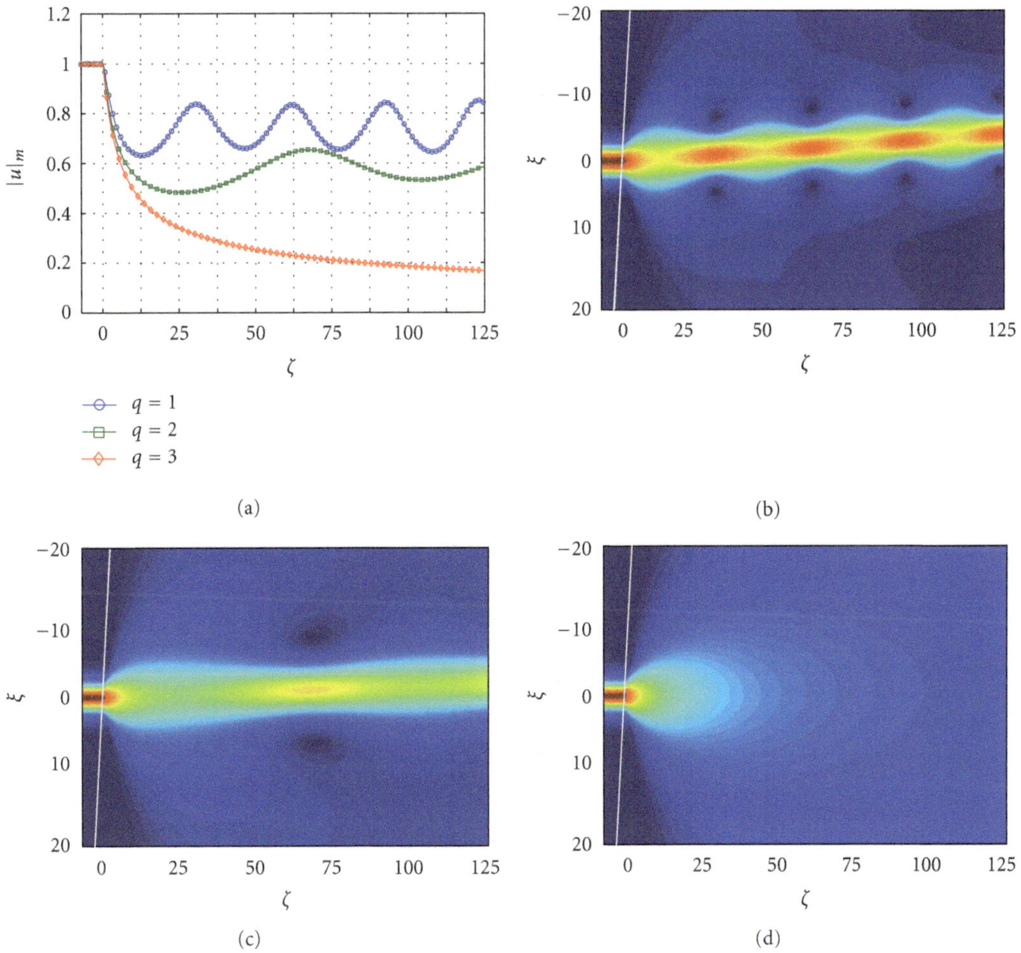

FIGURE 6: Internal refraction of a unit-amplitude ($\eta_0 = 1.0$) spatial soliton at a nonlinear interface with $\alpha = 0.5$ and a nonparaxial incidence angle $\theta_{\text{inc}} = 30^\circ$ when $\kappa = 2.5 \times 10^{-3}$. (a) Evolution in ζ of the peak amplitude $|u|_m$ of the beam. (b), (c), and (d) show the full numerical solution $|u(\xi,\zeta)|$ of (5) when the nonlinearity exponent is $q = 1, 2$, and 3, respectively.

the true critical angle slightly exceeds theory). However, it is worth noting that the qualitative behaviour predicted by (16), namely that θ_{crit} increases with q, is supported by numerics. Close to the (true) critical angle, simulations show that there is a strong divergence in the GH shift (which becomes highly sensitive to θ_{inc}). Two other general trends are that (i) GH shifts are larger (sometimes notably) for $q = 1$ than for $q = 3$; (ii) in nonlinear regimes, the GH shifts depend more strongly on q than for the case of linear interfaces (compare Figure 10 to Figure 9(a)).

Figure 10(b) reveals new types of behaviour at power-law interfaces when $q \neq 2$. In particular, for $q = 3$ one enters a regime wherein the GH shift no longer increases monotonically with θ_{inc}; instead, there is a marked decrease in the shift before the divergence at $\theta_{\text{inc}} \approx \theta_{\text{crit}}$ sets in. These results provide clear evidence that one can, quite reasonably, expect to find new qualitative phenomena in material regimes that deviate from the idealized Kerr-type response.

4. Helmholtz Nonlinear Surface Waves

Surface waves are well known in nonlinear photonics, being stationary localized light states that travel along the interface between different media. The transverse mode profiles are typically asymmetric due to the differences in dielectric properties defining the interface. We now derive the surface modes of model (5) using solitons (8a) and (10) as a nonlinear basis. These new solutions are most conveniently parameterized by β, which is related to the propagation constant in paraxial theory [27, 56].

4.1. Exact Analytical Solutions. To proceed, one seeks solutions to (5) of the form $u(\xi,\zeta) = F(\xi - \xi_j)\exp(ik_\zeta\zeta)\exp(-i\zeta/2\kappa)$, where k_ζ is the propagation constant and $F(\xi - \xi_j)$ is the (real) envelope profile that is centred on ξ_j. After substituting for u and defining $\kappa k_\zeta^2 - 1/4\kappa \equiv \beta$, it can be shown that in medium 1

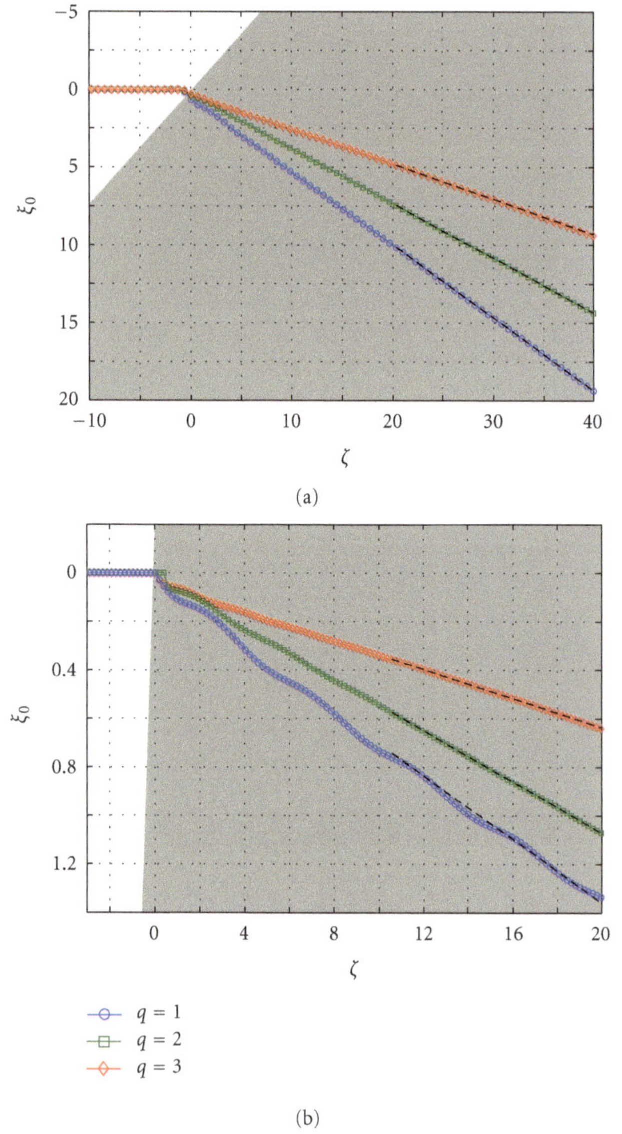

FIGURE 7: External refraction of a unit-amplitude ($\eta_0 = 1.0$) spatial soliton at a nonlinear interface with $\alpha = 2.0$ when the incidence angle is (a) quasi-paraxial ($\theta_{\text{inc}} = 3^\circ$) and (b) nonparaxial ($\theta_{\text{inc}} = 30^\circ$) for $\kappa = 2.5 \times 10^{-3}$. In (a), the trajectory of the beam in the second medium is essentially a straight line. In (b), the trajectory oscillates ("snakes") around the straight-line path predicted by the analysis in Section 2. Calculations of the beam centre ξ_0 were obtained by fitting the numerical solution at each longitudinal position to a trial function of the form $u_{\text{fit}}(\xi) = \eta(\zeta)\text{sech}^{2/q}\{a(\zeta)[\xi - \xi_0(\zeta)]\}$. Black dashed lines: best-fit trajectory.

$$u(\xi,\zeta) = \left(\frac{2+q}{2}\beta\right)^{1/q} \text{sech}^{2/q}\left[\frac{q}{\sqrt{2}}\beta^{1/2}(\xi - \xi_1)\right] \times \exp\left(\pm i\sqrt{1+4\kappa\beta}\frac{\zeta}{2\kappa}\right)\exp\left(-i\frac{\zeta}{2\kappa}\right), \tag{18a}$$

while in medium 2

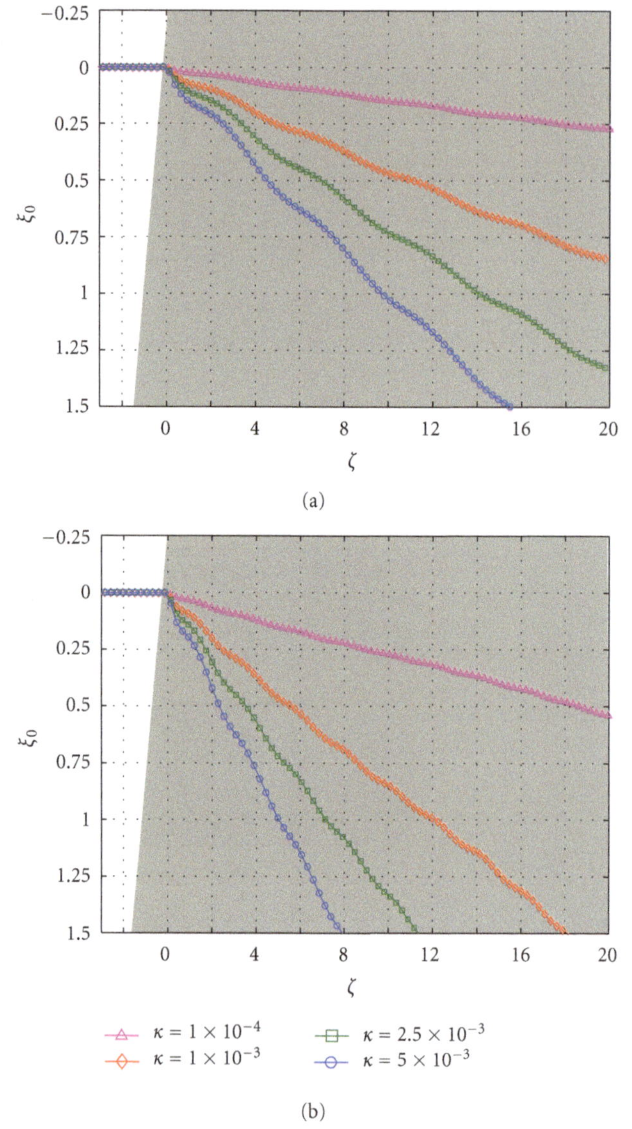

FIGURE 8: External refraction of spatial solitons at a nonlinear interface with $\alpha = 2.0$ for a nonparaxial angle $\theta_{\text{inc}} = 30^\circ$ for $q = 1$ and different values of κ. The peak amplitude of the incident beam in each case is (a) $\eta_0 = 1.0$ and (b) $\eta_0 = 2.0$.

$$u(\xi,\zeta) = \left[\left(\frac{2+q}{2}\beta\right)\left(\frac{1}{\alpha}\right)\left(1+\frac{\Delta}{4\kappa\beta}\right)\right]^{1/q} \times \text{sech}^{2/q}\left[\frac{q}{\sqrt{2}}\beta^{1/2}\left(1+\frac{\Delta}{4\kappa\beta}\right)^{1/2}(\xi - \xi_2)\right] \times \exp\left(\pm i\sqrt{1+4\kappa\beta}\frac{\zeta}{2\kappa}\right)\exp\left(-i\frac{\zeta}{2\kappa}\right). \tag{18b}$$

For a nonlinearity exponent q, the surface waves associated with any given interface are parameterized solely by β. The displacements ξ_1 and ξ_2, as yet undetermined, can be found by considering the auxiliary equations that arise from respecting continuity of u and its normal derivative (here

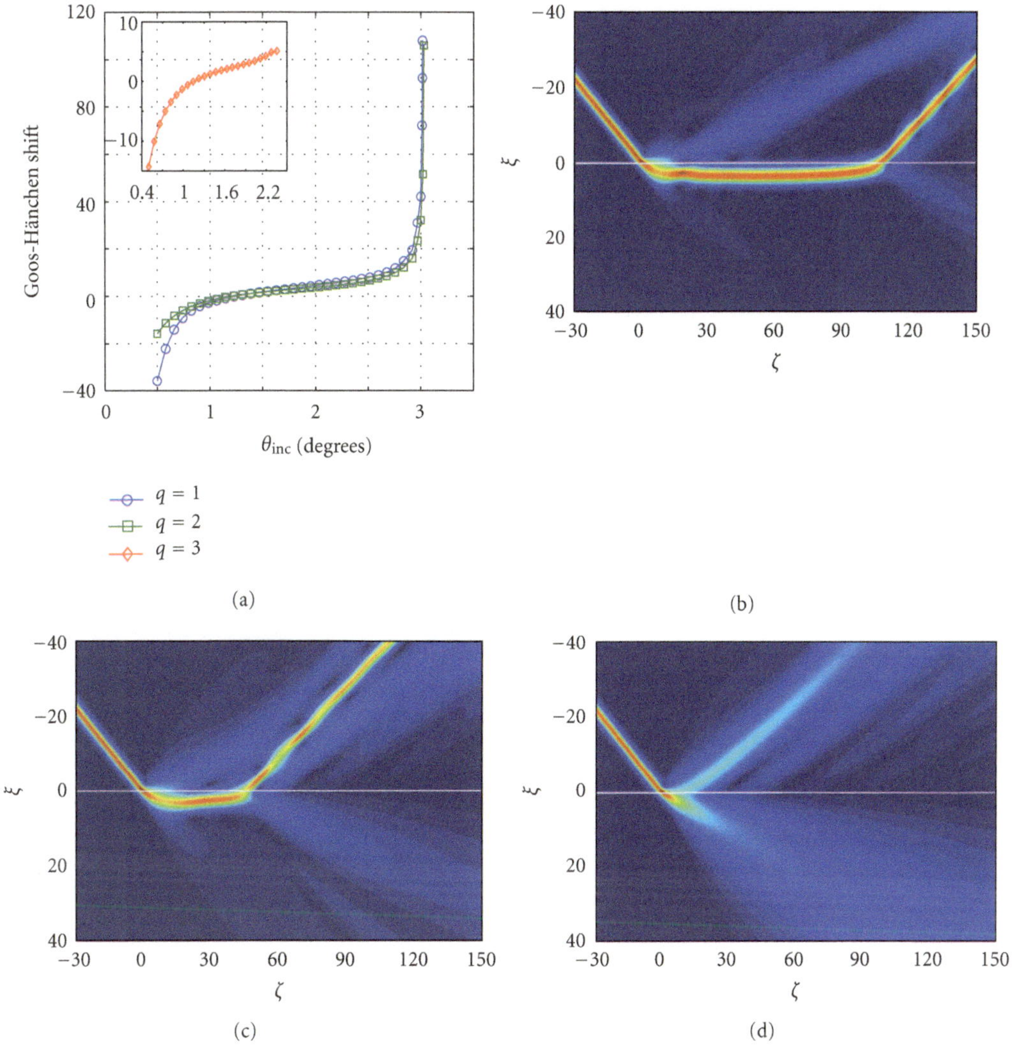

FIGURE 9: Demonstration of the GH shift for a unit-amplitude ($\eta_0 = 1.0$) spatial soliton at a linear interface with $\Delta = 0.0025$ and when $\kappa = 2.5 \times 10^{-3}$. (a) Variation of the GH shift with changing nonlinearity exponent q (the $q = 3$ results (inset) closely follow those for $q = 2$ until radiation effects come into play more strongly). (b), (c), and (d) show the full numerical solution $|u(\xi,\zeta)|$ of (5) when $q = 1$, 2 and 3, respectively (note that, over longer propagation lengths, the solution in (d) breaks up into radiation). The incidence angle in (b), (c), and (d) is $\theta_{inc} = 3.016°$, which exceeds the (almost q-independent) critical angle $\theta_{crit} \approx 2.86°$.

$\partial u/\partial\xi$ or, equivalently, $dF/d\xi$) across the interface. These conditions lead to

$$\operatorname{sech}^{2/q}\left(\frac{q}{\sqrt{2}}\beta^{1/2}\xi_1\right) = \left[\frac{1}{\alpha}\left(1+\frac{\Delta}{4\kappa\beta}\right)\right]^{1/q} \times \operatorname{sech}^{2/q}\left[\frac{q}{\sqrt{2}}\beta^{1/2}\left(1+\frac{\Delta}{4\kappa\beta}\right)^{1/2}\xi_2\right], \tag{19a}$$

$$\tanh\left(\frac{q}{\sqrt{2}}\beta^{1/2}\xi_1\right) = \left(1+\frac{\Delta}{4\kappa\beta}\right)^{1/2} \times \tanh\left[\frac{q}{\sqrt{2}}\beta^{1/2}\left(1+\frac{\Delta}{4\kappa\beta}\right)^{1/2}\xi_2\right], \tag{19b}$$

respectively. After some algebraic manipulation of (19a) and (19b), one finds that

$$\xi_1 = \frac{\sqrt{2}}{q}\beta^{-1/2}\ln\left(\frac{1\pm\sqrt{1-\delta^2}}{\delta}\right) \tag{20a}$$

$$\xi_2 = \frac{\sqrt{2}}{q}\beta^{-1/2}\left(1+\frac{\Delta}{4\kappa\beta}\right)^{-1/2}\ln\left(\frac{1\pm\sqrt{1-\mu^2}}{\mu}\right), \tag{20b}$$

where the parameters δ and μ are given by $\delta \equiv [\Delta/4\kappa\beta(\alpha - 1)]^{1/2}$ and $\mu \equiv [(\Delta/4\kappa\beta)(1+\Delta/4\kappa\beta)^{-1}(1-1/\alpha)^{-1}]^{1/2}$.

4.2. Surface Wave Existence Criterion. For displacements ξ_1 and ξ_2 to be real, it must be that $0 < \delta^2 < 1$ and $0 < \mu^2 < 1$. These two simultaneous requirements lead to a third

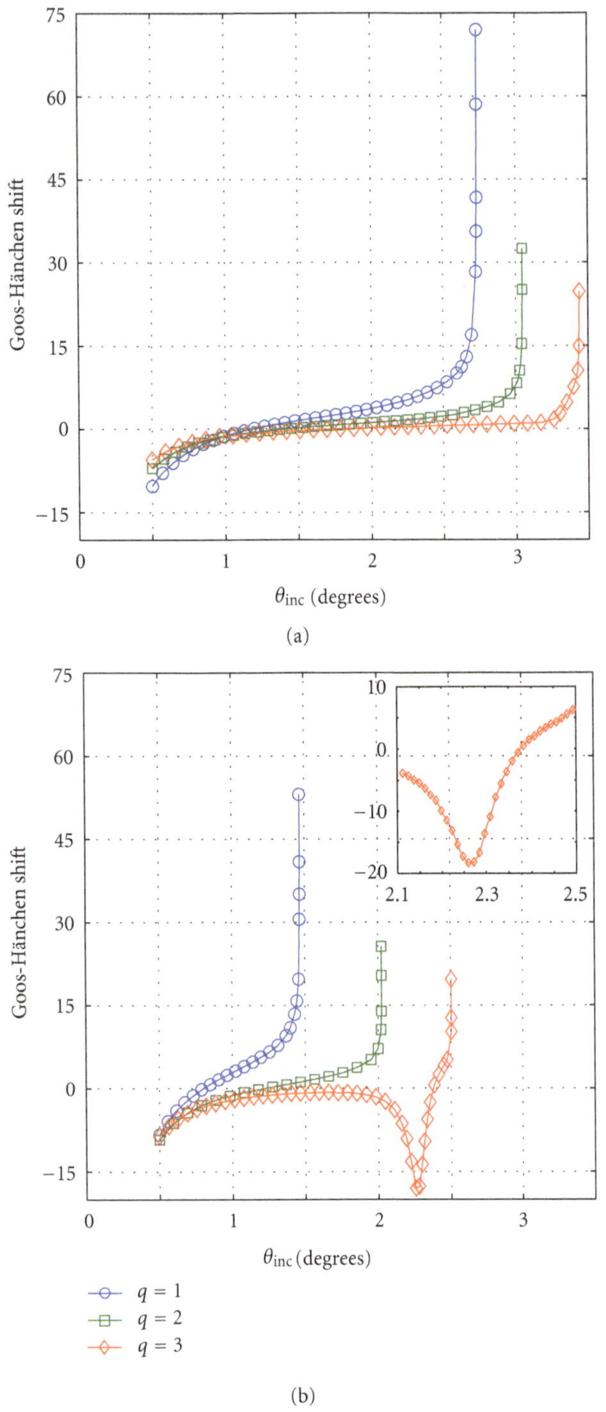

FIGURE 10: Numerical calculation of the GH shift for incident spatial solitons with $\eta_0 = 2.0$ at a nonlinear interface with $\alpha = 0.7$, (a) $\Delta = -0.001$, and (b) $\Delta = -0.0025$ when $\kappa = 2.5 \times 10^{-3}$ (inset shows the behaviour of the shift for $q = 3$ around the minimum).

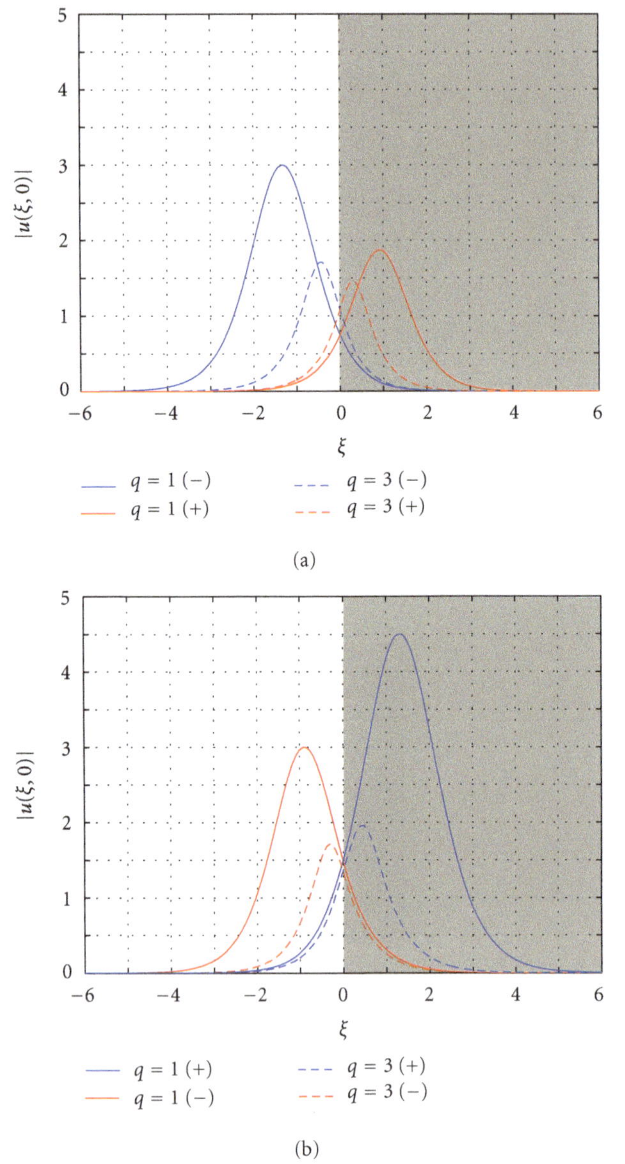

FIGURE 11: Nonlinear surface wave profiles for $\kappa = 2.5 \times 10^{-3}$ in (a) regime 1 (with $\Delta = 0.005$ and $\alpha = 2.0$) and (b) regime 2 (with $\Delta = -0.005$ and $\alpha = 0.5$). From (21), one has that $4\kappa\beta_{min} = 0.005$ and hence $\beta_{min} = 0.5$ for the solutions in (a), while $4\kappa\beta_{min} = 0.01$ and hence $\beta_{min} = 1.0$ in (b). In these profiles, $\beta = 2.0$ so that $\beta > \beta_{min}$ in each case. The + and − signs in the legends refer to the choice of sign solution in (20a) and (20b).

inequality placed on the product $4\kappa\beta$, namely, $4\kappa\beta > 4\kappa\beta_{min}$, where

$$4\kappa\beta_{min} = \frac{\Delta}{\alpha - 1} \tag{21}$$

(it is interesting to note that $4\kappa\beta_{min}$ is independent of q). Thus, existence criterion (21) for Helmholtz surface waves explicitly involves the (inverse) beam size through the appearance of κ. Since $4\kappa\beta$ must remain positive, it follows that surface modes are supported in two distinct parameter regimes: (i) regime 1: $\Delta > 0$ and $\alpha > 1$ (i.e., $n_{02}^2 < n_{01}^2$ and $\alpha_2 > \alpha_1$) and (ii) regime 2: $\Delta < 0$ and $0 < \alpha < 1$, (i.e., $n_{02} > n_{01}, \alpha_2 < \alpha_1$). We mention, in passing, that (21) is reminiscent of the existence criterion derived by Aceves et al. [8]; it differs through the explicit appearance of κ. Typical surface wave profiles are shown in Figure 11.

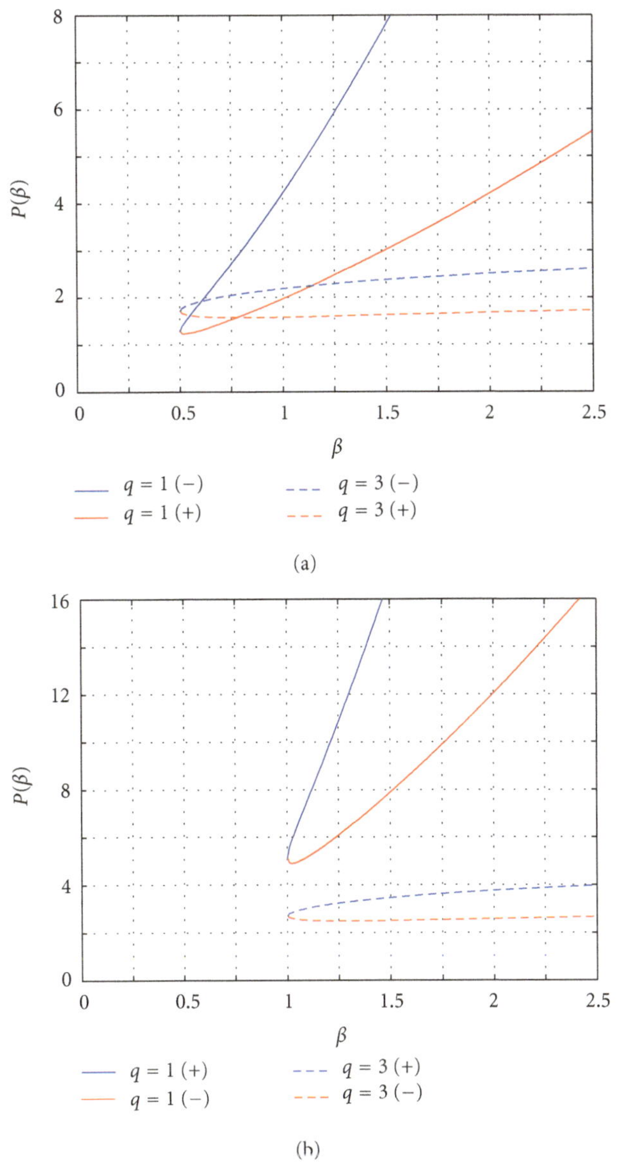

Figure 12: Power curves as a function of the propagation constant β, obtained from (22) with $\kappa = 2.5 \times 10^{-3}$. (a) Regime 1 with $\Delta = 0.005$ and $\alpha = 2.0$ and (b) regime 2 with $\Delta = -0.005$ and $\alpha = 0.5$. The $+$ and $-$ signs in the legends refer to the choice of solution in (20a) and (20b). Lower (upper) solution branches appear as red (blue) lines, and each branch generally satisfies the VK stability criterion [61].

4.3. Solution Families and Wave Power. For both forward- and backward-propagating surface waves, there exist two solution families. The origin of this duality lies in solving simultaneous equations (19a) and (19b), where one is eventually obliged to find the roots of quadratic equations. Figure 11 reveals that, for fixed (Δ, α, β), the profile depends strongly on the nonlinearity exponent q. That is, the peak amplitude, width, and area all decrease with increasing q. The difference between the two peak amplitudes and the distance of each solution peak from the interface also decrease with increasing nonlinearity exponent.

Since the surface wave profiles differ, it is plausible that the two families will not share the same stability properties. We begin an analysis of Helmholtz solutions (18a) and (18b) by considering the power P, where

$$P(\beta; q) \equiv \int_{-\infty}^{+\infty} d\xi \,|u(\xi,\zeta)|^2, \tag{22}$$

as a function of the free parameter β for different values of the nonlinearity exponent q. The energy-flow invariant W [56] is related to P through $W(\beta) = \pm(1+4\kappa\beta)^{1/2}P(\beta)$, where the $\pm$ sign here corresponds to forward- or backward-propagating envelopes (being distinct from the sign choice in (20a) and (20b)). A representative set of curves is shown in Figure 12, where it can be seen that $P(\beta)$ comprises two branches. In regime 1 (where $\Delta > 0$ and $\alpha > 1$), the lower (upper) branch corresponds to the $-(+)$ sign in (20a) and (20b). This situation is reversed for regime 2 (where $\Delta < 0, 0 < \alpha < 1$), in which the lower (upper) branch corresponds to the $+(-)$ sign (see Figure 11). We note that for lower-branch solutions, the peak of the surface wave always resides in whichever medium has the lower linear refractive index.

Global trends in the parameter dependence of the modes profiles can be readily identified and discussed in the context of the two solution branches. For instance, one might fix Δ, β, and κ and consider the effect of varying α. In regime 1, one finds that upon increasing α, the upper-branch solutions tend to retain their shape while the lower-branch solutions experience a decrease in amplitude, width, and area. The separation between the pair of solutions also becomes greater, with each localized wave moving *away from* the interface. As α is increased in regime 2, the lower-branch solutions tend to retain their shape while the upper-branch solution exhibits decreases in amplitude, width, and area. Also, the separation between the solutions tends to decrease with increasing α (so that the solutions move *toward* the boundary).

4.4. Surface Wave Stability. Except near the intersection point (where $\beta \approx \beta_{\min}$), both $P(\beta)$ branches satisfy the classic Vakhitov-Kolokolov (VK) criterion for stability; namely, $dP/d\beta > 0$ [61]. Extensive simulations have revealed that lower-branch solutions always tend to remain self-trapped within the vicinity of the interface (so long as $dP/d\beta > 0$) evolving with a stationary profile over arbitrarily long distances.

Upper-branch solutions tend to display a spontaneous instability in finite ζ. A set of typical results is shown in Figure 13 for regime 1 with $\Delta = 0.005$ and $\alpha = 2.0$, where the input wave is localized predominantly in medium 1 (compare with Figure 11(a)). The initial stages of evolution appear to be stationary, but instability sets in after a finite propagation length. The unstable solution deviates spontaneously into medium 2, crossing the boundary and shedding radiation in the process. The beam in medium 2 undergoes narrowing since $\alpha > 1$. For fixed interface and solution parameters, the instability growth rate clearly increases with q. However, the angular deviation of the

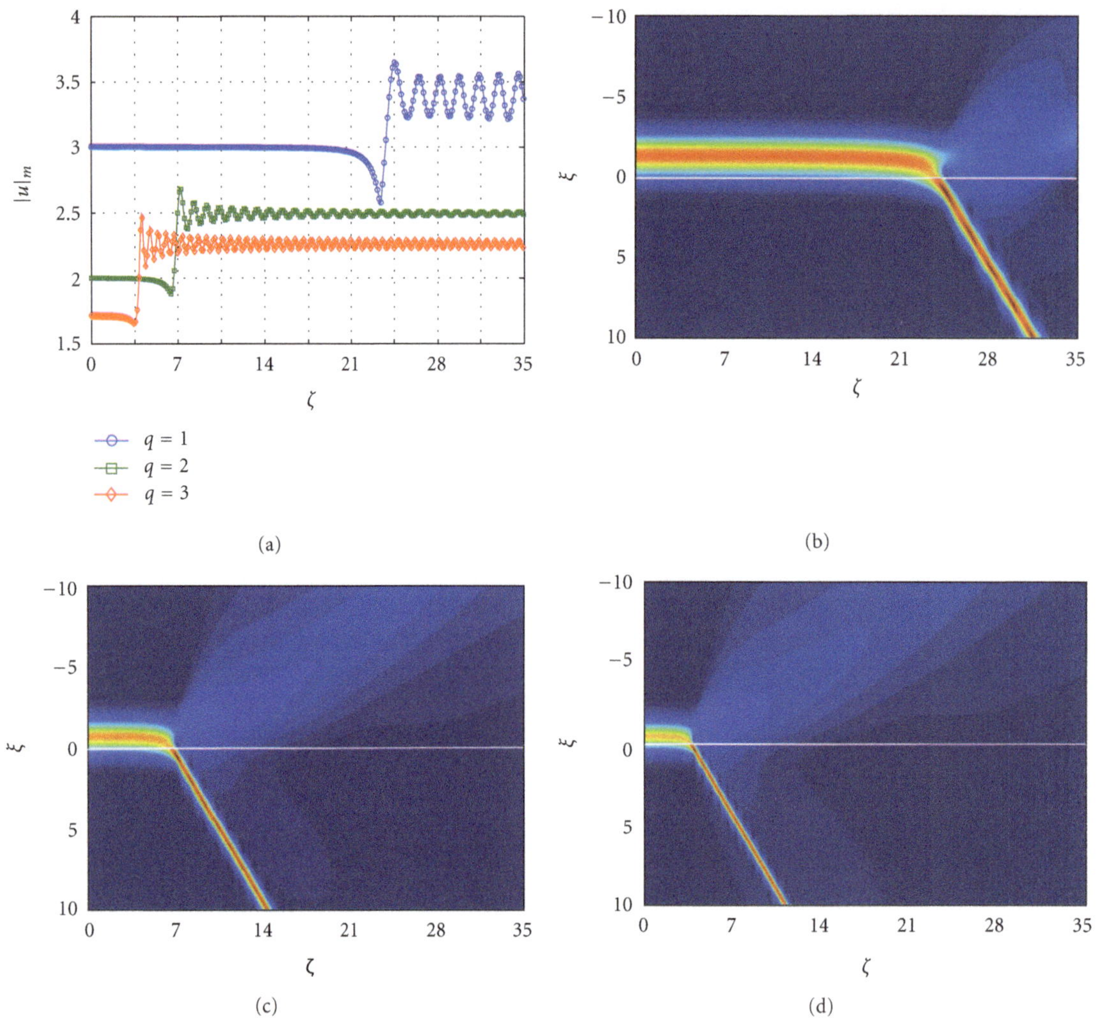

FIGURE 13: Spontaneous instability of nonlinear surface waves lying on the upper solution branch of Figure 12(a), where $\kappa = 2.5 \times 10^{-3}$ and $\beta = 2.0$ (interface mismatch parameters are $\Delta = 0.005$ and $\alpha = 2.0$). (a) Evolution in ζ of the peak amplitude $|u|_m$ of the beam. (b), (c), and (d) show the full numerical solution $|u(\xi, \zeta)|$ of (5) when the nonlinearity exponent is $q = 1$, 2, and 3, respectively. Note that the profiles of the input waves in (b) and (d) correspond to the upper-branch solutions shown in Figure 11(a).

(reshaping) daughter beam relative to the interface is largely insensitive to q.

Qualitatively different effects appear in regime 2 with $\Delta = -0.005$ and $\alpha = 0.5$; this time, the input wave is localized predominantly in medium 2 (compare with Figure 11(b)). After a finite propagation length, the surface wave bends smoothly away from the interface and is deflected deeper into medium 2. There is relatively little radiation shed in this process, and the localized wave suffers only a very small change to its shape (largely because the beam remains always on the same side of the interface, so does not encounter changes in refractive index). In common with regime 1, the instability growth rate increases with q.

4.5. Interactions between Solitons and Surface Waves. The stability of lower-branch surface waves is now investigated by considering their resilience against interactions with spatial solitons. Only a brief summary is presented here since the primary motivation is to uncover qualitatively new effects that depend upon the exponent q (detailed quantitative analyses are reserved for future works). For definiteness, we present simulation results for collisions between a unit-amplitude ($\eta_0 = 1.0$) soliton and surface waves in regimes 1 ($\Delta = 0.005, \alpha = 2.0$) and 2 ($\Delta = -0.005, \alpha = 0.5$) with $\beta = 2.0$ and $\kappa = 2.5 \times 10^{-3}$.

Regime 1 is considered first for a quasi-paraxial incidence angle of $\theta_{\text{inc}} = 3°$ (see Figure 14). When $q = 1$, the two distinct beams persist after the interaction. The path of the outgoing soliton has been deflected relative to its ingoing trajectory. The surface wave, on the other hand, survives as a localized spatial structure but can no longer be interpreted as a "surface wave" per se since it travels obliquely to (not along) the interface. This picture is qualitatively different for $q = 2$ and 3; there, the interaction results in the coalescence

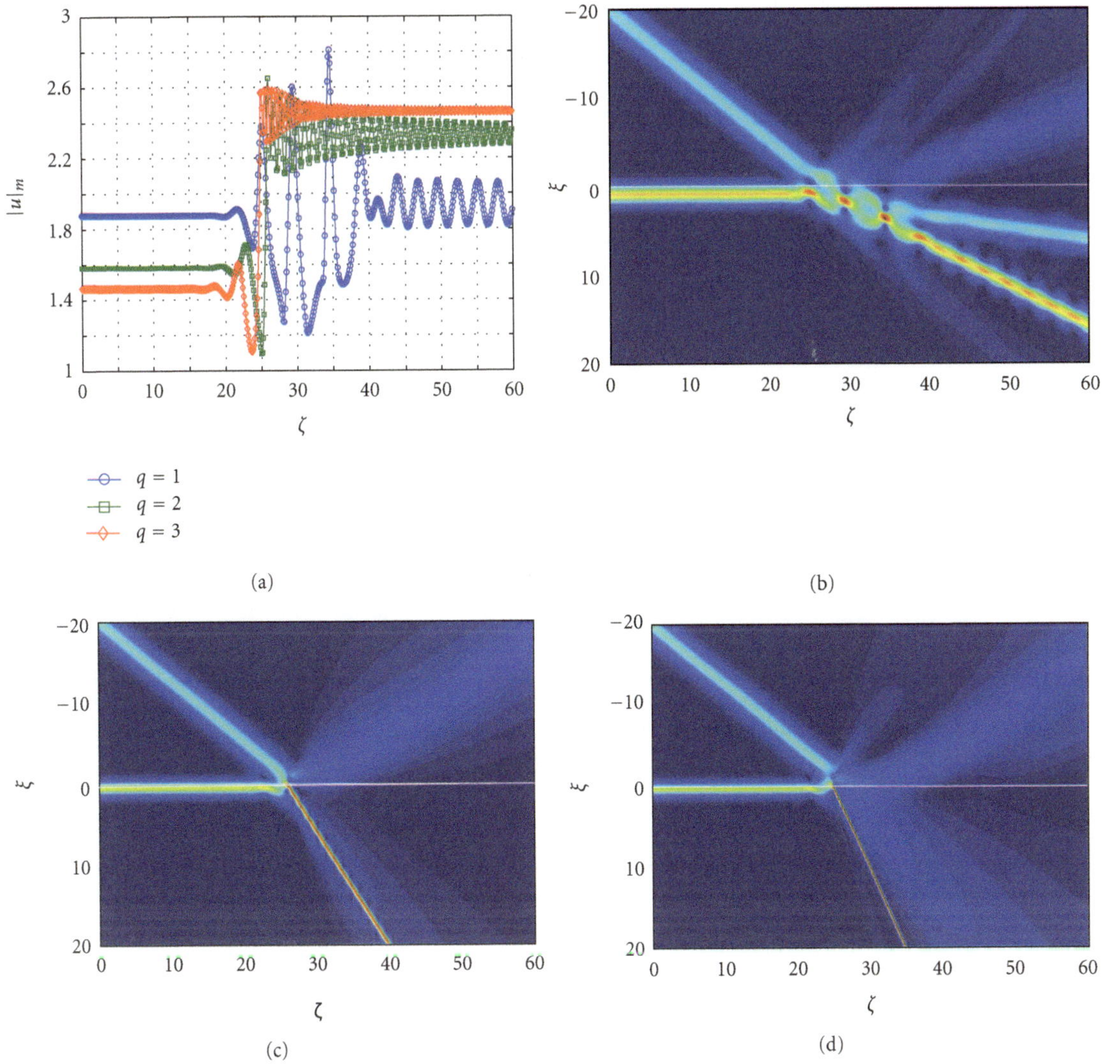

FIGURE 14: Quasi-paraxial interaction ($\theta_{\text{inc}} = 3°$) between a lower-branch nonlinear surface wave (with $\beta = 2.0$) and a unit-amplitude ($\eta_0 = 1$) soliton in regime 1 (mismatch parameters $\Delta = 0.005$ and $\alpha = 2.0$) with $\kappa = 2.5 \times 10^{-3}$. (a) Evolution in ζ of the peak amplitude $|u|_m$ of the solution. Parts (b), (c), and (d) show the full numerical solution $|u(\xi,\zeta)|$ of (5) when the nonlinearity exponent is $q = 1$, 2 and 3, respectively.

of the soliton and surface wave, producing a single higher-intensity narrow filament travelling obliquely to the interface (narrowing is to be expected for medium combinations with $\alpha > 1$). It is noteworthy that the propagation angle of the filament, relative to the interface, increases with q. Also, as one might expect, nonlinear beams interacting at quasi-paraxial angles tend to shed a large amount of radiation.

The qualitative behaviour can change dramatically at nonparaxial angles; a representative set of simulations for $\theta_{\text{inc}} = 30°$ is shown Figure 15. We have not observed coalescence phenomena; instead of this, individual beams retain their separate identities and can be clearly resolved. While the soliton often survives intact (and experiences a narrowing effect due to $\alpha > 1$), the evolution of the surface wave depends strongly on the nonlinearity exponent: (i) for $q = 1$, it acquires slow modulations in its shape but remains localized within the vicinity of the interface (i.e., it remains essentially a surface wave after the interaction); (ii) for $q = 2$, its path is deviated by the interaction so that it no longer travels along the interface (this obliquely-evolving self-trapped structure is, by definition, not a surface wave); (iii) for $q = 3$, the collision destroys it completely. It is interesting to note the general trend that larger-interaction angles generate far less radiation than their paraxial counterparts [62].

We now turn our attention to similar interaction scenarios in regime 2. For a quasi-paraxial incidence angle of 3°, the behaviour is strikingly different from that uncovered for the same angle in regime 1 (compare Figures 16 and 14, respectively). When $q = 1$, the soliton survives the interaction and the surface wave remains quasi-bound to the interface (but exhibiting a longitudinal "skimming" effect).

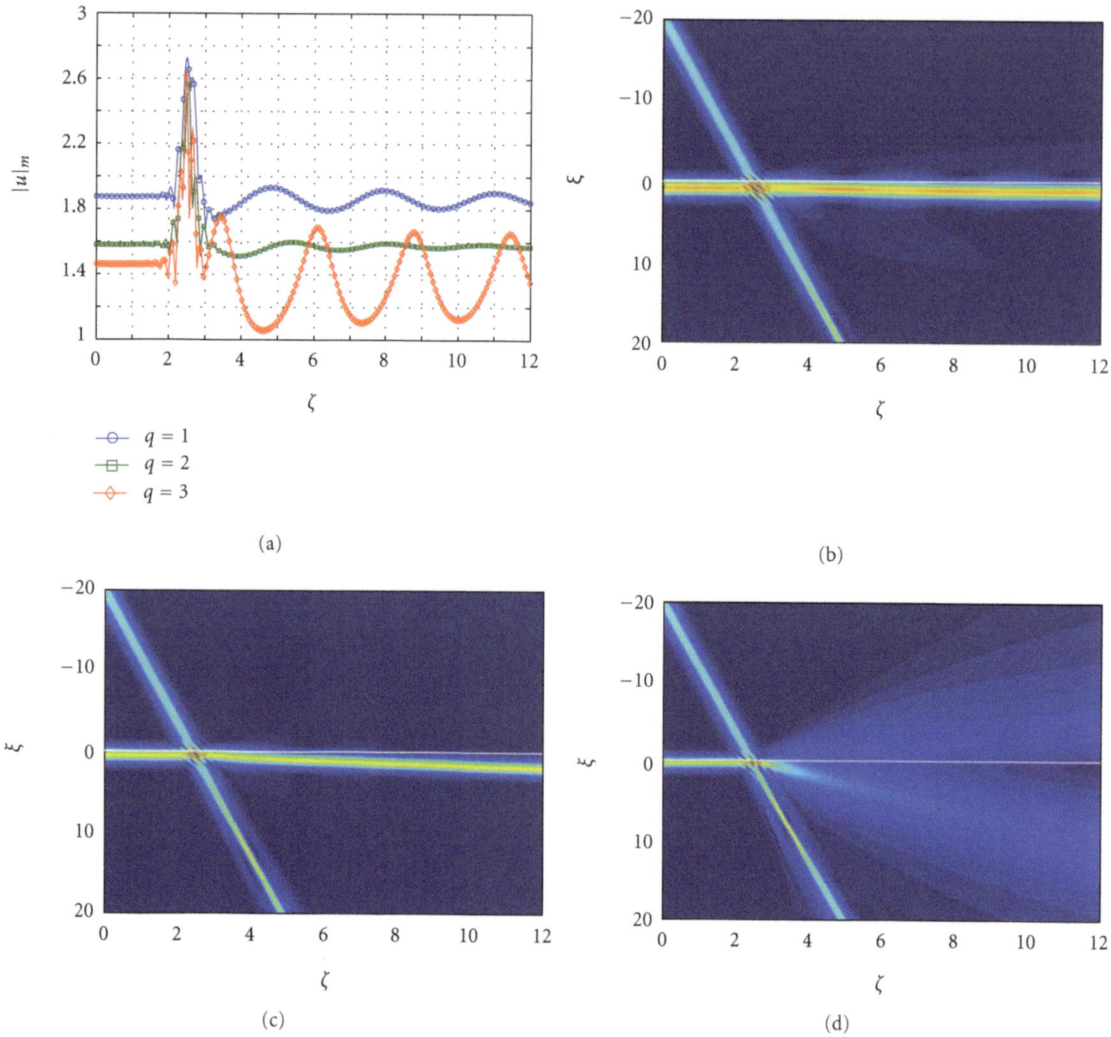

FIGURE 15: Nonparaxial interaction ($\theta_{\text{inc}} = 30°$) between a lower-branch nonlinear surface wave (with $\beta = 2.0$) and a unit-amplitude ($\eta_0 = 1$) soliton in regime 1 (mismatch parameters $\Delta = 0.005$ and $\alpha = 2.0$) with $\kappa = 2.5 \times 10^{-3}$. (a) Evolution in ζ of the peak amplitude $|u|_m$ of the solution. (b), (c), and (d) show the full numerical solution $|u(\xi, \zeta)|$ of (5) when the nonlinearity exponent is $q = 1$ (surface wave follows interface), 2 (surface wave deflected), and 3 (surface wave destroyed), respectively.

For $q = 2$ and 3, the interaction deflects the surface wave away from the boundary (i.e., the surface wave becomes an obliquely-evolving beam). However, the behaviour of the soliton is different for $q = 2$ and 3: it survives intact in the former case and breaks up into radiation in the latter (this effect is related to the threshold phenomenon discussed in Section 3.2 and is not a consequence of the interaction with the surface wave).

5. Conclusion

We have presented, to the best of our knowledge, the first investigation of the way spatial solitons behave at the planar interface between dissimilar materials whose refractive index has a power-law dependence on the electric field amplitude. This analysis has thus extended arbitrary angle refraction considerations beyond the ubiquitous Kerr-type case [17, 18, 25, 26]. Exact analytical solitons have been deployed as a nonlinear basis [56], permitting the derivation of a generalized Helmholtz-Snell law. Extensive numerical computations have tested its predictions, which are most accurate in regimes where only the linear refractive index changes across the boundary.

A range of new quantitative *and qualitative* effects that depend strongly upon the exponent q has been identified. For example, simulations have found that, at linear interfaces with $\Delta > 0$ and where $q = 1$ or 2, there is generally a well-defined transition (as θ_{inc} increases) from soliton reflection, through GH shifting, to soliton refraction. In contrast, systems with $q = 3$ are often far more complex: the reflection-to-refraction transition is generally obscured by radiation effects over a finite band of incidence angles around the (theoretical) critical angle: solitons interacting with

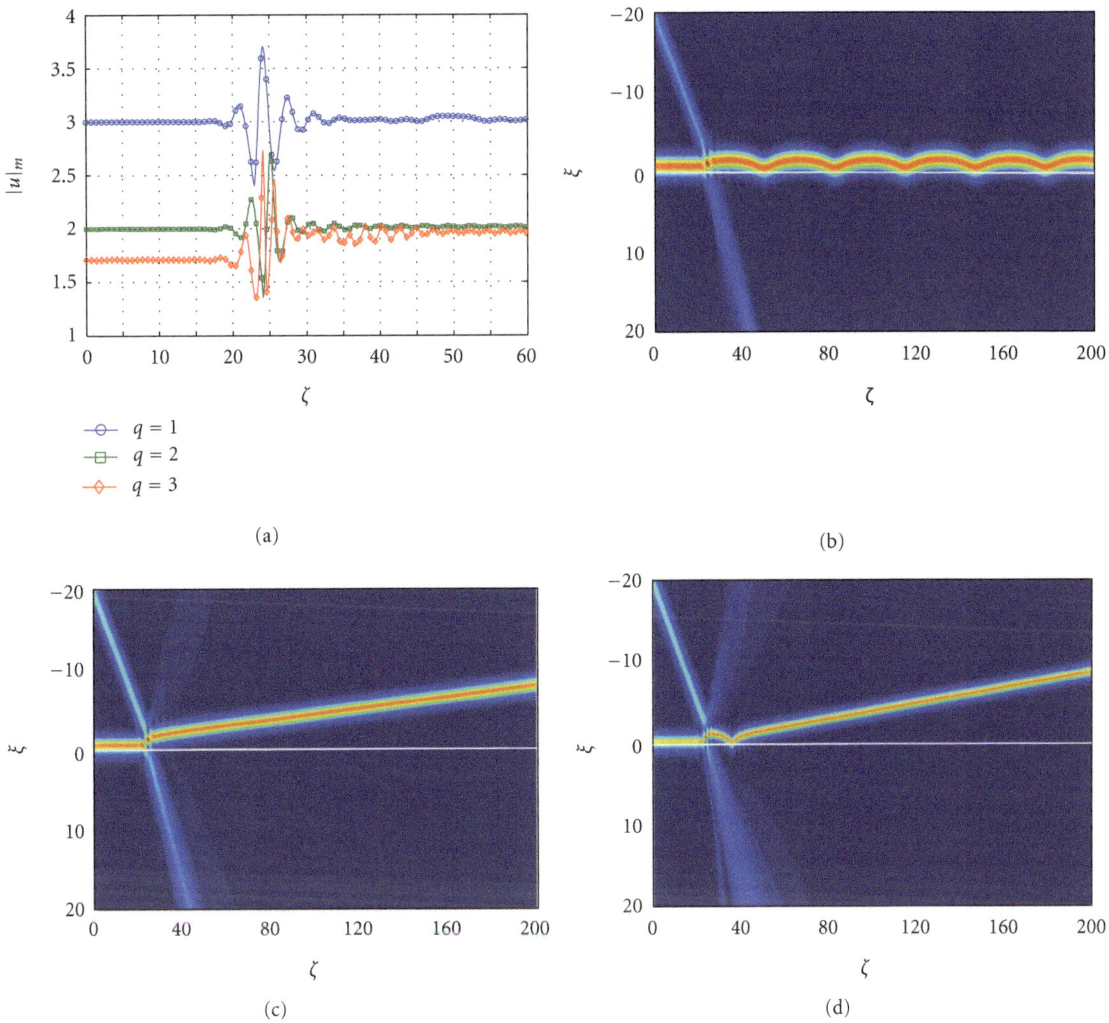

FIGURE 16: Quasi-paraxial interaction ($\theta_{\text{inc}} = 3^\circ$) between a lower-branch nonlinear surface wave (with $\beta = 2.0$) and a unit-amplitude ($\eta_0 = 1$) soliton in regime 2 (mismatch parameters $\Delta = -0.005$ and $\alpha = 0.5$) with $\kappa = 2.5 \times 10^{-3}$. (a) Evolution in ζ of the peak amplitude $|u|_m$ of the solution. (b), (c), and (d) show the full numerical solution $|u(\xi,\zeta)|$ of (5) when the nonlinearity exponent is $q = 1$ (surface wave "skimming"), 2 (deflection of the surface wave), and 3 (deflection of the surface wave and breakup of the soliton into radiation), respectively.

the interface may collapse into low-amplitude diffracting waves, with GH shifts becoming difficult to interpret or quantify in the absence of a well-defined reflected beam. However, strong supporting evidence has been obtained to confirm the validity of our Helmholtz-Snell modelling in arbitrary-angle non-Kerr regimes. In this way, the first steps have been taken towards understanding how (fully 2D) diffraction/nonlinearity interplays govern spatial soliton refraction in a much wider class of systems.

Nonlinear surface waves of model (5) have been derived, and we have performed the first numerical analysis of these types of solutions. Simulations have addressed the stability properties of the new surface waves, which tend to lie on one of two possible branches of the classic (β, P) curves. Solutions lying on the lower branch are predicted to behave as stable robust entities, while solutions on the upper branch are inherently unstable. Extensive computations have lent direct numerical support for this stability prediction in the more general Helmholtz context, and the growth rate of the upper-branch instability has been found to increase with q.

The stability properties of lower-branch Helmholtz surface waves have been further investigated by considering collisions with obliquely incident spatial solitons. A rich variety of behaviours, which depend crucially on both the nonlinearity exponent and the interaction angle, has been discovered. Finding analytical descriptions (e.g., through a perturbation theory [62]) of these phenomena seems a remote possibility since much of the behaviour is clearly non-adiabatic. Hence, computer simulations play a fundamental role in investigating solitons, surface waves, and their interactions in non-Kerr regimes.

The research presented in this paper provides a clear indication that deviating from the ideal Kerr-type nonlinearity ($q = 2$) can give rise to novel, interesting, and potentially exploitable phenomena. Each component of this paper (testing the Helmholtz-Snell law, calculating GH shifts, analyzing surface wave stability, and studying soliton-surface wave interactions) is a problem for detailed investigation in its own right. Our findings unpin analyses of other types of optical (and nonoptical) contexts involving solitons and surface waves where the power-law type of nonlinearity takes centre stage. One can expect other distinct classes of surface wave to exist when the interface comprises combinations of focusing/defocusing power-law nonlinearities [42, 43, 63]; the stability properties of these waves can, quite reasonably, be expected to differ from those reported here. Furthermore, the validity of our Helmholtz-Snell modelling in power-law regimes suggests that it may also be applicable to other material configurations, for example, to single- and multi-interface problems with cubic-quintic [64–67] and saturable [68–70] nonlinearities. Research is currently underway that investigates the generality of our findings in these other contexts, and preliminary results do suggest wider applicability.

Acknowledgment

This work was supported by the Engineering and Physical Sciences Research Council (EPSRC), grant number EP/H011595/1.

References

[1] P. W. Smith and W. J. Tomlinson, "Nonlinear optical interfaces: switching behaviour," *IEEE Journal of Quantum Electronics*, vol. 20, no. 1, pp. 30–36, 1984.

[2] W. J. Tomlinson, J. P. Gordon, P. W. Smith, and A. E. Kaplan, "Reflection of a Gaussian beam at nonlinear interface," *Applied Optics*, vol. 21, no. 11, pp. 2041–2051, 1982.

[3] P. W. Smith, J. P. Hermann, W. J. Tomlinson, and P. J. Maloney, "Optical bistability at a nonlinear interface," *Applied Physics Letters*, vol. 35, no. 11, pp. 846–848, 1979.

[4] A. E. Kaplan, "Theory of hysteresis reflection and refraction of light by a boundary of a nonlinear medium," *Soviet Physics Journal of Experimental and Theoretical Physics*, vol. 45, no. 1, pp. 896–905.

[5] A. E. Kaplan, "Hysteresis reflection and refraction by nonlinear boundary: a new class of effects in nonlinear optics," *Journal of Experimental and Theoretical Physics Letters*, vol. 24, no. 1, pp. 115–119, 1976.

[6] A. B. Aceves, J. V. Moloney, and A. C. Newell, "Reflection and transmission of self-focused channels at nonlinear dielectric interfaces," *Optics Letters*, vol. 13, no. 11, pp. 1002–1004, 1988.

[7] A. B. Aceves, J. V. Moloney, and A. C. Newell, "Snell's laws at the interface between nonlinear dielectrics," *Physics Letters A*, vol. 129, no. 4, pp. 231–235, 1988.

[8] A. B. Aceves, J. V. Moloney, and A. C. Newell, "Theory of light-beam propagation at nonlinear interfaces. I. Equivalent-particle theory for a single interface," *Physical Review A*, vol. 39, no. 4, pp. 1809–1827, 1989.

[9] A. B. Aceves, J. V. Moloney, and A. C. Newell, "Theory of light-beam propagation at nonlinear interfaces. II. Multiple-particle and multiple-interface extensions," *Physical Review A*, vol. 39, no. 4, pp. 1828–1840, 1989.

[10] A. B. Aceves, P. Varatharajah, A. C. Newell et al., "Particles aspects of collimated light channel propagation at nonlinear interfaces and waveguides," *Journal of the Optical Society of America B*, vol. 7, no. 6, pp. 963–974, 1990.

[11] P. Varatharajah, A. C. Newell, J. V. Moloney, and A. B. Aceves, "Transmission, reflection, and trapping of collimated light beams in diffusive Kerr-like nonlinear media," *Physical Review A*, vol. 42, no. 3, pp. 1767–1774, 1990.

[12] A. B. Aceves and J. V. Moloney, "Effect of two-photon absorption on bright spatial soliton switches," *Optics Letters*, vol. 17, no. 21, pp. 1488–1490, 1992.

[13] Y. M. Aliev, A. D. Boardman, A. I. Smirnov, K. Xie, and A. A. Zharov, "Spatial dynamics of solitonlike channels near interfaces between optically linear and nonlinear media," *Physical Review E*, vol. 53, no. 5, pp. 5409–5419, 1996.

[14] Y. M. Aliev, A. D. Boardman, K. Xie, and A. A. Zharov, "Conserved energy approximation to wave scattering by a nonlinear interface," *Physical Review E*, vol. 49, no. 2, pp. 1624–1633, 1994.

[15] A. D. Boardman, P. Bontemps, W. Ilecki, and A. A. Zharov, "Theoretical demonstration of beam scanning and switching using spatial solitons in a photorefractive crystal," *Journal of Modern Optics*, vol. 47, no. 11, pp. 1941–1957, 2000.

[16] I. V. Shadrivov and A. A. Zharov, "Dynamics of optical spatial solitons near the interface between two quadratically nonlinear media," *Journal of the Optical Society of America B*, vol. 19, no. 3, pp. 596–602, 2002.

[17] J. Sánchez-Curto, P. Chamorro-Posada, and G. S. McDonald, "Nonlinear interfaces: intrinsically nonparaxial regimes and effects," *Journal of Optics A*, vol. 11, no. 5, Article ID 054015, 2009.

[18] J. Sánchez-Curto, P. Chamorro-Posada, and G. S. McDonald, "Helmholtz solitons at nonlinear interfaces," *Optics Letters*, vol. 32, no. 9, pp. 1126–1128, 2007.

[19] T. A. Laine and A. T. Friberg, "Self-guided waves and exact solutions of the nonlinear Helmholtz equation," *Journal of the Optical Society of America B*, vol. 17, no. 5, pp. 751–757, 2000.

[20] S. Blair, "Nonparaxial one-dimensional spatial solitons," *Chaos*, vol. 10, no. 3, pp. 570–583, 2000.

[21] A. P. Sheppard and M. Haelterman, "Nonparaxiality stabilizes three-dimensional soliton beams in Kerr media," *Optics Letters*, vol. 23, no. 23, pp. 1820–1822, 1998.

[22] M. D. Feit and J. A. Fleck, "Beam nonparaxiality, filament formation, and beam breakup in the self-focusing of optical beams," *Journal of the Optical Society of America B*, vol. 5, no. 3, pp. 633–640, 1988.

[23] P. Chamorro-Posada, G. S. McDonald, and G. H. C. New, "Exact soliton solutions of the nonlinear Helmholtz equation: communication," *Journal of the Optical Society of America B*, vol. 19, no. 5, pp. 1216–1217, 2002.

[24] P. Chamorro-Posada, G. S. McDonald, and G. H. C. New, "Non-paraxial solitons," *Journal of Modern Optics*, vol. 45, no. 6, pp. 1111–1121, 1998.

[25] J. Sánchez-Curto, P. Chamorro-Posada, and G. S. McDonald, "Black and gray Helmholtz-Kerr soliton refraction," *Physical Review A*, vol. 83, no. 1, Article ID 013828, 2011.

[26] J. Sánchez-Curto, P. Chamorro-Posada, and G. S. McDonald, "Dark solitons at nonlinear interfaces," *Optics Letters*, vol. 35, no. 9, pp. 1347–1349, 2010.

[27] A. W. Snyder and D. J. Mitchell, "Spatial solitons of the power-law nonlinearity," *Optics Letters*, vol. 18, no. 2, pp. 101–103, 1993.

[28] A. Biswas, "Perturbation of solitons due to power law nonlinearity," *Chaos, Solitons and Fractals*, vol. 12, no. 3, pp. 579–588, 2001.

[29] S. Konar and A. Biswas, "Soliton-soliton interaction with power law nonlinearity," *Progress in Electromagnetics Research*, vol. 54, no. 1, pp. 95–108, 2005.

[30] E. W. Laedke and K. H. Spatschek, "Lifetime of spikons," *Physics Letters A*, vol. 74, no. 3-4, pp. 205–207, 1979.

[31] Y. V. Katyshev, N. V. Makhaldiani, and V. G. Makhankov, "On the stability of soliton solutions to the Schrödinger equation with nonlinear term of the form $\psi|\psi|^{\nu}$," *Physics Letters A*, vol. 66, no. 6, pp. 456–458, 1978.

[32] D. Mihalache, M. Bertolotti, and C. Sibilia, "Nonlinear wave propagation in planar structures," *Progress in Optics*, vol. 27, pp. 229–313, 1989.

[33] J. G. H. Mathew, A. K. Kar, N. R. Heckenberg, and I. Galbraith, "Time resolved self-defocusing in InSb at room temperature," *IEEE Journal of Quantum Electronics*, vol. 21, no. 1, pp. 94–99, 1985.

[34] D. S. Chemla, D. A. B. Miller, and P. W. Smith, "Nonlinear optical properties of GaAs/GaAlAs multiple quantum well material: Phenomena and applications," *Optical Engineering*, vol. 24, no. 4, pp. 556–564, 1985.

[35] R. K. Jain and R. C. Lind, "Degenerate four-wave mixing in semiconductor-doped glasses," *Journal of the Optical Society of America*, vol. 73, no. 5, pp. 647–653, 1983.

[36] S. S. Yao, C. Karaguleff, A. Gabel, R. Fortenberry, C. T. Seaton, and G. I. Stegeman, "Ultrafast carrier and grating lifetimes in semiconductor-doped glasses," *Applied Physics Letters*, vol. 46, no. 9, pp. 801–802, 1985.

[37] J. G. Ma, "Nonlinear surface waves on the interface of two non-Kerr-like nonlinear media," *IEEE Transactions on Microwave Theory and Techniques*, vol. 45, no. 6, pp. 924–930, 1997.

[38] A. W. Snyder and H. T. Tran, "Surface modes of power law nonlinearities," *Optics Communications*, vol. 98, no. 4–6, pp. 309–312, 1993.

[39] G. I. Stegeman, C. T. Seaton, J. Ariyasu, T. P. Shen, and J. V. Moloney, "Saturation and power law dependence of nonlinear waves guided by a single interface," *Optics Communications*, vol. 56, no. 5, pp. 365–368, 1986.

[40] L. Wu, "Examination of the core field uniformity for 3-layer power-law nonlinear slab waveguides," *Optics Communications*, vol. 224, no. 1–3, pp. 51–56, 2003.

[41] G. I. Stegeman, E. M. Wright, C. T. Seaton et al., "Nonlinear slab-guided waves in non-Kerr-like media," *IEEE Journal of Quantum Electronics*, vol. 22, no. 6, pp. 977–983, 1986.

[42] J. G. Ma and I. Wolff, "TE wave properties of slab dielectric guide bounded by nonlinear non-Kerr-like media," *IEEE Transactions on Microwave Theory and Techniques*, vol. 44, no. 5, pp. 730–738, 1996.

[43] J. G. Ma and I. Wolff, "Propagation characteristics of TE-waves guided by thin films bounded by nonlinear media," *IEEE Transactions on Microwave Theory and Techniques*, vol. 43, no. 4, pp. 790–795, 1995.

[44] P. Varatharajah, A. B. Aceves, J. V. Moloney, and E. M. Wright, "Stationary nonlinear surface waves and their stability in diffusive Kerr-like nonlinear media," *Journal of the Optical Society of America B*, vol. 7, no. 2, pp. 220–229, 1990.

[45] D. R. Andersen, "Surface-wave excitation at the interface between diffusive Kerr-like nonlinear and linear media," *Physical Review A*, vol. 37, no. 1, pp. 189–193, 1988.

[46] Y. V. Kartashov, F. Ye, V. A. Vysloukh, and L. Torner, "Surface waves in defocusing thermal media," *Optics Letters*, vol. 32, no. 15, pp. 2260–2262, 2007.

[47] P. J. Bradley and C. De Angelis, "Soliton dynamics and surface waves at the interface between saturable nonlinear dielectrics," *Optics Communications*, vol. 130, no. 1–3, pp. 205–218, 1996.

[48] F. Goos and H. Hänchen, "Ein neuer und fundamentaler Versuch zur Totalreflexion," *Annalyen Der Physik*, vol. 1, no. 1, pp. 333–346, 1947.

[49] J. Sánchez-Curto, P. Chamorro-Posada, and G. S. McDonald, "Giant Goos-Hänchen shifts and radiation-induced trapping of Helmholtz solitons at nonlinear interfaces," *Optics Letters*, vol. 36, no. 18, pp. 3605–3607, 2011.

[50] S. Chi and Q. Guo, "Vector theory of self-focusing of an optical beam in Kerr media," *Optics Letters*, vol. 20, no. 15, pp. 1598–1600, 1996.

[51] M. Lax, W. H. Louisell, and W. B. McKnight, "From Maxwell to paraxial wave optics," *Physical Review A*, vol. 11, no. 4, pp. 1365–1370, 1975.

[52] A. Ciattoni, B. Crosignani, S. Mookherjea, and A. Yariv, "Nonparaxial dark solitons in optical Kerr media," *Optics Letters*, vol. 30, no. 5, pp. 516–518, 2005.

[53] B. Crosignani, A. Yariv, and S. Mookherjea, "Nonparaxial spatial solitons and propagation-invariant pattern solutions in optical Kerr media," *Optics Letters*, vol. 29, no. 11, pp. 1254–1256, 2004.

[54] A. Ciattoni, P. Di Porto, B. Crosignani, and A. Yariv, "Vectorial nonparaxial propagation equation in the presence of a tensorial refractive-index perturbation," *Journal of the Optical Society of America B*, vol. 17, no. 5, pp. 809–819, 2000.

[55] B. Crosignani, P. Di Porto, and A. Yariv, "Nonparaxial equation for linear and nonlinear optical propagation," *Optics Letters*, vol. 22, no. 11, pp. 778–780, 1997.

[56] J. M. Christian, G. S. McDonald, R. J. Potton, and P. Chamorro-Posada, "Helmholtz solitons in power-law optical materials," *Physical Review A*, vol. 76, no. 3, Article ID 033834, 2007.

[57] J. M. Christian, G. S. McDonald, R. J. Potton, and P. Chamorro-Posada, "Erratum: Helmholtz solitons in power-law optical materials (Physical Review A (2007) 76 (033834))," *Physical Review A*, vol. 76, no. 4, Article ID 049905, 2007.

[58] J. M. Christian, G. S. McDonald, and P. Chamorro-Posada, "Helmholtz bright and boundary solitons," *Journal of Physics A*, vol. 40, no. 7, pp. 1545–1560, 2007.

[59] P. Chamorro-Posada, G. S. McDonald, and G. H. C. New, "Non-paraxial beam propagation methods," *Optics Communications*, vol. 192, no. 1-2, pp. 1–12, 2001.

[60] J. Sánchez-Curto, P. Chamorro-Posada, and G. S. McDonald, "Helmholtz bright and black soliton splitting at nonlinear interfaces," *Physical Review A*, vol. 85, no. 1, Article ID 013836, 2012.

[61] N. G. Vakhitov and A. A. Kolokolov, "Stationary solutions of the wave equation in a medium with nonlinearity saturation," *Radiophysics and Quantum Electronics*, vol. 16, no. 7, pp. 783–789, 1975.

[62] P. Chamorro-Posada and G. S. McDonald, "Spatial Kerr soliton collisions at arbitrary angles," *Physical Review E*, vol. 74, no. 3, Article ID 036609, 2006.

[63] Y. Chen, "Bright and dark surface waves at a nonlinear interface," *Physical Review A*, vol. 45, no. 7, pp. 4974–4978, 1992.

[64] J. M. Christian, G. S. McDonald, and P. Chamorro-Posada, "Bistable Helmholtz solitons in cubic-quintic materials," *Physical Review A*, vol. 76, no. 3, Article ID 033833, 2007.

[65] D. Mihalache, D. Mazilu, M. Bertolotti, and C. Sibilia, "Exact solutions for nonlinear thin-film guided waves in higher-order nonlinear media," *Journal of the Optical Society of America B*, vol. 5, no. 2, pp. 565–570, 1988.

[66] D. Mihalache and D. Mazilu, "Stability and instability of nonlinear guided waves in saturable media," *Solid State Communications*, vol. 63, no. 3, pp. 215–217, 1987.

[67] K. I. Pushkarov, D. I. Pushkarov, and I. V. Tomov, "Self-action of light beams in nonlinear media: soliton solutions," *Optical and Quantum Electronics*, vol. 11, no. 6, pp. 471–478, 1979.

[68] J. M. Christian, G. S. McDonald, and P. Chamorro-Posada, "Bistable Helmholtz bright solitons in saturable materials," *Journal of the Optical Society of America B*, vol. 26, no. 12, pp. 2323–2330, 2009.

[69] D. Mihalache and D. Mazilu, "Stability of nonlinear stationary slab-guided waves in saturable media: a numerical analysis," *Physics Letters A*, vol. 122, no. 6-7, pp. 381–384, 1987.

[70] D. Mihalache and D. Mazilu, "TM-polarized nonlinear slab-guided waves in saturable media," *Solid State Communications*, vol. 60, no. 4, pp. 397–399, 1986.

5

The Effect of Nonnative Interactions on the Energy Landscapes of Frustrated Model Proteins

Mark T. Oakley,[1] David J. Wales,[2] and Roy L. Johnston[1]

[1] *School of Chemistry, University of Birmingham, Edgbaston, Birmingham B15 2TT, UK*
[2] *University Chemical Laboratories, Lensfield Road, Cambridge CB2 1EW, UK*

Correspondence should be addressed to Mark T. Oakley, m.t.oakley@bham.ac.uk

Academic Editor: Jan Petter Hansen

The 46- and 69-residue BLN model proteins both exhibit frustrated folding to β-barrel structures. We study the effect of varying the strength of nonnative interactions on the corresponding energy landscapes by introducing a parameter λ, which scales the potential between the BLN ($\lambda = 1$) and Gō-like ($\lambda = 0$) limits. We study the effect of varying λ on the efficiency of global optimisation using basin-hopping and genetic algorithms. We also construct disconnectivity graphs for these proteins at selected values of λ. Both methods indicate that the potential energy surface is frustrated for the original BLN potential but rapidly becomes less frustrated as λ decreases. For values of $\lambda \leq 0.9$, the energy landscape is funnelled. The fastest mean first encounter time for the global minimum does not correspond to the Gō model: instead, we observe a minimum when the favourable nonnative interactions are still present to a small degree.

1. Introduction

Proteins are biopolymers constructed from a sequence of amino acid residues. The potential energy landscapes of proteins have many degrees of freedom and include important contributions between pairs of residues that are distant in sequence, but close to each other in space. Despite this complexity, many globular proteins fold to a well-defined the native state. According to the thermodynamic hypothesis, this structure is the global free energy minimum for a given sequence [1]. Frustration occurs when there are low-lying structures separated by high barriers [2]. All the favourable interactions between pairs of residues cannot be accommodated at the same time, which can lead to energetic frustration, where there are several low-lying structures with different patterns of contacts. Geometric frustration occurs when the interconversion of two low-lying structures requires the breaking of several favourable contacts.

A systematic way to simplify the potential energy surface for a protein is to include only attractive interactions between pairs of residues that are in contact in the native state, which constitutes a Gō model [3]. Various on- and off-lattice Gō models have been investigated by different authors to study a range of different proteins. In spite of the simplified potential, these models have proved capable of reproducing certain aspects of protein dynamics and thermodynamics [4–11]. Using a Gō model tends to lead to funnelled energy landscapes [12], with very little frustration. For some proteins, neglecting nonnative interactions can have a significant influence on the energy landscape [13].

United atom representations introduce a further level of coarse-graining, which can speed up simulations significantly, at the cost of atomistic detail. The simplest coarse-grained model is the HP model, in which each protein residue is represented by a single hydrophobic (H) or polar (P) bead and is constrained to lie on a regular lattice [14, 15]. The BLN model is an off-lattice generalisation of the HP model with three types of bead: hydrophobic (B), hydrophilic (L), and neutral (N). The 46-residue sequence [12, 16–33] $B_9N_3(LB)_4N_3B_9N_3(LB)_5L$ and the 69-residue sequence [34–38] $B_9N_3(LB)_4N_3B_9N_3(LB)_4N_3B_9N_3(LB)_5L$ were designed to exhibit frustrated folding and have several alternate β-barrel structures that are separated by large energy barriers. Disconnectivity graphs [39] for both of these proteins exhibit energy landscapes comprising several folding funnels [12, 38]. Using a Gō potential for these two proteins changes

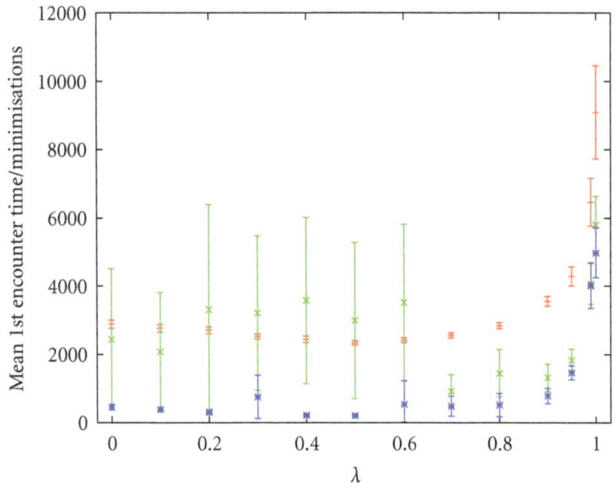

FIGURE 1: Mean first encounter times (number of minimisations) for 100 global optimisation runs initiated from random starting points for the 46-residue scaled BLN protein. The searches were run using a genetic algorithm (red), basin-hopping starting from random structures confined to a sphere (green), and basin-hopping starting from chain structures with randomised dihedral angles (blue). The error bars are the uncertainties calculated at the 95% level.

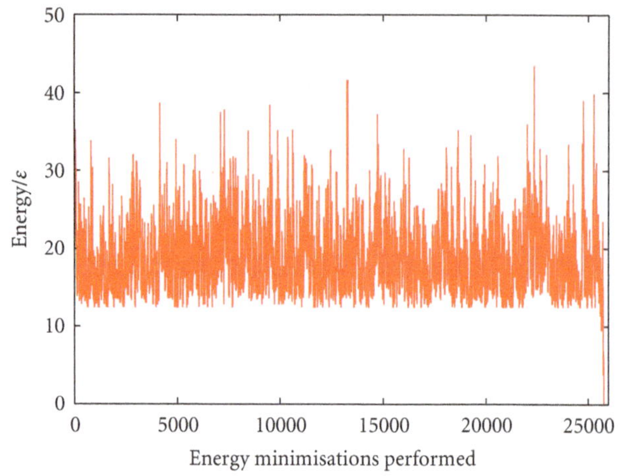

FIGURE 2: Energy of the minima in the Markov chain for a BH run where trapping occurs for the 46-residue scaled BLN protein with $\lambda = 0$.

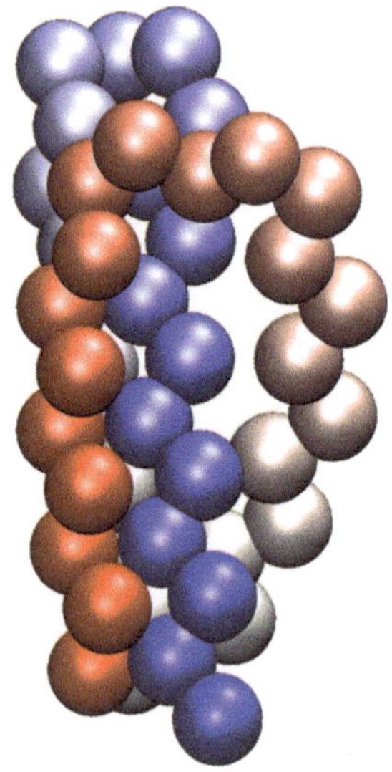

FIGURE 3: The most stable misfolded structure, which acts as a trap for global optimisation of the 46-residue BLN protein, illustrated using the VMD program [40] with a colouring scheme for the beads that varies from red to blue (N-terminus to C-terminus).

the nature of their energy landscapes, and they both exhibit single funnels with very little frustration [12, 38].

Intermediate potentials can be generated using a parameter, λ, which scales the strength of the nonnative interactions between the Gō ($\lambda = 0$) and BLN ($\lambda = 1$) limits. The folding thermodynamics of the 46-residue BLN protein have been investigated using this scaled BLN potential [23, 32, 33], showing that most of the frustration is only present for values of $\lambda \geq 0.9$. The introduction of salt bridges (gatekeepers) to the 46-residue protein also produces energy landscapes of intermediate character [27, 28].

In the present work, we study the effect of varying λ on the ease of global optimisation of the 46- and 69-residue BLN proteins using a basin-hopping algorithm and a genetic algorithm. We also construct disconnectivity graphs to compare the energy landscapes of the proteins for different values of λ.

2. Computational Methods

The protein structures were modelled using the following BLN potential [12, 21, 26, 28]:

$$\begin{aligned} V_{\text{BLN}} = {} & \frac{1}{2} K_r \sum_{i=1}^{N-1} (R_{i,i+1} - R_e)^2 + \frac{1}{2} K_\theta \sum_{i=1}^{N-2} (\theta_i - \theta_e)^2 \\ & + \epsilon \sum_{i=1}^{N-3} [A_i (1 + \cos\phi_i) + B_i (1 + 3\cos\phi_i)] \\ & + 4\epsilon \sum_{i=1}^{N-2} \sum_{j=i+2}^{N} C_{ij} \left[\left(\frac{\sigma}{R_{ij}} \right)^{12} - D_{ij} \left(\frac{\sigma}{R_{ij}} \right)^{6} \right], \end{aligned} \tag{1}$$

where R_{ij} is the distance between two beads i and j. The first term is a harmonic bond restraint with $K_r = 231.2\epsilon\sigma^{-2}$ and $R_e = \sigma$. The second term is a bond angle restraint with $K_\theta = 20\,\text{rad}^{-2}$ and $\theta_e = 1.8326$ rad. The third term involves torsional angles, ϕ, defined by four successive beads. If two or more of these beads are N, then $A = 0$ and $B = 0.2$. For all other sequences, $A = B = 1.2$. The final term introduces pairwise nonbonded interactions. If one residue is L and the other is L or B, then $C = 2/3$ and $D = -1$. If either of the residues is N, then $C = 1$ and $D = 0$. If both residues are B, then $C = 1$, but the value of D depends on the presence of the contact in the native state of the protein. For native contacts, $D = 1$. For nonnative contacts, $D = \lambda$, where $0 < \lambda < 1$. The case where $\lambda = 1$ is the original BLN potential and $\lambda = 0$ is the Gō potential.

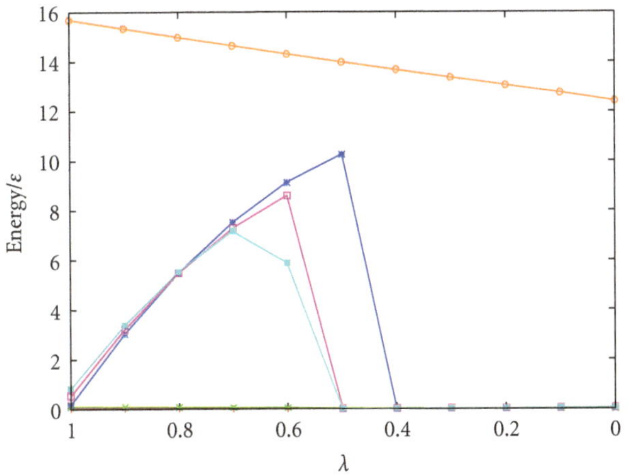

FIGURE 4: The energies of the five most stable BLN-46 structures relative to the global minimum as a function of λ. Also shown (orange) is the energy of the trap structure illustrated in Figure 3. The steep decreases mark the points at which structures cease to be local minima and collapse into the basin of attraction [41] of the global minimum.

TABLE 1: Parameters used for the two optimisation strategies.

	BLN-46	BLN-69
BH		
kT/ϵ	2.3	3.4
Step size/σ	0.65	0.70
GA		
Population size	140	200
Crossover rate	0.9	0.9
Mutation rate	0.05	0.05

Native contacts are defined as all pairs of residues where R_{ij} is less than a fixed cut-off distance in the native state (global minimum) of the protein. When $\lambda \neq 1$, the value of this cut-off radius will influence the energy landscape. Here, we use 1.167σ for consistency with previous work [12, 28, 38].

Global optimisation was performed using the basin-hopping approach [42–44] and a Lamarckian genetic algorithm [38, 45], which are both implemented in the `GMIN` program [46]. Each algorithm involves local energy minimisation after each structural perturbation. This minimisation transforms the potential energy surface into the basins of attraction of local minima [47] and removes downhill barriers. The search parameters for both algorithms were optimised in previous work for BLN proteins [38], and these parameters were used without adjustment for all searches presented here (Table 1). The `GMIN` input files used for these searches are included in the supplementary data (see Supplementary Material available online at doi:10.1155/2012/192613).

The genetic algorithm represents each structure with a genome consisting of the torsion angles in the backbone of the protein. Offspring structures are generated by one-point crossover from two parent structures. Mutants are generated by making a copy of an existing structure (parent or offspring) and replacing one of the torsion angles. To prevent stagnation of the genetic algorithm searches, a restart operator was used. If an entire generation of offspring contains no solutions that are fitter than any of the parent structures, a new epoch is started with a new random population. For the 69-residue protein, the fittest structure from each epoch survives into the next epoch.

All conformational searches were run until the global minimum structure was found. We report the mean time taken to encounter this structure in conformational searches from randomised starting points to compare the exploration of the energy landscape as a a function of λ. Searches were performed for values of λ between 0 and 1 in steps of 0.1, with additional points at $\lambda = 0.95$ and $\lambda = 0.99$. The initial structures for this benchmarking were generated using two alternative methods: either random placement of the residues inside a sphere of radius 3σ, or random assignment of the backbone dihedral angles. Full details of all of the global optimisation runs are available as supplementary data.

The disconnectivity graphs for the model proteins were constructed from databases of stationary points generated using the `PATHSAMPLE` program, [48] which organises independent pathway searches using `OPTIM` [49]. All the transition state searches in `OPTIM` were conducted in Cartesian coordinates [50] using a quasicontinuous interpolation scheme to avoid chain crossings, with local maxima accurately refined to transition states by hybrid eigenvector-following [51–53]. Successive pairs of local minima were selected for connection attempts within `OPTIM` using the missing connection algorithm [54]. Disconnectivity graphs [39] will be illustrated for both the 46- and 69-residue scaled BLN proteins with λ values of 0, 0.5, 0.9, and 1.

We also study the effect of λ on key structures of the BLN proteins. These structures were reminimised using values of λ between 0 and 1 in steps of 0.1. Pathways between pairs of interesting minima were studied by Dijkstra analysis [55] in `PATHSAMPLE` [48], with the discrete paths [56] that make the largest contribution to the steady-state rate constant [56, 57] presented here.

With a few exceptions, all of the stationary points of the BLN model proteins are chiral. However, the BLN potential includes no chiral terms, so each structure has an enantiomer with the same energy. When evaluating the optimisation algorithms, we accept convergence to either of the enantiomers of the global minimum. When looking at the pathways, it is important to use the same chirality for both structures, otherwise much longer paths result. For some of the trapped structures, pathways to both enantiomers of the global minimum can be viable.

3. Results

3.1. BLN-46. Searches for $\lambda = 0$ (Gō potential) find the global minimum much more rapidly than when $\lambda = 1$ (BLN potential), as one would expect for a more funnelled energy landscape [2, 58–61]. However, the number of steps required varies nonlinearly between these two extremes and behaves

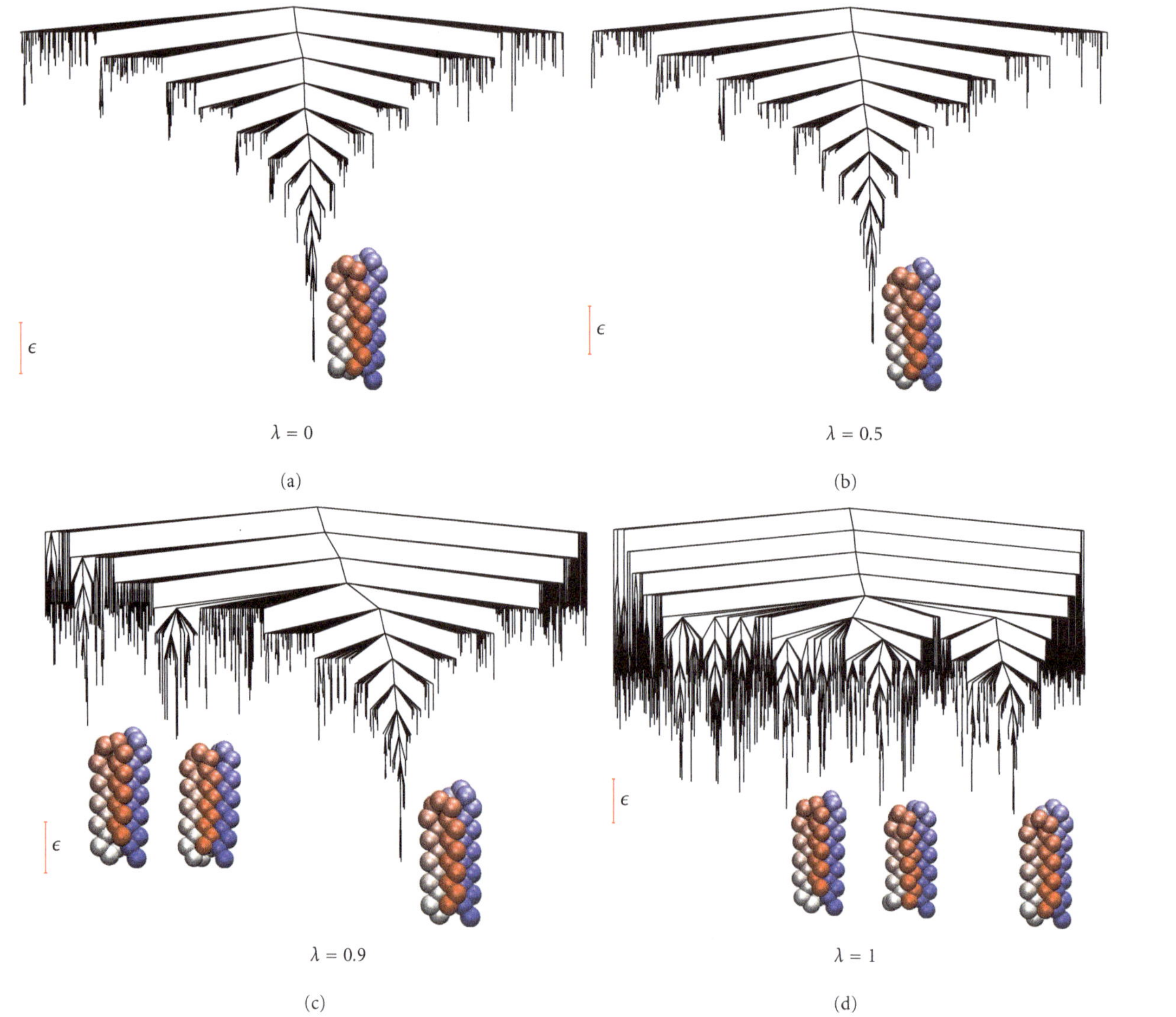

FIGURE 5: Disconnectivity graphs [39] showing the most stable minima accessible by transition states lower than 7ϵ from the global minimum of the 46-residue scaled BLN proteins. For $\lambda \geq 0.9$, only the 1000 most stable minima are shown. The structures of selected minima are illustrated close to the bottoms of the corresponding branches.

differently for each search algorithm. When optimising with the GA, the mean first encounter time decreases rapidly from $\lambda = 1$ to $\lambda = 0.9$ and then more slowly to a minimum at $\lambda = 0.5$ (Figure 1). After this minimum, there is a small increase in the required time as λ decreases to 0. This result is consistent with previous observations that the introduction of some nonnative interactions can assist the folding of some proteins [62]. Below $\lambda = 0.9$, almost all searches find the global minimum within the first epoch of the GA. For larger values of λ, several searches require two or more epochs, leading to much more variation in the first encounter time. The choice of the random starting configurations for the initial population of the GA makes little difference to the mean first encounter time.

In basin-hopping searches, the choice of starting structures makes a large difference to the efficiency of the optimisation. When starting from residues randomly distributed inside a sphere, for values of $\lambda < 0.7$, 95% of the searches find the global minimum rapidly. The remaining searches become trapped and require several thousand attempted Monte Carlo moves to escape (Figure 2). In this trap, the first, third, and fourth strands are correctly packed, but the second is wrapped around the outside of the protein (Figure 3). Searches with larger values of λ do not become trapped in this basin, which suggests that the nonnative interactions are important in stabilising the intermediates between this structure and the global minimum.

The trap configuration lies 12.4ϵ above the global minimum when $\lambda = 0$ and becomes more unfavourable for larger values of λ (Table 2). The fastest escape route from this trap involves unthreading of the N-terminus from the loop made by the second strand (Table 2). The energy of the highest transition state on this pathway relative to the trapped state increases from $\lambda = 0$ to $\lambda = 0.9$ before levelling off. The highest transition state on this pathway lies above the barrier to interconversion of the two enantiomers of

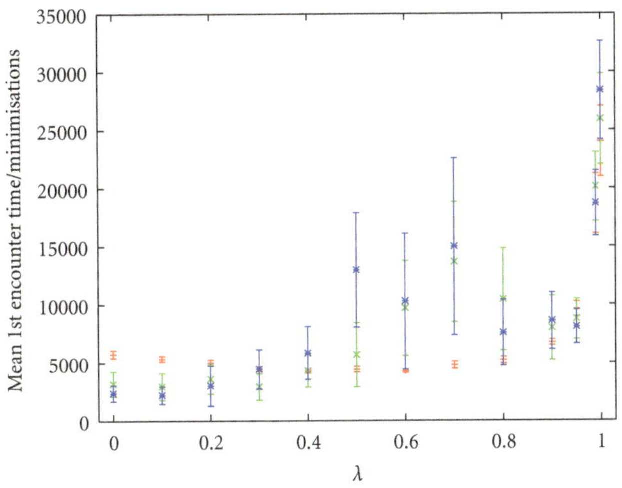

Figure 6: Mean first encounter times (number of minimisations) for 100 global optimisation runs initiated from random starting points for the 69-residue scaled BLN protein. The searches were run using a genetic algorithm (red), basin-hopping starting from random structures confined to a sphere (green), and basin-hopping starting from chain structures with randomised dihedral angles (blue). The error bars are the uncertainties calculated at the 95% level.

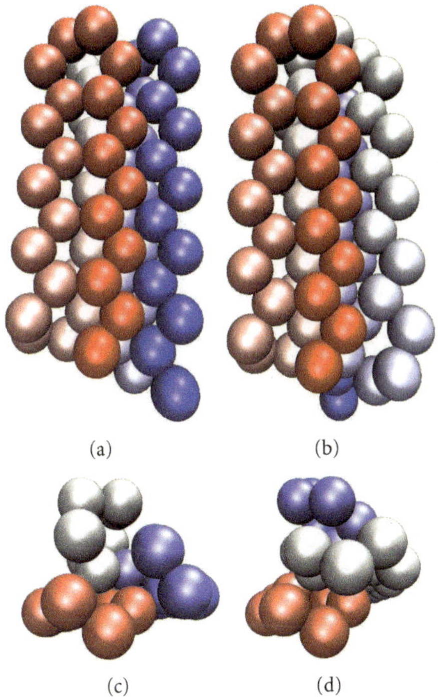

Figure 7: Side and top views of the global minimum (left) and trapped (right) structures of the 69-residue BLN protein illustrated using the VMD program [40] with a colouring scheme for the beads that varies from red to blue (N-terminus to C-terminus).

the global minimum. For searches starting from a random set of torsion angles, this trapping is much less frequent and is only seen in 3 of the 700 searches performed where $0 \leq \lambda \leq 0.6$. By retaining some notion of connectivity, these initial structures cover less of the configurational space than the entirely random starting points. However, the complete coverage of conformational space comes at the cost of including more unstable structures, such as the trap seen here.

The five lowest minima in the BLN-46 protein span an energy range of less than ϵ (Figure 4). The two most stable minima are in the same basin, and both have all of the BB contacts from the native state. Across the range of λ, the relative energies of these minima are within 0.1ϵ of each other, with the second-best minimum becoming slightly more stable as λ decreases and moving below the former global minimum when $\lambda < 0.3$ [12]. The next three minima are stabilised by some nonnative contacts and become less stable relative to the global minimum as λ decreases. In the region around $\lambda = 0.5$, these structures cease to be minima and fall into the basins of attraction [41] of the two lowest energy structures.

The disconnectivity graphs within 7ϵ of the global minimum for $\lambda = 0$ and $\lambda = 0.5$ are funnelled and almost indistinguishable (Figure 5). When $\lambda = 0.9$, some frustration appears in the low-energy regions of the energy landscape, but it is still mostly funnelled. Almost all of the frustration is introduced between $\lambda = 0.9$ and $\lambda = 1$, where several alternate β-barrel structures are separated by barriers of 4 to 5ϵ. This organisation is consistent with the increase in the mean first encounter times seen for global optimisation with $\lambda > 0.9$ and agrees with previous studies of the thermodynamics of the 46-residue protein [32, 33], where $\lambda = 0$ and $\lambda = 0.5$ were found to be good folders, $\lambda = 0.9$ an intermediate folder, and $\lambda = 1$ a poor folder.

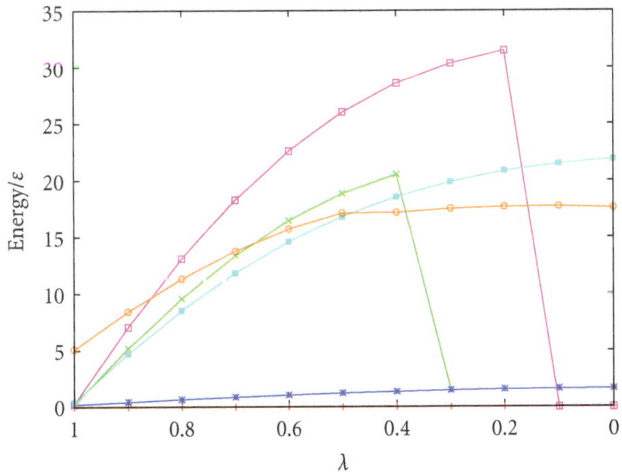

Figure 8: The energies of the five most stable BLN-69 structures relative to the global minimum as a function of λ. Also shown (orange) is the trap structure from Figure 7. The steep decreases in energy mark the points at which structures cease to be local minima and collapse into the basin of attraction [41] of the global minimum.

3.2. BLN-69. The behaviour of the GA for the 69-residue protein is similar to that for the 46-residue protein, with the fastest search time found at $\lambda = 0.5$. When optimising with basin-hopping on the 69-residue protein, there are several

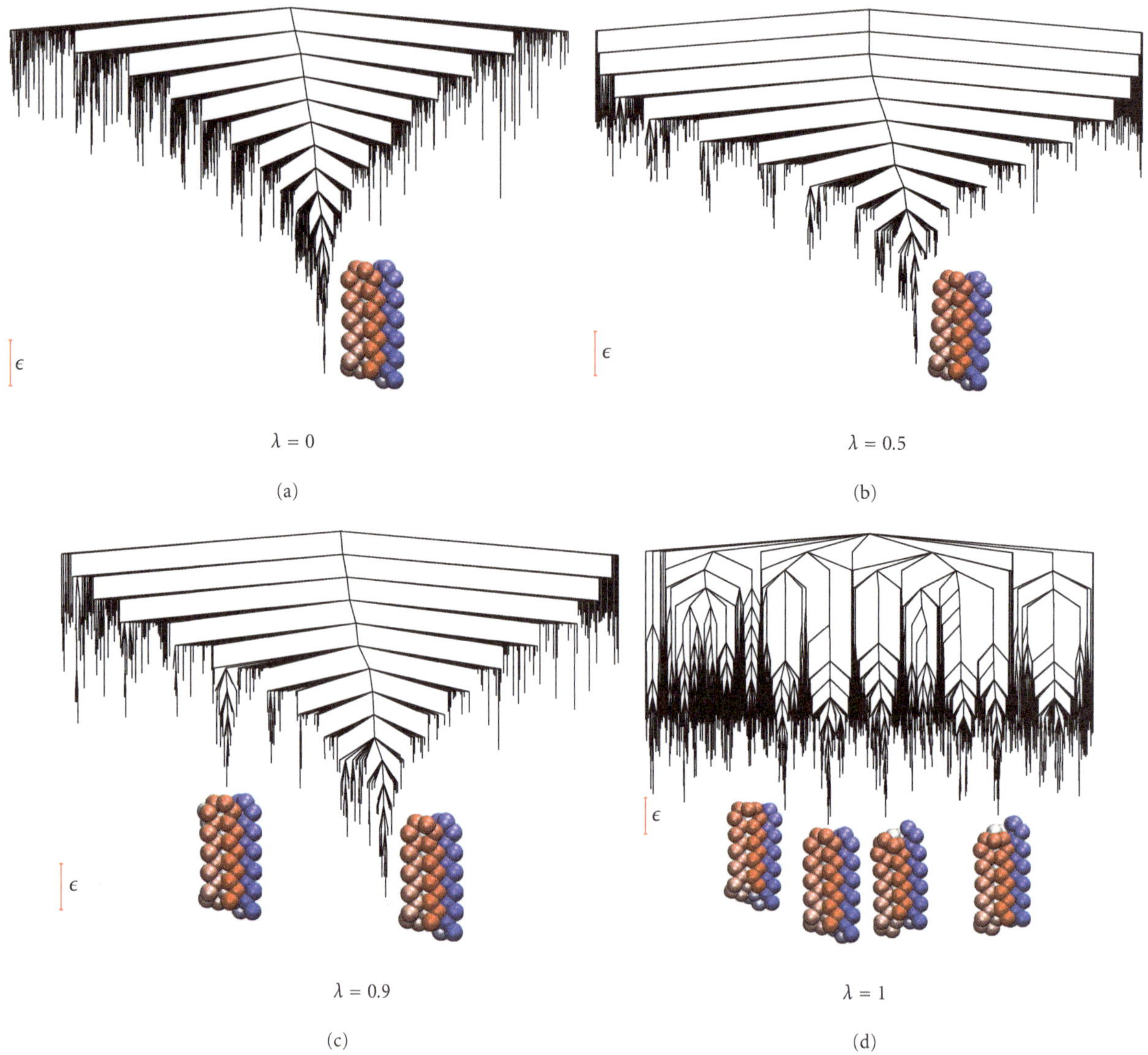

FIGURE 9: Disconnectivity graphs [39] showing the minima accessible by transition states lower than 8ϵ from the global minimum of the 69-residue scaled BLN proteins. For $\lambda \geq 0.5$, only the 1000 most stable minima are shown. The structures of selected minima are illustrated close to the bottoms of the corresponding branches.

TABLE 2: Energies of the trapped minimum and transition state for escape from the principal kinetic trap in the 46-residue scaled BLN protein. All energies are in units of ϵ and measured relative to the global minimum.

λ	E_{trap}	E_{untrap}
0.0	12.4	22.9
0.5	14.0	27.1
0.9	15.3	32.2
1.0	15.6	32.2

slow searches between $\lambda = 0.4$ and $\lambda = 0.8$ (Figure 6). There are multiple trap structures, and the one that is seen most frequently, which is responsible for the slowest searches, is formed from three strands from the left-handed barrel and three strands from right-handed barrel (Figure 7). This structure is a six-stranded β-barrel similar to the global minimum, but with two sets of interstrand contacts swapped (1–6 and 3-4 in the global minimum compared to 1–4 and 3–6 in the trap).

Conversion from the above structure to the global minimum proceeds either by inversion of the three strands at the N-terminus or of the three strands at the C-terminus. The barriers to these two mechanisms are different and vary with λ (Table 3). The barrier for the fastest pathway for inversion at the C-terminus becomes larger with increasing λ. However, the barrier for inversion of the N-terminus varies much less with λ. In the region where $0.5 \leq \lambda \leq 0.7$, the barriers to both routes out of the trap are relatively high, which is a possible explanation for the slow basin-hopping optimisation for these values of λ. This is doubtless an over-simplification when we consider that there are multiple trap structures.

TABLE 3: Energies of the trapped minimum and transition states for escape from the principal kinetic trap by inversion of the N- and C-termini in the 69-residue scaled BLN protein. All energies are in units of ϵ and measured relative to the global minimum.

λ	E_{trap}	$E_{untrap\text{-}C}$	$E_{untrap\text{-}N}$
0.0	17.6	30.2	38.2
0.5	16.7	43.8	33.2
0.7	13.8	30.2	31.1
0.9	8.4	25.7	26.9
1.0	5.1	25.8	23.9

For the 69-residue BLN protein, the energies of the five lowest minima span less than 0.4ϵ (Figure 8). One structure lies in the same funnel as the global minimum, and its relative energy increases from 0.2ϵ to 1.6ϵ when λ decreases from 1 to 0. The other three structures occupy different funnels from the global minimum, with several nonnative contacts, and their stability decreases steeply with decreasing λ. Unlike the 46-residue protein, the global minimum structure remains the same for all values of λ. The low-energy region of disconnectivity graphs for values of λ between 0 and 0.9 are mostly funnelled (Figure 9). Almost all of the frustration in this region of the potential energy surface appears for $\lambda > 0.9$.

4. Conclusions

Much of the energetic frustration in the BLN proteins is removed once the potential contains a 10% contribution from the Gō function. When looking at geometric frustration in higher-energy traps, the effect of λ is less predictable. The removal of nonnative interactions can stabilise or destabilise the transition states that must be crossed to escape from these traps. Measures of the landscape complexity [30] could provide a useful way to understand the influence of nonnative interactions and will be considered in future work.

Acknowledgments

The authors acknowledge the Engineering and Physical Sciences Research Council, UK (EPSRC), for funding under Programme Grant EP/I001352/1. The calculations described in this paper were performed using the University of Birmingham's BlueBEAR HPC service, which was purchased through HEFCE SRIF-3 funds (see http://www.bear.bham.ac.uk/).

References

[1] C. B. Anfinsen, "Principles that govern the folding of protein chains," *Science*, vol. 181, no. 4096, pp. 223–230, 1973.

[2] J. D. Bryngelson, J. N. Onuchic, N. D. Socci, and P. G. Wolynes, "Funnels, pathways, and the energy landscape of protein folding: a synthesis," *Proteins*, vol. 21, no. 3, pp. 167–195, 1995.

[3] Y. Ueda, H. Taketomi, and N. Gō, "Studies on protein folding, unfolding, and fluctuations by computer simulation. II. A three-dimensional lattice model of lysozyme," *Biopolymers*, vol. 17, no. 6, pp. 1531–1548, 1978.

[4] C. Micheletti, F. Seno, and A. Maritan, "Polymer principles of protein calorimetric two-state cooperativity," *Proteins*, vol. 40, no. 4, pp. 637–661, 2000.

[5] C. Clementi, H. Nymeyer, and J. N. Onuchic, "Topological and energetic factors: what determines the structural details of the transition state ensemble and "en-route" intermediates for protein folding? An investigation for small globular proteins," *Journal of Molecular Biology*, vol. 298, no. 5, pp. 937–953, 2000.

[6] J. W. H. Schymkowitz, F. Rousseau, and L. Serrano, "Surfing on protein folding energy landscapes," *Proceedings of the National Academy of Sciences of the United States of America*, vol. 99, no. 25, pp. 15846–15848, 2002.

[7] P. Das, C. J. Wilson, G. Fossati, P. Wittung-Stafshede, K. S. Matthews, and C. Clementi, "Characterization of the folding landscape of monomeric lactose repressor: quantitative comparison of theory and experiment," *Proceedings of the National Academy of Sciences of the United States of America*, vol. 102, no. 41, pp. 14569–14574, 2005.

[8] A. R. Lam, J. M. Borreguero, F. Ding et al., "Parallel folding pathways in the SH3 domain protein," *Journal of Molecular Biology*, vol. 373, no. 5, pp. 1348–1360, 2007.

[9] P. F. N. Faísca, R. D. M. Travasso, R. C. Ball, and E. I. Shakhnovich, "Identifying critical residues in protein folding: insights from ϕ-value and P_{fold} analysis," *Journal of Chemical Physics*, vol. 129, no. 9, Article ID 095108, 2008.

[10] R. D. Hills and C. L. Brooks, "Insights from coarse-grained go models for protein folding and dynamics," *International Journal of Molecular Sciences*, vol. 10, no. 3, pp. 889–905, 2009.

[11] P. O. Craig, J. Lätzer, P. Weinkam et al., "Prediction of native-state hydrogen exchange from perfectly funneled energy landscapes," *American Chemical Society*, vol. 133, no. 43, pp. 17463–17472, 2011.

[12] M. A. Miller and D. J. Wales, "Energy landscape of a model protein," *Journal of Chemical Physics*, vol. 111, no. 14, pp. 6610–6616, 1999.

[13] L. Sutto, J. Lätzer, J. A. Hegler, D. U. Ferreiro, and P. G. Wolynes, "Consequences of localized frustration for the folding mechanism of the IM7 protein," *Proceedings of the National Academy of Sciences of the United States of America*, vol. 104, no. 50, pp. 19825–19830, 2007.

[14] K. F. Lau and K. A. Dill, "A lattice statistical mechanics model of the conformational and sequence spaces of proteins," *Macromolecules*, vol. 22, no. 10, pp. 3986–3997, 1989.

[15] K. A. Dill, S. Bromberg, K. Yue et al., "Principles of protein folding—a perspective from simple exact models," *Protein Science*, vol. 4, no. 4, pp. 561–602, 1995.

[16] J. D. Honeycutt and D. Thirumalai, "Metastability of the folded states of globular proteins," *Proceedings of the National Academy of Sciences of the United States of America*, vol. 87, no. 9, pp. 3526–3529, 1990.

[17] J. D. Honeycutt and D. Thirumalai, "The nature of folded states of globular proteins," *Biopolymers*, vol. 32, no. 6, pp. 695–709, 1992.

[18] Z. Guo and D. Thirumalai, "Nucleation mechanism for protein folding and theoretical predictions for hydrogen-exchange labeling experiments," *Biopolymers*, vol. 35, no. 1, pp. 137–140, 1995.

[19] Z. Guo and D. Thirumalai, "Kinetics and thermodynamics of folding of a de novo designed four-helix bundle protein," *Journal of Molecular Biology*, vol. 263, no. 2, pp. 323–343, 1996.

[20] Z. Guo and C. L. Brooks III, "Thermodynamics of protein folding: a statistical mechanical study of a small all-β protein," *Biopolymers*, vol. 42, no. 7, pp. 745–757, 1997.

[21] R. S. Berry, N. Elmaci, J. P. Rose, and B. Vekhter, "Linking topography of its potential surface with the dynamics of folding of a protein model," *Proceedings of the National Academy of Sciences of the United States of America*, vol. 94, no. 18, pp. 9520–9524, 1997.

[22] H. Nymeyer, A. E. García, and J. N. Onuchic, "Folding funnels and frustration in off-lattice minimalist protein landscapes," *Proceedings of the National Academy of Sciences of the United States of America*, vol. 95, no. 11, pp. 5921–5928, 1998.

[23] J. E. Shea, Y. D. Nochomovitz, Z. Guo, and C. L. Brooks, "Exploring the space of protein folding Hamiltonians: the balance of forces in a minimalist β-barrel model," *Journal of Chemical Physics*, vol. 109, no. 7, pp. 2895–2903, 1998.

[24] N. Elmaci and R. S. Berry, "Principal coordinate analysis on a protein model," *Journal of Chemical Physics*, vol. 110, no. 21, pp. 10606–10622, 1999.

[25] J. E. Shea, J. N. Onuchic, and C. L. Brooks, "Energetic frustration and the nature of the transition state in protein folding," *Journal of Chemical Physics*, vol. 113, no. 17, pp. 7663–7671, 2000.

[26] D. A. Evans and D. J. Wales, "Free energy landscapes of model peptides and proteins," *Journal of Chemical Physics*, vol. 118, no. 8, pp. 3891–3897, 2003.

[27] A. D. Stoycheva, J. N. Onuchic, and C. L. Brooks, "Effect of gatekeepers on the early folding kinetics of a model β-barrel protein," *Journal of Chemical Physics*, vol. 119, no. 11, pp. 5722–5729, 2003.

[28] D. J. Wales and P. E. J. Dewsbury, "Effect of salt bridges on the energy landscape of a model protein," *Journal of Chemical Physics*, vol. 121, no. 20, pp. 10284–10290, 2004.

[29] T. Komatsuzaki, K. Hoshino, Y. Matsunaga, G. J. Rylance, R. L. Johnston, and D. J. Wales, "How many dimensions are required to approximate the potential energy landscape of a model protein?" *Journal of Chemical Physics*, vol. 122, no. 8, Article ID 084714, pp. 1–9, 2005.

[30] G. J. Rylance, R. L. Johnston, Y. Matsunaga, C.-B. Li, A. Baba, and T. Komatsuzaki, "Topographical complexity of multidimensional energy landscapes," *Proceedings of the National Academy of Sciences of the United States of America*, vol. 103, no. 49, pp. 18551–18555, 2006.

[31] J. Kim and T. Keyes, "Inherent structure analysis of protein folding," *Journal of Physical Chemistry B*, vol. 111, no. 10, pp. 2647–2657, 2007.

[32] J. Kim and T. Keyes, "Influence of go-like interactions on global shapes of energy landscapes inβ-barrel forming model proteins: inherent structure analysis and statistical temperature molecular dynamics simulation," *Journal of Physical Chemistry B*, vol. 112, no. 3, pp. 954–966, 2008.

[33] J. Kim, T. Keyes, and J. E. Straub, "Relationship between protein folding thermodynamics and the energy landscape," *Physical Review E*, vol. 79, no. 3, Article ID 030902, 2009.

[34] S. A. Larrass, L. M. Pegram, H. L. Gordon, and S. M. Rothstein, "Efficient generation of low-energy folded states of a model protein. II. Automated histogram filtering," *Journal of Chemical Physics*, vol. 119, no. 24, pp. 13149–13158, 2003.

[35] P. W. Pan, H. L. Gordon, and S. M. Rothstein, "Local-structural diversity and protein folding: application to all-beta off-lattice protein models," *The Journal of Chemical Physics*, vol. 124, no. 2, p. 024905, 2006.

[36] J. Kim, J. E. Straub, and T. Keyes, "Statistical temperature molecular dynamics: application to coarse-grained β-barrel-forming protein models," *Journal of Chemical Physics*, vol. 126, no. 13, Article ID 135101, 2007.

[37] S.-Y. Kim, "An off-lattice frustrated model protein with a six-stranded β-barrel structure," *Journal of Chemical Physics*, vol. 133, no. 13, Article ID 135102, 2010.

[38] M. T. Oakley, D. J. Wales, and R. L. Johnston, "Energy landscape and global optimization for a frustrated model protein," *Journal of Physical Chemistry B*, vol. 115, no. 39, pp. 11525–11529, 2011.

[39] O. M. Becker and M. Karplus, "The topology of multidimensional potential energy surfaces: theory and application to peptide structure and kinetics," *Journal of Chemical Physics*, vol. 106, no. 4, pp. 1495–1517, 1997.

[40] W. Humphrey, A. Dalke, and K. Schulten, "VMD: visual molecular dynamics," *Journal of Molecular Graphics*, vol. 14, no. 1, pp. 33–38, 1996.

[41] P. G. Mezey, *Potential Energy Hypersurfaces*, Elsevier, Amsterdam, The Netherlands, 1987.

[42] Z. Li and H. A. Scheraga, "Monte Carlo-minimization approach to the multiple-minima problem in protein folding," *Proceedings of the National Academy of Sciences of the United States of America*, vol. 84, no. 19, pp. 6611–6615, 1987.

[43] D. J. Wales and H. A. Scheraga, "Global optimization of clusters, crystals, and biomolecules," *Science*, vol. 285, no. 5432, pp. 1368–1372, 1999.

[44] D. J. Wales and J. P. K. Doye, "Global optimization by basin-hopping and the lowest energy structures of Lennard-Jones clusters containing up to 110 atoms," *Journal of Physical Chemistry A*, vol. 101, no. 28, pp. 5111–5116, 1997.

[45] R. L. Johnston, "Evolving better nanoparticles: genetic algorithms for optimising cluster geometries," *Dalton Transactions*, no. 22, pp. 4193–4207, 2003.

[46] D. J. Wales, "GMIN: A program for finding global minima and calculating thermodynamic properties from basin-sampling," http://www-wales.ch.cam.ac.uk/GMIN/.

[47] P. G. Mezey, "Catchment region partitioning of energy hypersurfaces, I," *Theoretica Chimica Acta*, vol. 58, no. 4, pp. 309–330, 1981.

[48] D. J. Wales, "PATHSAMPLE: A program for refining and analysing kinetic transition networks," http://www-wales.ch.cam.ac.uk/OPTIM/.

[49] D. J. Wales, "OPTIM: A program for characterising stationary points and reaction pathways," http://www-wales.ch.cam.ac.uk/PATHSAMPLE/.

[50] D. J. Wales, "Locating stationary points for clusters in cartesian coordinates," *Journal of the Chemical Society, Faraday Transactions*, vol. 89, no. 9, pp. 1305–1313, 1993.

[51] L. J. Munro and D. J. Wales, "Defect migration in crystalline silicon," *Physical Review B*, vol. 59, no. 6, pp. 3969–3980, 1999.

[52] G. Henkelman and H. Jónsson, "A dimer method for finding saddle points on high dimensional potential surfaces using only first derivatives," *Journal of Chemical Physics*, vol. 111, no. 15, pp. 7010–7022, 1999.

[53] Y. Kumeda, L. J. Munro, and D. J. Wales, "Transition states and rearrangement mechanisms from hybrid eigenvector-following and density functional theory: application to $C_{10}H_{10}$ and defect migration in crystalline silicon," *Chemical Physics Letters*, vol. 341, no. 1-2, pp. 185–194, 2001.

[54] J. M. Carr, S. A. Trygubenko, and D. J. Wales, "Finding pathways between distant local minima," *Journal of Chemical Physics*, vol. 122, no. 23, Article ID 234903, pp. 1–7, 2005.

[55] E. W. Dijkstra, "A note on two problems in connexion with graphs," *Numerische Mathematik*, vol. 1, no. 1, pp. 269–271, 1959.

[56] D. J. Wales, "Discrete path sampling," *Molecular Physics*, vol. 100, no. 20, pp. 3285–3305, 2002.

[57] D. J. Wales, "Energy landscapes: calculating pathways and rates," *International Reviews in Physical Chemistry*, vol. 25, no. 1-2, pp. 237–282, 2006.
[58] J. D. Bryngelson and P. G. Wolynes, "Spin glasses and the statistical mechanics of protein folding," *Proceedings of the National Academy of Sciences of the United States of America*, vol. 84, no. 21, pp. 7524–7528, 1987.
[59] J. N. Onuchic, P. G. Wolynes, Z. Luthey-Schulten, and N. D. Socci, "Toward an outline of the topography of a realistic protein-folding funnel," *Proceedings of the National Academy of Sciences of the United States of America*, vol. 92, no. 8, pp. 3626–3630, 1995.
[60] M. Karplus and A. Sali, "Theoretical studies of protein folding and unfolding," *Current Opinion in Structural Biology*, vol. 5, no. 1, pp. 58–73, 1995.
[61] J. N. Onuchic, H. Nymeyer, A. E. García, J. Chahine, and N. D. Socci, "The energy landscape theory of protein folding: insights into folding mechanisms and scenarios," *Advances in Protein Chemistry*, vol. 53, pp. 87–152, 2000.
[62] C. Clementi and S. S. Plotkin, "The effects of nonnative interactions on protein folding rates: theory and simulation," *Protein Science*, vol. 13, no. 7, pp. 1750–1766, 2004.

6

Proton Transfer Equilibria and Critical Behavior of H-Bonding

L. Sobczyk, B. Czarnik-Matusewicz, M. Rospenk, and M. Obrzud

Faculty of Chemistry, University of Wrocław, Joliot-Curie 14, 50-383 Wrocław, Poland

Correspondence should be addressed to L. Sobczyk, lucjan.sobczyk@chem.uni.wroc.pl

Academic Editor: Marek J. Wojcik

The aim of the present paper is an analysis of the hydrogen bond properties for the acid-base systems depending on the ability to the proton transfer in the formulation of the Brönsted approach. After definition of the proton transfer equilibrium expressed by using the equation $\log K_{\mathrm{PT}} = \xi \Delta pK_N$, various examples of different physical properties, such as dipole moments, IR spectra, and nuclear magnetic resonances, are presented which correlate with the ΔpK_N value. In such a way, a critical state of hydrogen bonding can be defined that corresponds to the potential of the proton motion for either single minimum or double minimum with low barrier. A particular attention in this paper found electronic spectra which have not been analysed so far and the quantitative analysis of the vibrational polarizability which can reach very high values of the order of electronic polarizability.

1. Introduction

The subject of our interest in the present review is hydrogen bonds which can be expressed as A–H· · · B. It is an acid-base system in the Brönsted formulation when the A–H group is treated as an acid while the B atom or group of atoms as proton acceptor (base). The potential energy curves for the proton motion can reach various shapes, as shown in Figure 1.

The extreme curves (1) and (6) correspond to states either without proton transfer (1) or to the complete ionization when the proton is attached to B while atom A is negatively charged (6). Among the intermediate states take place those when the proton is located in the central position either with two minima (3) and a low barrier or with one single minimum (4).

There is a rich literature [1–16] with various approaches to the hydrogen bonding corresponding to different definitions, showing an increase of systems analyzed with comprehensive theoretical treatments, and containing different rich chemical characteristic features of hydrogen bonds. Most actual comprehensive review was recently published by G. Gilli and P. Gilli [16].

From the point of view of the approach based on the acid-base interaction, the substantial, parameter is the proton transfer degree which evokes changes of further physico-chemical parameters. The main quantity is the ΔpK_a value which can be expressed in the form:

$$\Delta pK_a = pK_{\mathrm{B^+H}} - pK_{\mathrm{AH}}. \quad (1)$$

This quantity was introduced by Huyskens and Zeegers-Huyskens [17]. We introduced normalized parameter defined as

$$\Delta pK_N = \Delta pK_a - \Delta pK_a\,(\mathrm{crit}), \quad (2)$$

where $\Delta pK_a(\mathrm{crit})$ is related to ΔpK_a region when the proton transfer degree reaches 50% [18].

The dependence of proton transfer degree on the ΔpK_N value needs a correction connected with "softness/hardness" of interaction by using parameter $\xi < 1$ [19]. The value of this parameter is the higher, the harder is the interaction reaching maximal value equal to unity. As will be seen, this quantity is well correlated with the polarizability in the transition state of hydrogen bonds. The general equation presenting the dependence of proton transfer degree on ΔpK_N possesses the form:

$$\log K_{\mathrm{PT}} = \xi \Delta pK_N. \quad (3)$$

One should remember that physicochemical parameters measured depending on ΔpK_N and connected with the

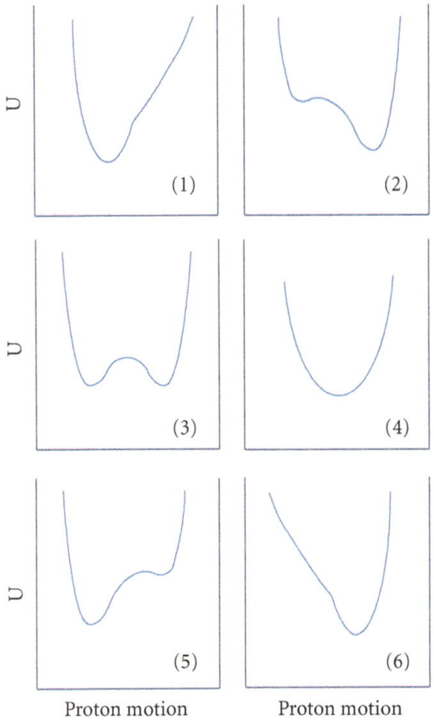

FIGURE 1: Postulated potential energy curves for the proton motion starting from nonproton-transfer state (1) up to fully ionized state (6).

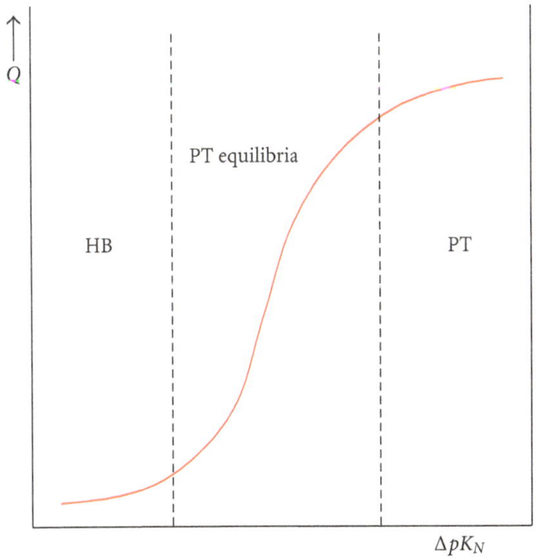

FIGURE 2: Three regions of physical properties depending on ΔpK_N: HB-related to nonproton-transfer states, PT-related to proton transfer state and HB + PT proton transfer equilibrium.

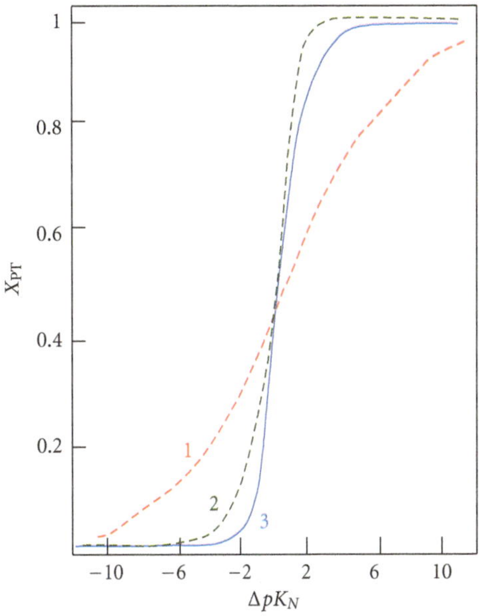

FIGURE 3: Proton transfer degree from NQR measurements for complexes composed of chlorine containing proton donors plotted versus ΔpK_N: (1) CCl_3COOH complexes ($\xi = 0.12$), (2) $CHCl_2COOH$ complexes ($\xi = 0.42$), and (3) C_6Cl_5OH complexes ($\xi = 0.74$) [20].

softness of interaction are related not only to ΔpK_N as has been shown in Figure 2.

There exist three regions; the central one with the equilibria of the proton transfer and side regions without proton transfer (HB) and with full ionization (PT).

Finally, as will be shown, it is necessary to mention the role of medium such as electric permittivity of the solvent and specific interaction between the solute and solvent molecules.

For characterization of the role of the ξ parameter we present in Figure 3 dependencies of the proton transfer degree deduced from the measurements of nuclear quadrupole resonance (NQR) for complexes of CCl_3COOH (1) ($\xi = 0.12$), $CHCl_2COOH$ (2) ($\xi = 0.42$), and C_6Cl_5OH (3) ($\xi = 0.74$) [20]. It is well seen the property of the curves in the critical region when approaching to $\Delta pK_N = 0$.

It is justified to mention in the introduction that curves expressing dependencies of physicochemical parameters on ΔpK_N possess various shapes [18]. One can distinguish two types of correlations between the physical quantity and ΔpK_N, namely, of the sigma and delta type. The examples of such correlations will be presented in the next chapter.

2. Examples of Correlation between Physicochemical Parameters and the ΔpK_N Quantity

So far a most precisely investigated phenomenon is the dependence of the increase of dipole moment $\Delta\mu$ for complexes of phenols with N-bases. In Figure 4, we present correlation between $\Delta\mu$ and ΔpK_N obtained for a number of systems in nonpolar solvents, particularly in benzene [18]. The experimental points are adjusted to the equation [21]:

$$\Delta\mu = \frac{\Delta\mu_{HB} + b_{HB}\Delta pK_N}{1 + \exp\left(2.303\xi\Delta pK_N\right)} + \frac{\left(\Delta\mu_{PT} + b_{PT}\Delta pK_N\right)\cdot\exp\left(2.303\xi\Delta pK_N\right)}{1 + \exp\left(2.303\xi\Delta pK_N\right)}, \quad (4)$$

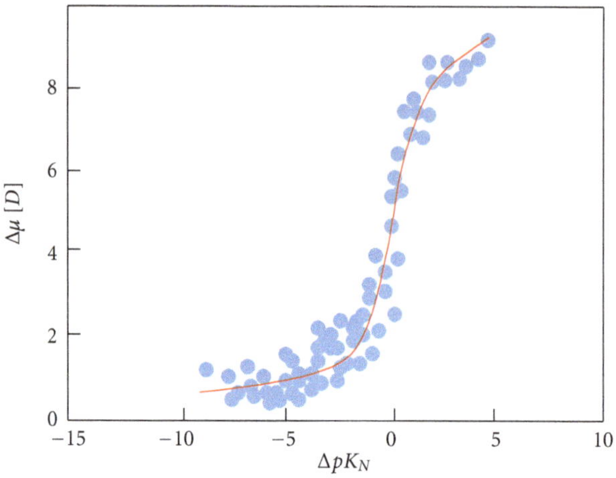

FIGURE 4: The increase of dipole moment $\Delta\mu$ plotted versus normalized parameter ΔpK_N [18].

where $\Delta\mu_{HB}$ and $\Delta\mu_{PT}$ mean the increase of the dipole moment without proton transfer (HB) and after the proton transfer (PT). These quantities depend nearly linearly on ΔpK_N with coefficients b_{HB} and b_{PT}. When approaching the critical region around $\Delta pK_N = 0$, a stepwise change of the dipole moment connected with the increase of the proton degree takes place. The proton transfer degree x_{PT} defines the equilibrium:

$$X_{PT} = \frac{\exp(2.303\xi\Delta pK_N)}{1+\exp(2.303\xi\Delta pK_N)}. \quad (5)$$

To obtain the agreement with the experiment, it is necessary to introduce the coefficient ξ which, as has been formulated, characterizes softness/hardness of interactions. It can be, on the other hand, connected with the barrier height for the proton transfer. The value of the ξ coefficient for the case of the situation in Figure 4 equals 0.65.

Very similar run of the dependence on ΔpK_N shows the value of the ^{15}N resonance chemical shift with the ξ value equal to 0.56 [22]. However, one should remember that the results are related to markedly different experimental conditions. Thus, the results obtained for ^{15}N chemical shift were obtained for complexes of carboxylic acids with pyridine in liquefied freons.

Sigmoidal type of the relationship of physical quantity on ΔpK_N is also observed for complexes of pentachlorophenol with amines by using the nuclear quadrupole resonance (NQR) [23] that is presented in Figure 5. In addition to experimental points, there are indicated values corresponding to neat pentachlorophenol, H-bis-phenolate, as well as to Na^+ and tributylamine salts. One should remember that NQR measurements are performed for solid state that reflects observed behavior.

The similar shape of the plot with that in Figure 5 is observed between geometrical parameters of complexes and ΔpK_N and particularly between C–O bond length and ΔpK_N [25].

An example of correlation between the measured quantity and ΔpK_N of the delta type relates first of all to the proton magnetic resonance δ^1H. It is presented for the

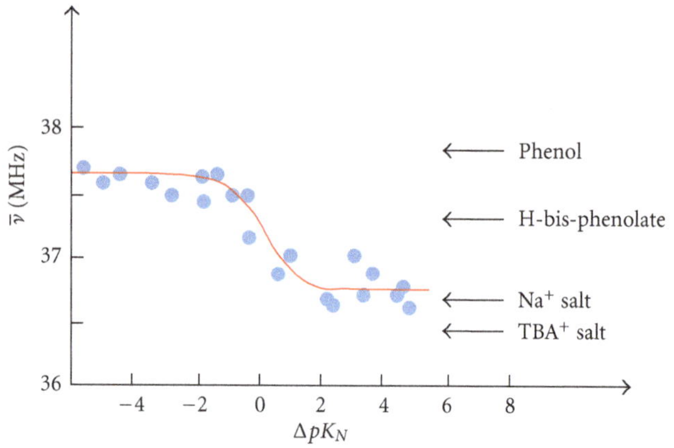

FIGURE 5: The dependence of average NQR ^{35}Cl frequency upon ΔpK_N for complexes of pentachlorophenol [24].

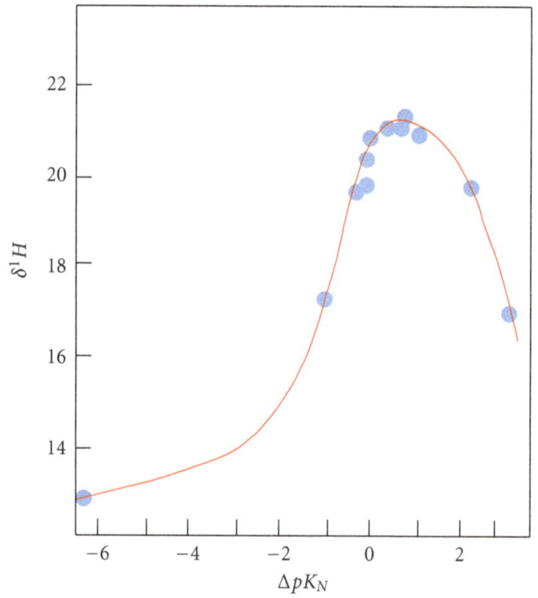

FIGURE 6: The dependence of δ^1H for complexes of carboxylic acids with pyridine in liquid freon [22].

systems analogues to the $\delta^{15}N$ resonance [18]. The experimental points of δ^1H presented in Figure 6 were obtained in the same conditions as for $\delta^{15}N$. The value of the ξ parameter is, however, somewhat lower (0.46) that we are not able to explain. From already done numerous experiments it follows that methods applied do not possess marked influence on the ξ value.

In the analysis of the correlation plots exhibiting an extremum in the critical region as in the case of δ^1H, a modified approach can be used. Thus, for the description of the dependence of given physical property Q showing an extremum, the following simple procedure can be employed. The reference value of a given physical property Q is its extremum; that is, maximum or minimum. In the case of δ^1H for the systems composed of carboxylic acids and pyridine in liquid freons the maximum value equals 21.5 ppm. The delta type correlation can be transformed to the sigmoidal one by assuming that $Q(\text{crit}) = 0$, while $\Delta Q_{HB} < 0$ and $\Delta Q_{PT} > 0$ as has been done in Figure 7.

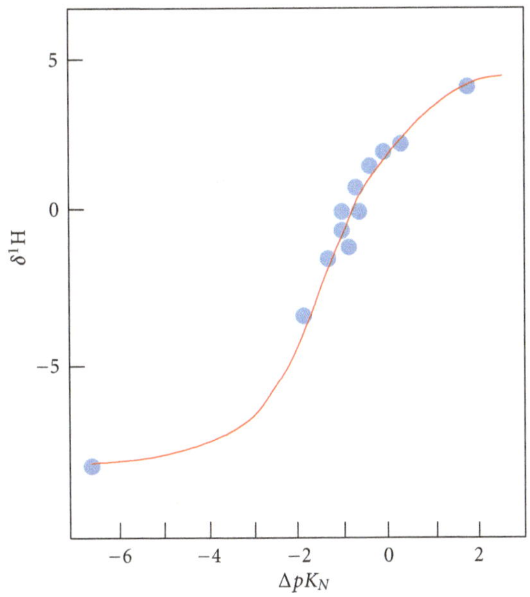

FIGURE 7: Correlation between δ^1H and ΔpK_N for complexes of carboxylic acids with pyridine in liquid freon according (6).

The correlation between ΔQ and ΔpK_N is presented in the following equation [18]:

$$\Delta Q = \frac{\Delta Q_{HB} + \Delta Q_{PT}\exp(2.303\xi\Delta pK_N)}{1 + \exp(2.303\xi\Delta pK_N)}. \quad (6)$$

The parameters for best fitting are Q_{max} = 21.5 ppm ΔQ_{HB} = −8.3 ppm while ΔQ_{PT} = 4.4 ppm and ξ = 0.46 as has been already mentioned.

The properties of infra-red spectra are commonly accepted for the hydrogen bonded systems. This relates first of all to the absorption band ascribed to the stretching vibrations of either AH group (HB state) or BH^+ group (PT state). The evolution of broad absorption ascribed to the ν(AH) or $\nu(NH^+)$ vibrations is illustrated in Figure 8 taking as an example complexes of pentachlorophenol with amines [26]. In the infra-red spectra the correlated quantity is the center of gravity of protonic vibrations (ν_{cg}) versus the ΔpK_N value. Figure 9 represents numerous data related to ν_{cg} collected for various O–H· · · N hydrogen bridges [27]. The scattering of experimental points is very large that seems to be understandable taking into account various experimental conditions and differences in the acid-base interaction for various components. One of the reasons of scattering is a difficulty connected with precise assessment of the position of broad bands. As follows from the results collected by Albrecht and Zundel [28] for the complexes of phenols with octylamine, the maximal absorbance in the range of continuous absorption corresponds to 50% of proton transfer that is shown in Figure 10.

3. Electronic Spectra and the Proton Transfer Degree

The UV-Vis spectroscopy is a very useful method of studies on the proton transfer degree in the Brönsted acid-base

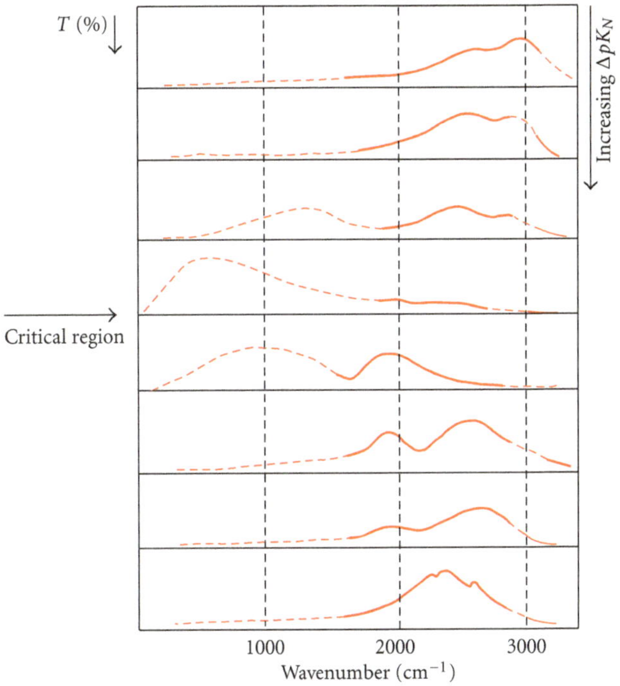

FIGURE 8: The evolution of infra-red absorption ascribed to ν(OH) when increasing ΔpK_N for complexes of pentachlorophenol with amines [26].

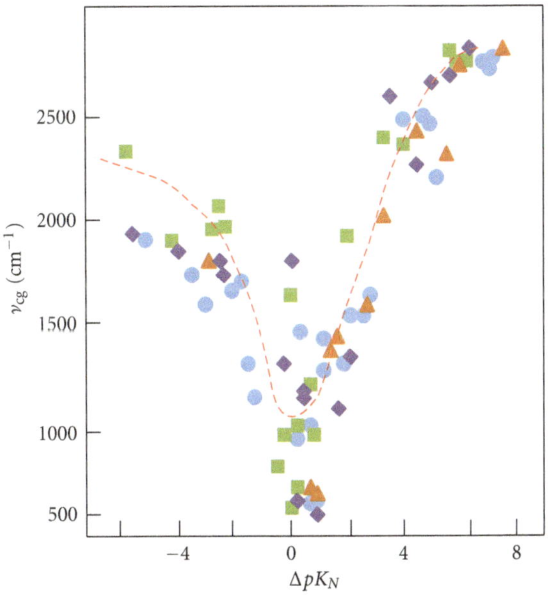

FIGURE 9: The center of gravity ν_{cg} for protonic vibrations as a function of ΔpK_N for various complexes of carboxylic acids [27].

system for the diluted solutions. The majority of quantitative data related to the proton transfer equilibria relates mainly to the complexes between phenols and amines [24, 29–35]. In the UV spectra, the tautomeric equilibrium is characterized by appearance of a new band corresponding to the $\pi \rightarrow \pi^*$ transition in the phenolate ion. After careful quantitative separation of the HB and PT bands the proton transfer equilibrium c_{PT}/c_{HB} can be evaluated. As an example of the UV spectra with the proton transfer equilibrium, we use

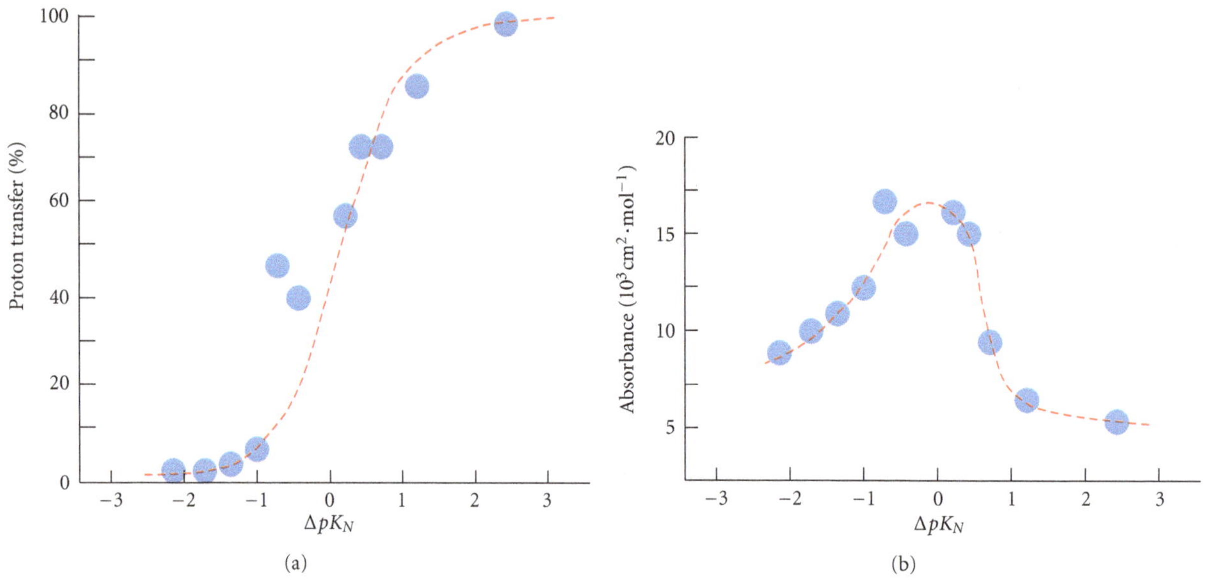

FIGURE 10: The proton transfer degree (a) and intensity of continuous absorption (b) for complexes of phenols with octylamine [28].

the system of 2,4,6-trichlorophenol in tributylamine (TBA) [29] presented in Figure 11 which shows the overlapping of HP and PT bands. From the equilibrium constant, other thermodynamic parameters can be determined according to equation:

$$\ln K = \frac{\Delta S^\circ}{R} - \frac{\Delta H^\circ}{RT}, \quad (7)$$

where K is calculated by using intensities of bands and molar absorption coefficients of corresponding forms:

$$K = \left(\frac{I_{\mathrm{PT}}}{I_{\mathrm{HB}}}\right)\left(\frac{\varepsilon_{\mathrm{HB}}}{\varepsilon_{\mathrm{PT}}}\right). \quad (8)$$

The first quantitative studies by using the electronic absorption spectra were performed by Baba et al. [30] for complex of 4-nitrophenol with triethylamine in 1,2-dichloroethane who found $\Delta H^\circ = -13\,\mathrm{kJ\cdot mol^{-1}}$ and $\Delta S^\circ = -49.8\,\mathrm{J\cdot mol^{-1}\cdot K^{-1}}$. Similarly, Crooks and Robinson [31] investigated complexes of bromophenol with methyl derivatives of pyridine in chlorobenzene. The obtained data correspond to $-\Delta H^\circ$ in the range 12–38 $\mathrm{kJ\cdot mol^{-1}}$ and $-\Delta S^\circ$ in the range 29–55 $\mathrm{J\cdot mol^{-1}\cdot K^{-1}}$. The values of thermodynamic parameters for the complexes of chlorophenols with TBA [29] are comparable with those of nitrophenol.

From the studies [29, 37–44] it follows that the concentration of the PT form, independently of the H-bonding type, increases with an increase of ΔpK_a value of interacting components, as well as with increase of the solvent activity and the drop of temperature.

For the systems with negative or close to zero ΔpK_a values, it was not possible to find traces of the PT band even in the most active solvents at temperatures as low as below −190°C [38]. Thus, for observation in UV spectrum participation of the PT form even in favorable conditions (low temperature and high polarity of solvent), some boundary ΔpK_a value is necessary.

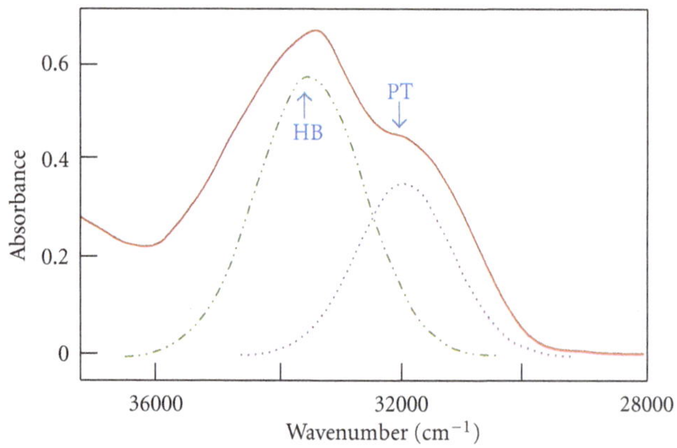

FIGURE 11: The plot of the absorbance versus wavenumber for 2,4,6-trichlorophenol in tributylamine at room temperature, $c = 5 \cdot 10^{-4}\,\mathrm{mol\cdot dm^{-3}}$; $d = 5\,\mathrm{mm}$.

Figure 12 shows the UV spectra for the series of complexes formed by TBA with various chlorophenols of increasing acidity. It can be seen that 2,4-dichlorophenol and 2,4,5-trichlorophenol do not show any contributions of PT species, only 2,6-dichlorophenol shows traces of the ionic PT form. For 2,4,6-trichlorophenol, a considerable amount (ca. 25%) of the PT form was estimated from the UV spectrum. Pentachlorophenol appears entirely in the zwitterionic state, whereas in a case of 2,6-dichloro- and 2,4,5-trichloro derivatives, characterized by almost the same ΔpK_a values, some contribution of the PT state shows only the former one. The ΔpK_a value is not, however, a completely satisfactory measure of the proton donor-acceptor properties in nonaqueous media.

In several papers, for example, [45–48] one considers the attention that one should apply another scale of proton donor and acceptor properties for defining the proton

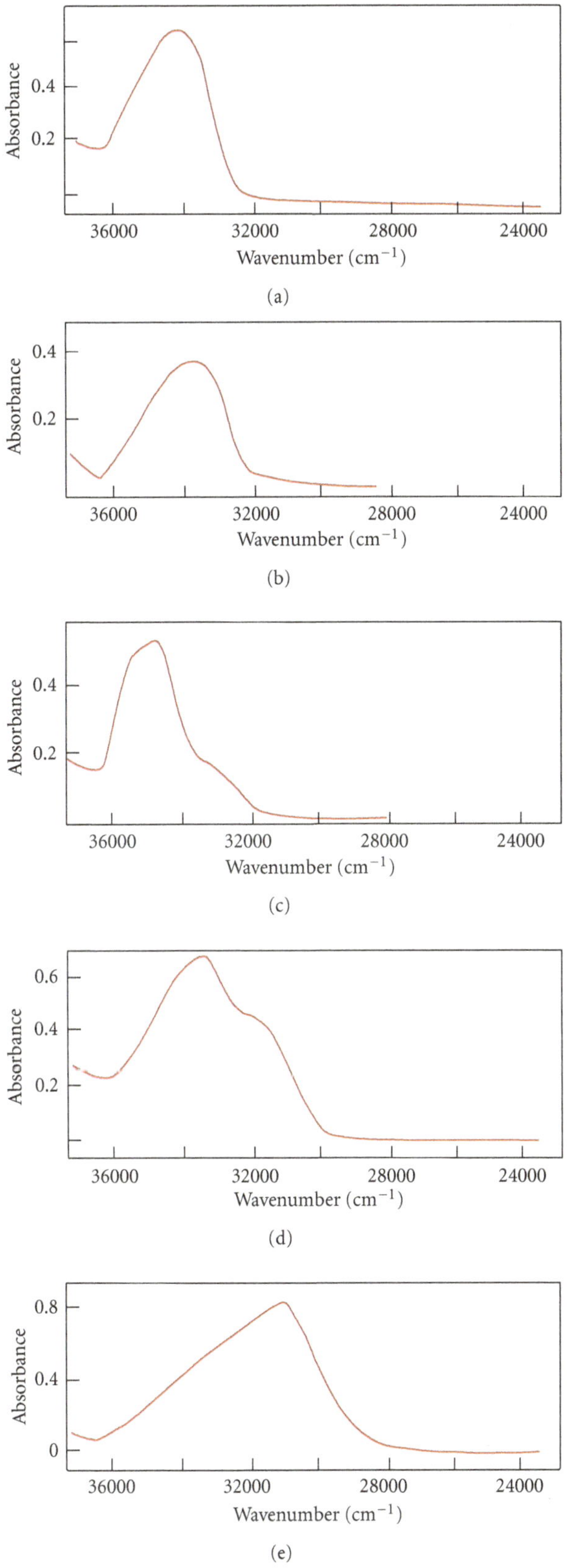

FIGURE 12: UV spectra of chlorophenols: (a) 2,4-dichlorophenol, (b) 2,4,5-trichlorophenol, (c) 2,6-dichlorophenol, (d) 2,4,6-trichlorophenol, (e) pentachlorophenol in TBA at room temperature, $c = 5 \cdot 10^{-4}$ mol$\cdot$dm^{-3}; $d = 5$ mm [29].

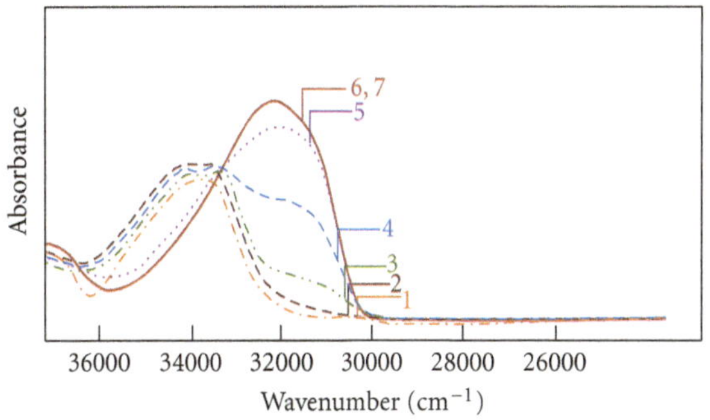

FIGURE 13: UV spectra of 2,4-dichlorophenol in TBA as a function of temperature: 298 K (1), 223 K (2), 203 K (3), 186 K (4), 165K (5), 143 K (6), 128 K (7). $C = 4 \times 10^{-4}$ mol dm^{-3}, $d = 5$ mm, wavenumber of PT form ≅ 32160 cm^{-1} and HB form ≅ 33840 cm^{-1} [29].

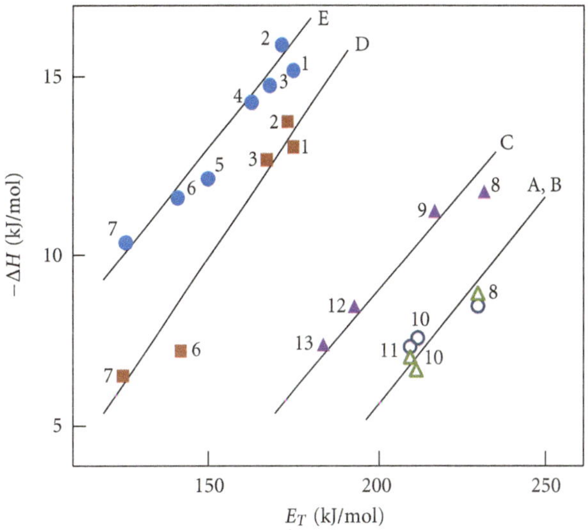

FIGURE 14: Comparison of the ΔH°_{PT} with E_T parameters for Mannich bases: (A) 2-(N,N-dimethylaminomethyl)-4,6-dibro- mo phenol, (B) 2-(N,N-diethylaminomethyl)-4-nitro-phenol, (C) 2-(N,N-diethylaminomethyl)-3,4,6-trichlorophenol, (D) 2-(N,N-diethylaminomethyl)-3,4,5,6-tetrachlorophenol, (E) 2-(N,N-diethylaminomethyl)-4-nitronaphthol-1, in 1,2-dichloroethane (1), dichloromethane (2), n-butylchloride (3), chloroform (4), 1,4-dioxane (5), isopropylbenzene (6) squalane (7), methanol (8), ethanol (9), butan-1-ol (10), propan-1-ol (11), acetonitrile (12), and N,N-dimethylformamide (13) [36].

position in hydrogen-bonded complexes. In the analysis, one takes into account the proton affinity and deprotonation enthalpy based on calculations by using DFT methods. However, in the present article, we limited our considerations to experimental methods leading to evaluation of the pK_a values.

A strong influence of cooling on the increase of concentration of the PT form indicates on negative change of enthalpy effect on the proton transfer process. In Figure 13, the UV spectra of 2,4-dichlorophenol in TBA are shown as a function of temperature [29]. The 2,4-dichlorophenol—TBA system at room temperature does not show any contribution of the PT state. Similar to other systems of this type, we observe a very strong influence of cooling on the contribution of the PT state. At the temperatures 203, 186, and 165 K, the values of K_{PT} are 0.33, 0.82, and 5.70, respectively. The complete proton-transfer state is reached at about 143 K and further cooling does not affect the intensity of the phenolate band.

By using electronic spectroscopy in the UV range, the PT equilibrium constants have been measured as a function of temperature in various solvents for various H-bonded systems. They allowed to determine the thermodynamic parameters of the PT process and correlate with various empirical parameters of the solvent activity. The results for Mannich bases [36, 39, 40] correlated with the Dimroth-Reichardt E_T parameter [49, 50] are presented in Figure 14. These correlations present individual straight lines with similar slope for particular Mannich bases. Such clear differentiation shows that the differences in the proton affinity of particular acid-base centers contribute essentially to the stabilization of both forms. The observed effect of solvent activity shows that the proton transfer process is characterized by two factors. Simultaneously, with previous ΔpK_a effect that can be classified as inter one, an additional factor, called an external, takes place, which correlates with the solvent activity expressed by the E_T parameter. Formally, one can express

$$\Delta H^{\circ}_{\mathrm{PT}} = \Delta H^{\circ}_{\mathrm{int}}\left(\Delta pK_a\right) + \Delta H^{\circ}_{\mathrm{ext}}(E_T), \tag{9}$$

however, quantitative estimation of both components is not an easy task.

The attempt has been undertaken to correlate the $\Delta H^{\circ}_{\mathrm{PT}}$ values with other parameters characterized the solvent activity, but the best correlation was obtained with E_T. Thus, the external factor contains two effects, that is, the electrostatic stabilization of the ionic form and the donor-acceptor interaction of solvent molecules with the free electron pair of the phenolate oxygen atom. So far, no proton transfer equilibrium was observed in the gas phase that prooves decisive role of the solvent for observation of the proton transfer. This is confirmed by relatively high values of entropy effect, ΔS°_{PT} from −30 up to −70 J K^{-1} mol^{-1} [36, 39, 40] that confirms a considerable redistribution of molecules and high increase of ordering of solvent molecules under influence of intramolecular proton transfer.

The UV spectra were used to locate the position of 50% proton transfer in chloranilic acid-amine complexes; the similar result was deduced from IR and NMR studies [51]. Chranina et al. [52] studied the proton transfer equilibria between hydroxyanthraquinone dyes and aliphatic amines in low-polarity solvents by UV spectroscopy. The shift of this equilibrium in an external electrical field has been observed by the method of electrochromism in the visible region. Also, the mechanism of proton transfer reactions between various acids and amines was studied kinetically by applying UV spectroscopy, when the order and the isotopic ratio effect were discussed [53, 54].

4. Vibrational Polarization of Hydrogen Bonded Systems

It has been broadly postulated by Zundel [55] that for the characteristic dependences of the important physical parameters on ΔpK_N, with the anomalous behavior in the critical region, the large proton polarizability of the hydrogen bonds is responsible. The extraordinary increase in proton polarizability with increased strength of the hydrogen bonds in heteroconjugated systems was the aim of detailed infrared studies conducted by Hawranek's group. For six systems of pentachlorophenol (PCPh) dissolved in different basis, the molar vibrational polarization (P^{vib}, called also atomic polarization as it arises from atomic motions) and molar electronic polarization were determined according to the procedure sketched below. Names of the basis are given in Table 1. The PCPh-base complexes were studied in binary solutions, that is, the proton donor (PCPh) was directly dissolved in an excess of the proton acceptor. Such conditions facilitated accurate determination of optical quantities necessary for calculations of the P^{vib} values, according to the following scheme.

Table 1 shows the P_2^{vib} values along with the position ($\nu_{\max}$) and the half width ($\Delta\nu_{1/2}$) of the ν_s(OH) band. The spectral parameters were obtained only for H-bonded systems related to the nonproton-transfer state, their values cannot be estimated with a sufficient accuracy for systems corresponding to other two states (see Figure 2). The plot of the P_2^{vib} values versus ΔpK_a, shown in Figure 15, possess the delta type character with a maximum.

It has to be mentioned here that the measurements in binary system have many advantages that facilitate the used procedure of determination of the molar vibrational polarization. However, there is also one disadvantage: the P^{vib} values are obtained for H-bonded systems differently polarized by their environment. The PCPh-base complexes are immersed in various media that have different macroscopic parameters and more or less strongly polarize the hydrogen bonds. For each system, the ξ and ΔpK_a (crit) parameters should be determined whenever the ΔpK_a values are subjected to the normalization procedure. Due to the lack of such data, the P^{vib} values on Figure 15 are plotted against ΔpK_a parameter. We can guess that the normalization and the different influence of solvents on the vibrational polarization should not meaningfully change the delta-type relation between P^{vib} and strength of the hydrogen-bonded systems.

According to Table 1, the molar vibrational polarization increases from a very small value for TMPh in inert CCl_4 solution, to a slightly larger for the OH group involved in a weak OH$\cdots$Cl intramolecular hydrogen bond in PCPh. Noticeable increase is observed for OH group engaged in a weak intermolecular hydrogen bonds in the PCPh-CH_3CN and PCPh-dioxane systems. Their P^{vib} values compared with that for the 2,4,6-TMPh-CCl_4 indicate on the 17- and 20-fold increase. The changes are strictly correlated with the typical spectral features of H-bond formation, that is, the shift of ν_s(OH) bands towards lower frequencies and the increase in its bandwidth. In relation to the system with intramolecular

Table 1: Spectral parameters related to the ν_s(OH) band and P^{vib} of the H-bond complexes of PCPh.

Acceptor	ν_{max} (cm^{-1})	$\Delta\nu_{1/2}$ (cm^{-1})	P^{vib} ($cm^3\ mol^{-1}$)	Reference
CCl_4	3525	21.6	0.048	[56]
CH_3CN	3322	275.5	0.294	[57]
Dioxane-D_8	3162	316.5	0.333	[58]
3-Chloropyridine	2737	945	1.182	[59]
Pyridine	—	—	12.5	[60]
2,4,6-Trimethylpyridine	—	—	17. 8	[61]
Tri-*n*-octylamine	—	—	9.0	[62]
2,4,6-TMPh-CCl_4	3622		0.017	[56]

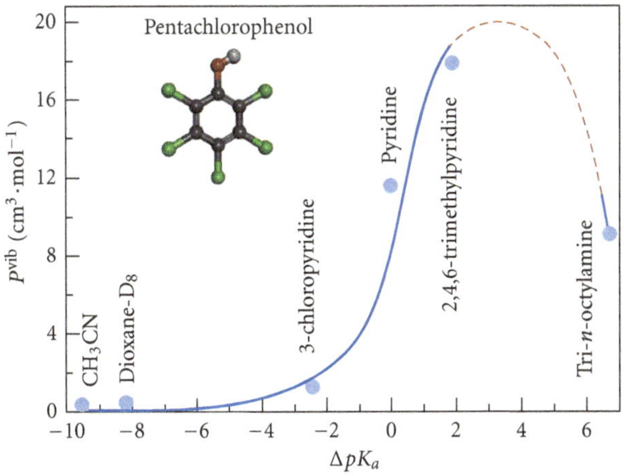

Figure 15: P^{vib} plotted versus ΔpK_a of complexes formed by PCPh with various proton acceptors.

hydrogen bonds (PCPh-CCl_4), the increase is 6-fold for the PCPh-CH_3CN and 7-fold for the PCPh-dioxane complex. It reveals that formation even rather weak intermolecular H-bond, when the proton is located in a relatively narrow single-minimum proton potential near the acid (Figure 1 (1)), leads to a drastic increase in P^{vib} of the OH group.

The PCPh-3-chloropyridine system, with still relatively asymmetrical hydrogen bond, is close to a border between the HB and the PT equilibrium states (see Figure 2). However, its P^{vib} value, compared with that obtained for the system with intramolecular H-bonded, shows almost 25- and 70-fold increase in comparison with the free OH-group in the 2,4,6-TMPh-CCl_4 system. Despite this, the molar vibrational polarization of the PCPh-3-chloropyridine system is still markedly less than its molar electronic polarization.

The complex of PCPh with pyridine with symmetrical O$\cdots$H$\cdots$N hydrogen bond is classified to the proton transfer state. The molar vibrational polarization of the OH group rises to 12.5 $cm^3 \cdot mol^{-1}$. This value compared with that obtained for free (2,4,6-TMPh-CCl_4) and for the intramolecularly bonded (PCPh-CCl_4) OH group shows almost 600- and 200-fold increase, respectively.

According to [28], the complex of PCPh with 2,4,5-trimethylpyridine is close to the border between the PT equilibrium and the PT states. Its molar vibrational polarization is more than 370 and 1000 times higher than in the PCPh-CCl_4 and 2,4,6-TMPh-CCl_4 system, respectively. For the PCPh-2,4,6-trimethylpyridine complex hydrogen bond possess largest proton polarizability. The last complex of PCPh with tri-*n*-octylamine belongs to the PT state. According to Figure 15, its P^{vib} value drops almost twice when compared with the previous system. For such large change of P^{vib}, a characteristic evolution of the infrared spectra corresponding to the PT state, shown in Figure 8, is responsible.

Summing up, the very large P^{vib} values determined for PCPh complexes with pyridine and 3-chloropyridine are excellent confirmation of the extraordinary properties of hydrogen bonds from the transition region with symmetrical potential. Moreover, they confirm very well Zundel's concept that an extreme broadening of the OH band occurs for hydrogen bonds showing the largest proton polarizability [55].

References

[1] D. Hadži and H. W. Thompson, Eds., *Hydrogen Bonding*, Pergamon Press, London, UK, 1959.

[2] L. Pauling, *The Nature of the Chemical Bond and the Structure of Molecules and Crystals: An Introduction to Modern Structural Chemistry*, Cornell University Press, Ithaca, NY, USA, 1960.

[3] G. C. Pimentel and A. L. McClellan, *The Hydrogen Bond*, W. H. Freeman, San Francisco, Calif, USA, 1960.

[4] S. N. Vinogradov and R. H. Linnel, *Hydrogen Bonding*, Van Nostrand-Reinhold, New York, NY, USA, 1971.

[5] M. D. Joesten and L. J. Schaad, *Hydrogen Bonding*, Marcel Dekker, New York, NY, USA, 1974.

[6] P. Schuster, G. Zundel, and C. Sandorfy, Eds., *The Hydrogen Bond. Recent Developments in Theory and Experiments*, vol. 1–3, North Holland, Amsterdam, The Netherlands, 1976.

[7] H. Ratajczak and W. J. Orwille-Thomas, Eds., *Molecular Interactions*, John Wiley & Sons, New York, NY, USA, 1980.

[8] P. L. Huyskens, W. A. P. Luck, and Th. Zeegers-Huyskens, Eds., *Intermolecular Forces: An Introduction to Modern Methods and Results*, Springer, Heidelberg, Germany, 1991.

[9] S. Scheiner, Ed., *Hydrogen Bonding. A Theoretical Perspective*, Oxford University Press, Oxford, UK, 1997.

[10] G. A. Jeffrey, *Introduction to Hydrogen Bonding*, Oxford University Press, Oxford, UK, 1997.

[11] D. Hadži, Ed., *Theoretical Treatments of Hydrogen Bonding*, Oxford University Press, Oxford, UK, 1997.

[12] G. R. Desiraju and T. Steiner, *The Weak Hydrogen Bond in Structural Chemistry and Biology*, Oxford University Press, Oxford, UK, 1999.

[13] Th. Elsaesser and H. J. Bakker, Eds., *Ultrafast Hydrogen Bonding Dynamics and Proton Transfer Processes in the Condensed Phase*, Kluwer Academic Publishers, Dordrecht, The Netherlands, 2002.

[14] S. J. Grabowski, Ed., *Hydrogen Bonding—New Insights*, Springer, Dordrecht, The Netherlands, 2006.

[15] Y. Maréchal, *The Hydrogen Bond and the Water Molecule: The Physics and Chemistry of Water Aqueous and Bio-Media*, Elsevier, Amsterdam, The Netherlands, 2007.

[16] G. Gilli and P. Gilli, *The Nature of The Hydrogen Bond: Outline of a Comprehensive Hydrogen Bond Theory*, Oxford University Press, Oxford, UK, 2009.

[17] P. L. Huyskens and Th. Zeegers-Huyskens, "Associations moleculaires et equilibres acide-base," *Journal de Chimie Physique*, vol. 61, aticle 84, 1964.

[18] P. Huyskens, L. Sobczyk, and I. Majerz, "On a hard/soft hydrogen bond interaction," *Journal of Molecular Structure*, vol. 615, no. 1–3, pp. 61–72, 2002.

[19] L. Sobczyk, "Softness of hydrogen bond interaction," *Khimicheskaya Fizika*, vol. 24, article 31, 2005.

[20] L. Sobczyk, "Quasi-symmetric O–H· · · N hydrogen bonds in solid state," *Molecular Physics Reports*, vol. 14, pp. 19–31, 1996.

[21] R. Nouwen and P. Huyskens, "Dipole moments and structure of the complexes of phenols with pyridines," *Journal of Molecular Structure*, vol. 16, no. 3, pp. 459–471, 1973.

[22] S. N. Smirnov, N. S. Golubev, G. S. Denisov, H. Benedict, P. Schah-Mohammedi, and H. H. Limbach, "Hydrogen/deuterium isotope effects on the NMR chemical shifts and geometries of intermolecular low-barrier hydrogen-bonded complexes," *Journal of the American Chemical Society*, vol. 118, no. 17, pp. 4094–4101, 1996.

[23] E. Grech, J. Kalenik, and L. Sobczyk, "^{35}Cl nuclear quadrupole resonance studies of pentachlorophenol-amine hydrogen-bonded complexes," *Journal of the Chemical Society, Faraday Transactions 1*, vol. 75, pp. 1587–1592, 1979.

[24] J. P. Castaneda, G. S. Denisov, and V. M. Schreiber, "Structure of 1 : 1 and 1 : 2 complexes formed by aromatic NH and OH proton donors with aliphatic amines. Possibility of homoconjugated NHN^+ cation formation," *Journal of Molecular Structure*, vol. 560, no. 1–3, pp. 151–159, 2001.

[25] I. Majerz, Z. Malarski, and L. Sobczyk, "Proton transfer and correlations between the C–O, O–H, N–H and O· · · N bond lengths in amine phenolates," *Chemical Physics Letters*, vol. 274, no. 4, pp. 361–364, 1997.

[26] Z. Malarski, M. Roepenk, E. Grech, and L. Sobczyk, "Dielectric and spectroscopic studies of pentachlorophenol-amine complexes," *Journal of Physical Chemistry*, vol. 86, no. 3, pp. 401–406, 1982.

[27] J. Kalenik, I. Majerz, L. Sobczyk, E. Grech, and M. M. M. Habeeb, "Infra-red and ^{35}Cl nuclear quadrupole resonance studies of hydrogen bonded adducts of 2-chlorobenzoic acid derivatives," *Collection of Czechoslovak Chemical Communications*, vol. 55, no. 1, pp. 80–90, 1990.

[28] G. Albrecht and G. Zundel, "Phenol–amine hydrogen bonds with large proton polarizabilities. Position of the OH· · · N $\rightleftharpoons$ O^{--} · · · H^+N equilibrium as a function of the donor and acceptor," *Journal of the Chemical Society, Faraday Transactions 1*, vol. 80, no. 3, pp. 553–561, 1984.

[29] V. M. Schreiber, A. Kulbida, M. Rospenk, L. Sobczyk, A. Rabold, and G. Zundel, "Temperature effect on proton-transfer equilibrium and IR spectra of chlorophenol-tributylamine systems," *Journal of the Chemical Society, Faraday Transactions*, vol. 92, no. 14, pp. 2555–2561, 1996.

[30] H. Baba, A. Matsuyama, and H. Kokubun, "Proton transfer in *p*-nitrophenol-triethylamine system in aprotic solvents," *Spectrochimica Acta Part A*, vol. 25, no. 10, pp. 1709–1722, 1969.

[31] J. E. Crooks and B. H. Robinson, "Hydrogen-bonded and ion-pair complexes in aprotic solvents," *Faraday Symposia of the Chemical Society*, vol. 10, pp. 29–40, 1975.

[32] H. Romanowski and L. Sobczyk, "Ultraviolet spectra and proton-transfer equilibria in 2,6-dichloro-4-nitrophenol-amine systems," *Journal of Physical Chemistry*, vol. 79, no. 23, pp. 2535–2542, 1975.

[33] M. M. Habeeb and M. A. Kharaba, "Intermolecular hydrogen bonds and proton transfer equilibrium in some nitro cresols-aliphatic amines-acetonitrile or methanol systems," *Journal of Molecular Liquids*, vol. 107, no. 1–3, pp. 205–219, 2003.

[34] M. M. Habeeb and R. M. Alghanmi, "Spectrophotometric study of intermolecular hydrogen bonds and proton transfer complexes between 1,2-dihydroxyanthraquinone and some aliphatic amines in methanol and acetonitrile," *Journal of Chemical and Engineering Data*, vol. 55, no. 2, pp. 930–936, 2010.

[35] Z. Dega-Szafran, E. Dulewicz, and M. Szafran, "Spectroscopic studies of N-methylpiperidine betaine complexes with phenols," *Journal of Molecular Structure*, vol. 704, no. 1–3, pp. 155–161, 2004.

[36] M. Rospenk, "The influence of steric effects of proton-transfer equilibrium in intramolecular hydrogen bonds," *Journal of Molecular Structure*, vol. 221, pp. 109–114, 1990.

[37] V. M. Schreiber, M. Rospenk, A. I. Kulbida, and L. Sobczyk, "Shaping of broad IR absorption in proton transfer equilibrating OH· · ·N hydrogen bonded systems," *Spectrochimica Acta—Part A*, vol. 53, no. 12, pp. 2067–2078, 1997.

[38] V. M. Schreiber, A. Koll, and L. Sobczyk, "Effect of temperature on the proton transfer equilibrium in the intramolecular hydrogen bond hydroxyl· · ·nitrogen," *Bulletin de l'Academie Polonaise des Sciences, Serie des Sciences Chimiques*, vol. 26, article 651, 1978.

[39] A. Koll, M. Rospenk, and L. Sobczyk, "Thermodynamic parameters for the proton-transfer reaction in Mannich bases," *Journal of the Chemical Society, Faraday Transactions 1*, vol. 77, no. 10, pp. 2309–2314, 1981.

[40] M. Rospenk, I. G. Ruminskaya, and V. M. Schreiber, "Elektronnye spektri i wnutrimolekularnij perekhod protona v osnovanyakh Mannikha v zhidkikh i tverdikh stekloobraznikh rastvorakh," *Journal of Applied Spectroscopy*, vol. 36, article 756, 1982.

[41] M. Rospenk, L. Sobczyk, A. Rabold, and G. Zundel, "Low temperature studies on ultraviolet and infrared spectra of ortho Mannich bases," *Spectrochimica Acta—Part A*, vol. 55, no. 4, pp. 855–860, 1999.

[42] I. Król-Starzomska, M. Rospenk, Z. Rozwadowski, and T. Dziembowska, "UV-visible absorption spectroscopic studies of intramolecular proton transfer in N-(R-salicylidene)-alkylamines," *Polish Journal of Chemistry*, vol. 74, no. 10, pp. 1441–1446, 2000.

[43] M. Rospenk, I. Król-Starzomska, A. Filarowski, and A. Koll, "Proton transfer and self-association of sterically modified Schiff bases," *Chemical Physics*, vol. 287, no. 1-2, pp. 113–124, 2003.

[44] A. Koll, M. Rospenk, L. Sobczyk, and T. Glowiak, "Properties of a strong intramolecular OHO hydrogen bond in 2-(N,N-diethylamino-N-oxymethyl)-4,6-dichlorophenol," *Canadian Journal of Chemistry*, vol. 64, no. 9, pp. 1850–1854, 1986.

[45] S. Kong, I. G. Shenderovich, and M. V. Vener, "Density functional study of the proton transfer effect on vibrations of strong (short) intermolecular O–H$\cdots$N/O$^-$$\cdots$H–N$^+$ hydrogen bonds in aprotic solvents," *Journal of Physical Chemistry A*, vol. 114, no. 6, pp. 2393–2399, 2010.

[46] T. Lankau and C. H. Yu, "Solubility of methane in water: The significance of the methane-water interaction potential," *Chemical Physics Letters*, vol. 424, article 264, 2006.

[47] P. Gilli, L. Pretto, and G. Gilli, "PA/pKa equalization and the prediction of the hydrogen-bond strength: a synergism of classical thermodynamics and structural crystallography," *Journal of Molecular Structure*, vol. 844-845, pp. 328–339, 2007.

[48] T. Lankau and C. H. Yu, "Correlated proton motion in hydrogen bonded systems: tuning proton affinities," *Physical Chemistry Chemical Physics*, vol. 9, no. 2, pp. 299–310, 2007.

[49] C. Reichardt, "Empirical parameters of the polarity of solvents," *Angewandte Chemie International Edition in English*, vol. 4, no. 1, pp. 29–40, 1965.

[50] C. Reichardt and K. Dimroth, "Solvents and empirical parameters for characterization of their polarity," *Fortschritte der Chemischen Forschung*, vol. 11, article 1, 1968.

[51] M. Habeeb, H. Alwakil, A. El-Dissouky, and H. Abdel-Fattah, "Spectroscopic studies of 1:1 chloranilic acid-amine complexes," *Polish Journal of Chemistry*, vol. 69, article 1428, 1995.

[52] O. V. Chranina, F. P. Czerniakowski, and G. S. Denisov, "UV-vis electrochromism due to proton transfer," *Journal of Molecular Structure*, vol. 177, pp. 309–315, 1988.

[53] W. Galezowski and A. Jarczewski, "Kinetics, isotope effects of the reaction of 1 (4 nitrophenyl)-1-nitroalkanes with DBU in tetrahydrofuran and chlorobenzene solvents," *Canadian Journal of Chemistry*, vol. 68, no. 12, pp. 2242–2248, 1990.

[54] A. Jarczewski, G. Schroeder, and K. T. Leffek, "The proton transfer reaction between bis(2,4-dinitrophenyl)methane and nitrogen bases in dimethyl sulfoxide and toluene solvents," *Canadian Journal of Chemistry*, vol. 69, no. 3, pp. 468–473, 1991.

[55] G. Zundel, "Hydrogen bonds with large proton polarizability and proton transfer processes in electrochemistry and biology," *Advances in Chemical Physics*, vol. 111, 2000.

[56] J. P. Hawranek and B. Czarnik-Matusewicz, "Infrared dispersion of H-bonded systems. The dielectric function for weak complexes," *Chemical Physics Letters*, vol. 109, no. 2, pp. 166–169, 1984.

[57] J. P. Hawranek and B. Czarnik-Matusewicz, "Infrared dispersion of the H-bonded pentachlorophenol-acetonitrile complex," *Chemical Physics Letters*, vol. 138, no. 5, pp. 397–400, 1987.

[58] J. P. Hawranek and B. Czarnik-Matusewicz, "IR dispersion of hydrogen bonded systems III. Pentachlorphenol—dioxane-D8 complex," *Journal of Molecular Structure*, vol. 143, no. C, pp. 337–340, 1986.

[59] B. Czarnik-Matusewicz and J. P. Hawranek, "Infrared dispersion of the hydrogen-bonded pentachlorophenol—3-chloropyridine complex," *Journal of Molecular Structure*, vol. 219, pp. 221–226, 1990.

[60] J. P. Hawranek, B. Czarnik-Matusewicz, and W. Wrzeszcz, "Infrared dispersion of the hydrogen-bonded pentachlorophenol-pyridine complex," *Journal of Molecular Structure*, vol. 322, pp. 181–186, 1994.

[61] J. P. Hawranek, J. Z. Flejszar-Olszewska, and A. S. Muszynski, "Infrared dispersion of the pentachlorophenol-sym-collidine complex," *Journal of Molecular Structure*, vol. 448, no. 2-3, pp. 149–159, 1998.

[62] J. P. Hawranek and A. S. Muszynski, "Infrared dispersion of the pentachlorophenol-trioctylamine complex," *Journal of Molecular Structure*, vol. 552, no. 1–3, pp. 205–212, 2000.

7

Polymorphism, Hydrogen Bond Properties, and Vibrational Structure of 1H-Pyrrolo[3,2-*h*]Quinoline Dimers

Alexandr Gorski,[1] Sylwester Gawinkowski,[1] Roman Luboradzki,[1] Marek Tkacz,[1] Randolph P. Thummel,[2] and Jacek Waluk[1]

[1] *Institute of Physical Chemistry, Polish Academy of Sciences, Kasprzaka 44/52, 01-224 Warsaw, Poland*
[2] *Department of Chemistry, University of Houston, Houston, TX 77204-5003, USA*

Correspondence should be addressed to Jacek Waluk, waluk@ichf.edu.pl

Academic Editor: Paul Blaise

Two forms of cyclic, doubly hydrogen-bonded dimers are discovered for crystalline 1H-pyrrolo[3,2-*h*]quinoline, a bifunctional molecule possessing both hydrogen bond donor and acceptor groups. One of the forms is planar, the other is twisted. Analysis of IR and Raman spectra, combined with DFT calculations, allows one to assign the observed vibrations and to single out vibrational transitions which can serve as markers of hydrogen bond formation and dimer structure. Raman spectra measured for samples submitted to high pressure indicate a transition from the planar towards the twisted structure. Formation of intermolecular hydrogen bonds leads to a large increase of the Raman intensity of the NH stretching band: it can be readily observed for the dimer, but is absent in the monomer spectrum.

1. Introduction

In studies of the intermolecular hydrogen bond (HB), an important class of model compounds consists of molecules which can form both H-bonded dimers and complexes with water or alcohols [1]. Such molecules are usually characterized by the simultaneous presence of HB donor and acceptor groups. Whether the strength of the intermolecular HBs is greater for dimers or complexes depends on the relative positions of the donor and acceptor in the molecular frame. Interestingly, different structures and stoichiometries are often encountered for the same molecule. A well-known example is 7-azaindole (7AI, Figure 1), which forms doubly hydrogen bonded dimers in solution [2], while the X-ray data reveal a tetrameric structure in the crystalline state [3]. Different stoichiometries and structures are possible for the complexes of 7AI with methanol and water: 1 : 1, 1 : 2, and 1 : 3 species have been detected [4–9].

The crystal structure of multiply H-bonded dimers/oligomers seems to be determined by the interplay of H-bonding and longer range intermolecular interactions. For instance, 1-azacarbazole (1AC), a molecule closely related to 7AI, exists in the crystal in the form of planar, doubly hydrogen bonded dimers [10] (Figure 2). While there is no doubt that 1AC also forms dimers in solutions, various possible structures have been discussed [11–14].

1H-pyrrolo[3,2-*h*]quinoline (PQ, Scheme 1) can be considered a counterpart to 7AI with regard to intermolecular HB characteristics. The NH group of PQ (HB-donor) and the pyridine nitrogen (HB-acceptor) are positioned three bonds apart, whereas in 7AI these groups are separated by two bonds. This change results in completely different excited state behaviour of complexes with water or alcohols [15–20]. Rapid photoinduced double proton transfer is observed for PQ in complexes of 1 : 1 stoichiometry. The process occurs on the time scale of single picoseconds and is not stopped by lowering of temperature or by increasing the viscosity of the medium. On the contrary, the reaction is slower and viscosity-dependent in 7AI complexes [21], since it requires a solvent rearrangement around an excited chromophore [22–27]. These different phototautomerization

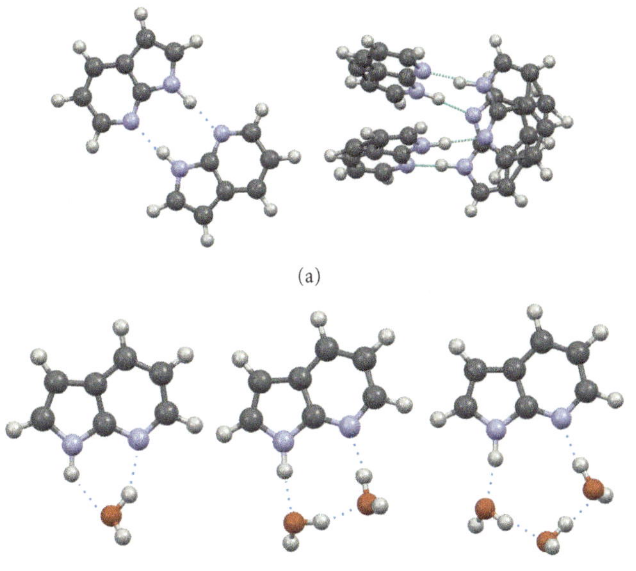

FIGURE 1: Various motifs of intermolecular HB formed by 7-azaindole. (a) Dimers and tetramers, (b) complexes with water.

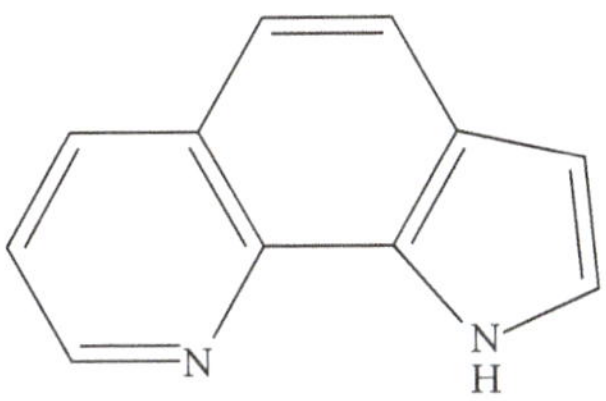

SCHEME 1

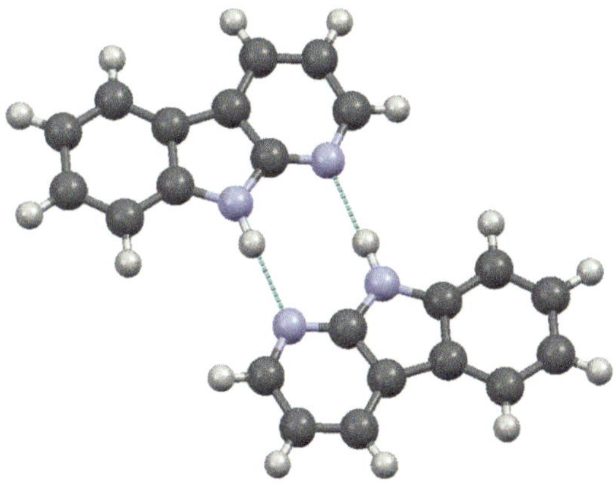

FIGURE 2: The structure of dimers of 1-azacarbazole in the solid phase.

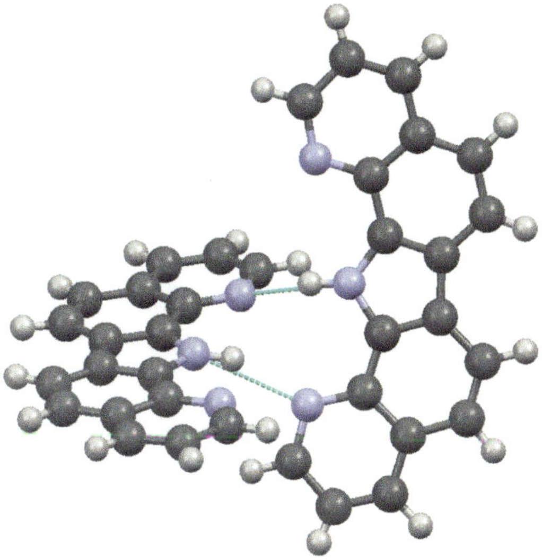

FIGURE 3: The X-ray structure of dipyrido[2,3-*a*:3,2-*i*]carbazole.

characteristics reflect different intermolecular HB strengths, imposed by molecular structure.

The HB characteristics, and, in consequence, tautomerization abilities in the dimeric species are expected to become reversed in PQ and 7AI. For the latter, a planar dimeric structure reveals two strong, linear, equivalent HBs. Therefore, it is not quite surprising that photoinduced double proton transfer in 7AI dimers has been observed at temperatures as low as 4 K [28]. In contrast, PQ dimers are predicted by theory to be nonplanar. This has been confirmed by X-ray studies, which reported an angle of 22.6° between the two monomeric units [29]. Our previous work on a similar structure, dipyrido[2,3-*a*:3,2-*i*]carbazole [30] demonstrated that in the crystalline phase this molecule forms cyclic, but strongly nonplanar doubly hydrogen-bonded dimers (Figure 3). No tautomeric fluorescence has been observed for such a dimer, but it could be readily detected when the crystalline sample was exposed to water vapor, prepared on a hydrophilic support, or embedded in a polymer containing hydroxyl groups. A general conclusion from this study was that HB-donor-acceptor molecules which readily form flat dimers should have a weak tendency for the formation of cyclic complexes, and *vice versa*.

In this work, we analyze structure and vibrational spectra of crystalline PQ dimers. Somewhat unexpectedly, our X-ray measurements of PQ reveal the existence of planar, doubly hydrogen-bonded dimeric species, and thus a structure very different than the one reported previously [29] (Figure 4). We analyze the experimental and theoretically predicted vibrational patterns, with particular interest regarding the vibrations involved in intermolecular hydrogen bonds. Finally, we show the influence of high pressure upon the HB strength, manifested by spectral shifts observed in the Raman spectra.

2. Experimental and Theoretical Details

Synthesis and purification of PQ have been described before [31].

The IR spectra were recorded on a Nicolet Magna 560 FTIR spectrometer, equipped with MCT/B liquid-nitrogen-cooled detector, with 1 cm^{-1} resolution. For the measurements of infrared spectra, thin polycrystalline PQ films were prepared on KBr or ZnSe windows by quick evaporation from a concentrated solution. The monomer IR

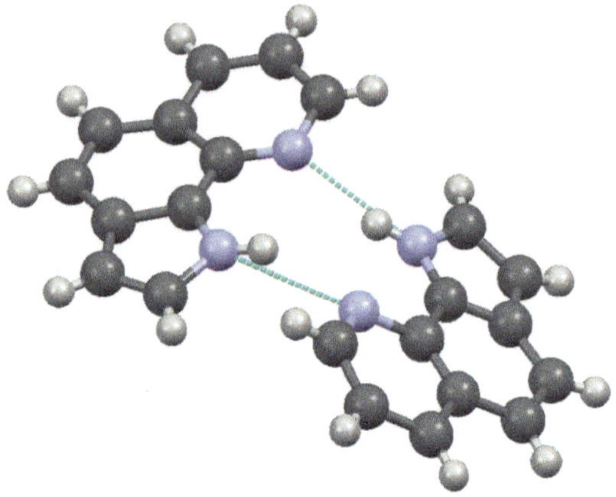

(a)

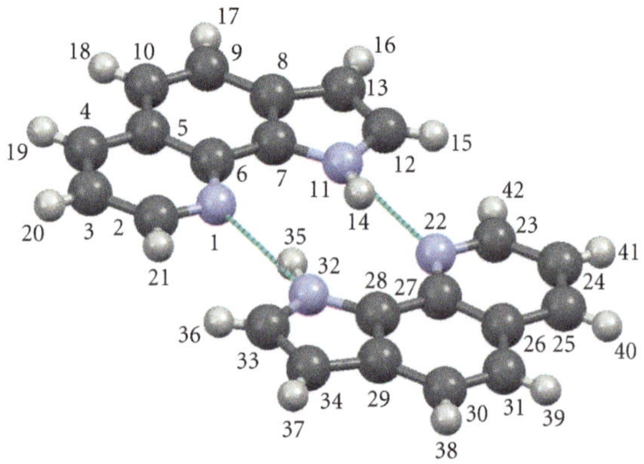

(b)

FIGURE 4: (a) X-ray structure of PQ reported in [29]; the geometrical positions of hydrogen atoms were inserted. (b) Our X-ray structure and atom numbering.

spectra have been recorded for PQ isolated in argon matrices using a closed-cycle helium cryostat (CSW-202N, Advanced Research Systems). The compound, contained in a glass tube, was heated to 350 K and codeposited with argon at a ratio of about 1 : 1000 onto a cold (20 K) KBr window mounted in a cryostat with 10^{-6} Torr background pressure. During spectral measurements the matrix temperature was kept at 10 K.

Renishaw inVia microscopic system was used for the measurements of Raman spectra. Ar^+ 514.5 nm (Stellar Pro Modu-Laser, LLC) laser line and a diode laser (HPNIR785) emitting 785 nm line were used as the excitation sources. With a $\times 100$ microscope objective the laser light was focused on a sample, the laser power at the sample being 5 mW or less. The Raman scattered light was collected by the same objective through a cut-off filter to block out Rayleigh scattering. Gratings of 1800 and 1200 grooves/mm were used for 514.5 and 785 nm laser lines, respectively. The resolution was 5 cm^{-1}, with the wavenumber accuracy of 2 cm^{-1}, both calibrated with the Rayleigh line and the 520.6 cm^{-1} line of silicon. The Raman scattered light was recorded by a 1024×256 pixel Peltier-cooled RenCam CCD detector.

High pressure experiments have been performed in Takemura type of diamond anvil cell [32]. The diameter of the diamond culet was 600 μm and a gasket made of stainless steel was used with 300 μm centrally drilled hole. Sample powder was loaded into the gasket hole without any pressure transmitting medium. Pressures were measured by recording the fluorescence spectrum of a small ruby chip embedded in the sample and converting the shift of the wavelength of the R_1 line to pressure, according to the scale proposed by Mao [33].

The samples of different polymorphs were prepared by quick crystallization by evaporation from concentrated PQ solutions in dichloromethane, diethyl ether, methanol, cyclohexane, and toluene.

For the X-ray studies, a colorless PQ crystal of approximate dimensions of $0.1 \times 0.2 \times 0.2\,mm^3$ was used. Diffraction data were collected at 100 K using a Bruker Kappa CCD diffractometer with graphite monochromated Mo Kα radiation. Structure was solved by direct methods (SHELXS-97) and refined on F^2 by full-matrix least-squares method (SHELXL-97) [34]. Formula is $C_{11}H_8N_2$, monoclinic, space group $P2_1/c$, $a = 9.0104(4)$, $b = 4.7302(1)$, $c = 19.3117(9)$ Å, $\beta = 103.1825(17)°$, $R_1 = 0.0449$ ($I > 2\sigma(I)$), $wR_2 = 0.1144$ for all data.

Unit cell parameters (but not the whole data) were also measured at room temperature, showing no significant differences compared with 100 K data ($a = 9.13$, $b = 4.87$, $c = 19.42$ Å, $\beta = 102.54°$, parameters not refined).

The crystallographic data have been deposited with the Cambridge Crystallographic Data Centre as a supplementary publication no. CCDC 868707. The data can be obtained free of charge at http://www.ccdc.cam.ac.uk or from the Cambridge Crystallographic Data Centre, 12 Union Road, Cambridge CB2 1EZ, UK.

Geometry optimizations were performed using density functional theory (DFT), with B3LYP functional and cc-PVTZ basis set, as implemented in Gaussian 09. This choice of functional/basis set was guided by extensive calculations for the PQ monomer which resulted in reliable assignments of nearly all of the vibrations.

In order to simulate the structure of PQ dimers in the crystalline environment, DFT-based quantum chemical calculations were performed using the CASTEP (Cambridge Serial Total Energy Package) computer code [35] in the framework of the generalized gradient approximation (GGA), as proposed by Perdew et al. [36], in combination with Vanderbilt ultrasoft pseudopotentials [37]. The plane wave basis set was truncated at a kinetic energy of 240 eV. Computations were performed over a range of k-points within the Brillouin zone as generated by the full Monkhorst-Pack scheme [38] with a $2 \times 2 \times 1$ mesh. A further increase of the cutoff energy and the number of k-points resulted in negligibly small changes in structure energies, indicating that the energy values are well converged. Two initial geometries of planar and twisted PQ dimers were taken from the X-ray data. In every case a slab including 16 molecules of PQ was constructed and repeated periodically.

3. Results

3.1. Dimer Structure. Geometry optimization performed for the PQ dimer yields a nonplanar, doubly H-bonded structure. The calculated nonplanar geometry agrees qualitatively with the X-ray data published in 1991 (Figure 4(a)). However, the quantitative differences are quite significant. The calculated twisting angle between the monomeric moieties, 45.6°, is much larger than the experimental one, 22.6°. For the separation of the H-bonded nitrogen atoms, the same value, 2.98 Å, is computed for both pairs, while the reported X-ray distances are very different, 2.92 and 2.99 Å. The calculations yield nearly planar monomeric units in the dimer, whereas the experiment clearly shows distortions. For instance, the experimental NCCN angles are 3.8° and 5.4°, while the calculations yield the same, smaller value of 2.5°. These results suggest that intermolecular interactions in the crystal may affect the dimer structure. We have therefore repeated the X-ray measurements, performing experiments both at 293 K and at lower temperatures. Surprisingly, a different structure than previously reported was obtained (Figure 4(b)), consisting of two doubly H-bonded PQ monomeric units in a planar arrangement. In order to obtain a theoretical model for the planar dimer, we imposed the planarity in the optimization procedure. This resulted in one negative frequency in the optimized structure, the vibration corresponding to mutual twisting of the planar moieties. Computationally, the planar geometry shows the NN distance of 3.12 Å, whereas the experimental value is 3.01 Å.

One can conclude that PQ forms polymorphs in the crystal, which differ in the structure of dimers, especially with regard to parameters usually considered important for the strength of intermolecular hydrogen bond. Therefore it seemed interesting to carry out vibrational spectroscopy studies in order to (i) determine how does the formation of a doubly H-bonded dimer affect the vibrational pattern and (ii) probe the possible differences in the vibrational structure between planar and nonplanar (but both doubly H-bonded) dimers.

3.2. IR Measurements. Figure 5 presents the IR spectra recorded for the monomeric PQ isolated in an Ar matrix and the spectra of polycrystalline PQ, corresponding to the planar dimeric structure, measured on a KBr window. The experimental data are compared with the results of calculations performed for the monomer and for the two forms of the dimer: a fully-optimized non-planar structure and a form with imposed planar geometry.

The spectra of monomeric PQ are very well reproduced by calculations with regard to both band positions and intensities. They will be treated in detail in a separate work, in which the combination of theoretical modelling, IR, Raman, and high resolution fluorescence spectra obtained for supersonic jet-isolated PQ allowed reliable assignments of nearly all of 57 vibrations of monomeric PQ. Here, we focus on the dimer, using the monomer vibrations as a starting point. Figure 5 shows that, while the general pattern of the IR spectrum of dimeric PQ roughly resembles that

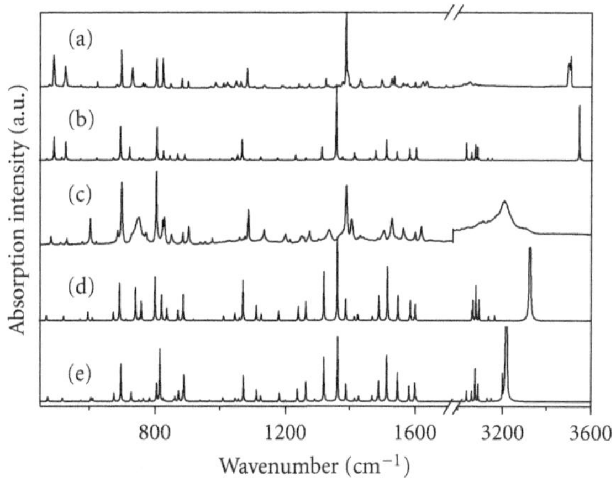

FIGURE 5: (a) IR spectrum of the monomer in Ar matrix at 15 K; (b) simulated monomer spectrum; (c) dimeric, polycrystalline PQ at 293 K; results of calculations performed for the planar (d) and twisted (e) dimer. The scaling factor of 0.9682 was used in calculations.

of the monomer, significant differences are observed in specific regions. The largest difference is observed for the NH stretching mode. The monomer peak observed around 3500 cm^{-1} (the observed triplet is due to argon site structure) disappears in the crystalline sample, where a broad band is detected, centered at 3210 cm^{-1}. This red shift of almost 300 cm^{-1} is characteristic for the formation of fairly strong NH· · ·N intermolecular hydrogen bonds. The calculations predict the shifts of 320 and 220 cm^{-1} for the twisted and planar forms, respectively. As expected, the larger shift is computed for a structure with a shorter N–N distance, and thus a stronger hydrogen bond. The better agreement with experiment for the larger value is somewhat misleading, since the X-ray measurements demonstrated that the sample corresponded to a planar dimer. Further arguments are provided by the analysis of the IR spectrum in the energy region corresponding to out-of-plane vibrations. For the monomeric PQ, calculations yield two modes that contain significant NH out-of-plane contributions. They can be readily identified in the experimental spectrum as the bands at 491 and 527 cm^{-1}. In the IR spectrum of a dimer these bands are still observed, but, in addition, a broad band appears at 743 cm^{-1}, in nice agreement with calculations, which predict for a planar structure a transition at 734 cm^{-1}. For the twisted dimer structure, there no longer exist pure "out-of-plane" modes. The mode which still retains much of that character is predicted to lie at 807 cm^{-1} and to have an intensity twice that of the planar structure. Comparison of the experimental and simulated IR spectra in the region of 650–950 cm^{-1} leaves no doubt that the observed spectrum originates from a planar species. The value of the blue shift of the NH out-of-plane bending mode, which exceeds 200 cm^{-1}, again points to a strong intermolecular HB in dimeric PQ.

There is no single particular vibration in the monomer which could be assigned to a pure NH in-plane bending mode. This is also true for the dimer. The IR transitions

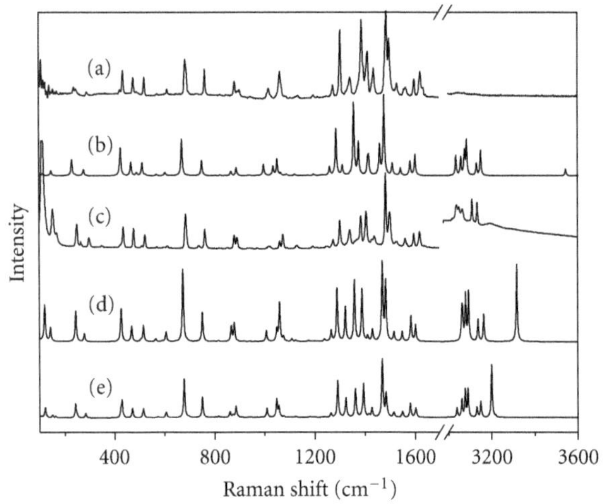

FIGURE 6: (a) Raman spectrum of the monomer in Ar matrix at 15 K; (b) simulated monomer spectrum; (c) dimeric, polycrystalline PQ, measured at 293 K; results of calculations performed for the planar (d) and twisted (e) dimer. The scaling factor of 0.9682 was used in calculations.

computed for the planar dimer consist of symmetric and antisymmetric combinations of the monomer modes. Only the latter are IR-active. A very similar pattern of IR transitions is obtained for the nonplanar dimer. Figure 5 shows that in the region above $1000\,cm^{-1}$ the predicted IR spectra for both planar and twisted dimer are almost identical.

The analysis of the IR spectra demonstrates that both NH stretching and out-of-plane bending modes are efficient markers for the HB formation. However, only the latter can be used to indicate the planar structure of the H-bonded dimer.

3.3. Raman Spectra. Comparison of Raman spectra simulated and measured for monomeric and dimeric PQ is presented in Figure 6. Contrary to the case of the IR spectra, the calculations now predict differences between monomeric and dimeric species in the region above $1200\,cm^{-1}$. Below that value, the simulated spectra are very similar for the three species. But even above $1200\,cm^{-1}$, the Raman spectra computed for planar and twisted dimers resemble each other very strongly, excluding their use for structure determination.

The calculations predict that the Raman activity of the NH stretching mode should be drastically increased, about 15 times, upon HB formation. This increase was confirmed by experiment. No band corresponding to the NH stretch was observed for monomeric PQ, but it could be readily detected at $3200\,cm^{-1}$ for the crystalline sample. Thus, formation of the intermolecularly H-bonded dimer enhances the polarizability to a degree that enables observation of a vibrational feature characteristic of the hydrogen bond.

3.4. Vibrational Assignments. Based on IR and Raman spectra and the results of calculations, we present in Table 1 the tentative assignments for the vibrations of dimeric PQ. The experimental data given in the Table correspond to the planar structure, whereas the calculations are given for both planar and twisted forms. Since the planarity was artificially imposed in the calculations, one might expect that the results in this case are less reliable. Still, as can be seen from Figures 5 and 6, the calculated vibrational patterns are very similar, both for IR and Raman spectra. The largest differences are observed for the NH stretching and out-of-plane bending modes, which were specifically discussed above.

3.5. Obtaining Different Polymorphic Forms. As already mentioned, the crystalline samples of PQ which we have examined by X-ray, IR and Raman techniques corresponded to planar dimers, and thus to a different polymorphic form than observed previously [29]. We have tried to obtain both forms by crystallization from different solvents, and then using Raman spectroscopy as a tool for structure determination. A trial and error approach was adopted, since no information about crystallization details was given in the work reporting the twisted structure [29]. Figure 7 presents the Raman spectra measured for samples crystallized from five different solvents. The spectra are similar, but significant differences can be detected in two regions. A peak of weak intensity appears at $738\,cm^{-1}$ for PQ crystallized from cyclohexane, toluene, and methanol, but not from diethyl ether and dichloromethane. The second region corresponds to two fairly strong peaks observed at 1062 and $1074\,cm^{-1}$. Their relative intensity patterns (a more intense feature lying at higher energy) are the same for the samples revealing the $738\,cm^{-1}$ transition. For two other samples, which lack the $738\,cm^{-1}$ peak, the intensity ratio changes: now the lower energy peak becomes higher. Such behavior strongly suggests that the PQ samples obtained from cyclohexane, toluene, and methanol correspond to planar dimers, whereas those crystallized from diethyl ether and dichloromethane, to the nonplanar ones. This is confirmed by the results of calculations, which predict exactly such reversal of the relative intensity pattern for the 1062 and $1074\,cm^{-1}$ peaks upon going from a planar to a twisted dimeric form (see Figure 6).

3.6. High-Pressure Experiments. The idea behind spectral measurements for samples submitted to high pressures was to observe pressure-induced changes in the strength, and possibly also of the structure of the intermolecular hydrogen bond. Figure 8 shows the Raman spectra recorded for PQ dimers under normal and elevated pressures. Nearly all peaks observed below $1700\,cm^{-1}$ evolve in a similar way with increasing pressure: the maxima shift to the blue by 5–$8\,cm^{-1}$. Much larger shifts towards higher transition energies are detected for the CH stretching bands, which shift by $30\,cm^{-1}$ or more. A reversal of the relative intensities is observed for the bands at 3114 and $3137\,cm^{-1}$. All these changes are reversible, as shown by comparison of the spectra recorded for the same sample before and after going through the high pressure cycle.

The effects most relevant to this work are related to changes in the HB strength and structure. Figure 8 shows that the NH stretching band, observed at $3200\,cm^{-1}$, moves to

TABLE 1: Comparison of the experimental IR and Raman spectra with the vibrational frequencies calculated for the twisted and planar forms of the PQ dimer.

	Calculated[a]						Observed				
	frequency[e] (cm^{-1})	frequency[f] (cm^{-1})	IR intensity[g] (km/mol)		Raman activity[g] (A^4/amu)		IR[b]		Raman[c]		Assignment[d]
							3362				
ν_1	3218.9	3218.8	2578.38	(1802.54)	28.41	(0)	3180[h]	s			as NH str
ν_2	3201.3	3211.2	48.28	(0.01)	1126.68	(1032.01)			3193[h]	s	s NH str
ν_3	3151.3	3065.3	4.97	(7.27)	49.50	(53.58)					as CH str
ν_4	3151.3	3065.3	0.52	(1.43)	281.36	(272.92)			3137	w	s CH str
ν_5	3133.0	3039.3	0.75	(0.64)	164.25	(237.43)			3112	w	s CH str
ν_6	3133.0	3039.3	5.02	(6.87)	39.47	(22.30)					as CH str
ν_7	3091.4	2996.7	4.88	(0.14)	453.54	(555.04)			3067	w	s CH str
ν_8	3091.4	2996.6	29.73	(36.36)	95.36	(2.32)					as CH str
ν_9	3078.3	2983.2	1.61	(0.01)	501.64	(515.94)			3052	w	s CH str
ν_{10}	3078.2	2983.1	63.42	(60.51)	27.11	(0.05)					as CH str
ν_{11}	3062.4	2970.2	0.17	(17.58)	302.47	(121.90)			3040	w	s CH str
ν_{12}	3062.4	2969.9	19.75	(17.35)	24.36	(125.43)					as CH str
ν_{13}	3059.7	2966.6	2.18	(8.51)	2.32	(113.18)					as CH str
ν_{14}	3059.6	2966.4	0.33	(6.59)	12.54	(140.11)			3019	sh	s CH str
ν_{15}	3039.2	2964.5	4.87	(0.27)	148.24	(30.94)			3000		s CH str
ν_{16}	3039.1	2964.5	15.73	(0.38)	28.75	(42.60)					as CH str
							1660	m			
							1632	m			
ν_{17}	1602.8	1551.4	6.25	(0)	50.40	(53.35)			1620	m	NH s b, CC str cr
ν_{18}	1598.7	1548.4	36.68	(27.79)	10.29	(0)	1615	s			NH as b,CC str cr
ν_{19}	1581.1	1534.2	27.87	(34.95)	13.16	(0.07)	1594	m			(CC, CN) as str pyridine
ν_{20}	1580.3	1533.4	4.33	(0.03)	66.13	(81.74)			1595	m	(CC, CN) s str pyridine
ν_{21}	1549.9	1499.8	0.99	(0.01)	28.48	(30.40)			1562	m	NH, CH s b pyridine
ν_{22}	1545.9	1497.7	57.03	(43.42)	6.11	(0.01)	1560	m			NH, CH as b pyridine
ν_{23}	1515.0	1468.1	4.14	(0)	21.16	(27.48)			1528	w	NH, CH18,20,41,39 s b
ν_{24}	1512.5	1466.6	90.78	(95.24)	5.73	(0)	1524	s			NH, CH18,20,41,39 as b
ν_{25}	1488.0	1441.3	40.20	(42.88)	17.42	(0)	1497	m			NH as b, CC str pyr
ν_{26}	1484.3	1436.4	4.55	(0)	111.20	(172.84)			1500	s	NH s b, CC str pyr
ν_{27}	1469.8	1423.5	1.32	(0)	243.81	(225.53)			1484	vs	CH20 s b, skel def CC
ν_{28}	1468.1	1422.0	9.96	(7.03)	56.54	(0)	1482	w			CH20 as b, skel def CC
ν_{29}	1427.9	1385.4	0.77	(0)	37.40	(32.30)			1440	w	NH, CH20,21,41,42 s b, skel def pyr
ν_{30}	1427.2	1379.4	9.25	(10.49)	6.91	(0)	1435	w			NH, CH20,21,41,42 as b, skel def pyr
ν_{31}	1414.9	1369.1	4.96	(6.27)	0.97	(0)	1428	w			CH17,18,19,20,38,39,40,41 as b, CC str cr
ν_{32}	1414.7	1366.9	0.84	(0)	9.32	(13.32)			1430	m	CH17,18,19,20,38,39,40,41 s b, CC str cr
ν_{33}	1394.6	1345.2	0.04	(0)	152.39	(137.61)			1407	m	NH, CH21,42 s b, CC str pyr
ν_{34}	1388.7	1343.6	33.68	(37.34)	17.26	(0.01)	1403	m			NH, CH21,42 as b, CC str pyr,
ν_{35}	1363.7	1318.7	124.83	(146.01)	22.36	(0)	1386	s			skel def, CH as b
ν_{36}	1361.7	1315.9	20.55	(0)	111.09	(157.34)			1386	s	skel def, CH s b
ν_{37}	1324.1	1280.6	6.01	(0)	78.48	(84.13)			1341	m	skel def, CH s b
ν_{38}	1320.2	1277.8	87.62	(85.48)	10.37	(0)	1333	m			skel def, CH as b
ν_{39}	1291.9	1249.4	3.58	(1.30)	32.85	(0.26)	1301	vw			skel def, CH as b
ν_{40}	1291.3	1248.8	0.77	(0)	119.86	(125.70)			1302	m	skel def, CH s b
ν_{41}	1263.8	1225.6	1.51	(0)	13.98	(24.15)			1275	m	CH s b, C7N11, C28N32, C8C9, C29C30 str
ν_{42}	1263.4	1222.7	37.25	(32.35)	1.74	(0)	1268	m			CH as b, C7N11, C28N32, C8C9, C29C30 str

Table 1: Continued.

	Calculated[a]						Observed				
	frequency[e] (cm^{-1})	frequency[f] (cm^{-1})	IR intensity[g] (km/mol)		Raman activity[g] (A^4/amu)		IR[b]		Raman[c]		Assignment[d]
ν_{43}	1239.0	1199.9	1.23	(24.33)	2.28	(0.02)			1251	w	NH and CH s b
ν_{44}	1236.5	1199.8	24.70	(0.11)	0.43	(4.81)	1243	m			NH and CH as b
ν_{45}	1199.6	1160.3	1.67	(0.57)	0.25	(0)	1210	w			CH as b, CC str cr
ν_{46}	1199.5	1159.7	0.03	(0)	1.27	(1.08)			1210	w	CH s b, CC str cr
ν_{47}	1180.7	1140.7	0.58	(16.59)	2.52	(0)			1193	w	CH s b, CC str
ν_{48}	1180.6	1138.8	15.28	(0)	0.09	(2.07)	1193	m			CH s b, CC str
ν_{49}	1123.7	1089.2	8.81	(10.81)	0.66	(0.01)	1133	m			CH17,18,19,20 and CH38,39,40,41 as b
ν_{50}	1123.5	1089.1	1.71	(0.03)	2.24	(2.15)			1132	w	CH17,18,19,20 and CH38,39,40,41 s b
ν_{51}	1112.3	1074.6	0.50	(25.90)	1.01	(0)					CH15,17,18,20 and CH36,38,39,41 and NH s b
ν_{52}	1110.6	1074.1	22.51	(0)	0.23	(4.96)	1121	m			CH15,17,18,20 and CH36,38,39,41 and NH as b
ν_{53}	1075.6	1041.7	0.14	(0)	3.42	(8.06)			1090	w	skel def, CH s b
ν_{54}	1071.0	1035.3	51.41	(69.84)	0.35	(0)	1082	s			skel def, CH as b
ν_{55}	1056.8	1026.7	0.02	(0)	30.98	(69.48)			1075	s	CH15,16,36,37 s b
ν_{56}	1055.9	1024.6	4.03	(3.50)	2.00	(0)	1065	w			CH15,16,36,37 s b
ν_{57}	1047.8	1015.9	0.86	(0)	53.49	(21.34)			1062	m	skel def, CH as b
ν_{58}	1046.4	1012.3	5.31	(11.68)	2.32	(0)	1058	w			skel def, CH as b
ν_{59}	1008.9	978.5	5.64	(6.69)	3.18	(0)	1025	w			skel def
ν_{60}	1008.6	975.5	0.60	(0)	24.42	(17.84)			1019	w	skel def
ν_{61}	968.6	931.8	1.25	(1.27)	0.15	(0)	972	vw	973	vw	CH19-21 and CH40-42 s "oop" twisting
ν_{62}	968.6	931.7	0.32	(0)	0.04	(0.24)	965	vw	969	vw	CH19-21 and CH40-42 as "oop" twisting
ν_{63}	944.2	909.4	0.03	(0)	0.26	(0.60)			951	w	CH17,18,21 and CH38,39,42 as "oop" wag
ν_{64}	944.1	909.4	0.11	(0.38)	0.13	(0)					CH17,18,21 and CH38,39,42 as "oop" wag
ν_{65}	938.9	887.5	0.86	(1.71)	0.40	(0.11)	946	w			CH17-19,21 and CH38-40,42 as "oop" twisting
ν_{66}	938.8	887.3	0.76	(0.28)	0.41	(0.66)			942	w	CH17-19,21 and CH38-40,42 as "oop" twisting
ν_{67}	888.4	857.1	51.13	(44.91)	2.84	(0)	899	m			as skel def pyr (N11-C12-C13)
ν_{68}	885.0	851.4	16.47	(0)	26.78	(25.25)			890	m	s skel def pyr (N11-C12-C13)
ν_{69}	872.9	841.8	10.51	(17.27)	4.86	(0)	882	m			skel def, NH s twisting
ν_{70}	871.3	839.8	16.74	(0)	1.46	(20.52)					skel def, NH as twisting
ν_{71}	860.9	809.0	9.07	(0)	13.06	(0.03)			853	w	CH15,16,36,37 and s "oop" wag
ν_{72}	857.2	808.8	2.82	(21.39)	0.83	(0)	860	vw			CH15,16,36,37 and s "oop" wag
ν_{73}	829.1	793.1	1.31	(43.69)	0.63	(0)					s skel "oop" def, NH, CH wag
ν_{74}	823.8	789.6	5.18	(0)	0.65	(1.24)					as skel "oop" def, NH, CH wag
ν_{75}	814.6	774.0	103.69	(76.52)	2.72	(0)	823	m	826	w	s NH "oop"
ν_{76}	805.0	771.8	9.08	(0)	0.45	(0.73)					as cr "oop" def, CH as wag
ν_{77}	804.3	735.0	29.08	(0)	0.76	(0.29)	801	s			s cr "oop" def, CH s wag
ν_{78}	782.9	733.5	5.92	(33.11)	0.03	(0)					as NH "oop"
ν_{79}	764.2	728.2	5.35	(0)	0.23	(32.85)					CH17-21,38-42 s "oop" wag
ν_{80}	763.1	727.2	0.29	(2.50)	0.05	(0)					CH17-21,38-42 s "oop" wag
ν_{81}	751.2	716.8	1.01	(57.55)	36.36	(0)			763	m	s "ip" skel def
ν_{82}	750.8	699.1	1.33	(0)	3.39	(0.26)	773	m	770	sh	as "ip" skel def
ν_{83}	727.8	688.3	17.14	(1.95)	0.14	(0)	738	m	739	w	CH15-18 and CH36-39 "oop" s wag
ν_{84}	726.3	669.7	2.58	(65.15)	0.02	(0)					CH15-18 and CH36-39 "oop" as wag
ν_{85}	696.2	668.3	61.18	(0)	1.87	(0.88)					s "oop" skel def, CH s wag
ν_{86}	696.1	656.9	13.35	(0)	0.05	(0.71)					as "oop" skel def, CH s wag

Table 1: Continued.

	Calculated[a]						Observed				Assignment[d]
	frequency[e] (cm^{-1})	frequency[f] (cm^{-1})	IR intensity[g] (km/mol)		Raman activity[g] (A^4/amu)		IR[b]		Raman[c]		
ν_{87}	678.6	653.4	0.03	(0)	64.70	(70.42)	684	vw	686	s	s cr and pyridine ring b (sym along N1-C4, C7-C10 axis)
ν_{88}	675.2	651.1	15.94	(14.61)	1.87	(0)	696	m	696	sh	as cr and pyridine ring b (sym along N1-C4, C7-C10 axis)
ν_{89}	610.6	589.1	5.28	(4.30)	0.01	(0)					as cr and pyridine ring b (asym along N1-C4, C7-C10 axis)
ν_{90}	607.0	587.8	1.15	(0)	6.13	(8.11)	615	vw	614	w	s cr and pyridine ring b (asym along N1-C4, C7-C10 axis)
ν_{91}	605.3	575.9	0.89	(14.68)	0.58	(0)					as "oop" skel def pyr
ν_{92}	604.5	572.2	6.17	(0)	2.14	(0.68)			602	w	s "oop" skel def pyr
ν_{93}	573.7	552.4	0.75	(2.39)	0.53	(0)					as "oop" skel def pyridine and cr
ν_{94}	573.5	547.6	0.55	(0)	0.98	(2.04)			571	w	s "oop" skel def pyridine and cr
ν_{95}	516.2	502.9	5.97	(7.19)	1.51	(0)					as "ip" skel def pyridine and cr
ν_{96}	516.1	500.6	0.74	(0)	8.04	(10.89)			524	m	s "ip" skel def pyridine and cr
ν_{97}	513.7	490.9	0.53	(0)	1.51	(1.03)			513	w	s "oop" skel def pyridine and cr
ν_{98}	511.6	485.2	0.17	(0.81)	0.72	(0)					as "oop" skel def pyridine and cr
ν_{99}	472.8	455.0	8.28	(0)	1.82	(9.13)					as "ip" skel def cr
ν_{100}	471.5	453.5	0.17	(8.39)	8.11	(0)			479	m	s "ip" skel def cr
ν_{101}	430.1	414.5	0.54	(8.03)	11.61	(0)			438	m	s "ip" skel def cr, CH17,19,38,40 "oop"
ν_{102}	429.9	414.3	7.54	(0)	2.59	(15.89)			432	sh	as "ip" skel def cr, CH17,19,38,40 "oop"
ν_{103}	426.3	411.4	0.52	(0)	1.64	(2.25)					as "oop" skel def cr, CH17,19,38,40 "oop"
ν_{104}	425.1	410.3	0.58	(2.00)	4.49	(0)			424	w	s "oop" skel def cr, CH17,19,38,40 "oop"
ν_{105}	284.4	273.3	1.89	(1.93)	0.81	(0)					as "oop" pyridine and cr rock
ν_{106}	283.5	270.9	1.37	(0)	1.34	(2.06)			300	w	s "oop" pyridine and cr rock
ν_{107}	252.8	244.7	0.65	(6.73)	0.04	(0)					as "oop" pyr and cr rock
ν_{108}	249.5	238.0	6.20	(6.29)	0.17	(0)					s "oop" pyr and cr rock
ν_{109}	248.8	237.5	3.53	(0)	0.52	(6.82)			267	w	as pyr and pyridine rings "ip" bend
ν_{110}	243.3	231.8	0.07	(0)	5.18	(0.05)			251	m	s pyr and pyridine rings "ip" bend
ν_{111}	163.0	141.8	0.01	(0.04)	0.27	(0)			169	w	as pyr and pyridine tor
ν_{112}	150.4	139.1	0.03	(0)	0.43	(1.24)			154	m	s pyr and pyridine tor
ν_{113}	122.3	117.0	6.27	(0)	0.23	(2.44)					as "oop" pyridine and pyr rock
ν_{114}	121.9	112.3	3.70	(6.00)	0.91	(0)					s "oop" pyridine and pyr rock
ν_{115}	80.2	75.9	3.58	(1.02)	3.54	(0)					dim rock
ν_{116}	78.2	71.4	0.02	(0)	1.54	(1.80)					dim b
ν_{117}	67.0	59.6	0.00	(0)	1.15	(3.12)					dim b
ν_{118}	43.4	21.3	0.18	(0.12)	9.82	(0)					dim rock
ν_{119}	26.1	9.2	0.27	(0)	4.96	(15.09)					dim rock
ν_{120}	21.3	−35.3	0.04	(0)	11.69	(0)					dim tor

[a]B3LYP/cc-pVTZ, C_2 symmetry group, scaling factor = 0.9682, as recommended in the literature [39].
[b]Polycrystalline sample, 293 K.
[c]Polycrystalline sample, 293 K, 785 nm laser (633 nm was used in the NH region).
[d]Abbreviations. s: symmetric; as: antisymmetric; str: stretch; b: bend; ip: in-plane; oop: out-of-plane; skel def: skeletal deformation; tor: torsion; pyrid: pyridine; pyr: pyrrole; cr: central ring.
[e]Twisted dimer.
[f]Planar dimer.
[g]In parentheses: values computed for the planar dimer.
[h]Very broad (~200 cm^{-1}).

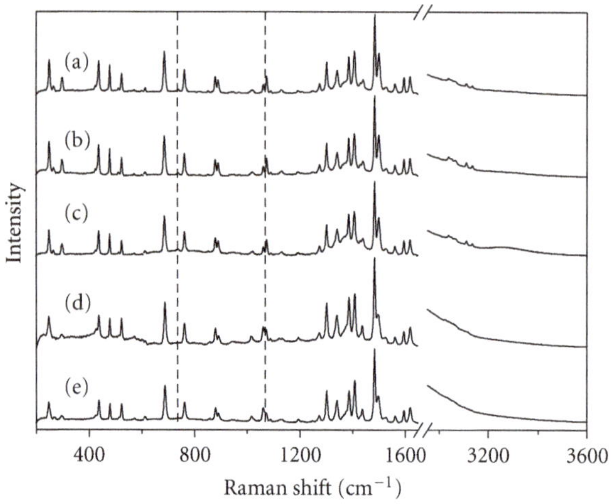

FIGURE 7: Raman spectra measured for samples crystallized from five different solvents: cyclohexane (a), methanol (b), toluene (c), dichloromethane (d), diethyl ether (e). Dashed vertical lines indicate regions with structure-sensitive transitions (see text).

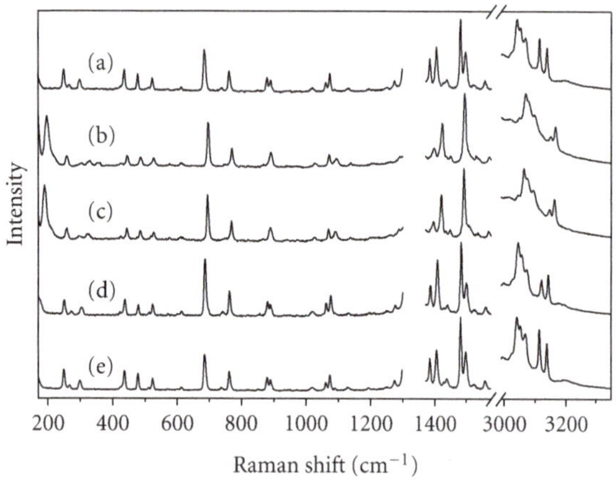

FIGURE 8: Raman spectra of crystalline PQ as a function of pressure: normal pressure, 1 atm (a); 2×10^3 atm (b); 22×10^3 atm (c); 35×10^3 atm (d); 1 atm (e), at the end of pressure cycle. The low and high frequency regions are normalized separately to their highest bands. A region between 1300–1370 cm^{-1}, exhibiting a strong Raman peak from diamond culets, was removed.

the red with increasing pressure. Such behavior is opposite to that of other modes and indicates the increase of the HB strength, most probably due to a shorter NH· · ·N distance. Unfortunately, the exact amount of the shift cannot be determined, as the band becomes buried under the transitions corresponding to CH stretches. Experiments are planned with either N- or C-deuterated PQ, to avoid interferences of NH/ND vibrations with other modes.

The second effect is the change in the relative intensity pattern with increasing pressure, observed for the peaks at 1062 and 1074 cm^{-1}. As discussed above, such behaviour can indicate a transition from a cyclic toward a twisted structure. For another mode diagnostic in this respect, 738 cm^{-1}, we observe decreasing intensity. However, it can still be detected at the highest pressures applied. It may be that what is

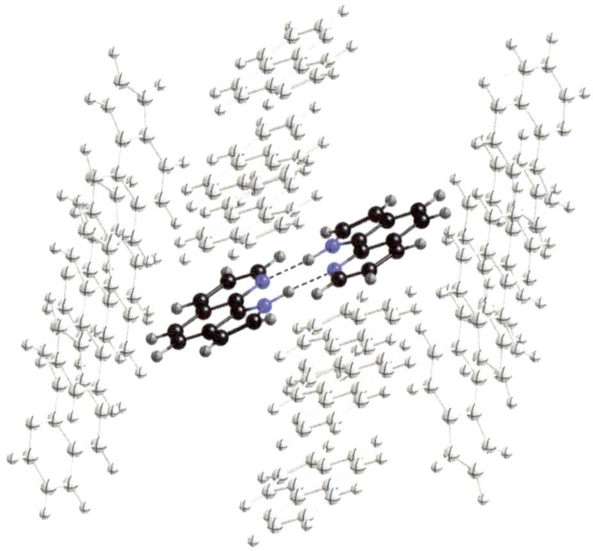

FIGURE 9: PQ dimer surrounded by identical neighbors (taken from X-ray data). The dimer in the middle was being distorted along the twisting coordinate, and then the whole structure was optimized.

observed is gradual twisting, not necessarily leading to the same angle between the monomeric units as observed for the nonplanar polymorph under normal pressure. More detailed investigations are planned once both planar and twisted dimeric samples are available. The experiments described in the previous section bode well for such studies.

3.7. Simulations of Polymorphic Structures. The existence of both planar and twisted dimers leads to the question of the energy barrier separating the two phases. Theoretical simulations have been carried out, in order to check the local minimum character of each structure and to estimate their relative stabilities. In this procedure, a dimer, surrounded by 14 identical neighbours (Figure 9), was distorted towards the structure of the other polymorph (twisted for the initially planar form, and vice versa). The whole ensemble was then optimized. Both planar and twisted structures relaxed back to the initial form, showing that they correspond to the minimum and providing additional independent confirmation of the existence of two crystal polymorphic forms of the PQ dimer. These results indicate that a collective rather than local distortion of the crystal is required for the phase change in PQ.

In agreement with the high pressure experiments, comparison of energies calculated for the slab consisting of 16 molecules for both planar and twisted dimers revealed a lower energy for the latter.

4. Summary and Conclusions

A combination of X-ray, IR and Raman spectroscopy, high pressure techniques, and quantum chemical calculations resulted in the detection of two polymorphic forms of dimeric PQ. Both types of dimer reveal a cyclic, doubly hydrogen-bonded structure, but differ in the planar versus twisted arrangement of the monomeric units. The calculations predict a twisted dimer structure, whereas imposing planarity results in one negative vibrational frequency, corresponding to the twisting coordinate. These results show that the isolated dimer should be nonplanar and thus the polymorphism is due to the interplay of interactions between the two monomeric units forming the hydrogen bond and dimer-dimer interactions in the crystal. The experiments indicate that upon applying pressure the planar form can be converted into the twisted one.

The NH stretching and out-of-plane bending modes observed in the IR spectra were shown to be clear indicators of the HB formation. The analysis of the position of the latter could be used to determine the structure of the H-bonded dimer. With respect to the influence of HB formation on the Raman spectra, a large increase of the intensity was observed for the NH stretching band in the H-bonded dimers, indicating increase of polarizability. The Raman spectra were also diagnostic for structural assignments: even though the spectra are quite similar, the intensity ratio of two peaks observed at 1062 and 1074 cm^{-1} provides information whether the PQ dimer is planar or not.

Our future plans include testing a possibility of photoinduced double proton transfer in both forms of crystalline PQ. Both kinetics and thermodynamics of such a process should be strongly structure-sensitive. Moreover, we have selected PQ as one of the objects in the investigations of the influence of plasmonic structures on the spectral and photophysical characteristics of chromophores located in the vicinity of metallic environments. The results of vibrational and structural analysis presented in this work will provide a starting point for experiments in which monomers and dimers of PQ will be placed on, or close to metal surfaces.

Acknowledgments

The work was supported by the Grant 3550/B/H03/2011/40 from the Polish National Science Centre. The authors acknowledge the computing grant G17-14 from the Interdisciplinary Centre for Mathematical and Computational Modeling of the Warsaw University. They would like to thank bwGRiD (http://www.bw-grid.de), member of the German D-Grid initiative, funded by the Ministry for Education and Research (Bundesministerium für Bildung und Forschung) and the Ministry for Science, Research and Arts Baden-Württemberg (Ministerium für Wissenschaft, Forschung und Kunst Baden-Württemberg), for providing the opportunity to use parallel computing facilities and perform quantum chemical calculations. R. P. Thummel thanks the Robert A. Welch Foundation (E-621) and the National Science Foundation (CHE-0714751).

References

[1] J. Waluk, "Hydrogen-bonding-induced phenomena in bifunctional heteroazaaromatics," *Accounts of Chemical Research*, vol. 36, no. 11, pp. 832–838, 2003.

[2] J. A. Walmsley, "Self-association of 7-azaindole in nonpolar solvents," *The Journal of Physical Chemistry*, vol. 85, no. 21, pp. 3181–3187, 1981.

[3] P. Dufour, Y. Dartiguenave, M. Dartiguenave et al., "Crystal structures of 7-azaindole, an unusual hydrogen-bonded tetramer, and of two of its methylmercury(II)complexes," *Canadian Journal of Chemistry*, vol. 68, no. 1, pp. 193–201, 1990.

[4] H. Yokoyama, H. Watanabe, T. Omi, S. I. Ishiuchi, and M. Fujii, "Structure of hydrogen-bonded clusters of 7-azaindole studied by IR dip spectroscopy and ab initio molecular orbital calculation," *Journal of Physical Chemistry A*, vol. 105, no. 41, pp. 9366–9374, 2001.

[5] K. Sakota, Y. Kageura, and H. Sekiya, "Cooperativity of hydrogen-bonded networks in 7-azaindole$(CH_3OH)_n$ $(n = 2,3)$clusters evidenced by IR-UV ion-dip spectroscopy and natural bond orbital analysis," *Journal of Chemical Physics*, vol. 129, no. 5, Article ID 054303, 2008.

[6] K. Sakota, Y. Komure, W. Ishikawa, and H. Sekiya, "Spectroscopic study on the structural isomers of 7-azaindole$(ethanol)_n$ $(n = 1-3)$ and multiple-proton transfer reactions in the gas phase," *Journal of Chemical Physics*, vol. 130, no. 22, Article ID 224307, 2009.

[7] T. B. C. Vu, I. Kalkman, W. L. Meerts, Y. N. Svartsov, C. Jacoby, and M. Schmitt, "Rotationally resolved electronic spectroscopy of water clusters of 7-azaindole," *Journal of Chemical Physics*, vol. 128, no. 21, Article ID 214311, 2008.

[8] G. A. Pino, I. Alata, C. Dedonder, C. Jouvet, K. Sakota, and H. Sekiya, "Photon induced isomerization in the first excited state of the 7-azaindole-$(H_2O)_3$ cluster," *Physical Chemistry Chemical Physics*, vol. 13, no. 13, pp. 6325–6331, 2011.

[9] K. Sakota, C. Jouvet, C. Dedonder, M. Fujii, and H. Sekiya, "Excited-state triple proton transfer in 7-azaindole$(H_2O)_2$ and reaction path studied by electronic spectroscopy in the gas phase and quantum chemical calculations," *Journal of Physical Chemistry A*, vol. 114, no. 42, pp. 11161–11166, 2010.

[10] K. Suwińska, "Crystal structure communications," *Acta Crystallographica C*, vol. 41, pp. 973–975, 1985.

[11] J. Waluk and B. Pakuła, "Viscosity and temperature effects in excited state double proton transfer: luminescence of 1-azacarbazole dimers in solid state and solution," *Journal of Molecular Structure*, vol. 114, pp. 359–362, 1984.

[12] J. Waluk, A. Grabowska, B. Pakuła, and J. Sepioł, "Viscosity vs. temperature effects in excited-state double proton transfer. Comparison of 1-azacarbazole with 7-azaindole," *The Journal of Physical Chemistry*, vol. 88, no. 6, pp. 1160–1162, 1984.

[13] J. Waluk, J. Herbich, D. Oelkrug, and S. Uhl, "Excited-state double proton transfer in the solid state: the dimers of 1-azacarbazole," *Journal of Physical Chemistry*, vol. 90, no. 17, pp. 3866–3868, 1986.

[14] J. Catalán, "Photophysics of 1-azacarbazole dimers: a reappraisal," *The Journal of Physical Chemistry A*, vol. 111, no. 36, pp. 8774–8779, 2007.

[15] D. Marks, H. Zhang, P. Borowicz, J. Waluk, and M. Glasbeek, "(Sub)picosecond fluorescence upconversion studies of intermolecular proton transfer of dipyrido[2,3-a:3′,2′-i]carbazole and related compounds," *Journal of Physical Chemistry A*, vol. 104, no. 31, pp. 7167–7175, 2000.

[16] A. Kyrychenko, J. Herbich, M. Izydorzak, F. Wu, R. P. Thummel, and J. Waluk, "Role of ground state structure in photoinduced tautomerization in bifunctional proton donor-acceptor molecules: 1H-pyrrolo[3,2-*h*]quinoline and related

compounds," *Journal of the American Chemical Society*, vol. 121, no. 48, pp. 11179–11188, 1999.

[17] A. Kyrychenko and J. Waluk, "Excited-state proton transfer through water bridges and structure of hydrogen-bonded complexes in 1H-pyrrolo[3,2-*h*]quinoline: adiabatic time-dependent density functional theory study," *The Journal of Physical Chemistry A*, vol. 110, no. 43, pp. 11958–11967, 2006.

[18] Y. Nosenko, M. Kunitski, R. P. Thummel et al., "Detection and structural characterization of clusters with ultrashort-lived electronically excited states: IR absorption detected by femtosecond multiphoton ionization," *Journal of the American Chemical Society*, vol. 128, no. 31, pp. 10000–10001, 2006.

[19] Y. Nosenko, A. Kyrychenko, R. P. Thummel, J. Waluk, B. Brutschy, and J. Herbich, "Fluorescence quenching in cyclic hydrogen-bonded complexes of 1H-pyrrolo[3,2-*h*]quinoline with methanol: cluster size effect," *Physical Chemistry Chemical Physics*, vol. 9, no. 25, pp. 3276–3285, 2007.

[20] Y. Nosenko, M. Kunitski, C. Riehn et al., "Separation of different hydrogen-bonded clusters by femtosecond UV-ionization-detected infrared spectroscopy: 1H-pyrrolo[3,2-*h*] quinoline$\cdot(H_2O)_{n=1,2}$ complexes," *Journal of Physical Chemistry A*, vol. 112, no. 6, pp. 1150–1156, 2008.

[21] J. Herbich, J. Sepioł, and J. Waluk, "Determination of the energy barrier origin of the excited state double proton transfer in 7-azaindole: alcohol complexes," *Journal of Molecular Structure*, vol. 114, pp. 329–332, 1984.

[22] D. McMorrow and T. J. Aartsma, "Solvent-mediated proton transfer. The roles of solvent structure and dynamics on the excited-state tautomerization of 7-azaindole/alcohol complexes," *Chemical Physics Letters*, vol. 125, no. 5-6, pp. 581–585, 1986.

[23] J. Konijnenberg, A. H. Huizer, and C. A. G. O. Varma, "Solute-solvent interaction in the photoinduced tautomerization of 7-azaindole in various alcohols and in mixtures of cyclohexane and ethanol," *Journal of the Chemical Society, Faraday Transactions 2*, vol. 84, no. 8, pp. 1163–1175, 1988.

[24] R. S. Moog, S. C. Bovino, and J. D. Simon, "Solvent relaxation and excited-state proton transfer: 7-azaindole in ethanol," *Journal of Physical Chemistry*, vol. 92, no. 23, pp. 6545–6547, 1988.

[25] R. S. Moog and M. Maroncelli, "7-Azaindole in alcohols: solvation dynamics and proton transfer," *Journal of Physical Chemistry*, vol. 95, no. 25, pp. 10359–10369, 1991.

[26] A. V. Smirnov, D. S. English, R. L. Rich et al., "Photophysics and biological applications of 7-azaindole and its analogs," *Journal of Physical Chemistry B*, vol. 101, no. 15, pp. 2758–2769, 1997.

[27] S. Mente and M. Maroncelli, "Solvation and the excited-state tautomerization of 7-azaindole and 1-azacarbazole: computer simulations in water and alcohol solvents," *Journal of Physical Chemistry A*, vol. 102, no. 22, pp. 3860–3876, 1998.

[28] K. C. Ingham, M. Abu-Elgheit, and M. Ashraf El-Bayoumi, "Confirmation of biprotonic phototautomerism in 7-azaindole hydrogen-bonded dimers," *Journal of the American Chemical Society*, vol. 93, no. 20, pp. 5023–5025, 1971.

[29] S. N. Krasnokutskii, L. N. Kurkovskaya, T. A. Shibanova, and V. P. Shabunova, "Structure of 1H-pyrrolo[3,2-*h*]quinoline," *Zhurnal Strukturnoi Khimii*, vol. 32, p. 131, 1991.

[30] J. Herbich, M. Kijak, R. Luboradzki et al., "In search for phototautomerization in solid dipyrido[2,3-a:3′,2′-i]carbazole ," *Journal of Photochemistry and Photobiology A*, vol. 154, no. 1, pp. 61–68, 2002.

[31] F. Wu, C. M. Chamchoumis, and R. P. Thummel, "Bidentate ligands that contain pyrrole in place of pyridine," *Inorganic Chemistry*, vol. 39, no. 3, pp. 584–590, 2000.

[32] K. Takemura, S. Minomura, O. Shimomura, and Y. Fujii, "Observation of molecular dissociation of iodine at high pressure by X-ray diffraction," *Physical Review Letters*, vol. 45, no. 23, pp. 1881–1884, 1980.

[33] H. K. Mao, P. M. Bell, J. W. Shaner, and D. J. Steiberg, "Specific volume measurements of Cu, Mo, Pd, and Ag and calibration of the ruby R_1 fluorescence pressure gauge from 0.06 to 1 Mbar," *Journal of Applied Physics*, vol. 49, no. 6, pp. 3276–3283, 1978.

[34] G. M. Sheldrick, "Foundations of crystallography," *Acta Crystallographica A*, vol. 64, pp. 112–122, 2008.

[35] M. D. Segall, P. J. D. Lindan, M. J. Probert et al., "First-principles simulation: ideas, illustrations and the CASTEP code," *Journal of Physics Condensed Matter*, vol. 14, no. 11, pp. 2717–2744, 2002.

[36] J. P. Perdew, J. A. Chevary, S. H. Vosko et al., "Atoms, molecules, solids, and surfaces: applications of the generalized gradient approximation for exchange and correlation," *Physical Review B*, vol. 46, no. 11, pp. 6671–6687, 1992.

[37] D. Vanderbilt, "Soft self-consistent pseudopotentials in a generalized eigenvalue formalism," *Physical Review B*, vol. 41, no. 11, pp. 7892–7895, 1990.

[38] H. J. Monkhorst and J. D. Pack, "Special points for Brillouin-zone integrations," *Physical Review B*, vol. 13, no. 12, pp. 5188–5192, 1976.

[39] J. P. Merrick, D. Moran, and L. Radom, "An evaluation of harmonic vibrational frequency scale factors," *Journal of Physical Chemistry A*, vol. 111, no. 45, pp. 11683–11700, 2007.

8

Effective Potential for Ultracold Atoms at the Zero Crossing of a Feshbach Resonance

N. T. Zinner[1, 2]

[1] *Department of Physics, Harvard University, Cambridge, MA 02138, USA*
[2] *Department of Physics and Astronomy, University of Aarhus, 8000 Aarhus, Denmark*

Correspondence should be addressed to N. T. Zinner, zinner@phys.au.dk

Academic Editor: Ali Hussain Reshak

We consider finite-range effects when the scattering length goes to zero near a magnetically controlled Feshbach resonance. The traditional effective-range expansion is badly behaved at this point, and we therefore introduce an effective potential that reproduces the full T-matrix. To lowest order the effective potential goes as momentum squared times a factor that is well defined as the scattering length goes to zero. The potential turns out to be proportional to the background scattering length squared times the background effective range for the resonance. We proceed to estimate the applicability and relative importance of this potential for Bose-Einstein condensates and for two-component Fermi gases where the attractive nature of the effective potential can lead to collapse above a critical particle number or induce instability toward pairing and superfluidity. For broad Feshbach resonances the higher order effect is completely negligible. However, for narrow resonances in tightly confined samples signatures might be experimentally accessible. This could be relevant for suboptical wavelength microstructured traps at the interface of cold atoms and solid-state surfaces.

1. Introduction

Cold atomic gases have enjoyed many great successes since the first realizations of Bose-Einstein condensates in the mid nineties [1]. Ensembles of ultracold atomic gases can be manipulated in magnetic or optical trap geometries and in lattice setups, effectively mimicking the structure of real materials and teaching us about their properties. In particular, extreme control can be exercised over the atom-atom interactions through the use of Feshbach resonance [2]. Tuning the system into the regime of resonant two-body interactions provides a controlled way of studying strongly correlated dynamics which is believed to be crucial for material properties such as high-temperature superconductivity or giant magnetoresistance.

Recently there has been extended interest in weakly interacting Bose-Einstein condensates for use as an atomic interferometer [3] and also to probe magnetic dipolar interactions in condensates [4]. This work was based on ^{39}K atoms where a broad Feshbach resonance exists at a magnetic field strength of $B_0 = 402.4\,$G [5] which allows a large tunability of the atomic interaction in experiments [6]. Similar tunability has also been reported in a condensate of ^{7}Li [7]. The atomic interaction can be reduced by tuning the scattering length, a, to zero, also known as zero crossing. In a Gross-Pitaevskii mean-field picture we can thus neglect the usual nonlinear term proportional to a. The question is then what other interactions are relevant. As shown in [4], the magnetic dipole will contribute here.

In the Gross-Pitaevskii picture we might also ask whether higher order terms in the interaction can contribute around zero crossing. Recently it was shown that effective-range corrections can in fact influence the stability of condensates around zero crossing [8–10]. The Feshbach resonances used thus far in experiments have typically been very broad, and as a result the effective range, r_e, will be small, rendering the higher order terms negligible. However, around narrow resonances this is not necessarily the case and finite-range corrections are not necessarily negligible.

For the two-component Fermi gas, there has been increased interest in producing a cold atom analog of the celebrated Stoner model of ferromagnetism [11] which applies to repulsively interacting fermions. Theoretical proposals indicate that this should be possible [12–19], and an MIT experiment subsequently announced indications of the ferromagnetic transition [20]. The results caused controversy since the spin domains were not resolved [21–24]. A later experiment in the same group did not find evidence of the ferromagnetic transition [25]. However, these studies consider broad Feshbach resonances, and the situation with narrow resonances is less clear. One can imagine that finite-range corrections could play a role in driving the phase transition. In fact, a recent experiment in Innsbruck [26] has found increased lifetimes of the repulsive gas in the strongly imbalanced case, providing hope that decay into molecules can be controlled and ferromagnetism can be studied.

The systematic inclusion of finite-range effects through derivative terms in zero-range models was begun in the study of nuclear matter decades ago [27, 28]. Later on the intricacies of the cut-off problems that arise in this respect were considered by many authors both for the relativistic and nonrelativistic case (see [29] for discussion and references). In the context of cold atoms and Feshbach resonances, we need to use a two-channel model [30] in order to take the lowest order finite-range term into account. Similar models were already introduced in [31] and denoted resonance models (see f.x. [32] for a comprehensive review of scattering models for ultracold atoms). We note that whereas resonance models treat the closed-channel molecular state as a point boson the model of [30] treats the molecule more naturally as a composite object of two atoms. In the end the parameters of the two models turn out to be similarly related to the physical parameters of Feshbach resonances (see for instance the discussion of resonance models in [32]).

In Figure 1 we show calculations of scattering length and effective range for the Feshbach resonance at $B = 202.1$ G in ^{40}K in both a coupled-channel model [33] and in the zero-range model discussed here. We see the effective range being roughly constant at resonance and then start to diverge at zero crossing. The zero-range model provides a good approximation to the full calculations and for many-body purposes it is preferable due to its simplicity.

Whereas the earlier work of [31] considered the regime close to the resonance, we will be exclusively concerned with zero crossing. To our knowledge the intricacies of this region have not been addressed in the literature in the context of Feshbach resonances. Around zero crossing the Feshbach model turns out to have a badly behaved effective-range expansion. The parameters obtained from the effective-range expansion should therefore be used with extreme caution as the series is divergent at this point. However, as we show in this paper, the finite-range corrections obtained from the full T-matrix at low momenta via an effective potential turn out to be the same as one would naively expect based on the effective-range expansion. After introducing the effective potential we consider its applicability and importance in the case of Bose-Einstein condensates and for two-component Fermi gases where the attractive nature of the effective interaction at zero crossing could lead to collapse above a certain critical particle number or to pairing instability and superfluidity. In general, we find that tight external confinement is a necessary condition for the higher order effects to dominate the magnetic dipole interaction and be experimentally observable.

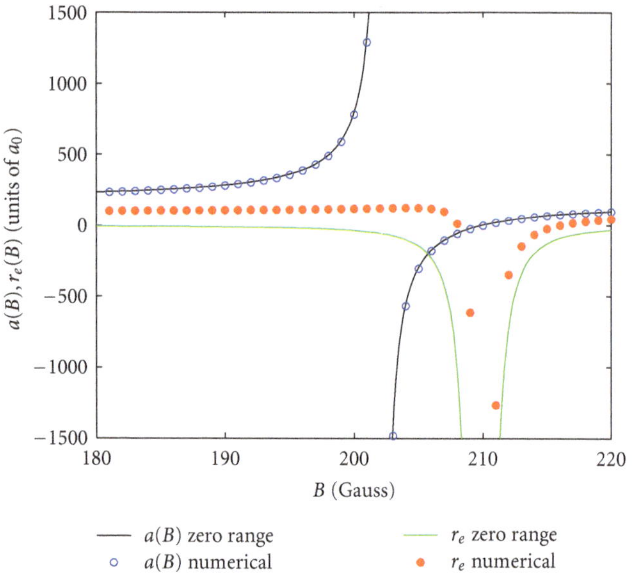

FIGURE 1: Scattering length and effective range for the s-wave scattering of fermionic ^{40}K atoms around the Feshbach resonance at $B_0 = 202.1$ G demonstrating the divergence in a coupled-channel calculation (symbols) [33] and in a zero-range model (full lines). The difference in the zero-range and coupled-channel models is caused by the presence of a bound state close to threshold in the open channel.

2. Two-Channel Model

We consider a two-channel s-wave Feshbach model with zero-range interactions [30] for which the on-shell open-open channel T-matrix as a function of magnetic field, B, is

$$T_{oo}(B) = \frac{(4\pi\hbar^2/m)a_{bg}}{\left(1 + \Delta\mu\Delta B/(\hbar^2 q^2/m - \Delta\mu(B - B_0))\right)^{-1} + ia_{bg}q}, \tag{1}$$

where $\Delta\mu$ is the difference between the magnetic moments in the open and closed channel, q is the relative momentum of the atoms of mass m, a_{bg} is the scattering length away from the resonance at magnetic field B_0, and ΔB is the width of the resonance. We can compare this to the standard vacuum expression for the T-matrix in terms of the phase-shift given by

$$T_v = \frac{(4\pi\hbar^2/m)a}{-qa\cot\delta(q) + iaq}. \tag{2}$$

Typically, one has the low-energy expression $-q\cot\delta(q) \to -1/a$ which implies that

$$T_v \longrightarrow \frac{(4\pi\hbar^2/m)a}{1 + iaq} \longrightarrow 0. \tag{3}$$

However, as we now discuss, for the realistic two-channel T-matrix for Feshbach resonances, the quantity $-q\cot\delta(q)$ is not well defined, and the conclusion that the T-matrix vanishes at zero crossing is only true for zero momentum, $q = 0$, as we now discuss.

From (1) and (2) we obtain the relation for the phase-shift:

$$q\cot\delta(q) = \frac{-1}{a_{bg}}\left(1+\frac{\Delta\mu\Delta B}{(\hbar^2q^2/m-\Delta\mu(B-B_0))}\right)^{-1}. \quad (4)$$

We now expand the right-hand side in powers of q as is usually done in an effective-range expansion. This yields

$$q\cot\delta(q) = \frac{-1}{a(B)} + \sum_{n=1}^{\infty}\frac{-1}{a_{bg}}\left[\frac{-a_{bg}r_{e0}}{2}\right]^n\left[\frac{a_{bg}}{a(B)}-1\right]^{n+1}q^{2n}, \quad (5)$$

where $a(B) = a_{bg}(1-\Delta B/(B-B_0))$ is the common parametrization from single-channel models and $r_{e0} = -2\hbar^2/(m\Delta B\Delta\mu a_{bg})$ is the background value of the effective range around the resonance. From (5) we can now read off all coefficients in an effective-range expansion with their full B-field dependence. For instance, the effective range is given simply by $r_e = r_{e0}[(a_{bg}/a)-1]^2$, which is divergent when $a(B) \to 0$. We also clearly see that all the other coefficients are divergent in that limit. This is signaled also before doing the full expansion in q as the first term in (5) diverges at zero crossing. However, in effective potentials derived from the T-matrix these problems are not transparent as the lowest order coefficient is proportional to $a(B)$ (see (12)). Below we will discuss what kind of constraints this introduces on the applicability of the effective-range expansion near zero crossing. We note that similar issues were briefly discussed in a different context in [34] where an equivalent to (7) below was obtained.

Let us first consider the low-q limit and compare the full T-matrix with the effective-range expansion as zero crossing is approached. Taking the low-q limit of (4) at zero crossing where $\Delta B/(B-B_0) = 1$, we find

$$q\cot\delta(q) \longrightarrow \frac{-1}{a_{bg}} - \frac{\Delta\mu\Delta B}{\hbar^2q^2/m}, \quad (6)$$

which diverges as q^{-2}. Therefore the coefficients of the expansion in (5) must necessarily diverge in order to retain any hope of describing the low-q behavior. Furthermore, since the expansion is an alternating series and therefore slowly converged, we also conclude that many terms must be retained for a fair approximation at very small but nonzero q. The same conclusion can be reached by considering the radius of convergence of (5), which we find by locating the pole in (4) at $\hbar^2q^2/m = \Delta\mu(B-B_0-\Delta B)$. This radius indeed goes to zero at zero crossing. We are thus forced to conclude that the effective-range expansion breaks down near zero crossing.

2.1. Effective Potential at Zero Crossing. Since the effective-range expansion is insufficient we consider the full T-matrix in the low-q limit at zero crossing. To lowest order we have

$$T_{oo}(B = B_0+\Delta B) = -\frac{4\pi\hbar^2a_{bg}}{m}\frac{\hbar^2q^2}{m\Delta\mu\Delta B} + O(q^4). \quad (7)$$

Using the expression for r_{e0}, this can be written

$$\frac{4\pi\hbar^2}{m}\frac{a_{bg}^2r_{e0}}{2}q^2. \quad (8)$$

Knowing the T-matrix at low q we can now proceed to find an effective low-q potential through the Lippmann-Schwinger equation:

$$V = T - TG_0V, \quad (9)$$

where $G_0 = (E-H_0+i\delta)^{-1}$ is the free space Green's function [35]. This equation can be solved for $T(q,q') \propto q^2+q'^2$ (the symmetrized version of the full T-matrix) in an explicit cut-off approach [29, 35] and then be expanded to order q^2 for consistence with the input T-matrix. In the long-wavelength limit we can take the cut-off to zero [35] and for the on-shell effective potential we then obtain the obvious answer:

$$V(q) = \frac{4\pi\hbar^2}{m}\frac{a_{bg}^2r_{e0}}{2}q^2 \quad (10)$$

in momentum space. The effective potential in real-space is now easily found by canonical substitution ($\mathbf{q} \to -i\nabla$) and appropriate symmetrization [36]. We have

$$V(\mathbf{r}) = -\frac{4\pi\hbar^2}{m}\frac{a_{bg}^2r_{e0}}{2}\frac{1}{2}\left[\overleftarrow{\nabla}_{\mathbf{r}}^2\delta(\mathbf{r})+\delta(\mathbf{r})\overrightarrow{\nabla}_{\mathbf{r}}^2\right]. \quad (11)$$

Notice that the Lippmann-Schwinger approach is nonperturbative as opposed to the perturbative energy shift method [36, 37].

2.2. Comparison to Effective-Range Expansion and Energy-Shift Method. Away from zero crossing one can easily relate the effective-range expansion to an effective potential through the perturbative energy shift method [18, 25, 26]. To second order the s-wave effective potential is

$$V(\mathbf{r}) = \frac{4\pi\hbar^2a}{m}\left[\delta(\mathbf{r})+\frac{g_2}{2}\left(\overleftarrow{\nabla}_{\mathbf{r}}^2\delta(\mathbf{r})+\delta(\mathbf{r})\overrightarrow{\nabla}_{\mathbf{r}}^2\right)\right], \quad (12)$$

where the first term is the effective interaction usually employed in mean-field theories of cold atoms [35]. In terms of a and r_e, we have $g_2 = a^2/3 - ar_e/2$ [36, 37] with the field-dependent $a = a(B)$ and $r_e = r_e(B)$.

At zero-crossing the first term in (12) vanishes and one might expect the second term to vanish as well. However, in the naive effective-range expansion of the two-channel model discussed above we saw that r_e diverges as a^{-2} and we therefore have

$$\lim_{a\to 0} ag_2 = -\frac{a_{bg}^2r_{e0}}{2}. \quad (13)$$

In particular, if we for a moment ignore q^4 terms in the effective-range expansion, we recover exactly the same effective potential as in (11) at zero crossing. The finite limiting result in (13) shows that the potential in (12) is well defined as $a \to 0$, provided that appropriate regularization and renormalization are performed. Equation (12) thus applies equally well at resonance ($a \to \infty$) where the gradient terms are small and at zero crossing where the lowest order delta function term is unimportant. It is thus a well-defined effective potential over the entire range of a Feshbach resonance.

We therefore see that even though the effective-range expansion has divergent coefficients at zero crossing, the lowest order does in fact give the same effective potential as the full T-matrix if we apply it naively. The effective-range expansion should thus be viewed as an asymptotic series. However, we cannot use the effective-range expansion to estimate the validity of the second-order effective potential since the radius of convergence goes to zero at zero crossing as discussed above.

The two-channel model in (1) compares well with a coupled-channel calculation [33] as shown in Figure 1. It also compares well to other scattering models [38, 39] that include finite-range effects. In fact, the model used here compares well with the analytical models of [38] when $a(B)$ and $r_e(B)$ have the field-dependence introduced above. This can be seen for instance in Figure 12 of [38], although a difference is that our $a(B)$ and $r_e(B)$ are parametrization and not taken from coupled-channels values as in [38] (our Figure 1 quantifies the difference which is largest on $r_e(B)$). However, here we are concerned with the behavior when $a(B) \to 0$ in the context of Feshbach resonances which is not addressed in [38, 39]. We note that the resonance models of [31] and the two-channel and resonance models in [32] are very similar to the model employed here, but again those references do not consider the specific problems arising when $a(B) \to 0$. In addition and in contrast to previous discussions, here we construct appropriate zero-range pseudo potentials that work around zero crossing.

3. Relation to Experiments

Above we only retained terms of order q^2 in the full T-matrix. We now estimate the energy regime in which this expression is valid. Demanding that the q^4 term be smaller than the q^2 term gives the criterion:

$$\frac{\hbar^2 q^2}{m} \ll \frac{\hbar^2}{m\left|a_{bg} r_{e0}\right|}. \tag{14}$$

We relate this condition to recent experiments with bosonic condensates of ^{39}K working around zero crossing [3]. The resonance used there is very broad ($\Delta B = -52\,\mathrm{G}$) with $a_{bg} = -29a_0$ and $r_{e0} = -58a_0$ (a_0 is the Bohr radius). The right-hand side of (14) is $2.3 \cdot 10^{-7}$ eV, corresponding to a temperature of about 3 mK. Since the experiments are performed at much lower temperatures the approximation above is certainly valid. However, as a_{bg} and particularly r_{e0} are small, the front factor in (11) is also small. The relevant scale of comparison is the outer trap parameter b [9] which is typically of order 1 μm, yielding a vanishing ratio $|a_{bg}^2 r_{e0}|/b^3 \sim 10^{-9}$. For broad Feshbach resonances the higher order interactions can thus be safely ignored. For very narrow resonances the situation potentially changes as r_{e0} can be very large and make the potential in (11) important. As an example, we consider the narrow resonance in ^{39}K at $B_0 = 25.85\,\mathrm{G}$ with $\Delta B = 0.47\,\mathrm{G}$, $a_{bg} = -33a_0$, and $r_{e0} = -5687a_0$ [5]. The right-hand side of (14) is now $2 \cdot 10^{-9}$ eV, corresponding to 24 μK. This is again much higher than experimental temperatures. A more careful argument can be made from the energy per particle of the noncondensed cloud. Ignoring the trap, we have $E/N = 0.770 k_B T_c (T/T_c)^{5/2}$ (T_c is the critical temperature) [35]. For a sample of $3 \cdot 10^4$ a critical temperature of 100 nK was reported in [6]. Using this T_c we find that $T \ll 900$ nK for (14) to hold. Again this is within the experimental regime. The effective potential approach should therefore be applicable around zero crossing for narrow resonances. However, even with this narrow resonance we find $|a_{bg}^2 r_{e0}|/b^3 \sim 10^{-7}$, and the effect is still completely negligible.

In order to increase the relevance of the higher order term, we now consider some very narrow resonances that have been found in ^{87}Rb. In particular, the resonance at $B_0 = 9.13\,\mathrm{G}$ [40] which was recently utilized in nonlinear atom interferometry [41]. We have $\Delta B = 0.015\,\mathrm{G}$, $a_{bg} = 99.8a_0$, and $\Delta\mu = 2.00\,\mu_B$ [42], which gives $r_{e0} = -19.8 \cdot 10^3 a_0$ and a ratio $|a_{bg}^2 r_{e0}|/b^3 = 2.92 \cdot 10^{-5}(1\,\mu\mathrm{m}/b)^3$. A trap length of $b \sim 0.5\,\mu$m as used in [41] would thus yield 10^{-4} and demonstrates that higher order corrections can safely be neglected. For a ratio of 1 we need $b \sim 0.03\,\mu$m which is unrealistically small in current traps or optical lattices. However, a resonance of width $\Delta B = 0.0004\,\mathrm{G}$ is known in the same system at $B_0 = 406.2\,\mathrm{G}$ [43] with $a_{bg} = 100a_0$ and $\Delta\mu = 2.01\mu_B$ [42]. In this case we find $r_{e0} = -7.4 \cdot 10^5 a_0$ and a much more favorable ratio of $|a_{bg}^2 r_{e0}|/b^3 = 0.001(1\,\mu\mathrm{m}/b)^3$. Here we see that a ratio of 1 is achieved already for $b \sim 0.1\,\mu$m which not far off from tight traps or optical lattice dimensions. In terms of temperature we still have to be in the ultralow regime of $T \lesssim 30$ nK according to (14) for the latter resonance.

Consider now a fermionic two-component system where s-wave interactions are dominant. Since we have $r_{e0} < 0$ for all Feshbach resonances [42], the effective potential in (10) is attractive, and the system could potentially be unstable toward a paired state or become unstable to collapse above a critical particle number. For simplicity we will use the semiclassical Thomas-Fermi approach to describe a gas with equal population of the two components and estimate the critical particle number. Assuming an isotropic trapping potential with length scale $b = \sqrt{\hbar/m\omega}$ where ω is the trap frequency, the ground-state density, $\rho(\mathbf{x})$, can be found by minimization and satisfies

$$\left[\frac{\mu}{\hbar\omega} - \frac{1}{2}\left(\frac{\mathbf{x}}{b}\right)^2\right] = \frac{1}{2}(k_F(\mathbf{x})b)^2 - \frac{4}{30\pi}\alpha(k_F(\mathbf{x})b)^5, \tag{15}$$

where $\rho(\mathbf{x}) = k_F(\mathbf{x})/6\pi^2$ and $\alpha = a_{bg}^2|r_{e0}|/b^3$. The maximum allowed momentum and chemical potential, μ, is found by solving for the turning point of the right-hand side of (15) which gives

$$k_{\max} b = \left[\frac{3\pi}{2\alpha}\right]^{1/3}, \qquad \mu_{\max} = \frac{3}{10}\hbar\omega(k_{\max} b)^2. \tag{16}$$

We can now compare this $k_{\max}$ to the value obtained from the noninteracting density within the Thomas-Fermi approximation at the center of the trap. In terms of the number of particles in each component, N, at the center of the trap we have $k_F(0)b \approx 1.906N^{1/6}$ [35]. By equating these two expression we obtain an estimate for the critical number of particles, $N_{\max}$. Inserting the relevant units, we have

$$N_{\max} = 2 \cdot 10^{25}\left(\frac{a_0}{a_{bg}}\right)^4\left(\frac{a_0}{r_{e0}}\right)^2\left(\frac{b}{1\mu\text{m}}\right)^6, \tag{17}$$

where a_0 is the Bohr radius. We note that the scaling $N_{\max} \propto \alpha^{-2}$ can also be obtained by considering the point at which the monopole mode becomes unstable.

Typical numbers for common fermionic species ^{6}Li or ^{40}K in the lowest hyperfine states [42] lead to $N_{\max} \sim 10^{12}$ for $b = 1\,\mu$m. This is of course a huge number and experiments are well within this limit. Even if one reduced the trap length by a factor of ten and made the presumably unrealistic assumption that the particle number remains the same we still have $N \ll N_{\max}$. The reason is that the s-wave Feshbach resonances utilized in the two-component gases are generally broad in order to study the universal regime. If we consider the narrow resonance at $B_0 = 543.25$ G in ^{6}Li [44] with $\Delta B = 0.1$ G, $a_{bg} = 60a_0$, and $\Delta\mu = 2.00\,\mu_B$ [42], we have $N_{\max} \sim 2 \cdot 10^{13}(b/1\,\mu\text{m})^6$. This is somewhat better but we still need $b \sim 0.06\,\mu$m to get to an experimentally relevant $N_{\max} \sim 10^6$. We have to conclude that higher order s-wave interactions are highly unlikely to be observable through monopole instabilities. In light of this it seems better to consider p-wave resonances which are much more narrow in general. However, also here extremely small trap sizes appear necessary [45].

The instability toward Cooper pairing around zero crossing can also be estimated in simple terms. In general the critical temperature is $T_c \sim T_F \exp(-1/N_0|U|)$, where $N_0 = mk_F(0)/2\pi^2\hbar^2$ is the density of states at the Fermi energy in the trap center and $U < 0$ is a measure of the attraction. For the latter we use the effective potential in momentum space from (10) and make the assumption that $q \sim k_F(0)$. Using the expression for $k_F(0)$ in terms of N above, we find

$$\frac{1}{N_0|U|} = \frac{1.5 \cdot 10^{12}}{\sqrt{N}}\left(\frac{b}{1\mu\text{m}}\right)^3\left(\frac{a_0}{a_{bg}}\right)^2\frac{a_0}{|r_{e0}|}. \tag{18}$$

For broad resonances in ^{6}Li or ^{40}K this exponent is of order 10^3 and T_c is thus vanishingly small. However, the scaling with trap size can help and if we imagine reducing to $b = 0.1\,\mu$m, we find $T_c \lesssim 0.5T_F$ for $N = 10^6$ atoms. For the narrow resonance in ^{6}Li discussed above, we find that $T_c \sim 0.5T_F$ with $N = 10^6$ can be achieved for $b \sim 0.5\,\mu$m and $T_c \sim 0.1T_F$ for $N = 10^5$. Thus there may be a possibility to reach the pairing instability near zero crossing if high particle numbers can be cooled in tight traps and narrow resonances are used.

While the suboptical wavelength trapping sizes needed for the above effects to be large are not achievable with typical optical or magnetic traps or optical lattice setups, they could potentially be reached via hybrid setups where atoms are trapped near a surface. Inspired by surface plasmon subwavelength optics [46], nanoscale trapping for neutral atoms has been studied [47, 48], and micropotential traps with width less than 100 nanometer (<0.1 μm) are within reach [49]. In these very tightly confined systems, it is very likely that finite-range effects could be enhanced. Devices that provide an interface between atoms and solid-state systems are under intense study at the moment, and our considerations here imply that finite-range corrections should be considered when the scattering length is tuned close to zero.

3.1. Dipole-Dipole Interactions. The discussion above ignores the dipole-dipole interaction discussed in the introduction which will compete against the higher order effective potential from the Feshbach resonance. A simple estimate can be made along the lines of the discussion in [35]. The external trapping potential is the characteristic scale of spatial variations and we thus find a ratio, r, of magnetic dipole-dipole, U_{md}, to higher-order s-wave zero-range interaction strength, U_2, which can be written as

$$r = \frac{U_{\text{md}}}{U_2} = \frac{a_0 b^2}{a_{bg}^2|r_{e0}|} = 35.7\left[\frac{b}{1\mu\text{m}}\right]^2\left[\frac{100a_0}{a_{bg}}\right]^2\frac{1000a_0}{|r_{e0}|}. \tag{19}$$

For $r < 1$ the higher order interaction term will therefore dominate the magnetic dipole term. For the case of narrow resonances in ^{87}Rb discussed above we find $r \sim 0.11(b/1\,\mu\text{m})^2$ for the resonance at $B_0 = 9.13$ G and $r \sim 0.05(b/1\,\mu\text{m})^2$ for the one at $B_0 = 406.2$ G. For the narrow resonance in ^{6}Li at $B_0 = 543.25$ G we find $r \sim 1.4(b/1\,\mu\text{m})^2$. These ratios clearly indicate that magnetic dipole-dipole interactions can be suppressed relative to higher order zero-range terms for narrow Feshbach resonances and standard trap sizes. This dominance becomes even stronger for the tight traps needed for the realization of the effects discussed above, and we thus conclude that interference of the magnetic dipole-dipole term is not a major concern.

4. Conclusions and Outlook

In this paper we have discussed the effective potential around a Feshbach resonances as the scattering length is tuned to zero and finite-range corrections become important. We showed that the effective-range expansion is badly behaved and the effective potential most be defined from the T-matrix. We have demonstrated that the low momenta effective potential obtained from the full T-matrix agrees with one obtained naively from the effective-range expansion when the scattering length goes to zero. Thus even though the effective-range expansion has divergent coefficients at zero-crossing the first terms of the associated effective potential yield consistent results. We then estimated the effects of the terms on different condensates. Since the effective potential at zero crossing is attractive it may induce various instabilities which we considered for the case of a two-component Fermi gas under harmonic confinement.

For the broad Feshbach resonances used in current experiments the effective potential discussed here are negligible and the dipole-dipole interaction dominates completely at zero crossing. However, for narrow resonances in very tightly confined systems some of the effects might be detectable. In particular, future generations of microtraps with suboptical wavelength trap sizes using surface plasmons could be small enough to make finite-range effects important. The competing dipole interaction is small for narrow resonances in tight confinement. However, it is conceivable that effects of spherically symmetric higher order terms could be separated from dipolar effects which change with system geometry [4].

Small trapped Fermi systems have recently become an experimental reality with particle numbers ranging from two to ten [50]. For two atomic fermions with different internal states, the system turns out to be well described by the analytic zero-range model of Busch et al. [51–56], and similarly for three fermions [57, 58]. Effective-range corrections to these results have also been studied [59–62]. Mesoscopic Fermi systems (less than about 50 particles) have been studied in harmonic traps using a number of numerical methods, [63–79] with particular emphasis on the unitary regime where the scattering length diverges. It would be interesting to investigate the situation also around zero-crossing of a narrow resonance where the effective range is sizable. A preliminary study along this line for three bosons is discussed in [80].

Another interesting direction of future work is the study of the contact introduced by Tan [81–90] to describe the universal behavior of strongly interacting quantum gases at a broad resonance where the range corrections are negligible, for instance through the tail of the momentum distribution which is predicted to behave as C/k^4, where C is the contact and k the momentum of a single particle. The relations found by Tan [81, 82] have subsequently been confirmed experimentally in three dimensions [91–93]. While the contact originally pertains to two-body correlations, signatures of three-body physics in momentum distributions have also been studied both theoretically [86, 94–98] and experimentally [99]. While a few studies have considered the universal behavior when including the effective range term [100, 101], it would be very interesting to consider the regime around zero crossing for a narrow resonance where the background effective range parameter.

Acknowledgments

The author would like to thank Martin Thøgersen for very fruitful collaborations. Correspondence with Georg Bruun about two-channel models is highly appreciated. The author is grateful to Nicolai Nygaard for discussions and for producing Figure 1. The author acknowledges the hospitality of the Niels Bohr Institute, Blegdamsvej 17, 2100 Copenhagen Ø, Denmark. This work was supported by the Villum Kann Rasmussen foundation.

References

[1] I. Bloch, J. Dalibard, and W. Zwerger, "Many-body physics with ultracold gases," *Reviews of Modern Physics*, vol. 80, no. 3, pp. 885–964, 2008.

[2] C. Chin, R. Grimm, P. Julienne, and E. Tiesinga, "Feshbach resonances in ultracold gases," *Reviews of Modern Physics*, vol. 82, no. 2, pp. 1225–1286, 2010.

[3] M. Fattori, C. D'Errico, G. Roati et al., "Atom interferometry with a weakly interacting bose-Einstein condensate," *Physical Review Letters*, vol. 100, no. 8, Article ID 080405, 4 pages, 2008.

[4] M. Fattori, G. Roati, B. Deissler et al., "Magnetic dipolar interaction in a Bose-Einstein condensate atomic interferometer," *Physical Review Letters*, vol. 101, no. 19, Article ID 190405, 4 pages, 2008.

[5] C. D'Errico, M. Zaccanti, M. Fattori et al., "Feshbach resonances in ultracold ^{39}K," *New Journal of Physics*, vol. 9, article 223, 2007.

[6] G. Roati, M. Zaccanti, C. D'Errico et al., "^{39}K bose-Einstein condensate with tunable interactions," *Physical Review Letters*, vol. 99, no. 1, Article ID 010403, 4 pages, 2007.

[7] S. E. Pollack, D. Dries, M. Junker, Y. P. Chen, T. A. Corcovilos, and R. G. Hulet, "Extreme tunability of interactions in a ^{7}Li Bose-Einstein condensate," *Physical Review Letters*, vol. 102, no. 9, Article ID 090402, 4 pages, 2009.

[8] H. Fu, Y. Wang, and B. Gao, "Beyond the Fermi pseudopotential: a modified Gross-Pitaevskii equation," *Physical Review*, vol. 67, no. 5, Article ID 053612, 6 pages, 2003.

[9] N. T. Zinner and M. Thøgersen, "Stability of a Bose-Einstein condensate with higher-order interactions near a Feshbach resonance," *Physical Review*, vol. 80, no. 2, Article ID 023607, 4 pages, 2009.

[10] M. Thøgersen, N. T. Zinner, and A. S. Jensen, "Thomas-Fermi approximation for a condensate with higher-order interactions," *Physical Review A*, vol. 80, no. 4, Article ID 043625, 8 pages, 2009.

[11] E. Stoner, "LXXX. Atomic moments in ferromagnetic metals and alloys with non-ferromagnetic elements," *Philosophical Magazine*, vol. 15, no. 101, pp. 1018–1034, 1933.

[12] M. Houbiers, R. Ferwerda, H. T. C. Stoof, W. I. McAlexander, C. A. Sackett, and R. G. Hulet, "Superfluid stateof atomic ^{6}Li in a magnetic trap," *Physical Review A*, vol. 56, no. 6, pp. 4864–4878, 1997.

[13] Y. Zhang and S. Das Sarma, "Exchange instabilities in electron systems: bloch versus Stoner ferromagnetism," *Physical Review B*, vol. 72, no. 11, Article ID 115317, 9 pages, 2005.

[14] R. A. Duine and A. H. MacDonald, "Itinerant ferromagnetism in an ultracold atom Fermi gas," *Physical Review Letters*, vol. 95, no. 23, Article ID 230403, 4 pages, 2005.

[15] G. J. Conduit and B. D. Simons, "Itinerant ferromagnetism in an atomic Fermi gas: influence of population imbalance," *Physical Review A*, vol. 79, no. 5, Article ID 053606, 9 pages, 2009.

[16] J. Conduit, A. G. Green, and B. D. Simons, "Inhomogeneous phase formation on the border of itinerant ferromagnetism," *Physical Review Letters*, vol. 103, no. 20, Article ID 207201, 4 pages, 2009.

[17] G. J. Conduit and B. D. Simons, "Repulsive atomic gas in a harmonic trap on the border of itinerant ferromagnetism,"

Physical Review Letters, vol. 103, no. 20, Article ID 200403, 4 pages, 2009.
[18] L. J. Leblanc, J. H. Thywissen, A. A. Burkov, and A. Paramekanti, "Repulsive Fermi gas in a harmonic trap: ferromagnetism and spin textures," *Physical Review A*, vol. 80, no. 1, Article ID 013607, 2009.
[19] S. Zhang, H.-H. Hung, and C. Wu, "Proposed realization of itinerant ferromagnetism in optical lattices," *Physical Review*, vol. 82, no. 5, Article ID 053618, 5 pages, 2010.
[20] G. B. Jo, Y. R. Lee, J. H. Choi et al., "Itinerant ferromagnetism in a fermi gas of ultracold atoms," *Science*, vol. 325, no. 5947, pp. 1521–1524, 2009.
[21] H. Zhai, "Correlated versus ferromagnetic state in repulsively interacting two-component Fermi gases," *Physical Review A*, vol. 80, no. 5, Article ID 051605, 4 pages, 2009.
[22] X. Cui and H. Zhai, "Stability of a fully magnetized ferromagnetic state in repulsively interacting ultracold Fermi gases," *Physical Review A*, vol. 81, no. 4, Article ID 041602, 4 pages, 2010.
[23] D. Pekker, M. Babadi, R. Sensarma et al., "Competition between pairing and ferromagnetic instabilities in ultracold Fermi gases near Feshbach resonances," *Physical Review Letters*, vol. 106, no. 5, Article ID 050402, 4 pages, 2011.
[24] V. B. Shenoy and T.-L. Ho, "Nature and properties of a repulsive Fermi gas in the upper branch of the energy spectrum," *Physical Review Letters*, vol. 107, no. 21, Article ID 210401, 5 pages, 2011.
[25] C. Sanner, E. J. Su, W. Huang, A. Keshet, J. Gillen, and W. Ketterle, "Correlations and pair formation in a repulsively interacting Fermi gas," *Physical Review Letters*, vol. 108, no. 24, Article ID 240404, 5 pages, 2012.
[26] C. Kohstall, M. Zaccanti, M. Jag et al., "Metastability and coherence of repulsive polarons in a strongly interacting Fermi mixture," *Nature*, vol. 485, pp. 615–618, 2012.
[27] T. H. R. Skyrme, "CVII. The nuclear surface," *Philosophical Magazine*, vol. 1, no. 11, pp. 1043–1054, 1956.
[28] T. H. R. Skyrme, "The effective nuclear potential," *Nuclear Physics*, vol. 9, no. 4, pp. 615–634, 1959.
[29] D. R. Phillips, S. R. Beane, and T. D. Cohen, "Nonperturbative regularization and renormalization: simple examples from nonrelativistic quantum mechanics," *Annals of Physics*, vol. 263, no. 2, pp. 255–275, 1998.
[30] G. M. Bruun, A. D. Jackson, and E. E. Kolomeitsev, "Multichannel scattering and Feshbach resonances: effective theory, phenomenology, and many-body effects," *Physical Review A*, vol. 71, no. 5, Article ID 052713, 10 pages, 2005.
[31] S. J. J. M. F. Kokkelmans, J. N. Milstein, M. L. Chiofalo, R. Walser, and M. J. Holland, "Resonance superfluidity: rsenormalization of resonance scattering theory," *Physical Review A*, vol. 65, no. 5, Article ID 536171, 4 pages, 2002.
[32] E. Braaten, M. Kusunoki, and D. Zhang, "Scattering models for ultracold atoms," *Annals of Physics*, vol. 323, no. 7, pp. 1770–1815, 2008.
[33] N. Nygaard, B. I. Schneider, and P. S. Julienne, "Two-channel R-matrix analysis of magnetic-field-induced Feshbach resonances," *Physical Review A*, vol. 73, no. 4, Article ID 042705, 10 pages, 2006.
[34] P. Massignan and Y. Castin, "Three-dimensional strong localization of matter waves by scattering from atoms in a lattice with a confinement-induced resonance," *Physical Review A*, vol. 74, no. 1, Article ID 013616, 2006.
[35] C. J. Pethick and H. Smith, *Bose-Einstein Condensation in Dilute Gases*, Cambridge University Press, Cambridge, Mass, USA, 2002.
[36] R. Roth and H. Feldmeier, "Effective s- and p-wave contact interactions in trapped degenerate Fermi gases," *Physical Review A*, vol. 64, no. 4, Article ID 043603, 17 pages, 2001.
[37] A. Collin, P. Massignan, and C. J. Pethick, "Energy-dependent effective interactions for dilute many-body systems," *Physical Review A*, vol. 75, no. 1, Article ID 013615, 2007.
[38] B. Marcelis, E. G. M. van Kempen, B. J. Verhaar, and S. J. J. M. F. Kokkelmans, "Feshbach resonances with large background scattering length: interplay with open-channel resonances," *Physical Review A*, vol. 70, no. 1, Article ID 012701, 15 pages, 2004.
[39] B. Marcelis and S. Kokkelmans, "Fermionic superfluidity with positive scattering length," *Physical Review A*, vol. 74, no. 2, Article ID 023606, 2006.
[40] A. Widera, O. Mandel, M. Greiner, S. Kreim, T. W. Hänsch, and I. Bloch, "Entanglement interferometry for precision measurement of atomic scattering properties," *Physical Review Letters*, vol. 92, no. 16, Article ID 160406, 2004.
[41] C. Gross, T. Zibold, E. Nicklas, J. Estève, and M. K. Oberthaler, "Nonlinear atom interferometer surpasses classical precision limit," *Nature*, vol. 464, no. 7292, pp. 1165–1169, 2010.
[42] C. Chin, R. Grimm, P. Julienne, and E. Tiesinga, "Feshbach resonances in ultracold gases," *Reviews of Modern Physics*, vol. 82, no. 2, pp. 1225–1286, 2010.
[43] A. Marte, T. Volz, J. Schuster et al., "Feshbach resonances in rubidium 87: precision measurement and analysis," *Physical Review Letters*, vol. 89, no. 28, Article ID 283202, 4 pages, 2002.
[44] K. E. Strecker, G. B. Partridge, and R. G. Hulet, "Conversion of an atomic Fermi gas to a long-lived molecular bose gas," *Physical Review Letters*, vol. 91, no. 8, Article ID 080406, 4 pages, 2003.
[45] N. T. Zinner, "Stability of a fully polarized ultracold Fermi gas near zero-crossing of a p-wave Feshbach resonance," *European Physical Journal D*, vol. 57, no. 2, pp. 235–240, 2010.
[46] W. L. Barnes, A. Dereux, and T. W. Ebbesen, "Surface plasmon subwavelength optics," *Nature*, vol. 424, no. 6950, pp. 824–830, 2003.
[47] B. Murphy and L. V. Hau, "Electro-optical nanotraps for neutral atoms," *Physical Review Letters*, vol. 102, no. 3, Article ID 033003, 4 pages, 2009.
[48] D. E. Chang, J. D. Thompson, H. Park et al., "Trapping and manipulation of isolated atoms using nanoscale plasmonic structures," *Physical Review Letters*, vol. 103, no. 12, Article ID 123004, 4 pages, 2009.
[49] C. Stehle, H. Bender, C. Zimmermann, D. Kern, M. Fleischer, and S. Slama, "Plasmonically tailored micropotentials for ultracold atoms," *Nature Photonics*, vol. 5, no. 8, pp. 494–498, 2011.
[50] F. Serwane, G. Zürn, T. Lompe, T. B. Ottenstein, A. N. Wenz, and S. Jochim, "Deterministic preparation of a tunable few-fermion system," *Science*, vol. 332, no. 6027, pp. 336–338, 2011.
[51] T. Busch, B. G. Englert, K. Rzazewski, and M. Wilkens, "Two cold atoms in a harmonic trap," *Foundations of Physics*, vol. 28, no. 4, pp. 549–559, 1998.
[52] T. Stöferle, H. Moritz, K. Günter, M. Köhl, and T. Esslinger, "Molecules of fermionic atoms in an optical lattice," *Physical Review Letters*, vol. 96, no. 3, Article ID 030401, 4 pages, 2006.
[53] T. Volz, N. Syassen, D. M. Bauer, E. Hansis, S. Dürr, and G. Rempe, "Preparation of a quantum state with one molecule

at each site of an optical lattice," *Nature Physics*, vol. 2, no. 10, pp. 692–695, 2006.

[54] G. Thalhammer, K. Winkler, F. Lang, S. Schmid, R. Grimm, and J. Hecker Denschlag, "Long-lived Feshbach molecules in a three-dimensional optical lattice," *Physical Review Letters*, vol. 96, no. 5, Article ID 050402, 4 pages, 2006.

[55] C. Ospelkaus, S. Ospelkaus, L. Humbert, P. Ernst, K. Sengstock, and K. Bongs, "Ultracold heteronuclear molecules in a 3D optical lattice," *Physical Review Letters*, vol. 97, no. 12, Article ID 120402, 4 pages, 2006.

[56] D. Blume and C. H. Greene, "Fermi pseudopotential approximation: two particles under external confinement," *Physical Review A*, vol. 65, no. 4, Article ID 043613, 6 pages, 2002.

[57] F. Werner and Y. Castin, "Unitary quantum three-body problem in a. Harmonic trap," *Physical Review Letters*, vol. 97, Article ID 150401, 2006.

[58] F. Werner and Y. Castin, "Unitary gas in an isotropic harmonic trap: symmetry properties and applications," *Physical Review A*, vol. 74, no. 5, Article ID 053604, 2006.

[59] Z. Idziaszek and T. Calarco, "Analytical solutions for the dynamics of two trapped interacting ultracold atoms," *Physical Review A*, vol. 74, no. 2, Article ID 022712, 2006.

[60] A. Suzuki, Y. Liang, and R. K. Bhaduri, "Two-atom energy spectrum in a harmonic trap near a Feshbach resonance at higher partial waves," *Physical Review A*, vol. 80, no. 3, Article ID 033601, 6 pages, 2009.

[61] S. G. Peng, S. Q. Li, P. D. Drummond, and X. J. Liu, "High-temperature thermodynamics of strongly interacting s-wave and p-wave Fermi gases in a harmonic trap," *Physical Review A*, vol. 83, no. 6, Article ID 063618, 10 pages, 2011.

[62] N. T. Zinner, "Universal two-body spectra of ultracold harmonically trapped atoms in two and three dimensions," *Journal of Physics A*, vol. 45, no. 20, Article ID 205302, 2012.

[63] J. L. DuBois and H. R. Glyde, "Bose-Einstein condensation in trapped bosons: a variational Monte Carlo analysis," *Physical Review A*, vol. 63, no. 2, Article ID 023602, 2001.

[64] J. Carlson, S. Y. Chang, V. R. Pandharipande, and K. E. Schmidt, "Superfluid Fermi gases with large scattering length," *Physical Review Letters*, vol. 91, no. 5, Article ID 050401, 4 pages, 2003.

[65] S. Y. Chang and G. F. Bertsch, "Unitary Fermi gas in a harmonic trap," *Physical Review A*, vol. 76, no. 2, Article ID 021603, 2007.

[66] D. Blume, J. Von Stecher, and C. H. Greene, "Universal properties of a trapped two-component fermi gas at unitarity," *Physical Review Letters*, vol. 99, no. 23, Article ID 233201, 2007.

[67] M. Thøgersen, D. V. Fedorov, and A. S. Jensen, "Trapped Bose gases with large positive scattering length," vol. 79, no. 4, Article ID 40002, 6 pages, 2007.

[68] J. von Stecher, C. H. Greene, and D. Blume, "Energetics and structural properties of trapped two-component Fermi gases," *Physical Review A*, vol. 77, no. 4, Article ID 043619, 20 pages, 2008.

[69] D. Lee, "Lattice simulations for few- and many-body systems," *Progress in Particle and Nuclear Physics*, vol. 63, no. 1, pp. 117–154, 2009.

[70] W. C. Haxton and T. Luu, "Perturbative effective theory in an oscillator basis?" *Physical Review Letters*, vol. 89, no. 18, Article ID 182503, 4 pages, 2002.

[71] I. Stetcu, B. R. Barrett, U. van Kolck, and J. P. Vary, "Effective theory for trapped few-fermion systems," *Physical Review A*, vol. 76, no. 6, Article ID 063613, 7 pages, 2007.

[72] Y. Alhassid, G. F. Bertsch, and L. Fang, "New effective interaction for the trapped fermi gas," *Physical Review Letters*, vol. 100, no. 23, Article ID 230401, 2008.

[73] N. T. Zinner, K. M. Mølmer, C. Özen, D. J. Dean, and K. Langanke, "Shell-model Monte Carlo simulations of the BCS-BEC crossover in few-fermion systems," *Physical Review A*, vol. 80, no. 1, Article ID 013613, 5 pages, 2009.

[74] I. Stetcu, J. Rotureau, B. R. Barrett, and U. van Kolck, "An effective field theory approach to two trapped particles," *Annals of Physics*, vol. 325, no. 8, pp. 1644–1666, 2010.

[75] T. Luu, M. J. Savage, A. Schwenk, and J. P. Vary, "Nucleon-nucleon scattering in a harmonic potential," *Physical Review C*, vol. 82, no. 3, Article ID 034003, 2010.

[76] J. Rotureau, I. Stetcu, B. R. Barrett, M. C. Birse, and U. Van Kolck, "Three and four harmonically trapped particles in an effective-field-theory framework," *Physical Review A*, vol. 82, no. 3, Article ID 032711, 2010.

[77] J. R. Armstrong, N. T. Zinner, D. V. Fedorov, and A. S. Jensen, "Analytic harmonic approach to the N-body problem," *Journal of Physics B*, vol. 44, no. 5, Article ID 055303, 2011.

[78] J. R. Armstrong, N. T. Zinner, D. V. Fedorov, and A. S. Jensen, "Quantum statistics and thermodynamics in the harmonic approximation," *Physical Review E*, vol. 85, no. 2, Article ID 021117, 10 pages, 2012.

[79] J. R. Armstrong, N. T. Zinner, D. V. Fedorov, and A. S. Jensen, "Virial expansion coefficients in the harmonicapproximation," http://arxiv.org/abs/1205.2574.

[80] N. T. Zinner, "Efimov trimers near the zero-crossing of a Feshbach resonance," http://arxiv.org/abs/1112.6358.

[81] S. Tan, "Energetics of a strongly correlated Fermi gas," *Annals of Physics*, vol. 323, no. 12, pp. 2952–2970, 2008.

[82] S. Tan, "Large momentum part of a strongly correlated Fermi gas," *Annals of Physics*, vol. 323, no. 12, pp. 2971–2986, 2008.

[83] E. Braaten and L. Platter, "Exact relations for a strongly interacting fermi gas from the operator product expansion," *Physical Review Letters*, vol. 100, no. 20, Article ID 205301, 2008.

[84] S. Zhang and A. J. Leggett, "Universal properties of the ultracold Fermi gas," *Physical Review A*, vol. 79, no. 2, Article ID 023601, 2009.

[85] R. Combescot, F. Alzetto, and X. Leyronas, "Particle distribution tail and related energy formula," *Physical Review A*, vol. 79, no. 5, Article ID 053640, 2009.

[86] F. Werner and Y. Castin, "Exact relations for quantum-mechanical few-body and many-body problems with short-range interactions in two and three dimensions," http://arxiv.org/abs/1001.0774.

[87] M. Barth and W. Zwerger, "Tan relations in one dimension," *Annals of Physics*, vol. 326, no. 10, pp. 2544–2565, 2011.

[88] M. Valiente, N. T. Zinner, and K. M. Mølmer, "Universal relations for the two-dimensional spin-1/2 Fermi gas with contact interactions," *Physical Review A*, vol. 84, no. 6, Article ID 063626, 4 pages, 2011.

[89] M. Valiente, "Tan's distributions and Fermi-Huang pseudopotential in momentum space," *Physical Review A*, vol. 85, no. 1, Article ID 014701, 4 pages, 2012.

[90] M. Valiente, N. T. Zinner, and K. M. Mølmer, "Universal properties of Fermi gases in arbitrary dimensions," http://arxiv.org/abs/1205.6388.

[91] J. T. Stewart, J. P. Gaebler, T. E. Drake, and D. S. Jin, "Rification of universal relations in a strongly interacting fermi gas," *Physical Review Letters*, vol. 104, no. 23, Article ID 235301, 2010.

[92] E. D. Kuhnle, H. Hu, X. J. Liu et al., "Universal behavior of pair correlations in a strongly interacting fermi gas," *Physical Review Letters*, vol. 105, no. 7, Article ID 070402, 2010.

[93] E. D. Kuhnle, S. Hoinka, P. Dyke, H. Hu, P. Hannaford, and C. J. Vale, "Temperature dependence of the universal contact parameter in a unitary Fermi gas," *Physical Review Letters*, vol. 106, no. 17, Article ID 170402, 2011.

[94] E. Braaten, D. Kang, and L. Platter, "Universal relations for identical bosons from three-body physics," *Physical Review Letters*, vol. 106, no. 15, Article ID 153005, 2011.

[95] Y. Castin and F. Werner, "Single-particle momentum distribution of an Efimov trimer," *Physical Review A*, vol. 83, no. 6, Article ID 063614, 2011.

[96] K. Helfrich and H. W. Hammer, "Resonant three-body physics in two spatial dimensions," *Physical Review A*, vol. 83, no. 5, Article ID 052703, 7 pages, 2011.

[97] F. F. Bellotti, T. Frederico, M. T. Yamashita, D. V. Fedorov, A. S. Jensen, and N. T. Zinner, "Scaling and universality in two dimensions: three-body bound states with short-ranged interactions," *Journal of Physics B*, vol. 44, no. 20, Article ID 205302, 2011.

[98] F. F. Bellotti, T. Frederico, M. T. Yamashita, D. V. Fedorov, A. S. Jensen, and N. T. Zinner, "Supercircle description of universal three-body states in two dimensions," *Physical Review A*, vol. 85, no. 2, Article ID 025601, 5 pages, 2012.

[99] R. J. Wild, P. Makotyn, J. M. Pino, E. A. Cornell, and D. S. Jin, "Measurements of Tan's contact in an atomic bose-Einstein condensate," *Physical Review Letters*, vol. 108, no. 14, Article ID 145305, 5 pages, 2012.

[100] E. Braaten, D. Kang, and L. Platter, "Universal relations for a strongly interacting Fermi gas near a Feshbach resonance," *Physical Review A*, vol. 78, no. 5, Article ID 053606, 2008.

[101] F. Werner, "Virial theorems for trapped cold atoms," *Physical Review A*, vol. 78, no. 2, Article ID 025601, 4 pages, 2008.

Transition Parameters for Doubly Ionized Lanthanum

Betül Karaçoban and Leyla Özdemir

Department of Physics, Sakarya University, 54187 Sakarya, Turkey

Correspondence should be addressed to Betül Karaçoban, bkaracoban@sakarya.edu.tr

Academic Editor: Alan Migdall

The transition parameters such as the wavelengths, weighted oscillator strengths, and transition probabilities (or rates) for the nd $(n = 5-9)-\text{nf}(n = 4-8)$, nd $(n = 5-9)-\text{np}\,(n = 6-9)$, np $(n = 6-9)-\text{ns}\,(n = 6-10)$, and ng $(n = 5-8)-\text{nf}(n = 4-8)$ electric dipole ($E1$) transitions of doubly ionized lanthanum (La III, $Z = 57$) have been calculated using the relativistic Hartree-Fock (HFR) method. In this method, configuration interaction and relativistic effects have been included in the computations combined with a least squares fitting of the Hamiltonian eigenvalues to the observed energy levels. We have compared the results obtained from this work with the previously available calculations and experiments in literature. We have also reported new transitions with the weighted transition probabilities greater than or equal to 10^5.

1. Introduction

The radiative properties of the lanthanides and their ions have been rather little considered. This can be explained by the fact that these atoms or ions are characterized by complex electronic structures with an unfilled 4f subshell, which makes the calculations very difficult, and that the laboratory analyses are still extremely fragmentary or even missing for many ions. Owing to the importance of rare earth elements in astrophysics, especially in relation to nucleosynthesis and star formation (notably the lanthanides in chemically peculiar (CP) stars) [7], there is a growing need for accurate spectroscopic data, that is, wavelengths, radiative transition rates, oscillator strengths, branching fractions, radiative lifetimes, hyperfine structure, and isotope shift data for lanthanide atoms and ions.

The lanthanum atom is the first member of the rare earth elements. Doubly ionized lanthanum (La III) is characterized by a simple atomic structure with core [Xe] and only one outer electron. There is substantial spectroscopic literature concerning La III, though less than the neutral or singly ionized species. The available theoretical and experimental works on energy levels, radiative lifetimes, and transition parameters for La III can be found in the literature [1–3, 5, 6, 8–13]. These works were reported in our previous work in detail [14].

Up till now the wavelengths, oscillator strengths, and transition probabilities available for La III were obtained by experimental, semiempirical, or pure theoretical approaches. Sixty-five spectral lines of La III in the 2000–12000 Å interval were reported by Odabasi [2]. Sugar and Kaufman [13] observed forty-five La III spectral lines in the interval from 700 to 2000 Å. Johansson and Litzén [5] recorded wavelengths of 5d–4f lines of La III. Relativistic single-configuration Hartree-Fock oscillator strengths for 6s–6p transitions in La III were reported by Migdalek and Baylis [4]. Migdalek and Wyrozumska [3] have calculated oscillator strengths obtained using the relativistic model-potential approach in there different versions: a model-potential without valence-core electron exchange but with core-polarization included (RMP + CP), with semiclassical exchange and core-polarization (RMP + SCE + CP), and with empirically adjusted exchange and core-polarization (RMP + EX + CP) for the 6s–6p, 5d–6p, 5d–4f, 5d–5f, 5d–6f, 6p–6d, and 6p–7d transition arrays. The single-configuration relativistic Hartree-Fock ionization potentials of La III were computed by Migdalek and Bojara [9]. Biémont et al. [1] have performed oscillator strengths and transition probabilities in La III by relativistic Hartree-Fock method with core-polarization.

Our aim here is to determine the transition parameters, such as the wavelengths, oscillator strengths, and transition

Table 1: Wavelengths λ(Å), weighted oscillator strengths gf, and weighted transition probabilities gA_{ki} (s^{-1}) for electric dipole ($E1$) transitions in La III.

Transition		λ		gf		gA_{ki}	
Lower level	Upper level	This work	Other works	This work	Other works	This work	Other works
6p $^{2}P^{o}_{1/2}$	7s $^{2}S_{1/2}$	2479.41[A,B]	2478.66[a]	0.489[A]	0.475[a]	5.31×10^{8}[A]	5.16×10^{8}[a]
			2478.652[c]	0.463[B]		5.03×10^{8}[B]	
6p $^{2}P^{o}_{3/2}$	7s $^{2}S_{1/2}$	2685.55[A,B]	2684.75[a]	0.904[A]	0.878[a]	8.36×10^{8}[A]	8.12×10^{8}[a]
			2684.757[c]	0.855[B]		7.91×10^{8}[B]	
6p $^{2}P^{o}_{1/2}$	6d $^{2}D_{3/2}$	2477.36[A]	2476.60[a]	2.474[A]	2.365[a]	2.69×10^{9}[A]	2.57×10^{9}[a]
		2477.35[B]	2476.599[c]	2.651[B]	2.142[d]	2.88×10^{9}[B]	2.27×10^{9}[b]
6p $^{2}P^{o}_{3/2}$	6d $^{2}D_{5/2}$	2652.29[A]	2651.50[a]	4.168[A]	3.976[a]	3.95×10^{9}[A]	3.77×10^{9}[a]
		2652.28[B]	2651.501[c]	4.457[B]	3.972[d]	4.23×10^{9}[B]	3.66×10^{9}[b]
6p $^{2}P^{o}_{3/2}$	6d $^{2}D_{3/2}$	2683.14[A]	2682.34[a]	0.458[A]	0.437[a]	4.24×10^{8}[A]	4.05×10^{8}[a]
		2683.13[B]	2682.345[c]	0.489[B]	0.444[d]	4.54×10^{8}[B]	4.01×10^{8}[b]
6p $^{2}P^{o}_{1/2}$	8s $^{2}S_{1/2}$	1466.39[A,B]	1466.39[a]	0.071[A]	0.069[a]	2.20×10^{8}[A]	2.15×10^{8}[a]
				0.064[B]		2.00×10^{8}[B]	
6p $^{2}P^{o}_{3/2}$	8s $^{2}S_{1/2}$	1536.13[A,B]	1536.13[a]	0.136[A]	0.132[a]	3.83×10^{8}[A]	3.73×10^{8}[a]
				0.123[B]		3.48×10^{8}[B]	
6p $^{2}P^{o}_{1/2}$	7d $^{2}D_{3/2}$	1459.45[A,B]	1459.45[a]	0.158[A]	0.156[a]	4.96×10^{8}[A]	4.88×10^{8}[a]
				0.216[B]	0.137[d]	6.76×10^{8}[B]	4.26×10^{8}[b]
6p $^{2}P^{o}_{3/2}$	7d $^{2}D_{5/2}$	1523.75[A,B]	1523.75[a]	0.277[A]	0.269[a]	7.95×10^{8}[A]	7.71×10^{8}[a]
				0.372[B]	0.211[d]	10.70×10^{8}[B]	6.02×10^{8}[b]
6p $^{2}P^{o}_{3/2}$	7d $^{2}D_{3/2}$	1528.51[A,B]	1528.51[a]	0.031[A]	0.030[a]	8.71×10^{7}[A]	8.49×10^{7}[a]
				0.041[B]	0.022[d]	11.80×10^{7}[B]	6.23×10^{7}[b]
6p $^{2}P^{o}_{1/2}$	9s $^{2}S_{1/2}$	1212.28[A,B]	1212.28[a]	0.026[A]	0.026[a]	1.19×10^{8}[A]	1.16×10^{8}[a]
				0.023[B]		1.06×10^{8}[B]	
6p $^{2}P^{o}_{3/2}$	9s $^{2}S_{1/2}$	1259.55[A,B]	1259.55[a]	0.051[A]	0.049[a]	2.13×10^{8}[A]	2.07×10^{8}[a]
				0.045[B]		1.89×10^{8}[B]	
6p $^{2}P^{o}_{1/2}$	8d $^{2}D_{3/2}$	1208.79[A,B]	1208.79[a]	0.040[A]	0.041[a]	1.80×10^{8}[A]	1.86×10^{8}[a]
				0.065[B]		2.95×10^{8}[B]	
6p $^{2}P^{o}_{3/2}$	8d $^{2}D_{5/2}$	1254.00[A,B]	1254.00[a]	0.070[A]	0.071[a]	2.98×10^{8}[A]	3.00×10^{8}[a]
				0.112[B]		4.76×10^{8}[B]	
6p $^{2}P^{o}_{3/2}$	8d $^{2}D_{3/2}$	1255.79[A,B]	1255.79[a]	0.008[A]	0.008[a]	3.27×10^{7}[A]	3.32×10^{7}[a]
				0.012[B]		5.27×10^{7}[B]	
6p $^{2}P^{o}_{1/2}$	10s $^{2}S_{1/2}$	1101.01[A,B]	1101.01[a]	0.013[A]	0.013[a]	7.26×10^{7}[A]	7.07×10^{7}[a]
				0.012[B]		6.39×10^{7}[B]	
6p $^{2}P^{o}_{3/2}$	10s $^{2}S_{1/2}$	1139.87[A,B]	1139.87[a]	0.026[A]	0.025[a]	1.31×10^{8}[A]	1.27×10^{8}[a]
				0.022[B]		1.15×10^{8}[B]	
6p $^{2}P^{o}_{1/2}$	9d $^{2}D_{3/2}$	1099.00[A,B]	1099.00[a]	0.015[A]	0.017[a]	8.54×10^{7}[A]	9.21×10^{7}[a]
				0.029[B]		16.10×10^{7}[B]	
6p $^{2}P^{o}_{3/2}$	9d $^{2}D_{5/2}$	1136.80[A,B]	1136.80[a]	0.028[A]	0.029[a]	1.44×10^{8}[A]	1.50×10^{8}[a]
				0.050[B]		2.61×10^{8}[B]	
6p $^{2}P^{o}_{3/2}$	9d $^{2}D_{3/2}$	1137.71[A,B]	1137.71[a]	0.003[A]	0.003[a]	1.58×10^{7}[A]	1.66×10^{7}[a]
				0.006[B]		2.89×10^{7}[B]	
6d $^{2}D_{3/2}$	5f $^{2}F^{o}_{5/2}$	9926.70[A]	9924.04[a]	2.549[A]	2.370[a]	1.73×10^{8}[A]	1.60×10^{8}[a]
		9926.74[B]	9923.989[c]	2.574[B]		1.74×10^{8}[B]	
6d $^{2}D_{5/2}$	5f $^{2}F^{o}_{7/2}$	10287.59[A,B]	10284.790[c]	3.515[A]	—	2.21×10^{8}[A]	—
				3.548[B]		2.24×10^{8}[B]	
6d $^{2}D_{5/2}$	5f $^{2}F^{o}_{5/2}$	10373.15[A]	10370.335[c]	0.174[A]	—	1.08×10^{7}[A]	—
		10373.12[B]		0.176[B]		1.09×10^{7}[B]	
6d $^{2}D_{3/2}$	7p $^{2}P^{o}_{3/2}$	8277.67[A,B]	8275.41[a]	0.243[A]	0.240[a]	2.37×10^{7}[A]	2.34×10^{7}[a]
			8275.388[c]	0.250[B]		2.43×10^{7}[B]	
6d $^{2}D_{5/2}$	7p $^{2}P^{o}_{3/2}$	8585.81[A]	8583.42[a]	2.115[A]	2.081[a]	1.91×10^{8}[A]	1.88×10^{8}[a]
		8585.76[B]	8583.453[c]	2.165[B]		1.96×10^{8}[B]	

Table 1: Continued.

Transition		λ		gf		gA_{ki}	
Lower level	Upper level	This work	Other works	This work	Other works	This work	Other works
$6d\ ^2D_{3/2}$	$7p\ ^2P^o_{1/2}$	9215.20[A]	9212.68[a]	1.094[A]	1.077[a]	8.59×10^{7}[A]	8.46×10^{7}[a]
		9215.23[B]	9212.628[c]	1.121[B]		8.80×10^{7}[B]	
$6d\ ^2D_{3/2}$	$6f\ ^2F^o_{5/2}$	3076.05[A]	3075.17[a]	0.733[A]	0.822[a]	5.17×10^{8}[A]	5.80×10^{8}[a]
		3076.06[B]	3075.173[c]	0.790[B]		5.57×10^{8}[B]	
$6d\ ^2D_{5/2}$	$6f\ ^2F^o_{7/2}$	3112.88[A,B]	3111.97[a]	1.037[A]	1.161[a]	7.14×10^{8}[A]	7.99×10^{8}[a]
			3111.969[c]	1.116[B]		7.68×10^{8}[B]	
$6d\ ^2D_{5/2}$	$6f\ ^2F^o_{5/2}$	3117.63[A,B]	3116.74[a]	0.052[A]	0.058[a]	3.55×10^{7}[A]	3.98×10^{7}[a]
			3116.738[c]	0.056[B]		3.82×10^{7}[B]	
$6d\ ^2D_{3/2}$	$8p\ ^2P^o_{3/2}$	2954.63[A]	2953.77[a]	0.009[A]	0.008[a]	6.82×10^{6}[A]	5.87×10^{6}[a]
		2954.64[B]		0.008[B]		6.18×10^{6}[B]	
$6d\ ^2D_{5/2}$	$8p\ ^2P^o_{3/2}$	2992.97[A,B]	2992.10[a]	0.079[A]	0.068[a]	5.89×10^{7}[A]	5.08×10^{7}[a]
			2992.098[c]	0.072[B]		5.35×10^{7}[B]	
$6d\ ^2D_{3/2}$	$8p\ ^2P^o_{1/2}$	3010.10[A]	3009.22[a]	0.044[A]	0.038[a]	3.23×10^{7}[A]	2.77×10^{7}[a]
		3010.11[B]	3009.223[c]	0.040[B]		2.92×10^{7}[B]	
$6d\ ^2D_{3/2}$	$7f\ ^2F^o_{5/2}$	2239.04[A]	2238.35[a]	0.339[A]	0.358[a]	4.51×10^{8}[A]	4.76×10^{8}[a]
		2239.05[B]		0.356[B]		4.74×10^{8}[B]	
$6d\ ^2D_{5/2}$	$7f\ ^2F^o_{7/2}$	2259.31[A,B]	2258.61[a]	0.480[A]	0.507[a]	6.28×10^{8}[A]	6.62×10^{8}[a]
			2258.609[c]	0.504[B]		6.59×10^{8}[B]	
$6d\ ^2D_{5/2}$	$7f\ ^2F^o_{5/2}$	2261.00[A,B]	2260.30[a]	0.024[A]	0.025[a]	3.13×10^{7}[A]	3.30×10^{7}[a]
			2260.295[c]	0.025[B]		3.29×10^{7}[B]	
$6d\ ^2D_{3/2}$	$9p\ ^2P^o_{3/2}$	2195.18[A]	2194.50[a]	0.003[A,B]	0.002[a]	4.15×10^{6}[A]	3.28×10^{6}[a]
		2195.19[B]				3.58×10^{6}[B]	
$6d\ ^2D_{3/2}$	$9p\ ^2P^o_{1/2}$	2213.95[A,B]	2213.26[a]	0.015[A]	0.012[a]	2.02×10^{7}[A]	1.60×10^{7}[a]
				0.013[B]		1.75×10^{7}[B]	
$6d\ ^2D_{5/2}$	$9p\ ^2P^o_{3/2}$	2216.28[A,B]	2215.58[a]	0.027[A]	0.021[a]	3.62×10^{7}[A]	2.87×10^{7}[a]
				0.023[B]		3.13×10^{7}[B]	
$6d\ ^2D_{3/2}$	$8f\ ^2F^o_{5/2}$	1923.33[A]	1923.33[a]	0.178[A]	0.185[a]	3.22×10^{8}[A]	3.33×10^{8}[a]
		1923.34[B]		0.188[B]		3.39×10^{8}[B]	
$6d\ ^2D_{5/2}$	$8f\ ^2F^o_{7/2}$	1938.53[A,B]	1938.53[a]	0.253[A]	0.262[a]	4.49×10^{8}[A]	4.65×10^{8}[a]
				0.267[B]		4.73×10^{8}[B]	
$6d\ ^2D_{5/2}$	$8f\ ^2F^o_{5/2}$	1939.51[A,B]	1939.51[a]	0.013[A,B]	0.013[a]	2.24×10^{7}[A]	2.32×10^{7}[a]
						2.36×10^{7}[B]	
$6f\ ^2F^o_{5/2}$	$6g\ ^2G_{7/2}$	8290.18[A]	8287.76[a]	8.903[A]	8.527[a]	8.64×10^{8}[A,B]	8.28×10^{8}[a]
		8290.16[B]	8287.752[c]	8.904[B]			
$6f\ ^2F^o_{7/2}$	$6g\ ^2G_{9/2}$	8323.43[A]	8321.16[a]	11.495[A]	11.009[a]	1.11×10^{9}[A,B]	1.06×10^{9}[a]
		8323.34[B]	8321.107[c]	11.496[B]			
$6f\ ^2F^o_{7/2}$	$6g\ ^2G_{7/2}$	8323.98[A]	8321.63[a]	0.328[A,B]	0.315[a]	3.16×10^{7}[A,B]	3.03×10^{7}[a]
		8323.97[B]					
$6f\ ^2F^o_{5/2}$	$7g\ ^2G_{7/2}$	5147.17[A,B]	5145.72[a]	1.282[A]	1.239[a]	3.23×10^{8}[A,B]	3.12×10^{8}[a]
			5145.729[c]	1.283[B]			
$6f\ ^2F^o_{7/2}$	$7g\ ^2G_{9/2}$	5159.84[A]	5158.39[a]	1.658[A]	1.602[a]	4.15×10^{8}[A]	4.01×10^{8}[a]
		5159.82[B]	5158.410[c]	1.659[B]		4.16×10^{8}[B]	
$6f\ ^2F^o_{7/2}$	$7g\ ^2G_{7/2}$	5160.18[A]	5158.76[a]	0.047[A,B]	0.046[a]	1.19×10^{7}[A,B]	1.15×10^{7}[a]
		5160.19[B]					
$6f\ ^2F^o_{5/2}$	$8g\ ^2G_{7/2}$	4130.43[A,B]	4129.24[a]	0.411[A,B]	0.394[a]	1.61×10^{8}[A,B]	1.54×10^{8}[a]
$6f\ ^2F^o_{7/2}$	$8g\ ^2G_{9/2}$	4138.59[A]	4137.43[a]	0.531[A]	0.509[a]	2.07×10^{8}[A,B]	1.98×10^{8}[a]
		4138.58[B]	4137.428[c]	0.532[B]			
$6f\ ^2F^o_{7/2}$	$8g\ ^2G_{7/2}$	4138.80[A]	4137.64[a]	0.015[A,B]	0.015[a]	5.91×10^{6}[A,B]	5.67×10^{6}[a]
		4138.81[B]					
$6f\ ^2F^o_{5/2}$	$9d\ ^2D_{3/2}$	5519.77[A]	5518.19[a]	0.132[A]	0.130[a]	2.89×10^{7}[A]	2.85×10^{7}[a]
		5519.75[B]	5518.187[c]	0.128[B]		2.79×10^{7}[B]	

Table 1: Continued.

Transition		λ		gf		gA_{ki}	
Lower level	Upper level	This work	Other works	This work	Other works	This work	Other works
$6f\,^2F^o_{5/2}$	$9d\,^2D_{5/2}$	5498.45[A] 5498.43[B]	5496.88[a]	0.009[A,B]	0.009[a]	2.08×10^{6} [A] 2.02×10^{6} [B]	2.06×10^{6} [a]
$6f\,^2F^o_{7/2}$	$9d\,^2D_{5/2}$	5513.29[A,B]	5511.76[a] 5511.721[c]	0.188[A] 0.182[B]	0.186[a]	4.14×10^{7} [A] 4.01×10^{7} [B]	4.09×10^{7} [a]
$6s\,^2S_{1/2}$	$6p\,^2P^o_{3/2}$	3172.60[A,B]	3171.69[a] 3171.735[c]	1.673[A] 1.935[B]	1.527[a] 1.418[d] 1.868[e]	1.11×10^{9} [A] 1.28×10^{9} [B]	1.01×10^{9} [a] 9.40×10^{8} [b]
$6s\,^2S_{1/2}$	$6p\,^2P^o_{1/2}$	3518.16[A] 3518.15[B]	3517.16[a] 3517.217[c]	0.754[A] 0.872[B]	0.689[a] 0.640[d] 0.850[e]	4.06×10^{8} [A] 4.70×10^{8} [B]	3.71×10^{8} [a] 3.45×10^{8} [b]
$6s\,^2S_{1/2}$	$7p\,^2P^o_{3/2}$	1236.55[A,B]	1236.55[a]	0.002[A] 0.001[B]	0.006[a]	0.70×10^{7} [A] 0.27×10^{7} [B]	2.39×10^{7} [a]
$6s\,^2S_{1/2}$	$7p\,^2P^o_{1/2}$	1255.63[A,B]	1255.63[a]	0.001[A] 0.0003[B]	0.003[a]	0.33×10^{7} [A] 0.13×10^{7} [B]	1.14×10^{7} [a]
$7s\,^2S_{1/2}$	$7p\,^2P^o_{3/2}$	8254.85[A] 8254.77[B]	8252.53[a] 8252.603[c]	2.418[A] 2.424[B]	2.279[a]	2.37×10^{8} [A,B]	2.23×10^{8} [a]
$7s\,^2S_{1/2}$	$7p\,^2P^o_{1/2}$	9186.87[A] 9186.92[B]	9184.34[a] 9184.380[c]	1.086[A] 1.089[B]	1.024[a]	8.59×10^{7} [A] 8.61×10^{7} [B]	8.09×10^{7} [a]
$7s\,^2S_{1/2}$	$8p\,^2P^o_{3/2}$	2951.72[A,B]	2950.843[c]	0.002[A,B]	—	1.81×10^{6} [A] 1.54×10^{6} [B]	—
$7s\,^2S_{1/2}$	$8p\,^2P^o_{1/2}$	3007.07[A] 3007.08[B]	3006.186[c]	0.001[A,B]	—	8.57×10^{5} [A] 7.27×10^{5} [B]	—
$7p\,^2P^o_{1/2}$	$8s\,^2S_{1/2}$	5890.23[A] 5890.25[B]	5888.63[a] 5888.620[c]	0.716[A] 0.714[B]	0.718[a]	1.38×10^{8} [A] 1.37×10^{8} [B]	1.38×10^{8} [a]
$7p\,^2P^o_{3/2}$	$8s\,^2S_{1/2}$	6349.93[A] 6349.97[B]	6348.21[a] 6348.213[c]	1.329[A] 1.324[B]	1.331[a]	2.20×10^{8} [A] 2.19×10^{8} [B]	2.20×10^{8} [a]
$7p\,^2P^o_{1/2}$	$7d\,^2D_{3/2}$	5779.74[A] 5779.71[B]	5778.14[a] 5778.138[c]	3.095[A] 3.045[B]	2.967[a]	6.18×10^{8} [A] 6.08×10^{8} [B]	5.92×10^{8} [a]
$7p\,^2P^o_{3/2}$	$7d\,^2D_{5/2}$	6143.64[A] 6143.71[B]	6141.99[a] 6141.987[c]	5.238[A] 5.157[B]	5.024[a]	9.26×10^{8} [A] 9.11×10^{8} [B]	8.88×10^{8} [a]
$7p\,^2P^o_{3/2}$	$7d\,^2D_{3/2}$	6221.70[A] 6221.69[B]	6219.99[a] 6219.999[c]	0.575[A] 0.566[B]	0.551[a]	9.91×10^{7} [A] 9.75×10^{7} [B]	9.50×10^{7} [a]
$7p\,^2P^o_{1/2}$	$9s\,^2S_{1/2}$	3197.77[A] 3197.78[B]	3196.85[a] 3196.844[c]	0.089[A,B]	0.092[a]	5.77×10^{7} [A,B]	5.98×10^{7} [a]
$7p\,^2P^o_{3/2}$	$9s\,^2S_{1/2}$	3328.60[A] 3328.61[B]	3327.64[a] 3327.655[c]	0.170[A,B]	0.176[a]	1.02×10^{8} [A,B]	1.06×10^{8} [a]
$7p\,^2P^o_{1/2}$	$8d\,^2D_{3/2}$	3173.60[A] 3173.61[B]	3172.69[a] 3172.689[c]	0.274[A] 0.284[B]	0.260[a]	1.81×10^{8} [A] 1.89×10^{8} [B]	1.72×10^{8} [a]
$7p\,^2P^o_{3/2}$	$8d\,^2D_{5/2}$	3290.05[A] 3290.07[B]	3289.11[a] 3289.110[c]	0.476[A] 0.494[B]	0.452[a]	2.94×10^{8} [A] 3.05×10^{8} [B]	2.79×10^{8} [a]
$7p\,^2P^o_{3/2}$	$8d\,^2D_{3/2}$	3302.41[A] 3302.43[B]	3301.47[a] 3301.481[c]	0.053[A] 0.055[B]	0.050[a]	3.22×10^{7} [A] 3.35×10^{7} [B]	3.06×10^{7} [a]
$7p\,^2P^o_{1/2}$	$10s\,^2S_{1/2}$	2524.74[A,B]	2523.98[a]	0.031[A,B]	0.033[a]	3.25×10^{7} [A] 3.27×10^{7} [B]	3.44×10^{7} [a]
$7p\,^2P^o_{3/2}$	$10s\,^2S_{1/2}$	2605.59[A] 2605.60[B]	2604.82[a] 2604.827[c]	0.060[A,B]	0.064[a]	5.91×10^{7} [A] 5.94×10^{7} [B]	6.25×10^{7} [a]
$7p\,^2P^o_{1/2}$	$9d\,^2D_{3/2}$	2514.19[A,B]	2513.43[a] 2513.432[c]	0.083[A] 0.088[B]	0.077[a]	8.72×10^{7} [A] 9.27×10^{7} [B]	8.10×10^{7} [a]
$7p\,^2P^o_{3/2}$	$9d\,^2D_{5/2}$	2589.64[A,B]	2588.86[a] 2588.867[c]	0.145[A] 0.153[B]	0.134[a]	1.44×10^{8} [A] 1.53×10^{8} [B]	1.33×10^{8} [a]
$7p\,^2P^o_{3/2}$	$9d\,^2D_{3/2}$	2594.36[A,B]	2593.58[a]	0.016[A] 0.017[B]	0.015[a]	1.59×10^{7} [A] 1.69×10^{7} [B]	1.47×10^{7} [a]

TABLE 1: Continued.

Transition		λ		gf		gA_{ki}	
Lower level	Upper level	This work	Other works	This work	Other works	This work	Other works
7d $^2D_{3/2}$	7f $^2F^o_{5/2}$	6057.51[A]	6055.85[a]	0.518[A]	0.604[a]	0.94×10^{8}[A]	1.10×10^{8}[a]
		6057.56[B]	6055.838[c]	0.578[B]		1.05×10^{8}[B]	
7d $^2D_{5/2}$	7f $^2F^o_{7/2}$	6121.04[A]	6119.27[a]	0.735[A]	0.854[a]	1.31×10^{8}[A]	1.52×10^{8}[a]
		6121.00[B]	6119.254[c]	0.818[B]		1.46×10^{8}[B]	
7d $^2D_{5/2}$	7f $^2F^o_{5/2}$	6133.39[A]	6131.67[a]	0.037[A]	0.043[a]	6.51×10^{6}[A]	7.55×10^{6}[a]
		6133.35[B]		0.041[B]		7.24×10^{6}[B]	
7d $^2D_{3/2}$	8f $^2F^o_{5/2}$	4194.71[A]	4193.51[a]	0.260[A]	0.292[a]	0.99×10^{8}[A]	1.11×10^{8}[a]
		4194.72[B]		0.283[B]		1.07×10^{8}[B]	
7d $^2D_{5/2}$	8f $^2F^o_{7/2}$	4226.33[A]	4225.12[a]	0.370[A]	0.414[a]	1.38×10^{8}[A]	1.54×10^{8}[a]
		4226.29[B]		0.402[B]		1.50×10^{8}[B]	
7d $^2D_{5/2}$	8f $^2F^o_{5/2}$	4230.95[A]	4229.73[a]	0.018[A]	0.021[a]	6.89×10^{6}[A]	7.70×10^{6}[a]
		4230.92[B]		0.020[B]		7.47×10^{6}[B]	
7d $^2D_{3/2}$	9p $^2P^o_{3/2}$	5746.85[A]	5745.26[a]	0.008[A]	0.007[a]	1.67×10^{6}[A]	1.49×10^{6}[a]
		5746.90[B]		0.007[B]		1.48×10^{6}[B]	
7d $^2D_{5/2}$	9p $^2P^o_{3/2}$	5815.10[A]	5813.45[a]	0.073[A]	0.066[a]	1.45×10^{7}[A]	1.30×10^{7}[a]
		5815.07[B]	5813.447[c]	0.065[B]		1.29×10^{7}[B]	
7d $^2D_{3/2}$	9p $^2P^o_{1/2}$	5877.26[A]	5875.63[a]	0.040[A]	0.036[a]	7.83×10^{6}[A]	6.98×10^{6}[a]
		5877.31[B]	5875.632[c]	0.036[B]		6.92×10^{6}[B]	
7f $^2F^o_{5/2}$	8g $^2G_{7/2}$	8293.36[A]	8291.04[a]	1.513[A,B]	1.485[a]	1.47×10^{8}[A,B]	1.44×10^{8}[a]
		8293.35[B]					
7f $^2F^o_{7/2}$	8g $^2G_{9/2}$	8315.16[A]	8312.96[a]	1.956[A,B]	1.919[a]	1.89×10^{8}[A,B]	1.85×10^{8}[a]
		8315.11[B]					
7f $^2F^o_{7/2}$	8g $^2G_{7/2}$	8316.03[A]	8313.81[a]	0.056[A,B]	0.055[a]	5.39×10^{6}[A,B]	5.29×10^{6}[a]
		8316.04[B]					
5f $^2F^o_{5/2}$	7d $^2D_{5/2}$	5469.30[A]	5467.81[a]	0.037[A]	0.035[a]	8.26×10^{6}[A]	7.89×10^{6}[a]
		5469.35[B]	5467.812[c]	0.035[B]		7.77×10^{6}[B]	
5f $^2F^o_{5/2}$	7d $^2D_{3/2}$	5531.09[A]	5529.54[a]	0.514[A]	0.490[a]	1.12×10^{8}[A]	1.07×10^{8}[a]
		5531.07[B]	5529.542[c]	0.482[B]		1.05×10^{8}[B]	
5f $^2F^o_{7/2}$	7d $^2D_{5/2}$	5493.40[A]	5491.90[a]	0.738[A]	0.704[a]	1.63×10^{8}[A]	1.56×10^{8}[a]
		5493.43[B]	5491.902[c]	0.693[B]		1.53×10^{8}[B]	
5f $^2F^o_{5/2}$	5g $^2G_{7/2}$	4484.21[A]	4482.98[a]	8.886[A]	8.277[a]	2.95×10^{9}[A,B]	2.75×10^{9}[a]
		4484.25[B]	4482.967[c]	8.889[B]			
5f $^2F^o_{7/2}$	5g $^2G_{9/2}$	4500.32[A]	4499.06[a]	11.478[A]	10.692[a]	3.78×10^{9}[A,B]	3.52×10^{9}[a]
		4500.34[B]	4499.050[c]	11.482[B]			
5f $^2F^o_{7/2}$	5g $^2G_{7/2}$	4500.39[A]	4499.15[a]	0.328[A,B]	0.306[a]	1.08×10^{8}[A,B]	1.01×10^{8}[a]
		4500.43[B]					
5f $^2F^o_{5/2}$	8d $^2D_{5/2}$	3086.28[A]	3085.38[a]	0.004[A,B]	0.005[a]	2.98×10^{6}[A]	3.13×10^{6}[a]
		3086.29[B]	3085.379[c]			2.80×10^{6}[B]	
5f $^2F^o_{5/2}$	8d $^2D_{3/2}$	3097.15[A]	3096.26[a]	0.059[A]	0.062[a]	4.13×10^{7}[A]	4.33×10^{7}[a]
		3097.16[B]	3096.255[c]	0.056[B]		3.88×10^{7}[B]	
5f $^2F^o_{7/2}$	8d $^2D_{5/2}$	3093.93[A]	3093.03[a]	0.085[A]	0.089[a]	5.91×10^{7}[A]	6.21×10^{7}[a]
		3093.94[B]	3093.028[c]	0.080[B]		5.56×10^{7}[B]	
5f $^2F^o_{5/2}$	6g $^2G_{7/2}$	2898.73[A]	2897.88[a]	0.866[A,B]	0.786[a]	6.87×10^{8}[A,B]	6.24×10^{8}[a]
		2898.74[B]	2897.875[c]				
5f $^2F^o_{7/2}$	6g $^2G_{9/2}$	2905.42[A]	2904.57[a]	1.120[A,B]	1.016[a]	8.85×10^{8}[A,B]	8.03×10^{8}[a]
		2905.41[B]	2904.576[c]				
5f $^2F^o_{7/2}$	6g $^2G_{7/2}$	2905.49[A,B]	2904.63[a]	0.032[A,B]	0.029[a]	2.53×10^{7}[A,B]	2.29×10^{7}[a]
5f $^2F^o_{5/2}$	9d $^2D_{5/2}$	2461.70[A,B]	2460.95[a]	0.001[A,B]	0.002[a]	1.51×10^{6}[A]	1.67×10^{6}[a]
						1.41×10^{6}[B]	
5f $^2F^o_{5/2}$	9d $^2D_{3/2}$	2465.97[A,B]	2465.22[a]	0.019[A]	0.021[a]	2.11×10^{7}[A]	2.33×10^{7}[a]
				0.018[B]		1.96×10^{7}[B]	

Table 1: Continued.

Transition		λ		gf		gA_{ki}	
Lower level	Upper level	This work	Other works	This work	Other works	This work	Other works
$5f\,^2F^o_{7/2}$	$9d\,^2D_{5/2}$	2466.59[A]	2465.82[a]	0.027[A]	0.030[a]	3.01×10^{7} [A]	3.33×10^{7} [a]
		2535.57[B]		0.025[B]		2.80×10^{7} [B]	
$5f\,^2F^o_{5/2}$	$7g\,^2G_{7/2}$	2388.71[A]	2387.99[a]	0.226[A,B]	0.196[a]	2.64×10^{8} [A,B]	2.29×10^{8} [a]
		2388.72[B]	2387.988[c]				
$5f\,^2F^o_{7/2}$	$7g\,^2G_{9/2}$	2393.23[A]	2392.49[a]	0.292[A,B]	0.254[a]	3.41×10^{8} [A,B]	2.96×10^{8} [a]
		2393.22[B]	2392.492[c]				
$5f\,^2F^o_{7/2}$	$7g\,^2G_{7/2}$	2393.30[A,B]	2392.57[a]	0.008[A,B]	0.007[a]	9.73×10^{6} [A,B]	8.45×10^{6} [a]
$5f\,^2F^o_{5/2}$	$8g\,^2G_{7/2}$	2143.81[A,B]	2143.13[a]	0.089[A,B]	0.074[a]	1.29×10^{8} [A,B]	1.08×10^{8} [a]
$5f\,^2F^o_{7/2}$	$8g\,^2G_{9/2}$	2147.44[A,B]	2146.77[a]	0.115[A,B]	0.096[a]	1.67×10^{8} [A,B]	1.39×10^{8} [a]
$5f\,^2F^o_{7/2}$	$8g\,^2G_{7/2}$	2147.50[A,B]	2146.83[a]	0.003[A,B]	0.003[a]	4.77×10^{6} [A,B]	3.98×10^{6} [a]
$5g\,^2G_{9/2}$	$7f\,^2F^o_{7/2}$	8116.78[A]	8114.48[a]	0.018[A,B]	0.021[a]	1.79×10^{6} [A,B]	2.11×10^{6} [a]
		8116.75[B]	8114.415[c]				
$5g\,^2G_{7/2}$	$7f\,^2F^o_{5/2}$	8138.27[A]	8136.00[a]	0.014[A,B]	0.016[a]	1.37×10^{6} [A,B]	1.62×10^{6} [a]
		8138.19[B]	8135.964[c]				
$5g\,^2G_{9/2}$	$8f\,^2F^o_{7/2}$	5090.54[A]	5089.12[a]	0.003[A,B]	0.003[a]	6.83×10^{5} [A]	7.82×10^{5} [a]
		5090.51[B]				6.86×10^{5} [B]	
$5g\,^2G_{7/2}$	$8f\,^2F^o_{5/2}$	5097.17[A]	5095.70[a]	0.002[A,B]	0.002[a]	5.25×10^{5} [A]	6.01×10^{5} [a]
		5097.11[B]				5.27×10^{5} [B]	
$5d\,^2D_{3/2}$	$4f\,^2F^o_{5/2}$	13898.50[A]	13894.47[f]	0.072[A]	0.031[d]	2.51×10^{6} [A]	—
		13898.06[B]		0.074[B]		2.54×10^{6} [B]	
$5d\,^2D_{5/2}$	$4f\,^2F^o_{7/2}$	14099.97[A]	14096.18[f]	0.102[A]	0.046[d]	3.44×10^{6} [A]	—
		14100.19[B]		0.104[B]		3.48×10^{6} [B]	
$5d\,^2D_{5/2}$	$4f\,^2F^o_{5/2}$	17882.04[A]	17878.09[f]	0.004[A,B]	0.002[d]	8.41×10^{4} [A]	—
		17883.69[B]				8.52×10^{4} [B]	
$5d\,^2D_{5/2}$	$6p\,^2P^o_{3/2}$	2298.44[A,B]	2297.74[a]	1.172[A]	1.120[a]	1.48×10^{9} [A]	1.41×10^{9} [a]
			2297.737[c]	1.315[B]	1.050[d]	1.66×10^{9} [B]	1.33×10^{9} [b]
			2298.44[g]				
$5d\,^2D_{3/2}$	$6p\,^2P^o_{3/2}$	2216.76[A]	2216.07[a]	0.135[A]	0.129[a]	1.83×10^{8} [A]	1.75×10^{8} [a]
		2216.75[B]	2216.067[c]	0.151[B]	0.115[d]	2.06×10^{8} [B]	1.56×10^{8} [b]
$5d\,^2D_{3/2}$	$6p\,^2P^o_{1/2}$	2380.10[A]	2379.37[a]	0.629[A]	0.601[a]	7.41×10^{8} [A]	7.07×10^{8} [a]
		2380.09[B]	2379.374[c]	0.705[B]	0.576[d]	8.31×10^{8} [B]	6.78×10^{8} [b]
			2380.10[g]				
$5d\,^2D_{3/2}$	$5f\,^2F^o_{5/2}$	1081.61[A,B]	1081.61[a]	1.649[A]	1.377[a]	9.40×10^{9} [A]	7.85×10^{9} [a]
				1.640[B]	1.604[d]	9.35×10^{9} [B]	9.06×10^{9} [b]
$5d\,^2D_{5/2}$	$5f\,^2F^o_{7/2}$	1099.73[A,B]	1099.73[a]	2.317[A]	1.935[a]	1.28×10^{10} [A]	1.07×10^{10} [a]
				2.304[B]	2.325[d]	1.27×10^{10} [B]	1.28×10^{10} [b]
$5d\,^2D_{5/2}$	$5f\,^2F^o_{5/2}$	1100.70[A,B]	1100.70[a]	0.115[A,B]	0.097[a]	6.37×10^{8} [A]	5.32×10^{8} [a]
					0.119[d]	6.34×10^{8} [B]	6.49×10^{8} [b]
$5d\,^2D_{3/2}$	$7p\,^2P^o_{3/2}$	1058.63[A,B]	1058.63[a]	0.013[A,B]	0.010[a]	7.84×10^{7} [A]	5.72×10^{7} [a]
						8.02×10^{7} [B]	
$5d\,^2D_{3/2}$	$7p\,^2P^o_{1/2}$	1072.59[A,B]	1072.59[a]	0.067[A,B]	0.048[a]	3.77×10^{8} [A]	2.75×10^{8} [a]
						3.86×10^{8} [B]	
$5d\,^2D_{5/2}$	$7p\,^2P^o_{3/2}$	1076.91[A,B]	1076.91[a]	0.116[A]	0.085[a]	6.70×10^{8} [A]	4.89×10^{8} [a]
				0.119[B]		6.86×10^{8} [B]	
$5d\,^2D_{3/2}$	$6f\,^2F^o_{5/2}$	870.40[A,B]	870.40[a]	0.614[A,B]	0.446[a]	5.41×10^{9} [A]	3.93×10^{9} [a]
					0.480[d]	5.40×10^{9} [B]	4.21×10^{9} [b]
$5d\,^2D_{5/2}$	$6f\,^2F^o_{7/2}$	882.34[A,B]	882.34[a]	0.865[A,B]	0.629[a]	7.42×10^{9} [A]	5.39×10^{9} [a]
					0.696[d]	7.41×10^{9} [B]	5.95×10^{9} [b]
$5d\,^2D_{5/2}$	$6f\,^2F^o_{5/2}$	882.72[A,B]	882.72[a]	0.043[A,B]	0.031[a]	3.71×10^{8} [A]	2.69×10^{8} [a]
					0.035[d]	3.70×10^{8} [B]	2.99×10^{8} [b]
$5d\,^2D_{3/2}$	$8p\,^2P^o_{3/2}$	860.39[A,B]	860.39[a]	0.005[A,B]	0.003[a]	4.24×10^{7} [A]	2.76×10^{7} [a]
						4.36×10^{7} [B]	

Table 1: Continued.

Transition		λ		gf		gA_{ki}	
Lower level	Upper level	This work	Other works	This work	Other works	This work	Other works
$5d\ ^2D_{3/2}$	$8p\ ^2P^o_{1/2}$	865.04[A,B]	865.04[a]	0.024[A,B]	0.015[a]	2.08×10^{8} [A] 2.15×10^{8} [B]	1.36×10^{8} [a]
$5d\ ^2D_{5/2}$	$8p\ ^2P^o_{3/2}$	872.43[A,B]	872.43[a]	0.043[A,B]	0.027[a]	3.65×10^{8} [A] 3.77×10^{8} [B]	2.38×10^{8} [a]
$5d\ ^2D_{3/2}$	$7f\ ^2F^o_{5/2}$	787.14[A,B]	787.14[a]	0.303[A] 0.305[B]	0.203[a]	3.24×10^{9} [A] 3.28×10^{9} [B]	2.18×10^{9} [a]
$5d\ ^2D_{5/2}$	$7f\ ^2F^o_{7/2}$	796.99[A,B]	796.99[a]	0.430[A,B]	0.286[a]	4.51×10^{9} [A,B]	3.00×10^{9} [a]
$5d\ ^2D_{5/2}$	$7f\ ^2F^o_{5/2}$	797.20[A,B]	797.20[a]	0.021[A,B]	0.014[a]	2.26×10^{8} [A,B]	1.50×10^{8} [a]
$5d\ ^2D_{3/2}$	$9p\ ^2P^o_{3/2}$	781.65[A,B]	781.65[a]	0.002[A,B]	0.001[a]	2.52×10^{7} [A] 2.62×10^{7} [B]	1.54×10^{7} [a]
$5d\ ^2D_{3/2}$	$9p\ ^2P^o_{1/2}$	784.01[A,B]	784.01[a]	0.012[A,B]	0.007[a]	12.50×10^{7} [A] 13.01×10^{7} [B]	7.61×10^{7} [a]
$5d\ ^2D_{5/2}$	$9p\ ^2P^o_{3/2}$	791.57[A,B]	791.57[a]	0.021[A,B]	0.013[a]	2.19×10^{8} [A] 2.27×10^{8} [B]	1.33×10^{8} [a]
$5d\ ^2D_{3/2}$	$8f\ ^2F^o_{5/2}$	744.19[A,B]	744.19[a]	0.176[A,B]	0.110[a]	2.12×10^{9} [A,B]	1.33×10^{9} [a]
$5d\ ^2D_{5/2}$	$8f\ ^2F^o_{7/2}$	753.03[A,B]	753.03[a]	0.249[A] 0.248[B]	0.155[a]	2.92×10^{9} [A,B]	1.83×10^{9} [a]
$5d\ ^2D_{5/2}$	$8f\ ^2F^o_{5/2}$	753.18[A,B]	753.18[a]	0.012[A,B]	0.008[a]	1.46×10^{8} [A,B]	9.14×10^{7} [a]
$4f\ ^2F^o_{5/2}$	$6d\ ^2D_{5/2}$	1322.42[A,B]	1322.42[a]	0.000[A,B]	0.002[a]	0.12×10^{6} [A] 0.17×10^{6} [B]	7.00×10^{6} [a]
$4f\ ^2F^o_{5/2}$	$6d\ ^2D_{3/2}$	1330.04[A,B]	1330.04[a]	0.001[A,B]	0.026[a]	0.22×10^{7} [A] 0.23×10^{7} [B]	9.64×10^{7} [a]
$4f\ ^2F^o_{7/2}$	$6d\ ^2D_{5/2}$	1349.18[A,B]	1349.18[a]	0.001[A,B]	0.036[a]	3.07×10^{6} [A] 3.16×10^{6} [B]	1.32×10^{8} [a]
$4f\ ^2F^o_{5/2}$	$5g\ ^2G_{7/2}$	929.72[A,B]	929.72[a]	0.058[A] 0.060[B]	0.040[a]	4.49×10^{8} [A] 4.63×10^{8} [B]	3.06×10^{8} [a]
$4f\ ^2F^o_{7/2}$	$5g\ ^2G_{9/2}$	942.86[A,B]	942.86[a]	0.074[A] 0.077[B]	0.051[a]	5.58×10^{8} [A] 5.76×10^{8} [B]	3.81×10^{8} [a]
$4f\ ^2F^o_{7/2}$	$5g\ ^2G_{7/2}$	942.87[A,B]	942.87[a]	0.002[A,B]	0.001[a]	1.60×10^{7} [A] 1.65×10^{7} [B]	1.09×10^{7} [a]
$4f\ ^2F^o_{5/2}$	$6g\ ^2G_{7/2}$	835.02[A,B]	835.02[a]	0.046[A] 0.048[B]	0.030[a]	4.44×10^{8} [A] 4.58×10^{8} [B]	2.87×10^{8} [a]
$4f\ ^2F^o_{7/2}$	$6g\ ^2G_{9/2}$	845.61[A,B]	845.61[a]	0.059[A] 0.061[B]	0.038[a]	5.54×10^{8} [A] 5.72×10^{8} [B]	3.58×10^{8} [a]
$4f\ ^2F^o_{7/2}$	$6g\ ^2G_{7/2}$	845.62[A,B]	845.62[a]	0.002[A,B]	0.001[a]	1.59×10^{7} [A] 1.63×10^{7} [B]	1.02×10^{7} [a]
$4f\ ^2F^o_{5/2}$	$7g\ ^2G_{7/2}$	786.64[A,B]	786.64[a]	0.033[A] 0.034[B]	0.021[a]	3.56×10^{8} [A] 3.68×10^{8} [B]	2.23×10^{8} [a]
$4f\ ^2F^o_{7/2}$	$7g\ ^2G_{9/2}$	796.03[A,B]	796.03[a]	0.042[A] 0.044[B]	0.026[a]	4.44×10^{8} [A] 4.60×10^{8} [B]	2.78×10^{8} [a]
$4f\ ^2F^o_{5/2}$	$8g\ ^2G_{7/2}$	758.12[A,B]	758.12[a]	0.023[A] 0.024[B]	0.014[a]	2.73×10^{8} [A] 2.82×10^{8} [B]	1.67×10^{8} [a]
$4f\ ^2F^o_{7/2}$	$8g\ ^2G_{9/2}$	766.83[A,B]	766.83[a]	0.030[A] 0.031[B]	0.018[a]	3.41×10^{8} [A] 3.54×10^{8} [B]	2.09×10^{8} [a]
$8s\ ^2S_{1/2}$	$9p\ ^2P^o_{3/2}$	5641.62[A,B]	5640.03[a]	0.005[A] 0.004[B]	0.003[a]	10.20×10^{5} [A] 9.34×10^{5} [B]	5.58×10^{5} [a]
$8s\ ^2S_{1/2}$	$9p\ ^2P^o_{1/2}$	5767.24[A] 5767.25[B]	5765.63[a]	0.002[A,B]	0.001[a]	4.76×10^{5} [A] 4.37×10^{5} [B]	2.61×10^{5} [a]
$8p\ ^2P^o_{1/2}$	$8d\ ^2D_{3/2}$	10940.83[A] 10940.86[B]	10937.898[c]	3.508[A] 3.439[B]	—	1.95×10^{8} [A] 1.92×10^{8} [B]	—
$8p\ ^2P^o_{1/2}$	$9d\ ^2D_{3/2}$	5745.68[A] 5745.66[B]	5744.08[a] 5744.088[c]	0.337[A] 0.347[B]	0.334[a]	6.81×10^{7} [A] 7.01×10^{7} [B]	6.75×10^{7} [a]

Table 1: Continued.

Transition		λ		gf		gA_{ki}	
Lower level	Upper level	This work	Other works	This work	Other works	This work	Other works
8p $^2P^o_{3/2}$	9d $^2D_{5/2}$	5934.37[A,B]	5932.73[a]	0.588[A]	0.582[a]	1.12×10^{8} [A]	1.10×10^{8} [a]
			5932.706[c]	0.604[B]		1.14×10^{8} [B]	
8p $^2P^o_{3/2}$	9d $^2D_{3/2}$	5959.22[A]	5957.57[a]	0.065[A]	0.064[a]	1.22×10^{7} [A]	1.21×10^{7} [a]
		5959.21[B]		0.067[B]		1.26×10^{7} [B]	
8p $^2P^o_{1/2}$	10s $^2S_{1/2}$	5801.06[A]	5799.48[a]	0.112[A]	0.115[a]	2.23×10^{7} [A]	2.27×10^{7} [a]
		5801.07[B]		0.113[B]		2.24×10^{7} [B]	
8p $^2P^o_{3/2}$	10s $^2S_{1/2}$	6018.81[A]	6017.18[a]	0.217[A]	0.221[a]	3.99×10^{7} [A]	4.06×10^{7} [a]
		6018.84[B]	6017.114[c]	0.218[B]		4.01×10^{7} [B]	

[a]Reference [1], [b]reference (in [1]), [c]reference [2], [d]reference [3, RMP + EX + CP], [e]reference [4, RHF + CP (a)], [f]reference [5], [g]reference [6].

probabilities, for electric dipole transitions ($E1$) in La III ($Z = 57$). These calculations have been performed by using code [15] developed Cowan for relativistic Hartree-Fock (HFR) [16] calculations. This code considers the correlation effects and relativistic corrections. These effects contribute importantly to the physical and chemical properties of atoms or ions, especially lanthanides. The ground-state level of doubly ionized lanthanum is [Xe] 5d $^2D_{3/2}$. We have taken into account $5p^6nd$, $5p^6ng$ (n = 5–10), $5p^6ns$ (n = 6–10), $5p^56s6p$, $5p^56s4f$, $5p^55d6p$, $5p^6nf$ (n = 4–10), $5p^6np$ (n = 6–10), $5p^54f^2$, and $5p^56p^2$ configurations outside the core [Cd], and nd, ng (n = 5–25), ns (n = 6–24), nf (n = 4–22), and np (n = 6–25) configurations outside the core [Xe] in La III. The configuration sets that we used have been denoted by A and B, respectively, and are given in tables and text. We presented the energies, the Landé g-factors, and the lifetimes for nd, ng (n = 5–25), ns (n = 6–24), nf (n = 4–22), and np (n = 6–25) excited levels of La III [14]. In addition, we have reported various atomic structure calculations such as energy levels, transition energies, hyperfine structure, lifetimes, and electric dipole transitions for some lanthanides (La I–III, Lu I III, and Yb I–III) [17–27].

2. Calculation Method

An electromagnetic transition between two states is characterized by the angular momentum and the parity of the corresponding photon. If the emitted or absorbed photon has angular momentum k and parity $\pi = (-1)^k$ then, the transition is an electric multipole transition (Ek). However, if the photon has parity $\pi = (-1)^{k+1}$ the transition is a magnetic multipole transition (Mk).

According to HFR method [16], the total transition probability from a state $\gamma' J' M'$ to all states M levels of γJ is given by

$$A = \frac{64\pi^4 e^2 a_0^2 \sigma^3}{3h}\mathbf{S}\sum_{Mq}\begin{pmatrix} J & 1 & J' \\ -M & q & M' \end{pmatrix}^2 = \frac{64\pi^4 e^2 a_0^2 \sigma^3}{3h(2J'+1)}\mathbf{S}, \tag{1}$$

and absorption oscillator strength is given by

$$f_{ij} = \frac{8\pi^2 mca_0^2\sigma}{3h(2J+1)}\mathbf{S} = \frac{(E_j - E_i)}{3(2J+1)}\mathbf{S}, \tag{2}$$

where, $\sigma = [(E_j - E_i)/hc]$ has units of kaysers (cm^{-1}) and $\mathbf{S} = |\langle \gamma J \| \mathbf{P}^{(1)} \| \gamma' J' \rangle|^2$ is the electric dipole line strength in atomic units of $e^2 a_0^2$. The strongest transition rate (or probability) is electric dipole ($E1$) radiation. For this reason, the $E1$ transitions are understood as being "allowed", whereas high-order transitions are understood as being "forbidden".

In HFR method, for an N electron atom of nuclear charge Z_0, the Hamiltonian is expanded as

$$H = -\sum_i \nabla_i^2 - \sum_i \frac{2Z_0}{r_i} + \sum_{i>j}\frac{2}{r_{ij}} + \sum_i \zeta_i(r_i)\mathbf{l}_i\cdot\mathbf{s}_i, \tag{3}$$

in atomic units, with r_i the distance of the ith electron from the nucleus and $r_{ij} = |\mathbf{r}_i - \mathbf{r}_j|$. $\zeta_i(R) = (\alpha^2/2)(1/r)(\partial V/\partial r)$ is the spin-orbit term, with α being the fine structure constant and V the mean potential field due to the nucleus and other electrons.

In this method, one calculates single-configuration radial functions for a spherically symmetrised atom (center-of-gravity energy of the configuration) based on Hartree-Fock method. The radial wave functions are also used to obtain the atom's total energy (E_{av}) including approximate relativistic and correlation energy corrections. Relativistic terms in the potential function give approximate relativistic corrections to the radial functions, as well as improved relativistic energy corrections in heavy atoms. In addition, a correlation term is included to make the potential function more negative, thereby helping to bind negative ions. These radial functions are also used to calculate Coulomb integrals F^k and G^k and spin-orbit integrals ζ_{nl}. After radial functions have been obtained based on Hartree-Fock model, the wave function $|\gamma JM\rangle$ of the M sublevel of a level labeled γJ is expressed in terms of LS basis states $|\alpha LSJM\rangle$ by the formula

$$|\gamma JM\rangle = \sum_{\alpha LS}|\alpha LSJM\rangle\langle \alpha LSJ \mid \gamma J\rangle. \tag{4}$$

If determinant wave functions are used for the atom, the total binding energy is given by

$$E = \sum_i\left(E_k^i + E_n^i + \sum_{j<i}E^{ij}\right), \tag{5}$$

where E_k^i is the kinetic energy, E_n^i is the electron-nuclear Coulomb energy, and E^{ij} is the Coulomb interaction energy

TABLE 2: New λ(Å), gf, and gA_{ki}(s^{-1}) for electric dipole ($E1$) transitions in La III.

Transition		λ	gf	gA_{ki}
Lower level	Upper level			
5d $^2D_{3/2}$	12f $^2F^o_{5/2}$	682.99^B	0.039^B	$5.62 \times 10^{8\,B}$
5d $^2D_{5/2}$	12f $^2F^o_{7/2}$	690.53^B	0.056^B	$7.77 \times 10^{8\,B}$
5d $^2D_{5/2}$	12f $^2F^o_{5/2}$	690.55^B	0.003^B	$3.89 \times 10^{7\,B}$
5d $^2D_{3/2}$	11f $^2F^o_{5/2}$	690.98^B	0.053^B	$7.46 \times 10^{8\,B}$
5d $^2D_{5/2}$	11f $^2F^o_{7/2}$	698.69^B	0.075^B	$1.03 \times 10^{9\,B}$
5d $^2D_{5/2}$	11f $^2F^o_{5/2}$	698.72^B	0.004^B	$5.15 \times 10^{7\,B}$
5d $^2D_{3/2}$	12p $^2P^o_{3/2}$	701.16^B	0.001^B	$8.21 \times 10^{6\,B}$
5d $^2D_{3/2}$	12p $^2P^o_{1/2}$	701.60^B	0.003^B	$4.10 \times 10^{7\,B}$
5d $^2D_{3/2}$	10f $^2F^o_{5/2}$	702.10^{A,B}	0.074^A	$1.00 \times 10^{9\,A}$
			0.075^B	$1.02 \times 10^{9\,B}$
5d $^2D_{5/2}$	12p $^2P^o_{3/2}$	709.13^B	0.005^B	$7.14 \times 10^{7\,B}$
5d $^2D_{5/2}$	10f $^2F^o_{7/2}$	710.05^{A,B}	0.104^A	$1.38 \times 10^{9\,A}$
			0.106^B	$1.41 \times 10^{9\,B}$
5d $^2D_{5/2}$	10f $^2F^o_{5/2}$	710.09^{A,B}	0.005^{A,B}	$6.92 \times 10^{7\,A}$
				$7.03 \times 10^{7\,B}$
5d $^2D_{3/2}$	11p $^2P^o_{3/2}$	716.83^B	0.001^B	$1.15 \times 10^{7\,B}$
5d $^2D_{3/2}$	11p $^2P^o_{1/2}$	717.49^B	0.004^B	$5.76 \times 10^{7\,B}$
5d $^2D_{5/2}$	11p $^2P^o_{3/2}$	725.17^B	0.008^B	$1.00 \times 10^{8\,B}$
4f $^2F^o_{5/2}$	10g $^2G_{7/2}$	729.02^A	0.013^{A,B}	$1.57 \times 10^{8\,A}$
		727.97^B		$1.67 \times 10^{8\,B}$
4f $^2F^o_{7/2}$	10g $^2G_{7/2}$	737.09^A	0.001^{A,B}	5.70×10^{6A}
		736.01^B		$5.97 \times 10^{6\,B}$
4f $^2F^o_{7/2}$	10g $^2G_{9/2}$	737.09^A	0.016^A	$1.97 \times 10^{8\,A}$
		736.01^B	0.017^B	$2.09 \times 10^{8\,B}$
4f $^2F^o_{5/2}$	9g $^2G_{7/2}$	742.58^A	0.017^A	$2.06 \times 10^{8\,A}$
		740.62^B	0.018^B	$2.15 \times 10^{8\,B}$
5d $^2D_{3/2}$	10p $^2P^o_{3/2}$	740.94^A	0.001^{A,B}	$1.60 \times 10^{7\,A}$
		740.93^B		$1.70 \times 10^{7\,B}$
5d $^2D_{3/2}$	10p $^2P^o_{1/2}$	742.00^A	0.007^{A,B}	$7.99 \times 10^{7\,A}$
		741.97^B		$8.44 \times 10^{7\,B}$
4f $^2F^o_{7/2}$	9g $^2G_{7/2}$	750.95^A	0.001^{A,B}	$7.44 \times 10^{6\,A}$
		748.94^B		7.72×10^{6B}
4f $^2F^o_{7/2}$	9g $^2G_{9/2}$	750.95^A	0.022^A	$2.58 \times 10^{8\,A}$
		748.94^B	0.023^B	$2.70 \times 10^{8\,B}$
5d $^2D_{5/2}$	10p $^2P^o_{3/2}$	749.86^A	0.012^{A,B}	$1.39 \times 10^{8\,A}$
		749.84^B		$1.47 \times 10^{8\,B}$
6p $^2P^o_{1/2}$	12d $^2D_{3/2}$	978.38^B	0.006^B	$4.54 \times 10^{7\,B}$
6p $^2P^o_{1/2}$	11d $^2D_{3/2}$	1002.77^B	0.010^B	$6.50 \times 10^{7\,B}$
6p $^2P^o_{3/2}$	12d $^2D_{5/2}$	1008.75^B	0.011^B	$7.46 \times 10^{7\,B}$
6p $^2P^o_{3/2}$	12d $^2D_{3/2}$	1008.94^B	0.001^B	8.29×10^{6B}
6p $^2P^o_{3/2}$	11d $^2D_{5/2}$	1034.62^B	0.017^B	$1.06 \times 10^{8\,B}$
6p $^2P^o_{3/2}$	11d $^2D_{3/2}$	1034.89^B	0.002^B	$1.18 \times 10^{7\,B}$
6p $^2P^o_{1/2}$	10d $^2D_{3/2}$	1061.80^A	0.008^A	$4.59 \times 10^{7\,A}$
		1039.29^B	0.016^B	$9.83 \times 10^{7\,B}$
6p $^2P^o_{3/2}$	10d $^2D_{5/2}$	1097.43^A	0.014^A	$7.83 \times 10^{7\,A}$
		1073.42^B	0.028^B	$1.61 \times 10^{8\,B}$
6p $^2P^o_{3/2}$	10d $^2D_{3/2}$	1097.89^A	0.002^A	8.55×10^{6A}
		1073.84^B	0.003^B	$1.78 \times 10^{7\,B}$
6d $^2D_{3/2}$	12f $^2F^o_{5/2}$	1561.65^B	0.036^B	$9.96 \times 10^{7\,B}$
6d $^2D_{5/2}$	12f $^2F^o_{7/2}$	1572.18^B	0.052^B	$1.39 \times 10^{8\,B}$
6d $^2D_{5/2}$	12f $^2F^o_{5/2}$	1572.29^B	0.002^B	6.97×10^{6B}

Table 2: Continued.

Transition		λ	gf	gA_{ki}
Lower level	Upper level			
6d $^2D_{3/2}$	11f $^2F^o_{5/2}$	1604.09[B]	0.050[B]	1.31×10^{8}[B]
6d $^2D_{5/2}$	11f $^2F^o_{7/2}$	1615.16[B]	0.072[B]	1.83×10^{8}[B]
6d $^2D_{5/2}$	11f $^2F^o_{5/2}$	1615.32[B]	0.004[B]	9.15×10^{6}[B]
6d $^2D_{3/2}$	12p $^2P^o_{1/2}$	1662.51[B]	0.002[B]	5.62×10^{6}[B]
6d $^2D_{3/2}$	10f $^2F^o_{5/2}$	1665.31[A]	0.080[A]	1.93×10^{8}[A]
		1665.37[B]	0.073[B]	1.76×10^{8}[B]
6d $^2D_{5/2}$	12p $^2P^o_{3/2}$	1672.06[B]	0.004[B]	9.94×10^{6}[B]
6d $^2D_{5/2}$	10f $^2F^o_{7/2}$	1677.17[A,B]	0.114[A]	2.70×10^{8}[A]
			0.104[B]	2.46×10^{8}[B]
6d $^2D_{5/2}$	10f $^2F^o_{5/2}$	1677.42[A,B]	0.006[A]	1.35×10^{7}[A]
			0.005[B]	1.23×10^{7}[B]
6d $^2D_{3/2}$	11p $^2P^o_{1/2}$	1754.58[B]	0.004[B]	7.85×10^{6}[B]
6d $^2D_{5/2}$	11p $^2P^o_{3/2}$	1764.04[B]	0.006[B]	1.39×10^{7}[B]
6d $^2D_{3/2}$	10p $^2P^o_{3/2}$	1901.87[A]	0.002[A]	2.94×10^{6}[A]
		1901.69[B]	0.001[B]	2.31×10^{6}[B]
6d $^2D_{3/2}$	10p $^2P^o_{1/2}$	1908.76[A]	0.008[A]	1.45×10^{7}[A]
		1908.57[B]	0.006[B]	1.14×10^{7}[B]
6d $^2D_{5/2}$	10p $^2P^o_{3/2}$	1917.68[A]	0.014[A]	2.57×10^{7}[A]
		1917.49[B]	0.011[B]	2.03×10^{7}[B]
5f $^2F^o_{5/2}$	10g $^2G_{7/2}$	1926.41[A]	0.025[A,B]	4.50×10^{7}[A]
		1919.09[B]		4.53×10^{7}[B]
5f $^2F^o_{7/2}$	10g $^2G_{7/2}$	1929.39[A]	0.001[A,B]	1.66×10^{6}[A]
		1922.05[B]		1.67×10^{6}[B]
5f $^2F^o_{7/2}$	10g $^2G_{9/2}$	1922.05[A]	0.032[A,B]	5.85×10^{7}[A]
		1929.39[B]		5.80×10^{7}[B]
5f $^2F^o_{5/2}$	12d $^2D_{3/2}$	1931.60[B]	0.003[B]	5.03×10^{6}[B]
5f $^2F^o_{7/2}$	12d $^2D_{5/2}$	1933.92[B]	0.004[B]	7.16×10^{6}[B]
7p $^2P^o_{1/2}$	12d $^2D_{3/2}$	1961.07[B]	0.014[B]	2.38×10^{7}[B]
7p $^2P^o_{3/2}$	12d $^2D_{5/2}$	2008.77[B]	0.024[B]	3.99×10^{7}[B]
7p $^2P^o_{3/2}$	12d $^2D_{3/2}$	2009.50[B]	0.003[B]	4.43×10^{6}[B]
5f $^2F^o_{5/2}$	9g $^2G_{7/2}$	2024.06[A]	0.044[A,B]	7.09×10^{7}[A]
		2009.57[B]		7.26×10^{7}[B]
5f $^2F^o_{7/2}$	9g $^2G_{7/2}$	2027.35[A]	0.002[A,B]	2.61×10^{6}[A]
		2012.81[B]		2.67×10^{6}[B]
5f $^2F^o_{7/2}$	9g $^2G_{9/2}$	2027.34[A]	0.056[A]	9.15×10^{7}[A]
		2012.81[B]	0.057[B]	9.36×10^{7}[B]
5f $^2F^o_{5/2}$	11d $^2D_{3/2}$	2029.02[B]	0.004[B]	7.36×10^{6}[B]
5f $^2F^o_{7/2}$	11d $^2D_{5/2}$	2031.29[B]	0.006[B]	1.05×10^{7}[B]
7p $^2P^o_{1/2}$	11d $^2D_{3/2}$	2061.56[B]	0.022[B]	3.47×10^{7}[B]
7p $^2P^o_{3/2}$	11d $^2D_{5/2}$	2114.03[B]	0.039[B]	5.80×10^{7}[B]
7p $^2P^o_{3/2}$	11d $^2D_{3/2}$	2115.16[B]	0.004[B]	6.43×10^{6}[B]
5f $^2F^o_{5/2}$	10d $^2D_{5/2}$	2284.24[A]	0.001[A,B]	7.40×10^{5}[A]
		2182.60[B]		8.22×10^{5}[B]
5f $^2F^o_{5/2}$	10d $^2D_{3/2}$	2286.23[A]	0.008[A,B]	1.03×10^{7}[A]
		2184.34[B]		1.15×10^{7}[B]
5f $^2F^o_{7/2}$	10d $^2D_{5/2}$	2288.43[A]	0.012[A,B]	1.47×10^{7}[A]
		2186.43[B]		1.63×10^{7}[B]
7p $^2P^o_{1/2}$	10d $^2D_{3/2}$	2327.62[A]	0.034[A]	4.21×10^{7}[A]
		2222.09[B]	0.040[B]	5.41×10^{7}[B]
7p $^2P^o_{3/2}$	10d $^2D_{5/2}$	2393.98[A]	0.060[A]	7.01×10^{7}[A]
		2282.58[B]	0.070[B]	8.98×10^{7}[B]
7p $^2P^o_{3/2}$	10d $^2D_{3/2}$	2396.17[A]	0.007[A]	7.73×10^{6}[A]

Table 2: Continued.

Transition				
Lower level	Upper level	λ	gf	gA_{ki}
		2284.49^{B}	0.008^{B}	9.95×10^{6} ^{B}
7d $^2D_{3/2}$	12f $^2F^o_{5/2}$	2786.97^{B}	0.045^{B}	3.86×10^{7} ^{B}
7d $^2D_{5/2}$	12f $^2F^o_{7/2}$	2802.54^{B}	0.064^{B}	5.42×10^{7} ^{B}
7d $^2D_{5/2}$	12f $^2F^o_{5/2}$	2802.90^{B}	0.003^{B}	2.71×10^{6} ^{B}
7d $^2D_{3/2}$	11f $^2F^o_{5/2}$	2925.08^{B}	0.064^{B}	4.98×10^{7} ^{B}
7d $^2D_{5/2}$	11f $^2F^o_{7/2}$	2942.09^{B}	0.091^{B}	7.00×10^{7} ^{B}
7d $^2D_{5/2}$	11f $^2F^o_{5/2}$	2942.64^{B}	0.004^{B}	3.50×10^{6} ^{B}
7d $^2D_{3/2}$	12p $^2P^o_{3/2}$	3116.57^{B}	0.001^{B}	4.10×10^{5} ^{B}
7d $^2D_{3/2}$	12p $^2P^o_{1/2}$	3125.35^{B}	0.003^{B}	2.03×10^{6} ^{B}
7d $^2D_{3/2}$	10f $^2F^o_{5/2}$	3135.24^{A,B}	0.103^{A}	6.99×10^{7} ^{A}
			0.096^{B}	6.53×10^{7} ^{B}
7d $^2D_{5/2}$	12p $^2P^o_{3/2}$	3136.51^{B}	0.005^{B}	3.62×10^{6} ^{B}
7d $^2D_{5/2}$	10f $^2F^o_{7/2}$	3154.57^{A}	0.146^{A}	9.78×10^{7} ^{A}
		3154.55^{B}	0.137^{B}	9.16×10^{7} ^{B}
7d $^2D_{5/2}$	10f $^2F^o_{5/2}$	3155.45^{A}	0.007^{A,B}	4.89×10^{6} ^{A}
		3155.42^{B}		4.58×10^{6} ^{B}
6f $^2F^o_{5/2}$	10g $^2G_{7/2}$	3392.75^{A}	0.098^{A}	5.70×10^{7} ^{A}
		3370.11^{B}	0.099^{B}	5.82×10^{7} ^{B}
6f $^2F^o_{7/2}$	10g $^2G_{7/2}$	3398.40^{A}	0.004^{A,B}	2.10×10^{6} ^{A}
		3375.69^{B}		2.14×10^{6} ^{B}
6f $^2F^o_{7/2}$	10g $^2G_{9/2}$	3398.41^{A}	0.127^{A}	7.35×10^{7} ^{A}
		3375.69^{B}	0.128^{B}	7.50×10^{7} ^{B}
6f $^2F^o_{5/2}$	12d $^2D_{5/2}$	3406.76^{B}	0.001^{B}	4.58×10^{5} ^{B}
6f $^2F^o_{5/2}$	12d $^2D_{3/2}$	3408.88^{B}	0.011^{B}	6.40×10^{6} ^{B}
6f $^2F^o_{7/2}$	12d $^2D_{5/2}$	3412.46^{B}	0.016^{B}	9.11×10^{6} ^{B}
7d $^2D_{3/2}$	11p $^2P^o_{3/2}$	3452.08^{B}	0.001^{B}	5.78×10^{5} ^{B}
7d $^2D_{3/2}$	11p $^2P^o_{1/2}$	3467.38^{B}	0.005^{B}	2.85×10^{6} ^{B}
7d $^2D_{5/2}$	11p $^2P^o_{3/2}$	3476.57^{B}	0.009^{B}	5.09×10^{6} ^{B}
8p $^2P^o_{1/2}$	12d $^2D_{3/2}$	3493.71^{B}	0.028^{B}	1.54×10^{7} ^{B}
8p $^2P^o_{3/2}$	12d $^2D_{5/2}$	3569.21^{B}	0.050^{B}	2.60×10^{7} ^{B}
8p $^2P^o_{3/2}$	12d $^2D_{3/2}$	3571.54^{B}	0.005^{B}	2.88×10^{6} ^{B}
6f $^2F^o_{5/2}$	9g $^2G_{7/2}$	3707.80^{A}	0.181^{A}	8.76×10^{7} ^{A}
		3659.44^{B}	0.183^{B}	9.13×10^{7} ^{B}
6f $^2F^o_{7/2}$	9g $^2G_{7/2}$	3714.55^{A}	0.007^{A,B}	3.23×10^{6} ^{A}
		3666.01^{B}		3.36×10^{6} ^{B}
6f $^2F^o_{7/2}$	9g $^2G_{9/2}$	3714.49^{A}	0.234^{A}	1.13×10^{8} ^{A}
		3666.01^{B}	0.237^{B}	1.18×10^{8} ^{B}
6f $^2F^o_{5/2}$	11d $^2D_{5/2}$	3720.97^{B}	0.001^{B}	6.83×10^{5} ^{B}
6f $^2F^o_{5/2}$	11d $^2D_{3/2}$	3724.47^{B}	0.020^{B}	9.54×10^{6} ^{B}
6f $^2F^o_{7/2}$	11d $^2D_{5/2}$	3727.77^{B}	0.028^{B}	1.36×10^{7} ^{B}
8p $^2P^o_{1/2}$	11d $^2D_{3/2}$	3825.97^{B}	0.050^{B}	2.30×10^{7} ^{B}
8p $^2P^o_{3/2}$	11d $^2D_{5/2}$	3915.63^{B}	0.089^{B}	3.86×10^{7} ^{B}
8p $^2P^o_{3/2}$	11d $^2D_{3/2}$	3919.50^{B}	0.010^{B}	4.28×10^{6} ^{B}
7d $^2D_{3/2}$	10p $^2P^o_{3/2}$	4093.95^{A}	0.003^{A}	1.05×10^{6} ^{A}
		4093.09^{B}	0.002^{B}	8.64×10^{5} ^{B}
7d $^2D_{3/2}$	10p $^2P^o_{1/2}$	4126.00^{A}	0.013^{A}	5.11×10^{6} ^{A}
		4125.12^{B}	0.011^{B}	4.22×10^{6} ^{B}
7d $^2D_{5/2}$	10p $^2P^o_{3/2}$	4128.47^{A}	0.023^{A}	9.13×10^{6} ^{A}
		4127.55^{B}	0.019^{B}	7.59×10^{6} ^{B}
6f $^2F^o_{5/2}$	10d $^2D_{5/2}$	4685.40^{A}	0.003^{A,B}	8.22×10^{5} ^{A}
		4276.87^{B}		1.10×10^{6} ^{B}
6f $^2F^o_{5/2}$	10d $^2D_{3/2}$	4693.78^{A}	0.038^{A}	1.14×10^{7} ^{A}

Table 2: Continued.

Transition: Lower level	Transition: Upper level	λ	gf	gA_{ki}
		4283.55[B]	0.042[B]	1.54×10^{7} [B]
$6f\,^2F^o_{7/2}$	$10d\,^2D_{5/2}$	4696.17[A]	0.054[A]	1.63×10^{7} [A]
		4285.85[B]	0.060[B]	2.19×10^{7} [B]
$8p\,^2P^o_{1/2}$	$10d\,^2D_{3/2}$	4856.14[A]	0.091[A]	2.57×10^{7} [A]
		4418.37[B]	0.110[B]	3.74×10^{7} [B]
$8p\,^2P^o_{3/2}$	$10d\,^2D_{5/2}$	4998.26[A]	0.160[A]	4.26×10^{7} [A]
		4536.05[B]	0.192[B]	6.22×10^{7} [B]
$8p\,^2P^o_{3/2}$	$10d\,^2D_{3/2}$	5007.81[A]	0.018[A]	4.69×10^{6} [A]
		4543.57[B]	0.021[B]	6.88×10^{6} [B]
$8d\,^2D_{3/2}$	$12f\,^2F^o_{5/2}$	4613.96[B]	0.058[B]	1.82×10^{7} [B]
$8d\,^2D_{5/2}$	$12f\,^2F^o_{7/2}$	4637.33[B]	0.083[B]	2.57×10^{7} [B]
$8d\,^2D_{5/2}$	$12f\,^2F^o_{5/2}$	4638.30[B]	0.004[B]	1.28×10^{6} [B]
$8d\,^2D_{3/2}$	$11f\,^2F^o_{5/2}$	5005.22[B]	0.086[B]	2.30×10^{7} [B]
$8d\,^2D_{5/2}$	$11f\,^2F^o_{7/2}$	5032.28[B]	0.123[B]	3.23×10^{7} [B]
$8d\,^2D_{5/2}$	$11f\,^2F^o_{5/2}$	5033.88[B]	0.006[B]	1.61×10^{6} [B]
$6g\,^2G_{9/2}$	$12f\,^2F^o_{7/2}$	5136.94[B]	0.001[B]	1.75×10^{5} [B]
$6g\,^2G_{7/2}$	$12f\,^2F^o_{5/2}$	5137.90[B]	0.001[B]	1.35×10^{5} [B]
$8d\,^2D_{3/2}$	$12p\,^2P^o_{3/2}$	5593.28[B]	0.001[B]	1.96×10^{5} [B]
$8d\,^2D_{3/2}$	$12p\,^2P^o_{1/2}$	5621.62[B]	0.005[B]	9.65×10^{5} [B]
$6g\,^2G_{9/2}$	$11f\,^2F^o_{7/2}$	5626.07[B]	0.001[B]	2.71×10^{5} [B]
$6g\,^2G_{7/2}$	$11f\,^2F^o_{5/2}$	5627.78[B]	0.001[B]	2.09×10^{5} [B]
$8d\,^2D_{5/2}$	$12p\,^2P^o_{3/2}$	5629.09[B]	0.008[B]	1.73×10^{6} [B]
$8d\,^2D_{3/2}$	$10f\,^2F^o_{5/2}$	5653.76[A]	0.144[A]	3.02×10^{7} [A]
		5653.69[B]	0.138[B]	2.87×10^{7} [B]
$8d\,^2D_{5/2}$	$10f\,^2F^o_{7/2}$	5687.53[A]	0.203[A]	4.23×10^{7} [A]
		5687.46[B]	0.196[B]	4.03×10^{7} [B]
$8d\,^2D_{5/2}$	$10f\,^2F^o_{5/2}$	5690.36[A]	0.010[A,B]	2.11×10^{6} [A]
		5690.29[B]		2.01×10^{6} [B]
$7f\,^2F^o_{5/2}$	$10g\,^2G_{7/2}$	5773.05[A]	0.247[A]	4.95×10^{7} [A]
		5707.79[B]	0.250[B]	5.13×10^{7} [B]
$7f\,^2F^o_{7/2}$	$10g\,^2G_{7/2}$	5784.03[A]	0.009[A,B]	1.82×10^{6} [A]
		5718.53[B]		1.89×10^{6} [B]
$7f\,^2F^o_{7/2}$	$10g\,^2G_{9/2}$	5784.05[A]	0.320[A]	6.38×10^{7} [A]
		5718.53[B]	0.324[B]	6.61×10^{7} [B]
$7f\,^2F^o_{5/2}$	$12d\,^2D_{5/2}$	5813.72[B]	0.002[B]	4.69×10^{5} [B]
$7f\,^2F^o_{5/2}$	$12d\,^2D_{3/2}$	5819.89[B]	0.033[B]	6.55×10^{6} [B]
$7f\,^2F^o_{7/2}$	$12d\,^2D_{5/2}$	5824.86[B]	0.047[B]	9.33×10^{6} [B]
$9p\,^2P^o_{1/2}$	$12d\,^2D_{3/2}$	5996.59[B]	0.061[B]	1.12×10^{7} [B]
$9p\,^2P^o_{3/2}$	$12d\,^2D_{5/2}$	6131.85[B]	0.107[B]	1.89×10^{7} [B]
$9p\,^2P^o_{3/2}$	$12d\,^2D_{3/2}$	6138.72[B]	0.012[B]	2.10×10^{6} [B]
$6g\,^2G_{9/2}$	$10f\,^2F^o_{7/2}$	6457.77[A]	0.003[A,B]	4.51×10^{5} [A]
		6457.76[B]		4.59×10^{5} [B]
$6g\,^2G_{7/2}$	$10f\,^2F^o_{5/2}$	6461.09[A]	0.002[A,B]	3.48×10^{5} [A]
		6461.03[B]		3.53×10^{5} [B]
$7f\,^2F^o_{5/2}$	$9g\,^2G_{7/2}$	6748.79[A]	0.512[A]	7.50×10^{7} [A]
		6590.26[B]	0.525[B]	8.07×10^{7} [B]
$7f\,^2F^o_{7/2}$	$9g\,^2G_{7/2}$	6763.80[A]	0.019[A,B]	2.76×10^{6} [A]
		6604.58[B]		2.97×10^{6} [B]
$7f\,^2F^o_{7/2}$	$9g\,^2G_{9/2}$	6763.63[A]	0.663[A]	9.66×10^{7} [A]
		6604.58[B]	0.679[B]	1.04×10^{8} [B]

TABLE 2: Continued.

Transition		λ	gf	gA_{ki}
Lower level	Upper level			
8d $^2D_{3/2}$	11p $^2P^o_{3/2}$	6775.04^B	0.002^B	$2.86\times10^{5\,B}$
7f $^2F^o_{5/2}$	11d $^2D_{5/2}$	6792.55^B	0.005^B	$7.23\times10^{5\,B}$
7f $^2F^o_{5/2}$	11d $^2D_{3/2}$	6804.23^B	0.070^B	$1.01\times10^{7\,B}$
7f $^2F^o_{7/2}$	11d $^2D_{5/2}$	6807.77^B	0.100^B	$1.44\times10^{7\,B}$
8d $^2D_{5/2}$	11p $^2P^o_{3/2}$	6827.66^B	0.017^B	$2.52\times10^{6\,B}$
8d $^2D_{3/2}$	11p $^2P^o_{1/2}$	6834.22^B	0.010^B	$1.39\times10^{6\,B}$
9p $^2P^o_{1/2}$	11d $^2D_{3/2}$	7046.99^B	0.130^B	$1.75\times10^{7\,B}$
9p $^2P^o_{3/2}$	11d $^2D_{5/2}$	7230.87^B	0.228^B	$2.91\times10^{7\,B}$
9p $^2P^o_{3/2}$	11d $^2D_{3/2}$	7244.09^B	0.025^B	$3.21\times10^{6\,B}$
9d $^2D_{3/2}$	12f $^2F^o_{5/2}$	7457.70^B	0.079^B	$9.47\times10^{6\,B}$
9d $^2D_{5/2}$	12f $^2F^o_{7/2}$	7494.42^B	0.112^B	$1.33\times10^{7\,B}$
9d $^2D_{5/2}$	12f $^2F^o_{5/2}$	7496.98^B	0.006^B	$6.66\times10^{5\,B}$
7g $^2G_{9/2}$	12f $^2F^o_{7/2}$	8263.95^B	0.003^B	$2.86\times10^{5\,B}$
7g $^2G_{7/2}$	12f $^2F^o_{5/2}$	8266.14^B	0.002^B	$2.21\times10^{5\,B}$
9d $^2D_{3/2}$	11f $^2F^o_{5/2}$	8536.25^B	0.124^B	$1.14\times10^{7\,B}$
9d $^2D_{5/2}$	11f $^2F^o_{7/2}$	8583.10^B	0.176^B	$1.60\times10^{7\,B}$
9d $^2D_{5/2}$	11f $^2F^o_{5/2}$	8587.74^B	0.009^B	$7.97\times10^{5\,B}$

between electrons i and j averaged over all possible magnetic quantum numbers.

In this method, relativistic corrections have been limited to calculations to the mass-velocity and the Darwin corrections by using the relativistic correction to total binding energy. The total binding energy can be given in by formulas (7.57), (7.58), and (7.59) in [16].

3. Results and Discussion

We calculated the radiative parameters (wavelengths, oscillator strengths, and transition probabilities) for electric dipole ($E1$) transitions in La III ($Z = 57$) using HFR code [15]. We have taken into account $5p^6nd$, $5p^6ng$ (n = 5–10), $5p^6ns$ (n = 6–10), $5p^56s6p$, $5p^56s4f$, $5p^55d6p$, $5p^6nf$ (n = 4–10), $5p^6np$ (n = 6–10), $5p^54f^2$, and $5p^56p^2$ configurations outside the core [Cd] for calculation A, and nd, ng (n = 5–25), ns (n = 6–24), nf (n = 4–22), and np (n = 6–25) configurations outside the core [Xe] for calculation B. Table 1 shows the wavelengths, λ (in Å); the weighted oscillator strengths, gf; the weighted transition rates (or probabilities), gA_{ki} (in s^{-1}), for nd (n = 5–9)–nf (n = 4–8), nd (n = 5–9)–np (n = 6–9), np (n = 6–9)–ns (n = 6–10), and ng (n = 5–8)–nf (n = 4–8) electric dipole ($E1$) transitions. The data obtained are too much. For this reason, we have here presented just a part of the results. The comparing values for these exist in literature. Therefore, it is also made a comparison with other calculations and experiments in Table 1. We have also reported the wavelengths, the weighted oscillator strengths, and the weighted transition probabilities that are greater than or equal to 10^5 for some new transitions (680 Å $\leq \lambda \leq$ 8600 Å) in Table 2. References for other comparison values are indicated below the tables with a lowercase superscript; odd-parity states are indicated by the superscript "o".

Electron correlation effects and relativistic effects play an important role in the spectra of heavy elements. To accurately predict the radiative atomic properties for heavy atoms such as La III, complex configuration interactions and relativistic effects must be considered simultaneously. Although Cowan's approach is based on Schrödinger's equation, it includes the most important relativistic effects like mass-velocity corrections and Darwin contributions. Also, for complex atoms, it is important to allow for spin-orbit interaction, which represents the magnetic interaction energy between electron's spin magnetic moment and the magnetic field that the electron sees due to its orbital motion through the electric field of the nucleus. These contributions are considered as perturbations. Thus, to solve the Schrödinger equation with this Hamiltonian, we define a new angular momentum operator in an intermediate coupling scheme.

In calculations, the eigenvalues of Hamiltonian were optimized to the observed energy levels via a least-squares fitting procedure using experimentally determined energy levels, specifically all of the levels from the NIST compilation [28]. The scaling factors of the Slater parameters (F^k and G^k) and of configuration interaction integrals (R^k), not optimized in the least-squares fitting, were chosen equal to 0.85, while the spin-orbit parameters were left at their initial values. This low value of the scaling factors has been suggested by Cowan for neutral heavy elements [15, 16].

We obtained 7785 and 4278 possible $E1$ transitions between odd- and even-parity levels in the calculations A and B, respectively. The results obtained are in excellent agreement with those of other works except some transitions. For some transitions, although the agreement is less in the weighted oscillator strengths and the weighted transition probabilities, it is very good in the wavelengths. Most of results related to low-lying levels obtained from this work are in agreement with literature [1–6]. The differences between our HFR results and other works for gf and gA_{ki} have been found in the 0–10% range for the transitions np ($n = 6$–8)–ns ($n = 6$–10), nd ($n = 6$–9), in the 0.5–9% range for the transitions nd ($n = 6$, 7)–nf ($n = 5$–8), np ($n = 7$–9), and in the 1.5–20% range for the transitions nf ($n = 5$–8)–nd ($n = 7$–9), ng ($n = 5$–8). But the agreement is less in the weighted oscillator strengths and the weighted transition probabilities for 5d and 4f transitions. In fact, except the transitions 6p $^2P^o_{3/2}$–9d $^2D_{3/2}$, 5d $^2D_{3/2}$–9p $^2P^o_{1/2}$, 4f $^2F^o_{7/2}$–5g $^2G_{7/2}$, 4f $^2F^o_{7/2}$–6g $^2G_{7/2}$, and 4f $^2F^o_{5/2}$–8g $^2G_{7/2,9/2}$, we found the values 1.064 (in calculation A) and 1.078 (in calculation B) for the mean ratio gf (this work)/gf [1]. Except the transitions 5d–9p, 4f–7g, 8g, 8s–9p, and 4f–6d, we found also the values 1.084 (in calculation A) and 1.126 (in calculation B) for the mean ratio gA_{ki} (this work)/gA_{ki} [1]. The transition results obtained from the calculation A agree with other works. This calculation includes core correlation (including excitation from 5p shell in core). These results obtained from HFR calculations may be better in case that the increasing number of configurations including the excitations from core. It is noted that there are no exist the works, especially experimental, on La III recently in available literature. A detailed comparison needs new experimental works. Most of our results are excellent in agreement, expect the transition results to 4f and 5d levels (for gf and gA_{ki} results), generally. It is well known that these levels interact strongly with core.

In conclusion, the main purpose of this paper was to perform HFR calculations for obtaining the description of La III spectrum. Accurate atomic structure data is an essential ingredient for a wide range of research fields. Areas from plasma research applications in nuclear fusion to lighting research, as well as astrophysics and cosmology, depend on such data. In spectrum synthesis works, particularly for CP stars, accurate data for transition probabilities (rates) and oscillator strengths for lanthanide atoms are needed to establish reliable abundances for these species. The agreement is excellent, especially for wavelengths, when our HFR results are compared with other available works in literature for the radiative transitions for La III. So, we may mention that new results presented in Table 2 for the transitions between some highly levels in this work are also reliable. There are a few experimental or theoretical radiative transition data for La III in literature. Consequently, we hope that our results, especially the new results in Table 2, which are obtained using the HFR method will be useful for research fields, technological applications, and other works in the future for La III spectra.

Acknowledgment

The authors are very grateful to the anonymous reviewer for stimulating comments and valuable suggestions, which resulted in improvements to this paper.

References

[1] E. Biémont, Z. S. Li, P. Palmeri, and P. Quinet, "Radiative lifetimes in La III and oscillator strengths in La III and Lu III," *Journal of Physics B*, vol. 32, no. 14, pp. 3409–3419, 1999.

[2] H. Odabasi, "Spectrum of doubly ionized lanthanum (La III)," *Journal of the Optical Society of America*, vol. 57, no. 12, pp. 1459–1463, 1967.

[3] J. Migdalek and M. Wyrozumska, "Relativistic oscillator strengths for the Cs isoelectronic sequence and collapse of f and d orbitals," *Journal of Quantitative Spectroscopy and Radiative Transfer*, vol. 37, no. 6, pp. 581–589, 1987.

[4] J. Migdalek and W. E. Baylis, "Relativistic Hartree-Fock oscillator strengths for the lowest $s \rightarrow p$ transitions in the first few members of the Rb(I) and Cs(I) isoelectronic sequences, with allowance for core polarization," *Journal of Quantitative Spectroscopy and Radiative Transfer*, vol. 22, no. 2, pp. 127–134, 1979.

[5] S. Johansson and U. Litzén, "Resonance lines of La III," *Journal of the Optical Society of America*, vol. 61, no. 10, pp. 1427–1428, 1971.

[6] Z. S. Li and J. Zhankui, "Lifetime measurements in La II and La III using time-resolved laser spectroscopy," *Physica Scripta*, vol. 60, no. 5, pp. 414–417, 1999.

[7] E. Biémont and P. Quinet, "Recent advances in the study of lanthanide atoms and ions," *Physica Scripta*, vol. T105, pp. 38–54, 2003.

[8] J. S. Badami, "The spectrum of trebly-ionized cerium (Ce IV)," *Proceedings of the Physical Society*, vol. 43, no. 1, pp. 53–58, 1931.

[9] J. Migdalek and A. Bojara, "Relativistic effects, core polarisation and relaxation in ionisation potentials along Rb and Cs isoelectronic sequences," *Journal of Physics B*, vol. 17, no. 10, pp. 1943–1951, 1984.

[10] P. Quinet and E. Biémont, "Landé g-factors for experimentally determined energy levels in doubly ionized lanthanides," *Atomic Data and Nuclear Data Tables*, vol. 87, no. 2, pp. 207–230, 2004.

[11] R. C. Gibbs and H. E. White, "Relations between doublets of stripped atoms in five periods of the periodic table," *Physical Review*, vol. 33, no. 2, pp. 157–162, 1929.

[12] H. N. Russell and W. F. Meggers, "An analysis of lanthanum spectra (La I, La II, La III)," *Journal of Research of the National Bureau of Standards*, vol. 9, no. 5, pp. 625–668, 1932.

[13] J. Sugar and V. Kaufman, "Spectrum of doubly ionized lanthanum (La III)," *Journal of the Optical Society of America*, vol. 55, no. 10, pp. 1283–1285, 1965.

[14] B. Karaçoban and L. Özdemir, "Energies, Landé g-factors, and lifetimes for some excited levels of doubly ionized lanthanum," *Central European Journal of Physics*, vol. 10, no. 1, pp. 124–131, 2012.

[15] http://www.tcd.ie/Physics/People/Cormac.McGuinness/Cowan/.

[16] R. D. Cowan, *The Theory of Atomic Structure and Spectra*, California, USA, 1981.

[17] B. Karaçoban and L. Özdemir, "Energies and lifetimes for some excited levels in La I," *Acta Physica Polonica A*, vol. 113, no. 6, pp. 1609–1618, 2008.

[18] B. Karaçoban and L. Özdemir, "Electric dipole transitions for La I (Z = 57)," *Journal of Quantitative Spectroscopy and Radiative Transfer*, vol. 109, no. 11, pp. 1968–1985, 2008.

[19] B. Karaçoban and L. Özdemir, "The hyperfine structure calculations of some excited levels for $^{(139)}$La I," *Acta Physica Polonica A*, vol. 115, no. 5, pp. 864–872, 2009.

[20] B. Karaçoban and L. Özdemir, "Transition energies of neutral and singly ionized lanthanum," *Indian Journal of Physics*, vol. 84, no. 3, pp. 223–230, 2010.

[21] B. Karaçoban and L. Özdemir, "Electric dipole transitions for Lu I (Z = 71)," *Arabian Journal for Science and Engineering*, vol. 36, no. 4, pp. 635–648, 2011.

[22] B. Karaçoban and L. Özdemir, "Energies and Landé factors for some excited levels in Lu I (Z = 71)," *Central European Journal of Physics*, vol. 9, no. 3, pp. 800–806, 2011.

[23] B. Karaçoban and L. Özdemir, "Energies, Landé factors, and lifetimes for some excited levels of neutral ytterbium (Z = 70)," *Acta Physica Polonica A*, vol. 119, no. 3, pp. 342–353, 2011.

[24] B. Karaçoban and L. Özdemir, "Electric dipole transitions for neutral ytterbium (Z = 70)," *Journal of the Korean Physical Society*, vol. 58, no. 3, pp. 417–428, 2011.

[25] B. Karaçoban and L. Özdemir, "Transition energies of ytterbium (Z = 70)," *Zeitschrift für Naturforschung A*, vol. 66, pp. 543–551, 2011.

[26] B. Karaçoban and L. Özdemir, "The level structure of atomic lutetium (Z = 71): a relativistic Hartree-Fock calculation," *Indian Journal of Physics*, vol. 85, no. 5, pp. 683–702, 2011.

[27] B. Karaçoban and L. Özdemir, "Transition energies of lutetium," *Chinese Journal of Physics*, vol. 50, no. 1, pp. 40–49, 2012.

[28] NIST, http://www.nist.gov/pml/data/asd.cfm.

10

Relativistic Time-Dependent Density Functional Theory and Excited States Calculations for the Zinc Dimer

Ossama Kullie

Laboratoire de Chimie Quantique, Institute de Chimie de Strasbourg, CNRS et Université de Strasbourg, 4 rue Blaise Pascal, 67070 Strasbourg, France

Correspondence should be addressed to Ossama Kullie, ossama.kullie@unistra.fr

Academic Editor: Jan Petter Hansen

I present a time-dependent density functional study of the 20 low-lying excited states as well the ground states of the zinc dimer Zn_2, analyze its spectrum obtained from all electrons calculations performed using time-depended density functional with a relativistic 4-component and relativistic spin-free Hamiltonian as implemented in Dirac-Package, and show a comparison of the results obtained from different well-known and newly developed density functional approximations, a comparison with the literature and experimental values as far as available. The results are very encouraging, especially for the lowest excited states of this dimer. However, the results show that long-range corrected functionals such as CAMB3LYP gives the correct asymptotic behavior for the higher states, and for which the best result is obtained. A comparable result is obtained from PBE0 functional. Spin-free Hamiltonian is shown to be very efficient for relativistic systems such as Zn_2.

1. Introduction

Zinc dimer Zn_2 is the first member of the group 12 (IIB) (Zn_2, Cd_2, Hg_2, and Cn_2) and has a representative character of these dimers. The interest in the dimers of the group IIB (12) is in part due to the possibility of laser applications in analogy with the rare gas dimers. A second point is the importance of the metallic complexes similar to the transition metal complexes [1–4] and some important application like the solar cell and renewable energy [5, 6] as well as electric battery for new cars technology [7, 8]. Zn_2, Cd_2, and Hg_2 are exciter with a shallow, predominantly Van der Waals ground state and low-lying covalent bound excited states. They are also interesting from a theoretical point of view due to the different character of the ground and excited states and consequently the different methodological demands for an accurate theoretical description of the spectrum. The dimer of group 12 has been studied both experimentally and theoretically. Relevant reviews have been provided by Morse [9] and more recently by Koperski [10, 11]. The covalent contributions to the ground state bonding in the group 12 dimers have been investigated in [12], it was concluded that the bond is a mixture of 3/4 Van der Waals and 1/4 covalent interactions. Bucinisky et al. [13] provides spectroscopic constants using the coupled cluster method (CCSD(T)) and different level of the theory 4-component relativistic Hamiltonian, using Dirac-Coulomb Hamiltonian, relativistic spin-free Hamiltonian and nonrelativistic (NR) Hamiltonian. Furthermore, they investigated the relativistic effects and found to be about 5, 8, 19% of the binding energies for Zn_2, Cd_2, and Hg_2, respectively. Finally the last member of the group Cn_2, copernicium, has an academic interest [14–16] due the chemical character of the bonding in comparison to Hg_2 (and the lighter dimers of the group), and the influence of the relativistic effects on the atomic orbitals providing a change of the boding character in the dimer to more covalent or Van der Waals type.

The paper presents all-electron calculations on the lowest-lying excited states as well as the ground state. The first 8 lowest exited states are discussed with a comparison to experimental and literature values, and several other higher excited states are presented and discussed. Earlier works investigated the lowest 8 excited states using different wave function methods. Ellingsen et al. [17] showed ab initio results for the ground and lowest 8 excited states of Zn_2, they performed all electron calculations and present NR as well as relativistic spin-free Douglas-Kroll result, the spin-orbit coupling was accounted perturbatively. The ground state is studied at ACPTF (averaged coupled pair functional, CCSD(T) and CASPT2 (complete active space second-order perturbation theory) level, and the excited states are studied at MR-ACPF (multireference ACPF) and CASPT2 level. Czuchaj et al. [18–20] performed their computations for Zn_2 (later for Cd_2 and Hg_2) using (NR) pseudopotential approach and MRCI (multireference configuration interaction), and the spin-orbit coupling was taken only approximately.

In this work, we use a relativistic spin-free Hamiltonian (SFH), without spin-orbit coupling, with a comparison to a relativistic 4-component Dirac-Coulomb Hamiltonian (DCH), spin-orbit coupling included, in the framework of time-dependent density functional theory (TDDFT) and its linear-response approximation (LRA). The calculations are performed using Dirac-Package (program for atomic and molecular direct iterative relativistic all-electron calculations) [21]. The relativistic effects for Zn_2 (and even for Cd_2) are small but visible and in some respects not negligible. To my experience, generally around zinc ($Z = 30$) the relativistic effects started to become important for chemical properties. For Hg_2, they are large enough (for Cn_2 expected to be very large) to make it necessary to incorporate them into any properties that are sensitive to the potential [13]. This is predominantly due to the contraction of 6s orbital, a well-known and important relativistic effects in heavy atoms [22–25]. We will follow this issue in future works on the group 12 (IIB).

The paper is organized as follows. Section 2 is devoted to the theory and method. We briefly introduce in Section 2.1 the key concepts of the static density functional (DFT) and discuss its extension to the relativistic domain. In Section 2.2, we introduce the key concepts of time-dependent density functional (TDDFT) and the linear response approximation. Section 3 is devoted to the computational details and Section 4 to the result and discussion, and finally we give a conclusion in Section 5. Some useful (well-known) notations used in this paper are collected in Table 1.

Table 1: Some of the acronyms used in this work.

HF	Hartree Fock method
NR	Nonrelativistic
DHF	Dirac or relativistic HF
DCH	Dirac-Coulomb Hamiltonian
MP2	Møller-Plesst 2nd-order perturbation theory
CCSD(T)	Coupled cluster singles-doubles (triples)
SFH	Relativistic spin-free Hamiltonian
(TD)DFT	(Time-depended) density functional theory
xc	Exchange-correlation
LR(A)	Linearresponse (approximation)
ALR	Adiabatic LR
srLDAMP2	Short-range LDA, long-range MP2

2. Theory and Methods

Time-dependent density functional theory (TDDFT) currently has a growing impact and intensive use in physics and chemistry of atoms, small and large molecules, biomolecules, finite systems, and solidstate. For excited states resulting from a single excitation that present a single jump from the ground state to an excited state, I used in this work the LRA as implemented in Dirac-Package [26–28] and well-known approximations of density functionals like LDA (SVWN5 correlation) [29, 30], PBE [31], PB86 [32–34], BPW91 (Becke exchange [32] and Perdew-Wang correlation [35]), long-range corrected PBE0 [36] and its gradient corrected functional GRAC-PBE0 [37, 38], BLYB and B3LYP [32, 39–41], or newly developed range-separated functionals such as CAMB3LYP [42]. Today's available DFT cannot describe the ground state of the group IIB dimers accurately due to a large contribution of dispersion in the bonding [12], despite this when calculating the covalently well-bound excited states the error is reduced considerably, quite possible accompanied with error cancellations.

The ground state of the group 12 dimer has a (closed-shell) valence orbitals configuration: $(ns^2 + ns^2) : \sigma_g^2, \sigma_u^2, n = 4, 5, 6$ for Zn_2-Hg_2. This configuration essentially arising from the interaction of atomic (ns) orbitals. It is weakly covalent and preponderantly dispersion interaction, well known especially in the rare gas dimers [43]. The potential curve displays a shallow van der Waals type of minimum. Exciting electrons from σ_g^2 or σ_u^2 to the lowest set of molecular orbitals spanned by the atomic orbitals Atom(ns^2) + Atom$(nsnp)$ or Atom(ns^2) + Atom$(ns(n+1)s)$, or Atom(ns^2) + Atom$(ns(n+1)p)$ gives rise to a manifold of states (see Table 2) among them states which strongly have covalent contributions as we will see in Section 4 Results and Discussion. This makes TDDFT using LRA and well-known functional approximations, adequate to describe these states [26].

We will discuss the lowest 20 excited states dissociating to the atomic asymptotes (NR notation) given in Table 2, resulting from exciting one electron from the ground state $(4s^2\ {}^1S + 4s^2\ {}^1S)^1\Sigma_g^+$. The concern will be in the first place on the 8 lowest excited states corresponding to the asymptote Atom(ns^2) + Atom$(nsnp)$. States corresponding to the higher asymptotes Atom(ns^2) + Atom$(ns(n+1)s)$ and Atom(ns^2) + Atom$(ns(n+1)p)$ are computed and some of them are well-bound states, we will discuss their quality in view of the limit of the validity of the known DFT approximations yielding inaccurate potential curves and causing a disturbance near the avoiding crossing with states of the same symmetry (see Section 4). To my best knowledge, there is no experimental or theoretical values from DFT or wave function methods available for the higher states to compare with, this makes it difficult to judge the result of the present work. It is expected that the result of the lowest states will show an excellent agreement with the experimental data [10, 11] (and the references therein), whereas for the higher states a satisfactory result is expected showing the important features of these states. The comparison between spin-free and 4-component results shows clearly the capability of SFH to deal with the computation of the properties of the Zn_2 dimer

Table 2: Lowest excited states and the corresponding asymptotes.

Equation (1) $((ns^2)\ ^1S + (nsnp)\ ^3P) : {}^3\Pi_g, {}^3\Pi_u, {}^3\Sigma_g^+, {}^3\Sigma_u^+$
Equation (2) $((ns^2)\ ^1S + (nsnp)\ ^1P) : {}^1\Pi_g, {}^1\Pi_u, {}^1\Sigma_g^+, {}^1\Sigma_u^+$
Equation (3) $((ns^2)\ ^1S + (ns(n+1)s)\ ^3S) : {}^3\Sigma_g^+, {}^3\Sigma_u^+$
Equation (4) $((ns^2)\ ^1S + (ns(n+1)s)\ ^1S) : {}^1\Sigma_g^+, {}^1\Sigma_u^+$
Equation (5) $((ns^2)\ ^1S + (ns(n+1)p)\ ^3P) : {}^3\Pi_g, {}^3\Pi_u, {}^3\Sigma_g^+, {}^3\Sigma_u^+$
Equation (6) $((ns^2)\ ^1S + (ns(n+1)p)\ ^1P) : {}^1\Pi_g, {}^1\Pi_u, {}^1\Sigma_g^+, {}^1\Sigma_u^+$

or similar systems. We also emphasize its importance for heavier relativistic systems [13], although spin-orbit effect is expected to be larger for Cd_2, Hg_2, and Cn_2. Pyper et al. [22] pointed out that the relativistic ground-state potential well depth of Hg_2 is 45% of the NR one and clearly it is stronger for Cn_2.

2.1. Density Functional Theory. Density functional theory [44–46] has become recently a very large popularity as a good compromise between accuracy and computational expediency. The Hohenberg-Kohn theorem [44] proves the existence of an unique (up to an additive constant) external potential $v_{\text{ext}}(\mathbf{r})$ for a given nondegenerate density $n(\mathbf{r})$ of interacting Fermions. The key point behind this scheme is the very useful simplification, namely, the transformation of the many-body quantum problem to a set of equations of one-particle Schrödinger (or Dirac) type of a noninteracting reference system with the density as a central ingredient quantity to carry all the relevant information of the system under consideration, instead of the many-body quantum wave function in which all the information of the system is stored:

$$\hat{H}\phi_i(\mathbf{r}) = E[n(\mathbf{r})]\phi_i(\mathbf{r}), \tag{1}$$

$$\hat{H} = \hat{T} + V_{\text{eff}}[n(\mathbf{r})] = \sum_i \hat{t}(\mathbf{r}_i) + v_{\text{eff}}(\mathbf{r}_i)[n(\mathbf{r})], \tag{2}$$

$$v_{\text{eff}}(\mathbf{r}_i) = v_{\text{ext}}(\mathbf{r}_i) + v_H(\mathbf{r}_i) + v_{\text{xc}}(\mathbf{r}_i) + v^{nn}, \tag{3}$$

$$n(\mathbf{r}) = \sum_{i=1}^{N} \|\phi_i(\mathbf{r})\|^2, \tag{4}$$

where $n(\mathbf{r})$ is the total density of the system and the sum is over N, that is, all occupied orbitals $\phi_i(\mathbf{r})$. $\hat{t}(\mathbf{r}_i)$ is the one-particle kinetic energy operator, $v_{\text{eff}}(\mathbf{r}_i)$ is the one-particle effective potential (also called Kohn-Sham potential $v_{\text{eff}}(\mathbf{r}_i) \equiv v_{\text{KS}}(\mathbf{r}_i)$), with $v_{\text{ext}}(\mathbf{r}_i)$ is the Coulombic interaction of the electron i with all the nuclei, called the external potential. $v_H(\mathbf{r}_i)$ is the Hartree and $v_{\text{xc}}(\mathbf{r}_i)$ exchange-correlation potential. And v^{nn} is the classical Coulombic repulsion of the nuclei in the system. $v_H(\mathbf{r}_i)$ is given by the usual expression, but the crucial part $v_{\text{xc}}(\mathbf{r}_i)$ in this scheme is the explicitly unknown $v_{\text{xc}}(\mathbf{r}_i)$:

$$v_H(\mathbf{r}_i) = \int d^3r \frac{n(\mathbf{r})}{|\mathbf{r}_i - \mathbf{r}|}, \qquad v_{\text{xc}}(\mathbf{r}_i) = \frac{\partial E_{\text{xc}}[n(\mathbf{r})]}{\partial n(\mathbf{r}_i)}, \tag{5}$$

for which an appropriate good approximation must be found. Experiences in DFT (and TDDFT) over the past decades shows that the density of atoms, molecules, finite systems, and solids have very complicated structures [47]. To find a good mathematical functionality form between the density (and its gradients) and an exchange-correlation potential with widely physical applications success is one of the most challenging problems in quantum physics and chemistry. Moreover, most of the problems arise when evaluating the results of the calculating systems can be tracked back to the limits of the validity of the today's known and employed approximations specially the long-range behavior leaving quite a room for improvements. One should note that that in many applications the usual approximations are quite reliable and give good results and acceptable accuracies. The present work is not an exception as we will see when analyzing the results of the ground state and excited states of the Zn_2 dimer.

2.1.1. Density Functional Theory in the Relativistic Domain. In the relativistic Dirac theory in absence of electromagnetic field, the DCH has the same generic form as the NR Hamiltonian (for molecules) [26, 48]:

$$\hat{H}_{\text{DC}} = \sum_i^N h_D(i) + \frac{1}{2}\sum_{i\neq j}^N \hat{g}^{\text{Coul}}(i,j) + \sum_{K\neq K'}^M V_{K,K'}^{nn},$$
$$h_D(i) = \left(c^2\hat{\beta} + c\,\hat{\boldsymbol{\alpha}} \cdot \hat{\mathbf{p}}(i) - c^2 \cdot \mathbf{I}_4\right) + \mathbf{I}_4 \cdot \sum_{K=1}^M V_K^{\text{ext}}(i),$$
$$\hat{\alpha}_j = \begin{pmatrix} 0 & \sigma_j \\ \sigma_j & 0 \end{pmatrix}, \quad j = x, y, z; \qquad \hat{\beta} = \begin{pmatrix} \mathbf{I}_2 & 0 \\ 0 & -\mathbf{I}_2 \end{pmatrix}, \tag{6}$$

where $h_D(i)$ is the one-particle DCH, and c is the speed of light in atomic units (atomic units are used throughout this work unless otherwise noted). V^{nn} is the classical nucleus-nucleus repulsion and $V_K^{\text{ext}}(i) = -Z_K/r_{iK}$ is the external Coulombic interaction of the electron i with the nucleus K, and the sum is over all nuclei M. $\mathbf{I}_2$ and $\mathbf{I}_4$ are the 2×2- and 4×4-unity matrix and the term $c^2 \cdot \mathbf{I}_4$ is a shift to align the relativistic and NR energy scales. $\hat{\beta}$ and $\hat{\boldsymbol{\alpha}} = (\alpha_x, \alpha_y, \alpha_z)$ are the Dirac matrices, with the well-known Pauli matrices σ's. The generic term

$$\hat{g}^{\text{Coul}}(i,j) = \frac{\mathbf{I}_4 \times \mathbf{I}_4}{r_{ij}} \tag{7}$$

is the Coulombic instantaneous two-electron i, j interaction operator, it contains in the relativistic theory the spin-own orbit interaction. The DCH approximation reduces the density functional theory in the relativistic domain to the usual density functional theory with the density as the central ingredient, and there is no need to introduce the current density [48]. A density functional theory in the relativistic domain can be constructed on the the basis of (1)–(4) with the density is constructed from the relativistic 4-component wave function. The total energy of the system is given by

$$E[n] = \sum_i^N \varepsilon_i - E_J[n] + E_{\mathrm{xc}}[n] - \int d^3r\, v_{\mathrm{xc}}(\mathbf{r}, n)n(\mathbf{r}) + E^{nn}, \tag{8}$$

where ε_i are the electronic eigenvalues of the system and are calculated iteratively in a self-consistent manner (SCF iterations) in an effective many-body potential v_{eff} given in (3). E^{nn} is the nuclear-nuclear repulsion energy, $E_J[n]$ is the Hartree energy equation (9), and $E_{\mathrm{xc}}[n]$ is the exchange-correlation energy, it can be further divided into exchange and correlation parts $E_{\mathrm{xc}}[n] = E_x[n] + E_c[n]$. At the (single, determinant) Hartree-Fock (HF) level, which in the relativistic calculations is usually called Dirac-Hartree-Fock (DHF), the two-particle interaction, the Hartree and exact exchange are given by (9) and (10) as follows:

$$E_J[n] = \frac{1}{2}\int\int d^3r_1 d^3r_2\, \frac{n(\mathbf{r}_1)n(\mathbf{r}_2)}{|\mathbf{r}_1 - \mathbf{r}_2|}, \tag{9}$$

$$E_x = -\frac{1}{4}\sum_{i,j}^N \int\int d^3r_1 d^3r_2\, \frac{\phi_i^\dagger(\mathbf{r}_1)\phi_j^\dagger(\mathbf{r}_2)\phi_j(\mathbf{r}_1)\phi_i(\mathbf{r}_2)}{|\mathbf{r}_1 - \mathbf{r}_2|}, \tag{10}$$

where $\mathbf{r}_1$ and $\mathbf{r}_2$ denote the coordinates of the electron one and two, respectively. $E_J[n]$ is a classical interaction between two one-particle densities $n(\mathbf{r}_1)$ and $n(\mathbf{r}_2)$, whereas E_x is a quantum mechanical nonlocal part of many-particle interaction. The $\phi(\mathbf{r})$s are the electronic one-particle HF-orbitals and the sum is over all the occupied orbitals N. A well-known approximation for the Hartree-Fock exchange energy is the (α-)Slater approximation [29] with remarkable performance for covalent bonding in covalently bound molecules with heavy atoms [49, 50]

$$E_x^\alpha[n] = -\frac{3}{2}\alpha C_x \int d^3r n^{4/3}(\mathbf{r}), \tag{11}$$

where $C_x = (3/4)(3/\pi)^{1/3}$ is a constant, in the Slater approximation the parameter $\alpha = 0.7$ is chosen. The missing of the correlation made the Slater approximation unpopular for chemical calculations. In the DFT, the exact $E_{\mathrm{xc}}[n]$ is unknown as a functional of the density (and its gradients). Many approximations exist with different performance and accuracy depending on their application area. In LDA one assumes a slowly varying local density dependence; hence, the Dirac-formula [51] of the exchange energy for an uniform electronic gas equation (11) with $\alpha = 2/3$ is applied and the Vosko-Wilk-Nusair correlation formula [29, 30] for the correlation energy (we use SVWN5). LDA depends only on the density, whereas in the generalized gradient approximation (GGA) the density and its gradient are involved, meta GGAs [52] include higher gradients, this systematic improvements is known in the DFT community under the term "Jacob's ladder." In hybrid functional, for example, BLYP and B3LYP [32, 39–41], one add a (fixed) suitable fraction of exact (Hartree-Fock) exchange (10) to the approximate x-energy part, which often improves the performance of the DFT approximation, whereas in the range-separated density functional [53] a parametric fraction of exchange (and possibly correlation) from wave function methods are added to the DFT exchange energy, with the parameter dictate the amount of exchange to be added, like CAMB3LYP [42], or of exchange-correlation like srLDAMP2 (see [43, 54–56] and the references therein), this improves the results considerably, unfortunately it is found that the optimum parameter value depends on the specific property of the system.

2.1.2. The Relativistic 4-Component and SFH. The Dirac equation with the Dirac-Coulomb Hamiltonian (DCH) describes the important relativistic effects for chemical calculation, which become large for systems with large Z. It is a firs-order differential equation(s), hence nonvariational "variational collapse" in contrast to the second-order differential Schrödinger equation in the NR case. The solutions to the Dirac equation describe both positrons (the "negative energy" states) and electrons (the "positive energy" states) as well as both spin orientations and a four-component wave function is involved called Dirac spinors:

$$|\psi\rangle = \begin{pmatrix}\Psi^L \\ \Psi^S\end{pmatrix}, \qquad \Psi^L = \begin{pmatrix}\phi^1 \\ \phi^2\end{pmatrix}, \qquad \Psi^S = \begin{pmatrix}\phi^3 \\ \phi^4\end{pmatrix}, \tag{12}$$

where Ψ^L is called the large and Ψ^S the small component. This notation originally comes from the well-known kinetic balance approximation and is justified by the relation $\sim 1/c$ between them, from which it follows the NR limit $\lim_{c\to\infty}\Psi_S = 0$ and one identify Ψ_L with the 2-component vector (spin up; down) of the Schrödinger equation. The full relativistic 4-component DCH is computationally demanding; therefore, it is desirable to reduce the computational effort in relativistic calculations by reducing the dimension of the involved quantities, normally by reducing or transforming the Hamiltonian to a new from, so that the calculations involving operators acting only on the large components and requiring a moderate computational effort by keeping the main physical features of the results. The relativistic SFH implemented in Dirac-Package uses the Dyall's formulation [57] to obtain results without spin-orbit coupling for the four-component Hamiltonian in the default restricted kinetic balance scheme. In Section 4, we show that the results obtained for the excited states of Zn_2 based on (relativistic) SFH are accurate similar and well comparable to those obtained from the 4-component DCH. For the deriving of this Hamiltonian, we kindly refer the reader to [57], see also [58] with advanced description in framework of second quantization formalism. The relativistic SFH permits factorization of the spin as in NR calculations so that standard NR post-SCF methods can be used for inclusion of electron correlation. The extension and implementation of relativistic SFH for many-body system or molecular calculation is straightforward see [21].

2.2. TDDFT and Linear Response. In this section, we briefly introduce TDDFT formulation with a special emphasis on the linear density-response function and its connection to the electronic excitation spectrum, a more extensive derivations and wide discussions can be found in refs [47, 59–78] and the references therein. TDDFT was pioneered by a work of Zangwill and Soven [78], but the fundamental step was done later by Runge and Gross [60, 61], the Runge-Gross theorem is a rigorous foundation for the formally extension of the Hohenberg-Kohn theorem [44] to the time-dependent phenomena. It results in a time-dependent Kohn-Sham equation:

$$\left[\hat{T} + v_{\text{ext},\sigma}[n](\mathbf{r}t) + v_H[n](\mathbf{r}t) + v_{\text{xc},\sigma}[n](\mathbf{r}t)\right]\psi_{j\sigma}(\mathbf{r}t) = i\frac{\partial}{\partial t}\psi_{j\sigma}(\mathbf{r}t), \tag{13}$$

where $\hat{T}$ is the kinetic energy, $v_{\text{ext},\sigma}(\mathbf{r}t)$, $v_H(\mathbf{r}t)$, are $v_{\text{xc},\sigma}(\mathbf{r}t)$ are the time-dependent external, Hartree, and exchange-correlation potential respectively, and we adopt the notation $(\mathbf{r}t) \equiv (\mathbf{r}, t)$. $\psi_{j\sigma}(\mathbf{r}t)$ is the wave function of a particle j with a spin σ. The external potential is unique determined via the total density:

$$n(\mathbf{r}t) = \sum_{\sigma} n_{\sigma}(\mathbf{r}t) = \sum_{\sigma}\sum_{j}^{N_\sigma}\left\|\psi_{j\sigma}(\mathbf{r}t)\right\| \tag{14}$$

of the interacting system, where the sum is taken over all occupied spin-orbitals N_σ of a spin possibility σ.

2.2.1. Linear Response. In the special case of the response of the ground-state density to a weak external field, that is, the case in the most optical applications, the slightly perturbed system, which can be written in a series expansion $v_{\text{ext}} = v^0_{\text{ext}} + v^1_{\text{ext}} + \cdots \approx v^0_{\text{ext}} + \delta v_{\text{ext}}$, see [72], starts its evolution slowly from its ground-state density n_0 corresponding to the ground-state external potential v^0_{ext}. The xc can be expressed in terms of the states of (unperturbed) system, and thus as a functional of the ground-state density. The interacting real system and the Kohn-Sham fictitious system are connected via the same infinitesimal density change $\delta n(rt)$. The infinitesimal change in the Hartree-xc-potential $\delta v_{H\text{xc}} = \delta v_H + \delta v_{\text{xc}}$ due to the infinitesimal change in the density can be expressed in its functional derivative:

$$\delta v_{H\text{xc}}(\mathbf{r}t) = \int d^3r' dt' f_{H\text{xc}}(\mathbf{r}\mathbf{r}', t - t')\delta n(\mathbf{r}'t'), \tag{15}$$

where $f_{H\text{xc}}$ is called the Hartree-xc-kernel and is given in LR regime by

$$f_{H\text{xc}}[n_0](\mathbf{r}\mathbf{r}'; t - t') = \frac{\delta(t - t')}{|\mathbf{r} - \mathbf{r}'|} + \left.\frac{\delta v_{\text{xc}}[n](\mathbf{r}t)}{\delta n(\mathbf{r}'t')}\right|_{n=n_0(\mathbf{r})}, \tag{16}$$

where $\delta(t - t')$ is the Dirac-delta function. The first term in (16) is the Hartree contribution, it is instantaneous, or local in time. The second term in (16), $f_{\text{xc}}[n_0]$, called the xc-kernel, is much simpler than $v_{\text{xc}}[n](\mathbf{r}t)$ since it is a functional of the ground-state density n_0, it is nonlocal in space and time [70].

In the adiabatic approximation which is the most common in TDDFT, one ignores all time-dependencies in the past and takes only the instantaneous density $n(t)$ being local in time. The adiabatic approach is a drastic simplification and a priori only justified for systems with a weak time-dependence, which are always locally close to equilibrium [72]. In practice, one takes a known ground-state functional approximation and insert $n_0(t)$ into it; thus, any ground-state approximation (LDA, GGA, ...) provides an adiabatic approximation for the TDDFT xc-functional. The most common one is the ALDA.

3. Computational Details

The reported results in this paper have been performed using a development version of the Dirac10-Package [21] based on the 4-component relativistic DCH and SFH. We would like to stress, though, that the present implementation allows the use of all Hamiltonians implemented in the Dirac-Package such as the eXact 2-component relativistic Hamiltonian (X2C) [79] and the 4-component NR Lévy-Leblond Hamiltonian [80]. The nuclear charge distribution was described by a Gaussian model using the recommended values of [81].

The values of the spectroscopic constants R_e, ω_e, and D_e were extracted from a Morse potential fit based on at least ten equidistant points of step length 0.05 a.u. around the equilibrium distance a second fit using polynomial fit procedure available in Dirac-Package is used too, the comparison between the two fits show that 5-order polynomial fit is rather equivalent to a Morse potential fit, provided that Morse potential fit is performed for small region around the minimum which is done throughout this work, the agreement between the two fits gives us an additional criterion for the safety and correctness of the calculated spectroscopic constants reported in the present result.

We employed the aug-cc-pVTZ (likewise aug-cc-pVQZ) Gaussian basis sets of Dunning and coworkers [82–84]. This basis set is widely used in the literature, thus simplifying the comparison between different works. The small components basis set for the 4-component relativistic calculations has been generated using restricted kinetic balance imposed in the canonical orthogonalization step [80]. All basis sets are used in uncontracted form. Test calculations with aug-cc-pVQZ basis sets indicate that the reported structures can be considered converged with respect to the chosen basis sets, see Section 4. The potential curves are generated with a bout 175 point densely chosen equidistant with of step length of 0.05 a.u. in the significant part of the potential curves 4.00–10.00 a.u. The asymptotic point is taken at 400 a.u., the value of this point is used to get the values $(D_e(R_i))$ at the point i.

4. Results and Discussion

In this section, we discuss our computational result based on our calculations with the linear response adiabatic TDDFT module in Dirac-Package. Our main concern will be (beside the correctness of our computational result) to compare the behavior of different density functional approximations (and in comparison to other methods) to draw conclusions on the performance, the quality, and the validity of the different functional approximations, also in regard to applications to similar systems and possibly enlighten improvements of the DFT approximations in future works. The comparison with the literature values is accompanying our discussion, where works with different computational methods are available and with experimental values as far as available to judge the quality of our result.

4.1. Ground State. As already mentioned, the ground-state bond of Zn_2 dimer is a mixture of 3/4 Van der Waals and 1/4 covalent interactions [85] and the DFT can hardly deal with it as seen in Table 3, where the spectroscopic constants of the ground state are given for different density functional approximations. We note that the effect of the basis set size, typically by DFT, is very small clearly seen in Table 3 from PBE values calculated with aug-cc-pVTZ and aug-cc-pVQZ basis set. In Table 3, one sees that a comparable result is obtained by MP2 and srLDAMP2 as expected [43]. Similar to the rare-gas dimers [43], the range-separated DFT improves the DFT result (here LDA) for Zn_2 and suitably cure the lack of correct long-range behavior known by pure DFT approximations because the long-range part of the exchange (and the long-range correlation in srLDAMP2) is treated by a wave function method (MP2). However, a crucial point is to determine a suitable value of the rage-separation parameter. Generally, a suitable range for this parameter is 0.2–0.5 a.u., for details and indepth discussion see [43] and the references therein. DFT approximations and CAMB3LYP, as well as srLDAMP2, do not yield a satisfactory result. Looking at the LDA, we see that the correction of the LDA by srLDA-MP2 is large; however, the improvement gives no advantage over the MP2 as they have similar computational coast. Dramatically behave the long-range corrected PBE0 and the hybrid functionals BLYP and B3LYP (contain a fixed fraction of exact HF-exchange only), they yield a dissociative ground state. BP86 is the only functional with accurate dissociation energy value, but its R_e and ω_e are not helpful. Although CAMB3LYP gives the best R_e value comparison to experiment, this is not sufficient as the bond energy and vibrational frequency are not helpful. It is worthwhile to mention at this point that CAMB3LYP gives the correct asymptotic behavior for the excited states, see Figure 2, in contrast to pure (LDA, PBE, BPW91, BP86, . . .), long-range corrected (PBE0,GARC-PBE0) or hybrid (BLYP, B3LYP) DFTs, as seen in Figures 2 and 3. Whether this means that CAMB3LYP potential curves has a correct shape (in all regions) is difficult to say at the moment. The shape of the potential curve is an important feature for the DFT accuracy as noted by Grüning et al. [38].

4.2. Excited States. The excited states shown in the pw are given in Table 2, where $n = 4$ for Zn atom. The results are given in the Tables 5–8. We first discuss the lowest 8 states given in the Tables 5–8, then we proceed to discuss the higher states given in Table 8.

Table 3: Ground-state $^1\Sigma_g^+$ of Zn_2 dimer.

	R_e (Å)	ω_e (cm^{-1})	D_e (eV)
exp[1]		25.7	0.034
exp[2]	4.19	25.9	0.035
HF-MP2[Q]	3.611	29	0.049
srLDAMP2[Q]	3.445	31	0.0459
PBE[Q]	3.157	48	0.678
PBE	3.156	49	0.683
PBE0	diss	diss	diss
BPW91	3.225	41	0.0154
BP86	3.181	46	0.036
BLYP	diss	diss	diss
B3LYP	diss	diss	diss
GRAC-PBE0	3.338	40.0	0.045
CAMB3LYP	4.219	11	0.001
LDA	2.846	85	0.225
[a]	3.959	22	0.024
[b]	3.96	22.5	0.030
[c1]	4.03	20.4	0.0205
[c2]	4.03	20.4	0.0205

pw using aug-cc-pVTZ basis set and SFH. [Q]aug-cc-pVQZ basis set, for PBE, HF-MP2 and srLDAMP2 (NR with parameter $\mu = 0.5$), see text: [1][86]; [2][85]. [a][12] using CCSD(T) in pseudopotential. [b][17] using NR-CCSD(T). [c1][13] CCSD(T) with 4-comp. DCH. [c2][13] CCSD(T) with SFH.

Table 4: Comparison between SFH (NR state assignment) and 4-component DCH of the spectroscopic constant. Above R_e (Å), middle ω_e (cm^{-1}), and below D_e (cm^{-1}), calculated with PBE functional and aug-cc-pVTZ basis set. For 4-component states assignment gerade, ungerade follow the symmetry of state in the first line.

	$^3\Pi_g$	$^3\Sigma_u^+$	$^3\Pi_u$	$^3\Sigma_g^+$
SFH	2.347	2.534	4.795	4.79
4-c. $0_g^-, 0_u^-$	2.345	—	4.874	—
4-c. $0_g^+, 0_u^+$	2.345	—	4.480	—
4-c. $0_u, 0_g$	—	2.534	—	4.553
4-c. $1_g(1_u)$	2.347	2.534	4.625	4.574
4-c. $2_g, 2_u$	2.349	—	4.945	—
SFH	219	172	7	27
4-c. $0_g^-, 0_u^-$	220	—	6	—
4-c. $0_g^+, 0_u^+$	220	—	13	—
4-c. $0_u, 0_g$	—	172	—	33
4-c. $1_g(1_u)$	219	172	13	34
4-c. $2_g, 2_u$	219	—	8	—
SFH	13097	10870	52	405
4-c. $0_g^-, 0_u^-$	12934	—	52	—
4-c. $0_g^+, 0_u^+$	13130	—	417	—
4-c. $0_u, 0_g$	—	10486	—	533
4-c. $1_g(1_u)$	12906	10680	235	550
4-c. $2_g, 2_u$	13068	—	53	—

At first we compare for PBE functional a 4-component and spin-free result for the four lowest states calculated in

TABLE 5: Bond lengths R_e (Å) of the lowest states corresponding to the lowest two asymptotes.

Method	$^3\Pi_g$	$^3\Sigma_u^+$	$^3\Pi_u$	$^3\Sigma_g^+$	$^1\Pi_g$	$^1\Sigma_u^+$	$^1\Pi_u$	$^1\Sigma_g^+$
P^Q	2.345	2.532	4.254	4.765	2.350	2.596	4.735	2.573
P^T	2.347	2.534	4.795	4.79	2.351	2.602	4.744	2.592
W91^T	2.343	2.517	diss	4.546	2.347	2.621	dis	5.158
P0^T	2.358	2.517	5.046	4.517	2.351	2.631	2.715	2.594
GP0^T	2.356	2.522	diss	5.806	2.345	2.780	2.929	4.755
CB3L^T	2.343	2.489	diss	diss	2.327	2.613	2.637	2.572
B3L^T	2.371	2.566	diss	5.525	2.366	2.655	2.807	2.624
BL^T	2.371	2.587	diss	4.882	2.376	2.648	diss	2.639
B86^T	2.337	2.534	diss	4.583	2.341	2.611	4.647	5.370
LDA^T	2.265	2.454	2.764	4.364	2.267	2.485	2.702	5.414
[17][a]	2.33	2.48	3.99	diss	2.30	2.64	2.40	2.74
[17][b]	2.35	2.50	4.11	diss	2.33	2.69	2.42	2.92
[87][c]	2.41	2.70	diss	diss	2.33	3.22	2.40	3.05
[19][d]	2.38	2.59	4.36	diss	2.38	2.64	2.65[f]	2.65[f]
[88][d]	2.53	2.74	diss	—	2.51	2.97	2.64	3.07
[89][d]	2.56	2.70	diss	diss	2.48	2.92	2.64	—
[90][e]	2.372	2.53						
exp	—	—	4.49[g]	—	—	3.0[g]	—	—

[T]Present work calculated with aug-cc-pVTZ and [Q]with aug-cc-pVQZ basis set. P, W91, P0, GP0, B86, BL, B3L, and CB3L denote PBE, BPW91, PBE0, GRAC-PBE0, BP86, BLYP, B3LYP, and CAMB3LYP, respectively. [a]With DK-CASPT2. [b]With DK-MRACPF. [c]With CI. [d]With MRCI. [e]With CCSD(T). [f]Value are ca. [g]From [85], for $^3\Pi_u$ [91] gives the value 3.30.

TABLE 6: Vibrational frequencies ω_e (cm^{-1}) of the lowest states corresponding to the lowest two asymptotes.

Method	$^3\Pi_g$	$^3\Sigma_u^+$	$^3\Pi_u$	$^3\Sigma_g^+$	$^1\Pi_g$	$^1\Sigma_u^+$	$^1\Pi_u$	$^1\Sigma_g^+$
P^Q	220	173	9	28	219	136	12	142
P^T	219	172	7	27	219	135	13	135
W91^T	218	177	diss	33	220	129	diss	21
P0^T	215	182	9	11	223	135	116	146
GP0^T	216	181	diss	11	226	106	78	38
CB3L^T	220	189	diss	diss	232	139	137	150
B3L^T	211	167	diss	13	215	126	90	139
BL^T	210	157	diss	27	207	122	diss	115
B86^T	222	172	diss	37	222	131	14	29
LDA^T	247	189	85	45	247	160	89	26
[17][a]	231	200	23	diss	250	131	211	58
[17][b]	220	208	32	diss	244	121	205	104
[87][c]	211	169	diss	diss	212	77	175	112
[88][d]	192	175	—	diss	210	134	178	—
[89][d]	175	150	diss	diss	202	107	166	104
exp	223 ± 5[e]	161 ± 5[f]	20.3 ± 0.2[g]	—	—	122 ± 10[h]	148 ± 6[i]	—

For the acronyms, see Table 5. [T,Q]as in Table 5. [a]With DK-CASPT2. [b]With DK-MRACPF. [c]With CI. [d]With MRCI. [e]From [92]. [f]From [93]. [g]From [85]. [h]From [94]. [i]From [95].

aug-cc-pVTZ basis set and demonstrate that SFH describes accurately the main relevant contributions of the relativistic effects. As seen in Table 4, the difference between SFH and 4-components DCH is rather small. To see the difference and the splitting in the 4 component precisely D_e is given in cm^{-1}. The splitting is very small or negligible clearly seen in Figure 1, where we compare visually the 8 lowest states of PBE functional using SFH and the corresponding 16 lowest excited states using 4-component DCH. We note that the CCSD(T) result of [13] for the ground state (see Table 3) using SFH and 4-components DCH confirms our result.

In Figure 2, we show the 20 lowest excited states corresponding to the 6 asymptotes given in Table 2, for the CAMB3LYP and B3LYP functionals. The overall behavior in Figure 2 for CAMB3LYP is satisfactory, it shows a better behavior for all states, and the states follow (at least)

Table 7: Dissociation energies (eV) of the lowest states corresponding to the lowest two asymptotes.

Method	$^3\Pi_g$	$^3\Sigma_u^+$	$^3\Pi_u$	$^3\Sigma_g^+$	$^1\Pi_g$	$^1\Sigma_u^+$	$^1\Pi_u$	$^1\Sigma_g^+$
P$^\mathrm{Q}$	1.626	1.347	0.0065	0.050	1.703	0.579	0.0180	0.112*
P$^\mathrm{T}$	1.624	1.348	0.0065	0.050	1.698	0.572	0.0175	0.08*
W91$^\mathrm{T}$	1.423	1.23	diss	0.031	1.654	0.541	diss	0.027
P0$^\mathrm{T}$	1.481	1.332	0.0034	0.0031	2.387	1.247	0.413	0.10
GP0$^\mathrm{T}$	1.43	1.316	diss	0.0148	2.385	1.111	0.270	0.279
CB3L$^\mathrm{T}$	1.436	1.281	diss	diss	2.298	1.126	0.099	0.292*
B3L$^\mathrm{T}$	1.45	1.189	diss	0.0033	2.226	1.125	0.393	0.148*
BL$^\mathrm{T}$	1.514	1.181	diss	0.468	1.426	0.361	diss	0.542
B86$^\mathrm{T}$	1.593	1.312	diss	0.0673	1.688	0.546	0.008	0.058
LDA$^\mathrm{T}$	2.119	1.704	1.456	1.902	2.089	0.798	0.145	0.788
[17][a]	1.502	1.225	0.026	diss	2.713	1.189	0.734	0.60
[17][b]	1.457	1.204	0.110	diss	2.694	1.292	0.718	0.204
[87][c]	0.91	0.90	diss	diss	2.35	0.71	—	—
[19][d]	1.21	0.95	0.016	diss	2.26	1.12	0.63	0.32
[88][d]	1.10	0.98	—	diss	2.43	1.13	0.66	—
[89][d]	1.05	0.87	diss	diss	2.42	1.06	0.83	0.44
[90][e]	1.41	1.21	—	—	—	—	—	—
exp	—	—	0.027[f]	—	—	1.117[g]	—	—

For the acronyms, see Table 5. T,Q As in Table 5. *See text. [a]With DK-CASPT2. [b]With DK-MRACPF. [c]With CI. [d] With MRCI. [e]With CCSD(T). [f]From [96]. [g]From [94] (1.117 ± 0.025), whereas [91] gives the value 1.30.

Table 8: Higher states corresponding to higher asymptotes see Table 2 and text.

State	R_e (Å)						ω_e (cm^{-1})						D_e (eV)					
	CB3L	P0	GP0	B3L	W91	B86	CB3L	P0	GP0	B3L	W91	B86	CB3L	P0	GP0	B3L	W91	B86
$^3\Sigma_u^+$	2.527	2.546	2.711	2.578	2.531	2.532	168	164	115	150	163	160	0.914	0.938	0.174	0.636*	0.555	0.644*
$^3\Sigma_g^+$	2.737	2.769	5.772	2.802	2.71	2.714	185	196	23	168	193	186	0.533	0.728	0.596	0.421	0.118	0.094
$^1\Sigma_u^+$	2.60	2.630	2.787	2.679	2.622	2.605	149	142	92	120	134	140	0.839	0.677	0.231	0.583	0.513	0.539
$^1\Sigma_g^+$	3.444	3.388	8.434	3.449	3.256	3.21	174	146	19	118	131	139	0.339	0.333	0.383	0.097	0.152	0.153
$^3\Pi_u$	2.919	3.080	3.162	3.352	3.323	3.451	99	82	72	59	51	45	1.416	0.95	0.90	0.646*	0.039	0.040*
$^3\Pi_g$	2.487	2.504	4.748	2.524	2.491	2.485	178	174	41	163	171	172	1.140	0.434	0.635	0.213*	0.143	0.20
$^3\Sigma_u^+$	2.519	2.532	diss	2.551	2.546	2.506	172	171	diss	158	166	164	0.905	0.270	diss	0.515	0.482	0.480
$^3\Sigma_g^+$	2.569	2.583	diss	2.603	2.513	2.563	153	150	diss	145	140	150	0.247	0.158*	diss	0.163*	0.150	0.157
$^1\Pi_u$	3.650	5.750	6.209	9.026	diss	diss	123	14	22	12	diss	diss	1.50	0.483	0.486	0.274	diss	diss
$^1\Pi_g$	2.459	2.482	6.317	2.495	2.472	2.465	190	184	26	174	180	182	1.417	0.344	0.482	0.46*	0.43	0.393
$^1\Sigma_u^+$	2.534	2.555	diss	2.585	2.537	2.533	169	167	diss	155	162	159	1.125	0.302*	diss	0.50*	0.561	0.560
$^1\Sigma_g^+$	2.704	2.682	diss	4.237	2.616	2.583	281	288	diss	296	244	210	0.517*	0.298	diss	0.218*	0.146	0.165

All values with SFH and aug-cc-pVTZ basis set. For the acronyms, see Table 5. *See text.

qualitatively to the correct asymptotes. In contrast to the B3LYP as seen in Figure 2(b), where similar result is obtained for all other functionals used in this work. These functionals show an incorrect asymptotic limit and only for the lowest 8 states give the correct (two) asymptotes, whereas most of the higher states follow to a wrong asymptotic limit. This is somehow unexpected since B3LYP includes a (fixed) fraction of exact exchange.

In Figure 3, a second example is presented for PBE0 and GARC-PBE0. GARC-PBE0 is supposed to give a better result than PBE0, but for Zn_2 dimer it does not show a correct description for the higher excited states. Indeed it is well known that pure DFT has incorrect long-range behavior which is the key point behind the range-separated DFT. It is clearly from this result that the separation of the two-electron interaction in short- and long-range parts as done in range-separated DFT like CAMB3LYP offers an advantage by treating the long-range part with a wave function method incorporating a suitable parametric amount of exact exchange. That only CAMB3LYP shows a better or a correct long-range behavior does not mean generally that a range-separated functional describes the excited states better in the short-range (or mid-range) region; however, its accuracy is satisfactory even it fails for the ground state (see Table 3) rather due to the lack of long-range correlation (in HF correlation is not present) important for dispersion interaction.

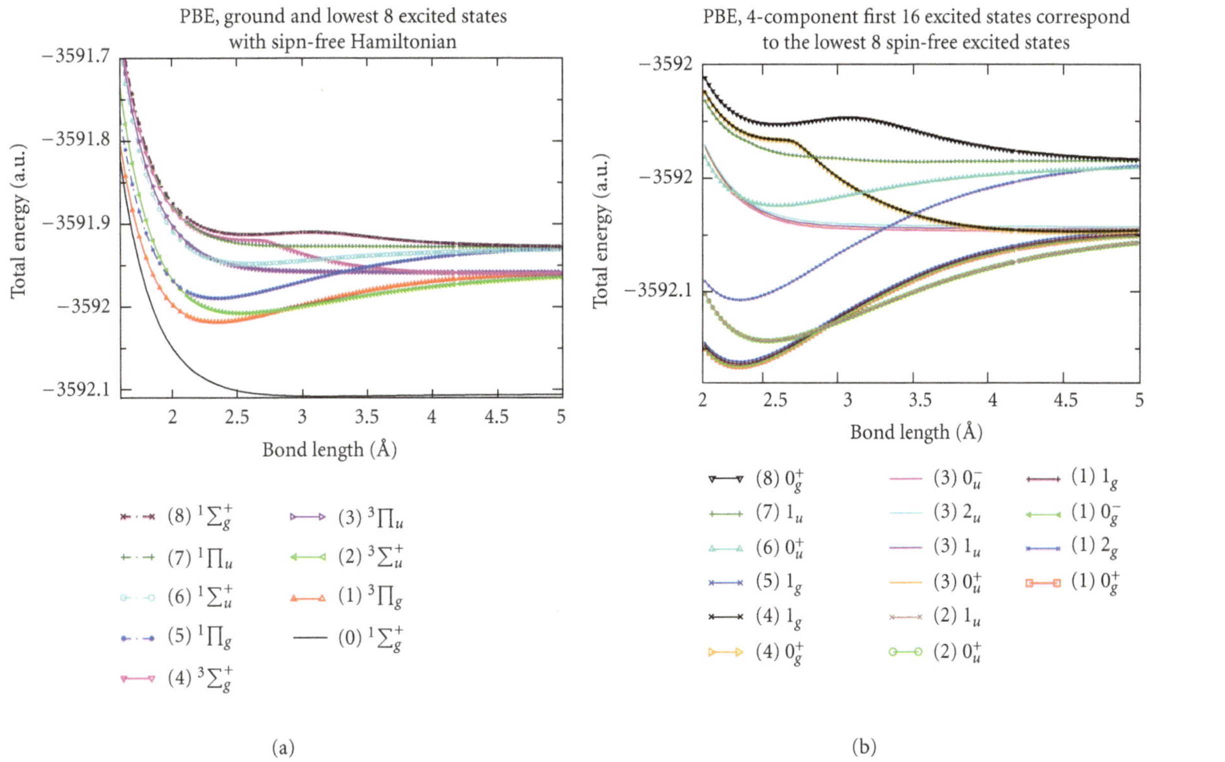

FIGURE 1: (a) Zn_2 PBE functional, with SFH (left) ground state (lowest curve) and 8 lowest excited state (corresponding to the two asymptotes $(4s^2S^1 + 4s4pP^1)$ lower ones, and $(4s^2S^1 + 4s4pP^v)$ upper ones. And (b) accordingly the 16 excited states with the same asymptotes using 4-component DCH. Numbering in brackets shows the correspondence between states of (a) and (b).

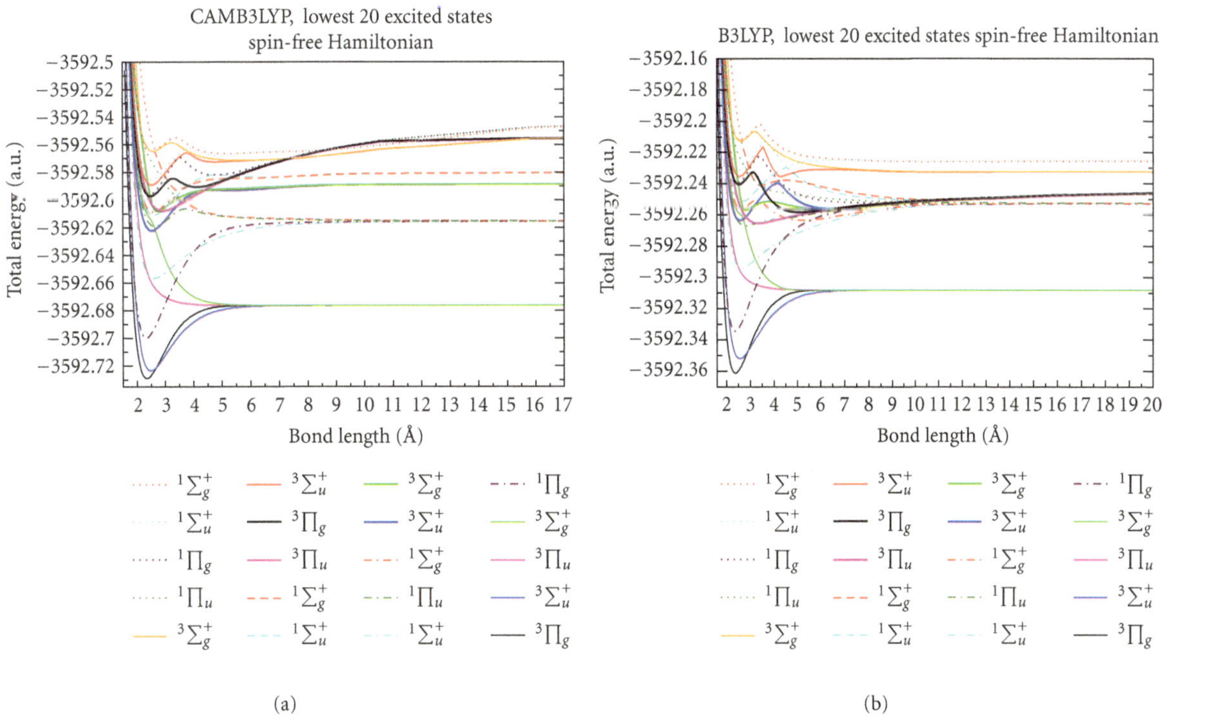

FIGURE 2: Zn_2, CAMB3LYP (a) and B3LYP (b) the 20 lowest states with SFH, corresponding to the asymptotes (from below) $4s^2\,^1S + 4s4p\,^3P$, $4s^2\,^1S + 4s4p\,^1P$, $4s^2\,^1S + 4s5s\,^3S$, $4s^2\,^1S + 4s5s\,^1S$, $4s^2\,^1S + 4s5p\,^3P$, and $4s^2\,^1S + 4s5p\,^1P$, respectively. Note some of the upper curves of B3LYP show incorrect asymptotes.

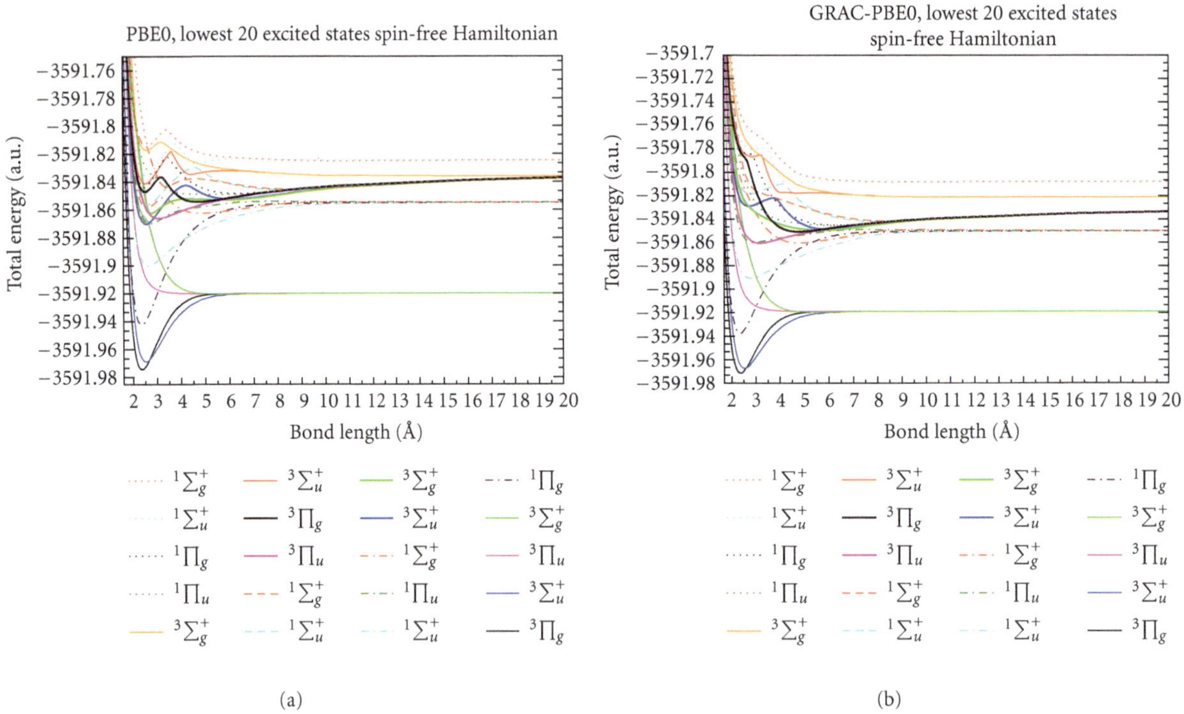

FIGURE 3: Zn_2, PBE0 (a) and GRAC-PBE0 (b) the 20 lowest states with SFH, corresponding to the same asymptotes as in Figure 2. Note that some of the upper curves show an incorrect asymptotes, compare CAMB3LYP Figure 2 and see text.

Obviously, a crucial point in calculating the excited states in TDDFT is that the most of the DFT approximations are semilocal, the long-range interaction is incorrectly described, consequently a disturbed potential curves is obtained, especially near the avoiding crossing point where the disturbed curves show enhanced effects. This can be clearly seen for the $^1\Sigma_g^+$, $^3\Sigma_g^+$, and $^1\Pi_u^+$ in Figure 4. For CAMB3LYP, we see every two states of the same symmetry push each other away and later both follow to the correct limit. For PBE0, as an example, the avoiding crossing is clear for $^1\Sigma_g^+$ and $^3\Sigma_g^+$ states but not for $^1\Pi_u^+$, most likely because it is disturbed by the incorrect long-range behavior. Similar behavior to PBE0 was found in all other DFT approximations used in this work, that is, an incorrect long-range behavior, with (or leading to) an incorrect asymptotic limit (and a disturbed avoiding crossing) is responsible for incorrect description of the higher excited states. We will discuss the accuracies in detail in the next sections.

4.2.1. Lowest 8 Excited States. In Tables 5–7, we give the evaluated spectroscopic constants for the lowest 8 excited states of Zn_2 using TDDFT, SFH, and aug-cc-pVTZ basis set. The lowest 8 excited states $^3\Pi_g$; $^3\Pi_u$; $^3\Sigma_g^+$; $^3\Sigma_u^+$ and $^1\Pi_g$; $^1\Pi_u$; $^1\Sigma_g^+$; $^1\Sigma_u^+$ are corresponding to the Atom($(4s^2)$ ^1S) + Atom($(4s4p)$ ^3P) and Atom($(4s^2)$ ^1S) + Atom($(4s4p)$ ^1P), respectively.

First, we look at the PBE values using aug-cc-pVTZ basis set and aug-cc-pVQZ basis set. As we see from Tables 3–5, the basis effect is small and only about $2 * 10^{-3}$ Å for R_e, about 1 unit for ω_e and between 2–6 meV in D_e. Following this we conclude that the SFH (see Table 4) with aug-cc-pVTZ basis set enable us to calculate the excited states of zinc dimer accurately. Our result is sufficiently accurate to compare with experimental values, wave function methods and compare the behavior of different functional approximations with each other for this dimer.

(a) The Lowest States $^3\Pi_g$, $^3\Pi_u$, $^3\Sigma_g^+$, $^3\Sigma_u^+$. Looking at the Tables 5–7, we see immediately that the best result is obtained for these states. For the lowest two state $^3\Pi_g$, $^3\Sigma_u^+$, all functionals give excellent agreement with wave function results giving in the literature, for example, [17] or the experimental value of ω_e, although the agreement for the first excited state, $^3\Pi_g$, is more pronounced. Recently, Determan et al. [90] have published accurate result for these two states using CCSD(T) and some density functional approximations, the excellent agreement with our values confirms our result. This is not surprising since these states are well bound and largely covalent in contrast to the ground state; moreover, the most known DFT approximations are more or less capable to describe (strong) covalent bonding due to its largely localized character in the bond region. It is also noticeable that all DFTs show for the eight lowest states asymptotically a correct behavior and the correct (two) asymptote, see Figures 2 and 3. For the lowest two states $^3\Pi_g$, $^3\Sigma_u^+$, only LDA strongly underestimates the dissociation energy and gives short bond lengths and large ω_e's. PBE gives larger bond energy for both states, likewise BP86 for the first one. BLYP and PBE0 give smaller values for ω_e. For R_e all these approximations give a similar result. For the

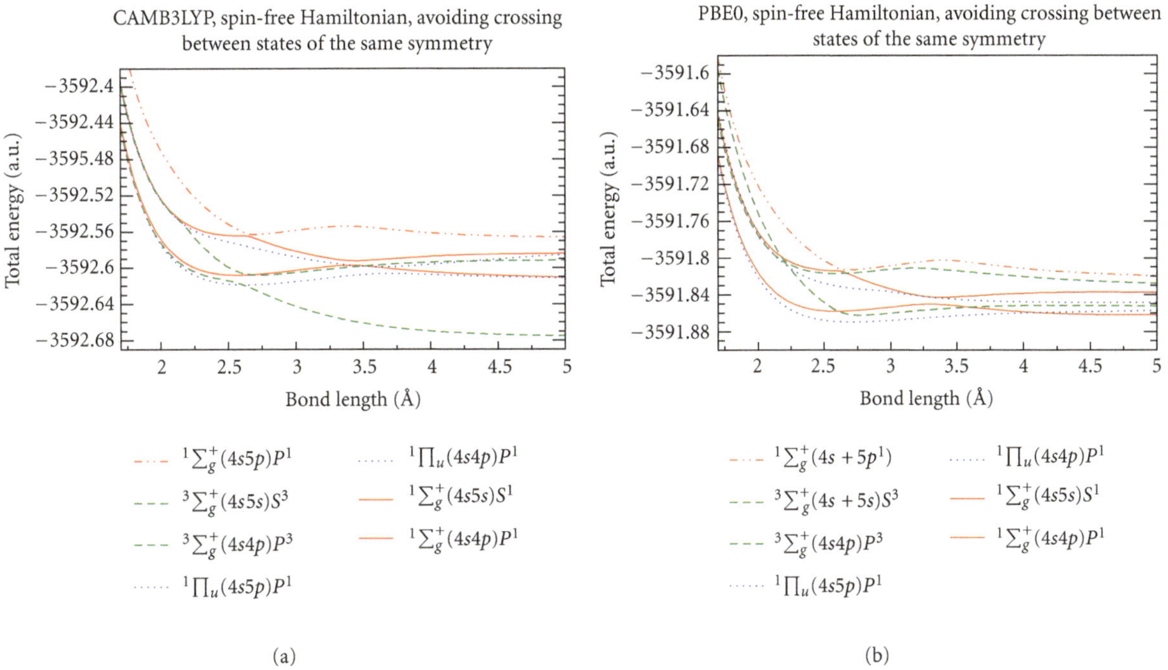

Figure 4: Zn_2, spinfree Hamiltonian avoiding crossing, CAMB3LYP (a) and PBE0 (b) between the two $^1\Sigma_g^+$ corresponding to the asymptotes $(4s^2S^1 + 4s4p\,^1P)$ and $(4s^2\,^1S + 4s5s\,^1S)$, the two $^1\pi_u$ corresponding to the asymptotes $(4s^2S^1 + 4s4p\,^1P)$ and $(4s^2\,S^1 + 4s5p\,^1P)$; the two $^3\Sigma_g^+$ states corresponding to the asymptotes $(4s^2\,S^1 + 4s4p\,^3P)$ and $(4s^2\,S^1 + 4s5s\,^3S)$. The highest $^1\Sigma_g^+$ is corresponding to the asymptotes $(4s^2\,S^1 + 4s5p\,^1P)$, see text.

next lowest two states, $^3\Pi_u$, $^3\Sigma_g^+$, the situation is somehow complicated. For $^3\Pi_u$ the experimental value shows a weak bound state, whereas wave function methods show different results, likewise in the DFT. PBE and PBE0 describe it as a weak bound state, but apart from LDA all other DFTs give a dissociative state. Whereas for the $^3\Sigma_u^+$ only CAMB3LYP shows a dissociative state in an agreement with the wave function methods. This is a first hint that CAMB3LYP gives a better long-range behavior and correct asymptotic limit for higher states than the other DFTs shown in the present work. This can be attributed to the fact that for high-quality response properties it is of primary importance for the potential curve to be accurate in the shape, rather than the condition to be met of being a functional derivative of a given density functional for the exchange-correlation energy [38]. For higher states, both the long-range behavior and the asymptotic limit in pure DFTs are incorrect and thus the shape of potential curves. BLYP gives $D_e \approx 0.47$eV for $^3\Sigma_g^+$ which somehow large comparing to other functional. The state $^3\Sigma_g^+$ (Atom$(4s+4s)$ ^1S + Atom$(4s+4p^1)$ ^3P) shows a hump around 2.5 Å clearly seen in Figure 4 due to an avoiding crossing with the higher state $^3\Sigma_g^+$ (Atom$(4s+4s)$ ^1S + Atom$(4s + 5p^1)$ ^3P), the later is well bound (see Table 8) and shows a small hump around 2.2 Å (hardly seen in Figure 4) presumably due to an avoiding crossing with a more higher state of the same symmetry.

(b) The States $^1\Pi_g$, $^1\Pi_u$, $^1\Sigma_g^+$, $^1\Sigma_u^+$. From Tables 5–7, we again see a good agreement, especially for $^1\Pi_g$ and $^1\Sigma_u^+$, between our result and the results of the wave function methods, where the agreement is less pronounced than the lowest two states. For $^1\Pi_g$ and $^1\Sigma_u^+$ bond lengths, apart from LDA, all functionals give comparable results. For the vibrational frequencies, BLYP and B3LYP give smaller values, for $^1\Sigma_u^+$ this is in excellent agreement with the experimental value of [94] or the value of [17]. CAMB3LYP gives the largest value of ω_e. For the dissociation energy D_e, B3LYP, CAMB3LYP, PBE0, and GRAC-PBE0 give reasonable values with a good agreement with the experiment for $^1\Sigma_u^+$. This remarkable result could be a hint that these three functionals have a correct mid-range behavior. From the agreement with the experiment and the wave function values, one concludes that the values of $^1\Pi_g$ of B3LYP, CAMB3LYP, PBE0, and GRAC-PBE0 should be close to the experiment. Next, we look at the two states $^1\Pi_u$, $^1\Sigma_g^+$, as mentioned above in Figure 4, these two states have avoided crossing with higher lying states of the same symmetry. From the tables, we now see a less agreement with the wave function method, and the lack of experimental values makes it more difficult to judge the result. If we take the values of [17], as a reference we see that reasonable DFTs values show larger bond lengths, smaller vibrational frequencies for $^1\Pi_u$ and for $^1\Sigma_g^+$ vice versa for the most of the functionals. For $^1\Pi_u$, the dissociation energies are smaller than the reference value. For $^1\Sigma_g^+$, the obtained bond energy values for some functionals denoting the depth of the minimum (marked with "∗") relating to the shallowest point after the minimum, otherwise the incorrect asymptotic point will show a dissociative state, which of course an artifact of the (quantitatively) incorrect tail of the

potential curve. We have seen in Figure 2 that CAMB3LYP has asymptotically a correct behavior specially for the higher states; however, it is quantitatively questionable and for some states seems to be inaccurate. In such cases, the spectroscopic constants are calculated relative to the shallowest point after the minimum, and not to the asymptotic point. This yields approximately the same R_e and ω_e, but the obtained value D_e will be definitely shallower than, or approximately equal to a value D_e relating to the "correct" asymptotic point. We note that all values marked with an "$*$" in Tables 7-8 are obtained this way. For the ${}^1\Sigma_g^+$ state, we see from Table 7 that CAMB3LYP has a good agreement with [19], likewise B3LYP with [17], whereas BLYP shows an agreement with DK-CASPT2 value of [17], but to conclude we see that the result(s) of ${}^1\Sigma_g^+$ are widely distributed; furthermore, the lack of any experimental value makes the situation more difficult.

4.2.2. Higher Excited States. To deal with more higher excited states is difficult because of the above-mentioned reasons. Available approximations do not describe the long-range behavior correctly and/or fail to offer the correct asymptotic limit or predict it accurately [97]. We will discuss the higher molecular states given in Table 6 corresponding to the last four asymptotes (3–6) in Table 2. The result is given for the functionals BPW91 and BP86 (pure), B3LYP (hybrid), CAMB3LYP (range-separated), PBE0 (long-range corrected), and its gradient corrected one GRAC-PBE0. GRAC is an interpolation scheme, it is an asymptotic correction and supposed to be able to deal with higher excited states [37, 38]. The pw shows that the best result is obtained for CAMB3LYP and a comparable result is obtained for PBE0. Indeed strictly only CAMB3LYP was able to deal with higher excited states, it shows (at least qualitatively) the correct asymptotic as can be clearly seen in Figure 2. Other functionals do not show a correct asymptotic behavior as expected [37], including the ones for which no data shown in Table 8. B3LYP is given in Figure 2 as an example, it mixes the asymptotic for higher states with lower states. Our conclusion based on analyzing the data of all functionals and comparing them with each other. It is clear that lacking to the correct long-range behavior is primarily the origin of the problem, CAMB3LYP is able to cure this although not accurately, the question is why other corrections like GRAC does not have the expected improvement? At one side important is the nonlocal part of exact exchange which improves the situation considerably when the two-electron interaction is separated in short- and long-range part such as in CAMB3LYP, and we notice that there is no long-range correlation present in CAMB3LYP because HF offers only (nonlocal) exchange. Another point is the wrong long-range behavior of the response function [72, 77] caused by the incorrect long-range behavior of the density functional approximation is more crucial than it might be believed. This is supported by the fact that the spatial nonlocality of f_{xc} is strongly frequency-dependent [98], in [98] Tokatly and Pankratov argued that not only any static approximation but also any LDA-based dynamic approximation (including any gradient corrections) for f_{xc} cannot provide consistent result. To my best knowledge, there is no calculated or experimental result reported for any of the higher states given in Table 8, this makes the situation more difficult to analyze and be clarified. In Table 8 surprisingly we see that PBE0 gives a better result for higher excited states than its asymptotic corrected one GRAC-PBE0 and better than B3LYP, BP86, or BPW91. Furthermore, it gives for all states a comparable result to CAMB3LYP for R_e and ω_e. This supports our view and stress the importance of the long-range correction. It is a clear evident that PBE0 has a correct shape in inner part of the potential curve, and only its asymptotic part (tail of the potential curve) is incorrect, unfortunately the applied correction of GRAC is not good. As seen in Table 8 our next four states, ${}^3\Sigma_u^+$, ${}^3\Sigma_g^+$, and ${}^1\Sigma_u^+$, ${}^1\Sigma_g^+$ corresponding to Atom($(4s^2)$ ${}^1S$ + Atom($(4s5s)$ ${}^3S$), and Atom($(4s^2)$ ${}^1S$ + Atom($(4s5s)$ ${}^1S$), have more or less a similar result for all functionals, only GRAC-PBE0 shows unexplainable result, since it is supposed to show asymptotically a better behavior. We think that the CAMB3LYP result is the most correct one although it might be not satisfactory accurate. It is worthwhile to mention that states with avoiding crossing get a second shallow minimum after the avoiding crossing at large internuclear distances, this is not reported and only the first minimum is presented. Next, we look to the states ${}^3\pi_u$, ${}^3\pi_g$, ${}^3\Sigma_u^+$, ${}^3\Sigma_g^+$ corresponding to the Atom($(4s^2)$ ${}^1S$ + Atom($(4s5p)$ ${}^3P$). Here, we see that the result is distributed, BPW91, BP86, and B3LYP show similar results, whereas GRAC-PBE0 differs considerably from all approximations given in Table 8. PBE0 result is close to CAMB3LYP when looking to R_e and ω_e, but its D_e values are different clearly due to its incorrect asymptotic limit. The last states treated in this work ${}^1\pi_u$, ${}^1\pi_g$, ${}^1\Sigma_u^+$, ${}^1\Sigma_g^+$ are corresponding to the Atom($(4s^2)$ ${}^1S$ + Atom($(4s5p)$ ${}^1P$). The results of ${}^1\pi_u$ are puzzling and presumably only the values of CAMB3LYP are reasonable, whereas for ${}^1\pi_g$ all functional apart from GRAC-PBE0 give comparable values for ω_e and R_e, which could be a hint that these values are reasonable. ${}^1\Sigma_u^+$, and ${}^1\Sigma_g^+$ follow the general trend that PBE0 result is close to CAMB3LYP. BPW91, BP86, and B3LYP show a similar result, GRAC-PBE0 shows unexplainable result.

The general conclusion of this section is that CAMB3LYP gives the best result due to its better treatment of the long-range part of the two-electron interaction and its asymptotically better behavior (tail of the potential curve) apparently due to including a suitable amount of exact exchange, PBE0 gives a comparable result, the main problem here is the tail of the potential curve. BPW91, BP86, and B3LYP are less satisfactory but still show acceptable result, whereas (most likely) the result of GRAC-PBE0 is not useful.

5. Conclusion

In the present work, we have studied the ground as well the 20 lowest exited states of the zinc dimer in the framework of DFT and TDDFT using well-known and newly developed functional approximations. We performed the calculations with Dirac-Package using relativistic 4-component DCH and SFH. First, we showed that SFH is capable to achieve the same accuracy as 4-components DCH and can describe quantitatively the main relevant contributions of the relativistic effects. In analyzing the results obtained from different functional approximations, comparing them with each other, with literature and experimental values as far as available, we drew some conclusions. The results show that the linear response in the adiabatic approximation with the known DFT approximations give good performance for the 8 lowest excited states of Zn_2. For higher excited states, we found, somehow as expected, that most of DFT approximations used in the pw did not show a correct long-range behavior and the correct asymptotic limit to perform a fair accuracy for these states, where we have to stress that the lack of experimental or other theoretical results makes a judgment difficult. Nevertheless, we can say that the best result is obtained with the range-separated CAMB3LYP functional, which was the only one able (at least qualitatively) to show the correct asymptotic behavior. This can be led back to the separation of the two-electron interaction in a suitable manner, short- and long-range part, where the former is handled by the DFT and the later by HF. Showing that including a suitable (parametric) amount of the exact exchange improves the result considerably. Moreover, the (long-range corrected) PBE0 was able to give a comparable result to CAMB3LYP for the higher states although it fails to give the correct asymptotes. The comparison between CAMB3LY and other functionals allows us to conclude that for higher states the lack of a correct long-range and a suitable amount of exact exchange is responsible for incorrect result rather than the linear response approximation and the adiabatic limit. In addition, it causes a wrong long-range behavior of the response function a crucial point for the long-range behavior in TDDFT. In future works, we will be concerned with the heavier members of the group 12, Cd_2, and Hg_2, where relativistic effects are expected to be more important than in zinc dimer. Furthermore, the superheavy dimer Cn_2 is under consideration, where the bonding character of its ground and excited states of academic interest due to the large relativistic effects and its influence on the atomic levels and hence on the molecular ground and excited states of the dimer.

Acknowledgments

The author gratefully acknowledges fruitful discussions with Dr. Trond Saue, Laboratoire de Chimie et Physique Quantique, Université de Toulouse (France), and the kindly support from him. Dr. Radovan Bast, Tromsø University (Norway), is acknowledged for his kindly support and the kindly support from the Laboratoire de Chimie Quantique, CNRS et Université de Strasbourg.

References

[1] K. G. Caulton and L. G. Hubert-Pfalzgraf, "Synthesis, structural principles, and reactivity of heterometallic alkoxides," *Chemical Reviews*, vol. 90, no. 6, pp. 969–995, 1990.

[2] M. C. Heitz, K. Finger, and C. Daniel, "Photochemistry of organometallics: quantum chemistry and photodissociation dynamics," *Coordination Chemistry Reviews*, vol. 159, pp. 171–193, 1997.

[3] L. Huebner, A. Kornienko, T. J. Emge, and J. G. Brennan, "Heterometallic lanthanide group 12 metal iodides," *Inorganic Chemistry*, vol. 43, no. 18, pp. 5659–5664, 2004.

[4] R. Kobayashia and R. D. Amos, "The application of CAM-B3LYP to the charge-transfer band problem of the zincbacteriochlorin-bacteriochlorin complex," *Chemical Physics Letters*, vol. 420, no. 1–3, pp. 106–109, 2006.

[5] G. Hua, Y. Zhang, J. Zhang, X. Cao, W. Xu, and L. Zhang, "Fabrication of ZnO nanowire arrays by cycle growth in surfactantless aqueous solution and their applications on dye-sensitized solar cells," *Materials Letters*, vol. 62, no. 25, pp. 4109–4111, 2008.

[6] J. H. Lee, Y. W. Chun, M. H. Hon, and I. C. Leu, "Density-controlled growth and field emission property of aligned ZnO nanorod arrays," *Applied Physics A*, vol. 97, no. 2, pp. 403–4408, 2009.

[7] T. Yamase, H. Gerischer, M. Lübke, and B. Pettinger, "Spectral sensitization of ZnO-electrodes by methylene blue," *Berichte der Bunsengesellschaft für physikalische Chemie*, vol. 83, no. 7, pp. 658–6663, 1979.

[8] D. K. Roe, L. Wenzhao, and H. Gerischer, "Electrochemical deposition of cadmium sulfide from DMSO solution," *Journal of Electroanalytical Chemistry*, vol. 136, no. 2, pp. 323–337, 1982.

[9] M. D. Morse, "Clusters of transition-metal atoms," *Chemical Reviews*, vol. 86, no. 6, pp. 1049–11109, 1986.

[10] J. Koperski, "Study of diatomic van der Waals complexes in supersonic beams," *Physics Reports*, vol. 369, no. 3, pp. 177–1326, 2002.

[11] J. Koperski, "Group-12 vdW dimers in free-jet supersonic beams: the legacy of Eugeniusz Czuchaj continues," *Europhysics Letters*, vol. 144, pp. 107–114, 2007.

[12] M. Yu and M. Dolg, "Covalent contributions to bonding in group 12 dimers M2 (Mn = Zn, Cd, Hg)," *Chemical Physics Letters*, vol. 273, no. 5-6, pp. 329–3336, 1997.

[13] L. Bucinisky, S. Biskupic, M. Ilcin, V. Lukeš, and V. Lauring, "On relativistic effects in ground state potential curves of Zn_2, Cd_2, and Hg_2 dimers. A CCSD(T) study," *Journal of Computational Chemistry*, vol. 30, no. 1, pp. 65–674, 2009.

[14] R. Eichler, N. V. Aksenov, A. V. Belozerov et al., "Chemical characterization of element 112," *Nature*, vol. 447, no. 7140, pp. 72–75, 2007.

[15] N. Gaston, I. Opahle, H. W. Góggeler, and P. Schwerdtfeger, "Is Eka-Mercury (element 112) a group 12 metal ?" *Angewandte Chemie International Edition*, vol. 46, pp. 1663–11666, 2007.

[16] V. Pershina, J. Anton, and T. Jacob, "Theoretical predictions of adsorption behavior of elements 112 and 114 and their homologs Hg and Pb," *Journal of Chemical Physics*, vol. 131, no. 8, Article ID 084713, 8 pages, 2009.

[17] K. Ellingsen, T. Saue, C. Puchan, and O. Groupen, "An *Ab initio* study of the electronic spectrum of Zn_2 including spin-orbit coupling," *Chemical Physics*, vol. 311, no. 1-2, pp. 35–344, 2005.

[18] E. Czuchaj, F. Rebentrost, H. Stoll, and H. Preuss, "Adiabatic potential curves for the Cd_2 dimer," *Chemical Physics Letters*, vol. 225, no. 1–3, pp. 233–239, 1994.

[19] E. Czuchaj, F. Rebentrost, H. Stoll, and H. Preuss, "Potential energy curves for the Zn_2 dimer," *Chemical Physics Letters*, vol. 255, no. 1–3, pp. 203–209, 1996.

[20] E. Czuchaj, F. Rebentrost, H. Stoll, and H. Preuss, "Calculation of ground- and excited-state potential energy curves for the Hg_2 molecule in a pseudopotential approach," *Chemical Physics*, vol. 214, no. 2-3, pp. 277–289, 1997.
[21] T. Saue, L. Visscher, H. J. Aa. Jensen et al., DIRAC, a relativistic *Ab initio* electronic structure program, Release DIRAC10, 2010, http://dirac.chem.vu.nl/.
[22] N. C. Pyper, I. Grant, and R. Gerber, "Relativistic effects on interactions between heavy atoms: the Hg_Hg potential," *Chemical Physics Letters*, vol. 49-, pp. 479–483, 1977.
[23] M. Seth, P. Schwerdtfeger, and M. Dolg, "The chemistry of the superheavy elements. I. Pseudopotentials for 111 and 112 and relativistic coupled cluster calculations for (112)H+, (112)F2, and (112)F4," *Journal of Chemical Physics*, vol. 106, no. 9, pp. 3623–3632, 1997.
[24] J. Antona, B. Fricke, and P. Schwerdtfeger, "Non-collinear and collinear four-component relativistic molecular density functional calculations," *Chemical Physics*, vol. 311, no. 1-2, pp. 97–103, 2005.
[25] L. Belpassi, L. Storchi, H. M. Quineyb, and F. Tarantelli, "Recent advances and perspectives in four-component Dirac-Kohn-Sham calculations," *Physical Chemistry Chemical Physics*, vol. 13, pp. 12368–12394, 2011.
[26] R. Bast, A. Heßelmann, P. Sałek, T. Helgaker, and T. Saue, "Static and frequency-dependent dipole-dipole polarizabilities of all closed-shell atoms up to radium: a four-component relativistic DFT study," *ChemPhysChem*, vol. 9, no. 3, pp. 445–453, 2008.
[27] R. Bast, H. J. A. A. Jensen, and T. Saue, "Relativistic adiabatic time-dependent density functional theory using hybrid functionals and noncollinear spin magnetization," *International Journal of Quantum Chemistry*, vol. 109, no. 10, pp. 2091–2112, 2009.
[28] T. Saue and H. J. A. Jensen, "Linear response at the 4-component relativistic level: application to the frequency-dependent dipole polarizabilities of the coinage metal dimers," *Journal of Chemical Physics*, vol. 118, no. 2, pp. 533–515, 2003.
[29] J. C. Slater, "A simplification of the Hartree-Fock method," *Physical Review*, vol. 81, no. 3, pp. 385–390, 1951.
[30] S. J. Vosko, L. Wilk, and M. Nusair, "Accurate spin-dependent electron liquid correlation energies for local spin density calculations: a critical analysis," *Canadian Journal of Physics*, vol. 58, no. 8, pp. 1200–11211, 1980.
[31] J. P. Perdew, K. Burke, and M. Ernzerhof, "Generalized gradient approximation made simple," *Physical Review Letters*, vol. 77, no. 18, pp. 3865–3868, 1996.
[32] A. D. Becke, "Density-functional exchange-energy approximation with correct asymptotic behavior," *Physical Review A*, vol. 38, no. 6, pp. 3098–3100, 1988.
[33] J. P. Perdew, "Density-functional approximation for the correlation energy of the inhomogeneous electron gas," *Physical Review B*, vol. 33, no. 12, pp. 8822–8824, 1986.
[34] J. P. Perdew, "Density-functional approximation for the correlation energy of the inhomogeneous electron gas," *Physical Review B*, vol. 34, no. 10, article 7406, 1986.
[35] J. P. Perdew and Y. Wang, "Accurate and simple analytic representation of the electron-gas correlation energy," *Physical Review B*, vol. 45, no. 23, pp. 13244–13249, 1992.
[36] M. Ernzerhof and G. E. Scuseria, "Assessment of the Perdew-Burke-Ernzerhof exchange-correlation functional," *Journal of Chemical Physics*, vol. 110, no. 11, pp. 5029–5036, 1999.
[37] R. van Leeuwen and E. J. Baerends, "Exchange-correlation potential with correct asymptotic behavior," *Physical Review A*, vol. 49, no. 4, pp. 2421–2431, 1994.
[38] M. Grüning, O. V. Gritsenko, S. J. A. van Gisbergen, and E. J. Baerends, "Shape corrections to exchange-correlation potentials by gradient-regulated seamless connection of model potentials for inner and outer region," *Journal of Chemical Physics*, vol. 114, no. 2, pp. 652–660, 2001.
[39] C. Lee, W. Yang, and R. G. Parr, "Development of the Colle-Salvetti correlation-energy formula into a functional of the electron density," *Physical Review B*, vol. 37, no. 2, pp. 785–789, 1988.
[40] A. D. Becke, "Density-functional thermochemistry. III. The role of exact exchange," *Journal of Chemical Physics*, vol. 98, no. 7, article 5648, 5 pages, 1993.
[41] P. J. Stephens, F. J. Devlin, C. F. Chabalowski, and M. J. Frisch, "*Ab initio* calculation of vibrational absorption and circular dichroism spectra using density functional force fields," *Journal of Physical Chemistry*, vol. 98, no. 45, pp. 11623–11627, 1994.
[42] T. Yanai, D. P. Tew, and N. C. Handy, "A new hybrid exchange-correlation functional using the Coulomb-attenuating method (CAM-B3LYP)," *Chemical Physics Letters*, vol. 393, no. 1–3, pp. 51–57, 2004.
[43] O. Kullie and T. Saue, "Range-separated density functional theory: a 4-component relativistic study of the rare gas dimers He_2, Ne_2, Ar_2, Kr_2, Xe_2, Rn_2 and Uuo_2," *Chemical Physics*, vol. 395, pp. 54–62, 2012.
[44] P. Hohenberg and W. Kohn, "Inhomogeneous electron gas," *Physical Review*, vol. 136, no. 3B, pp. B864–B871, 1964.
[45] W. Kohn and L. J. Sham, "Self-consistent equations including exchange and correlation effects," *Physical Review*, vol. 140, no. 4, pp. A1133–A1138, 1965.
[46] W. Kohn, "Nobel lecture: electronic structure of matter—wave functions and density functionals," *Reviews of Modern Physics*, vol. 71, no. 5, pp. A1133–A1266, 1999.
[47] W. Koch and M. C. Holthausen, *A Chemist's Guide to Density Functional Theory*, Willy-VCH, New York, NY, USA, 2001.
[48] T. Saue and T. Helgaker, "Four-component relativistic Kohn-Sham theory," *Journal of Computational Chemistry*, vol. 23, no. 8, pp. 814–823, 2002.
[49] O. Kullie, H. Zhang, and D. Kolb, "Relativistic and non-relativistic local-density functional, benchmark results and investigation on the dimers Cu_2,Ag_2,Au_2,Rg_2," *Chemical Physics*, vol. 351, no. 1–3, pp. 106–110, 2008.
[50] O. Kullie, E. Engel, and D. Kolb, "Accurate local density functional calculations with relativistic two-spinor minimax and finite element method for the alkali dimers," *Journal of Physics B*, vol. 42, no. 9, Article ID 095102, 2009.
[51] P. A. M. Dirac, "Note on exchange phenomena in the Thomas atom," *Mathematical Proceedings of the Cambridge Philosophical Society*, vol. 26, no. 3, pp. 376–385, 1930.
[52] J. P. Perdew, S. Kurth, A. Zupan, and P. Blaha, "Accurate density functional with correct formal properties: a step beyond the generalized gradient approximation," *Physical Review Letters*, vol. 82, no. 12, pp. 2544–2547, 1999.
[53] A. Savin, in *Recent Developments of Modern Density Functional Theory*, J. M. Seminario, Ed., pp. 327–357, Elsevier, Amsterdam, The Netherlands, 1996.
[54] E. Goll, H. J. Werner, and H. Stoll, "A short-range gradient-corrected density functional in long-range coupled-cluster calculations for rare gas dimers," *Physical Chemistry Chemical Physics*, vol. 7, pp. 3917–3923, 2005.
[55] I. C. Gerber and J. G. Ángyán, "Potential curves for alkaline-earth dimers by density functional theory with long-range correlation corrections," *Chemical Physics Letters*, vol. 416, no. 4–6, pp. 370–375, 2005.

[56] R. Baer, E. Livshits, and U. Salzner, "Tuned range-separated hybrids in density functional theory," *Annual Review of Physical Chemistry*, vol. 61, pp. 85–109, 2010.

[57] K. G. Dyall, "An exact separation of the spin–free and spin–dependent terms of the Dirac-Coulomb-Breit Hamiltonian," *Journal of Chemical Physics*, vol. 100, no. 3, article 2118, 10 pages, 1994.

[58] L. Cheng and J. Gauss, "Analytical evaluation of first-order electrical properties based on the spin-free Dirac-Coulomb Hamiltonian," *Journal of Chemical Physics*, vol. 134, no. 24, Article ID 244112, 11 pages, 2011.

[59] M. A. L. Marques, C. A. Urlich, F. Nogueira, A. Rubio, K. Burke, and E. K. Gross, Eds., *Time-Dependent Density Functional Theory*, Lecture Notes in Physics, Springer, New York, NY, USA, 2006.

[60] E. Runge and E. K. U. Gross, "Density-functional theory for time-dependent systems," *Physical Review Letters*, vol. 52, no. 12, pp. 997–1000, 84.

[61] E. Gross and W. Kohn, "Time-dependent density-functional theory," *Advances in Quantum Chemistry*, vol. 21, pp. 255–291, 1990.

[62] M. E. Casida, in *Recent Advances in Density Functional Methods*, D. P. Chong, Ed., p. 155, World Scientific, Singapore, 1995.

[63] E. Gross, J. Dobson, and M. Petersilka, "Density functional theory of time-dependent phenomena," *Topics in Current Chemistry*, vol. 181, pp. 81–172, 1996.

[64] M. Casida, "Time-dependent density functional response theory of molecular systems: theory, computational methods, and functionals," in *Recent Developments and Applications of Modern Density Functional Theory*, J. M. Seminario, Ed., chapter 11, p. 391, Elsevier, Amsterdam, The Netherlands, 1996.

[65] K. Burke and E. K. U. Gross, in *Density Functionals: Theory and Applications*, D. Joubert, Ed., vol. 500 of *Springer Lecture Notes in Physics*, p. 116, Springer, New York, NY, USA, 1998.

[66] R. van Leeuwen, "Key concepts in time-dependent density-functional theory," *International Journal of Modern Physics B*, vol. 15, no. 14, pp. 1969–2023, 2001.

[67] M. A. L. Marques and E. K. U. Gross, "Time dependent density functional theory," in *A Primer in Density Functional Theory*, M. M. C. Fiolhais and F. Nogueira, Eds., p. 144, Springer, New York, NY, USA, 2003.

[68] H. Appel, E. K. Gross, and K. Burke, "Excitations in time-dependent density-functional theory," *Physical Review Letters*, vol. 90, no. 4, Article ID 043005, 4 pages, 2003.

[69] M. A. L. Marques and E. K. U. Gross, "Time-dependent density functional theory," *Annual Review of Physical Chemistry*, vol. 55, pp. 427–455, 2004.

[70] K. Burke, J. Werschnik, and E. Gross, "Time-dependent density functional theory: past, present, and future," *Journal of Chemical Physics*, vol. 123, Article ID 062206, 12 pages, 2005.

[71] P. Elliott, F. Furche, and K. Burke, in *Reviews in Computational Chemistry*, K. B. Lipkowitz and T. R. Cundari, Eds., pp. 91–165, Wiley, Hoboken, NJ, USA, 2009.

[72] S. Botti, A. Schindlmayr, R. Del Sole, and L. Reining, "Time-dependent density-functional theory for extended systems," *Reports on Progress in Physics*, vol. 70, no. 3, pp. 357–407, 2007.

[73] O. V. Gritsenko and E. J. Baerends, "Double excitation effect in non-adiabatictime-dependent density functional theory with an analytic construction of the exchange-correlation kernel in the common energy denominator approximation," *Physical Chemistry Chemical Physics*, vol. 11, pp. 4640–4646, 2009.

[74] T. Ziegler, M. Seth, M. Krykunov, J. Autschbach, and F. Wangc, "Is charge transfer transitions really too difficult for standard density functionals or are they just a problem for time-dependent density functional theory based on a linear response approach," *Journal of Molecular Structure*, vol. 914, no. 1–3, pp. 106–109, 2009.

[75] M. E. Casida, "Time-dependent density-functional theory for molecules and molecular solids," *Journal of Molecular Structure*, vol. 914, no. 1–3, pp. 3–18, 2009.

[76] M. E. Casida and M. Huix-Rotllant, "Progress in time-dependent density-functional theory," *Annual Review of Physical Chemistry*, vol. 63, pp. 287–323, 2012.

[77] G. Onida, R. Reininger, and A. Rubio, "Electronic excitations: density-functional versus many-body Green's-function approaches," *Reviews of Modern Physics*, vol. 74, no. 2, pp. 601–659, 2002.

[78] A. Zangwill and P. Soven, "Resonant photoemission in barium and cerium," *Physical Review Letters*, vol. 45, no. 3, pp. 204–207, 1980.

[79] M. Iliaš and T. Saue, "An infinite-order two-component relativistic Hamiltonian by a simple one-step transformation," *Journal of Chemical Physics*, vol. 126, no. 6, Article ID 064102, 9 pages, 2007.

[80] L. Visscher and T. Saue, "Approximate relativistic electronic structure methods based on the quaternion modified Dirac equation," *Journal of Chemical Physics*, vol. 113, no. 10, pp. 3996–4002, 2000.

[81] L. Visscher and K. G. Dyall, "Dirac-fock atomic electronic structure calculations using different nuclear charge distributions," *Atomic Data and Nuclear Data Tables*, vol. 67, no. 2, pp. 207–224, 1997.

[82] T. Dunning, "Gaussian basis sets for use in correlated molecular calculations. I. The atoms boron through neon and hydrogen," *Journal of Chemical Physics*, vol. 90, no. 2, article 1007, 17 pages, 1989.

[83] D. Woon and T. Dunning, "Gaussian basis sets for use in correlated molecular calculations. III. The atoms aluminum through argon," *Journal of Chemical Physics*, vol. 98, no. 2, article 1358, 14 pages, 1993.

[84] A. K. Wilson, D. E. Woon, K. A. Peterson, and T. H. Dunning, "Gaussian basis sets for use in correlated molecular calculations. IX. The atoms gallium through krypton," *Journal of Chemical Physics*, vol. 110, no. 16, pp. 7667–7676, 1999.

[85] M. A. Czajkkowski and J. Koperski, "The Cd_2 and Zn_2 van der Waals dimers revisited. Correction for some molecular potential parameters," *Spectrochimica Acta*, vol. 55, no. 11, pp. 2221–2229, 1999.

[86] R. D. Van Zee, S. C. Blankespoor, and T. Z. Zweir, "Direct spectroscopic determination of the Hg_2 bond length and an analysis of the 2540 Å band," *Journal of Chemical Physics*, vol. 88, no. 8, article 4650, 5 pages, 1988.

[87] A. Aguado, J. de la Vega, and B. Miguel, "*Ab initio* configuration interactioncalculations of ground state and lower excited states of Zn_2 using optimized Slater-typewavefunctions," *Journal of the Chemical Society, Faraday Transactions*, vol. 93, no. 1, pp. 29–32, 1997.

[88] H. Tatewaki, M. Tomonari, and T. Nakamura, "The excited states of Zn_2 and Zn_3. Inclusion of the correlation effects," *The Journal of Chemical Physics*, vol. 82, no. 12, pp. 5608–5615, 1984.

[89] P. J. Hay, T. H. Dunning, and R. C. Raffenetti, "Electronic states of Zn_2. *Ab initio* calculations of a prototype for Hg_2," *The Journal of Chemical Physics*, vol. 65, no. 7, pp. 2679–2689, 1976.

[90] J. J. Determan, M. A. Omary, and A. K. Wilson, "Modeling the photophysics of Zn and Cd monomers, metallophilic dimers, and covalent excimers," *Journal of Physical Chemistry A*, vol. 115, no. 4, pp. 374–382, 2011.

[91] C. H. Su, P. K. Liao, Y. Huang, S. Liou, and R. F. Brebick, "A study of the symmetric charge transfer reaction H_2^++H_2 using the high resolution photoionization and crossed ion-neutral beam methods," *Journal of Chemical Physics*, vol. 81, no. 12, article 5672, 20 pages, 1984.

[92] W. Kedzierski, J. B. Atkinson, and L. Krause, "Laser-induced fluorescence from the $^3\Pi_u$ (4 ^{3}P, 4 ^{3}P) state of Zn_2," *Chemical Physics Letters*, vol. 215, no. 1–3, pp. 185–187, 1993.

[93] W. Kedzierski, J. B. Atkinson, and L. Krause, "The $\sum_g{}^+$ (4^3P, 4^3P) ← $\sum_u{}^+$ (4^3P, 4^1S) vibronic spectrum of Zn_2," *Chemical Physics Letters*, vol. 222, no. 1-2, pp. 146–148, 1994.

[94] G. Rodriguez and J. G. Eden, "Bound → free emission spectra and photoassociation of $^{114}Cd_2$ and $^{64}Zn_2$," *Journal of Chemical Physics*, vol. 95, no. 8, article 5539, 14 pages, 1991.

[95] W. Kedzierski, J. B. Atkinson, and L. Krause, "Laser-induced fluorescence of the Zn_2 excimer," *Optics Letters*, vol. 14, no. 12, pp. 607–608, 1989.

[96] M. Czajkkowski, R. Bobkowski, and L. Krause, *Physical Review A*, vol. 200, p. 103, 1990.

[97] T. Bally and G. N. Sastry, "Incorrect dissociation behavior of radical ions in density functional calculations," *The Journal of Physical Chemistry A*, vol. 101, no. 43, pp. 7423–7925, 1997.

[98] I. Tokatly and O. Pankratov, "Many-body diagrammatic expansion in a Kohn-Sham basis: implications for time-dependent density functional theory of excited states," *Physical Review Letters*, vol. 86, no. 10, pp. 2087–2081, 2001.

11

Millimeter-Wave Rotational Spectra of *trans*-Acrolein (Propenal) (CH_2CHCOH): A DC Discharge Product of Allyl Alcohol (CH_2CHCH_2OH) Vapor and DFT Calculation

A. I. Jaman and Rangana Bhattacharya

Experimental Condensed Matter Physics Division, Saha Institute of Nuclear Physics, Sector 1, Block AF, Bidhannagar, Kolkata 700 064, India

Correspondence should be addressed to A. I. Jaman, aismail.jaman@saha.ac.in

Academic Editor: Nigel J. Mason

Millimeter-wave rotational spectrum of *trans*-acrolein (propenal) (CH_2CHCOH) produced by applying a DC glow discharge through a low-pressure (~10–20 mTorr) flow of allyl alcohol (CH_2CHCH_2OH) vapor has been observed in the ground and several excited torsional states in the frequency region: 60.0–99.0 GHz. A least-square analysis of the measured and previously reported rotational transition frequencies has produced a set of rotational and centrifugal distortion constants for the ground as well as excited torsional states. Detailed DFT calculations were also carried out with various functional and basis sets to evaluate the spectroscopic constants, dipole moment, and various structural parameters of the *trans* conformer of propenal for the ground state and compared with their corresponding experimental values. A linear variation of the inertia defect values with torsional quantum number ($v = 0, 1, 2, 3$) demonstrates that the equilibrium configuration of *trans*-propenal is planar.

1. Introduction

The *trans* form of propenal (CH_2CHCOH) also known as *trans*-acrolein has been detected largely in absorption toward the star-forming region Sagittarius B2(N) by Hollis et al. [2] through the observation of rotational transitions using 100 m Green Bank Telescope (GBT) operating in the frequency range from 18.0 GHz to 26.0 GHz. Spectroscopic measurements in the microwave [1, 3], infrared [4], and near ultraviolet region [5, 6] have confirmed that the *trans*-form is the most abundant and stable conformer of acrolein. The first microwave study of *trans*-acrolein in the $J = 2 \leftarrow 1$, $3 \leftarrow 2$, and $4 \leftarrow 3$ a-type R-branch transitions was reported by Wagner et al. [3]. Later on, Cherniak and Costain [1] have measured both a- and b-type transitions for $J = 2 \leftarrow 1$ and $J = 3 \leftarrow 2$. First spectroscopic evidence of the existence of the less abundant *cis*-conformer of acrolein in the gas phase was found from studies of the near ultraviolet spectrum [7, 8]. Later on, *cis*-acrolein were detected in argon matrices [9, 10] and in the gas-phase Raman spectrum [11]. The first microwave detection of the *cis* form of acrolein in the gas phase was reported by Blom and Bauder [12]. They have reported the ground state rotational, quartic centrifugal distortion constants as well as dipole moment values. Blom et al. [13] also reported the complete substitution structures of both *trans* and *cis* conformers. The dipole moment values of the *trans* and *cis*-form of acrolein have been found to be $\mu = 3.117 \pm 0.004$ D [13] and $\mu = 2.552 \pm 0,003$ D [12] respectively. Winnewisser [14] have extended the analysis of the ground state of the *trans*-form of acrolein to the millimeter-wave region up to 180.0 GHz which has yielded a set of ground state rotational and centrifugal distortion

Table 1: Microwave and millimeter wave rotational transition frequencies of *trans*-propenal (CH_2CHCOH) in the ground and excited torsional states (in MHz.).

Transitions						Torsional levels							
						$v = 0$		$v = 1$		$v = 2$		$v = 3$	
J'	K'_{-1}	K'_{+1}	J''	K''_{-1}	K''_{+1}	Obs. Freq.	Obs. − cal.	Obs. Freq.	Obs. − cal.	Obs. Freq.	Obs. − cal.	Obs. Freq.	Obs. − cal.
1	0	1	0	0	0	8902.19	−.01						
2	0	2	1	0	1	17801.28	−.06						
2	1	1	1	1	0	18221.12	−.07	18258.18	−.12	18289.86	−.01		
3	0	3	2	0	2	26694.35	−.01			26824.88	.01	26895.18	.05
3	0	3	2	1	2			26765.43	−.09			27487.70	.06
3	1	2	2	1	1	27329.73	−.07	27385.58	.08	27432.89	.02	26322.50	−.07
3	1	3	2	1	2	26079.50	−.01	26165.84	−.05	26237.28	−.12		
3	2	1	2	2	0	26718.70	−.16						
3	2	2	2	2	1	26706.76	.04						
4	0	4	3	0	3			35673.16	−.09	35752.72	.17	35846.64	.15
4	1	4	3	1	3	34768.96	−.11						
7	0	7	6	0	6	62144.50[a]	.01	62312.22[a]	.01	62452.30[a]	−.11		
7	1	7	6	1	6	60816.60[a]	.04	61018.40[a]	.02				
7	2	6	6	2	5	62290.80[a]	−.06	62456.80[a]	−.03	62597.20[a]	.05	62760.10[a]	−.02
7	2	5	6	2	4	62460.30[a]	−.05	62624.20[a]	−.08	62762.20[a]	−.07	62922.30[a]	.03
7	3	4	6	3	3	62341.20[a]	−.01	62506.50[a]	−.08	62648.90[a]	.00		
7	3	5	6	3	4	62339.40[a]	−.07	62504.80[a]	−.04	62647.20[a]	.03		
7	4	3	6	4	2	62331.10[a]	.02	62496.50[a]	−.06				
7	5	3	6	5	2			62493.40[a]	.03				
8	0	8	7	0	7	70961.80[a]	.00	71154.30[a]	.10	71315.20[a]	.04	71506.60[a]	−.01
8	1	7	7	1	6	72820.00[a]	−.01	72969.40[a]	.03	73096.20[a]	.09	73243.30[a]	−.06
8	1	8	7	1	7	69489.80[a]	.06	69720.40[a]	−.12	69911.60[a]	.01	70139.80[a]	−.08
8	2	7	7	2	6	71179.00[a]	.07			71529.20[a]	.07		
8	2	6	7	2	5	71432.70[a]	.12	71619.40[a]	.08	71776.20[a]	−.03	71958.30[a]	.04
8	3	6	7	3	5	71251.50[a]	.06	71440.40[a]	.05				
8	3	5	7	3	4	71255.00[a]	.08	71443.90[a]	.07	71606.10[a]	−.17		
8	4	4	7	4	3	71239.10[a]	.05	71428.20[a]	.09				
8	5	4	7	5	3			71423.10[a]	.12				
8	6	3	7	6	2			71421.10[a]	−.06				
9	0	9	8	0	8							80371.70[a]	−.01
9	1	8	8	1	7	81901.20	.01	82069.50[a]	−.01	82212.30[a]	.06		
9	2	8	8	2	7	80062.80[a]	.04	80276.30[a]	−.11	80457.00[a]	.08	80667.10[a]	.09
9	2	7	8	2	6	80424.00[a]	.07	80633.30[a]	.06	80808.70[a]	−.05	81012.40[a]	−.08
9	3	6	8	3	5	80172.10[a]	.05	80384.50[a]	.04				
9	3	7	8	3	6	80165.70[a]	.03	80378.20[a]	.12				
9	4	5	8	4	4	80148.40[a]	−.04	80361.10[a]	.04				
9	5	5	8	5	4	80140.70[a]	.08	80353.40[a]	.06				
9	6	4	8	6	3	80137.50[a]	−.01	80350.40[a]	.06				
10	0	10	9	0	9	88523.30[a]	−.01	88765.90[a]	−.05			89211.90[a]	−.12
10	1	9	9	1	8	90974.40[a]	−.05						
10	1	10	9	1	9	86818.50[a]	−.02						
10	2	8	9	2	7	89436.20[a]	−.01	89667.83[a]	.03	89861.60[a]	−.02		
10	2	9	9	2	8	88941.80[a]	−.01	89179.48[a]	.10	89380.10[a]	.07	89613.80[a]	−.02
10	3	7	9	3	6	89093.20[a]	−.01	89329.07[a]	−.01				
10	3	8	9	3	7	89082.20[a]	−.08						
10	4	6	9	4	5	89059.35[a]	−.09	89295.47[a]	−.09				
10	5	5	9	5	4	89048.30[a]	−.01						

TABLE 1: Continued.

Transitions						Torsional levels							
						$v = 0$		$v = 1$		$v = 2$		$v = 3$	
J'	K'_{-1}	K'_{+1}	J''	K''_{-1}	K''_{+1}	Obs. Freq.	Obs. − cal.	Obs. Freq.	Obs. − cal.	Obs. Freq.	Obs. − cal.	Obs. Freq.	Obs. − cal.
10	6	4	9	6	3	89043.55[a]	−.04						
10	7	3	9	7	2	89041.90[a]	−.02						
10	9	1	9	9	0	89043.17[a]	−.02						
11	0	11	10	0	10	97263.13[a]	−.01	97531.40[a]	−.03	97756.50[a]	−.12	98025.70[a]	.05
11	1	11	10	1	10	95473.05[a]	−.03	95791.10[a]	.01	96054.40[a]	.14	96369.80[a]	.04
11	2	9	10	2	8	98470.83[a]	−.01	98724.30[a]	−.11	98936.00[a]	.02		
11	2	10	10	2	9	97815.59[a]	.03	98077.10[a]	.00			98555.60[a]	.08
11	3	8	10	3	7	98019.04[a]	−.02	98278.40[a]	.05	98500.60[a]	.03	98752.70[a]	−.03
11	3	9	10	3	8	98001.32[a]	.00	98260.60[a]	.01	98482.90[a]	−.06	98735.30[a]	−.01
11	4	8	10	4	7			98231.70[a]	−.08				
11	5	7	10	5	6	97957.00[a]	.00	98216.90[a]	−.05	98453.50[a]	.04		
11	6	5	10	6	4	97950.28[a]	.00			98456.20[a]	−.01		
11	7	4	10	7	3	97947.54[a]	.01						
11	8	3	10	8	2	97947.05[a]	.04						
1	1	1	2	0	2	24892.58	.01						
2	1	2	3	0	3	15585.86	.01						
6	0	6	5	1	5	13444.27	.02						
7	0	7	6	1	6	23450.46	.01						
8	0	8	7	1	7	33595.68	−.01						

[a] This work, rest are from [1].

constants. Analysis of the far-infrared spectrum of *trans* acrolein in the ν_{18} fundamental and $(\nu_{17} + \nu_{18}) - \nu_{18}$ hot bands were reported by McKellar et al. [15]. Very recently, 10 μm high-resolution rotational spectral analysis of the ν_{11}, ν_{16}, ν_{14} and $\nu_{16} + \nu_{18} - \nu_{18}$ bands of trans-acrolein were reported by Xu et al. [16]. In all the previous works *trans*-acrolein (propenal) was either procured commercially or prepared chemically.

Production, identification, and spectroscopic characterization of new stable and transient molecules by applying a DC glow discharge through a low-pressure flow of gas or a mixture of gases inside an absorption cell have become a well-established area of research in the field of molecular spectroscopy [21]. Recently, Jaman et. al have reported analysis of the millimeter-wave rotational spectra of propyne (CH_3CCH) [22] and propynal (HCCCOH) [23] produced by DC glow discharge technique and carried out detailed DFT calculations for both the molecules to evaluate the spectroscopic constants and molecular parameters and compared them with their respective experimental values. In the present communication, we report the analysis of the ground state ($v = 0$) as well as several torsional excited states ($v = 1, 2, 3$) rotational spectra of *trans*-propenal produced by a DC glow discharge through a low-pressure flow of allyl alcohol (CH_2CHCH_2OH) vapor in the frequency region 60.0–99.0 GHz. Asymmetric-top $K_{-1}K_{+1}$-structures of different $J{+}1{\leftarrow}J$ transitions which falls under this frequency range have been observed and measured. The measured rotational transition frequencies along with the previously reported frequencies were fitted to standard asymmetric-top Hamiltonian to determine the rotational and centrifugal distortion (CD) constants for the ground as well as excited torsional states. A detailed quantum chemical

TABLE 2: Ground state rotational and centrifugal distortion constants of *trans*-propenal (CH_2CHCOH).

Constants	Global fit using microwave and millimeter wave data	DFT calculation
A (MHz.)	47353.729 ± 0.009	47532.149
B (MHz.)	4659.4894 ± 0.0004	4635.391
C (MHz.)	4242.7034 ± 0.0004	4223.524
D_J (kHz)	1.031 ± 0.001	0.983
D_{JK} (kHz)	−8.684 ± 0.006	−9.099
D_K (kHz)	361.949 ± 0.963	346.316
d_1 (kHz)	−0.1197 ± 0.0002	−0.119
d_2 (kHz)	−0.0069 ± 0.0001	−0.006
H_{JK} (Hz)	0.014 ± 0.012	
H_{KJ} (Hz)	−0.490 ± 0.021	
σ^b	0.041	
κ^c	−0.9806	
Δ^d	−0.018	
N^e	224	

[b] Standard deviation of the overall fit.
[c] Asymmetry parameter.
[d] Inertia defect $\Delta = I_c - I_b - I_a$.
[e] Number of transitions used in the fit.

Table 3: Excited-state spectroscopic constants of *trans*-propenal (CH_2CHCOH).

Constants	$v = 1$		$v = 2$		$v = 3$	
	This work	Ref. [3]	This work	Ref. [3]	This work	Ref. [3]
A (MHz.)	45782.822 ± 3.231		44727.881 ± 3.873		43420.393 ± 5.888	
B (MHz.)	4666.210 ± 0.004	4666.19 ± 02	4672.056 ± 0.005	4672.10 ± 0.02	4678.661 ± 0.006	4678.69 ± 0.02
C (MHz.)	4259.668 ± 0.005	4259.66 ± 0.02	4273.558 ± 0.006	4273.56 ± 0.02	4290.297 ± 0.007	4290.29 ± 0.02
D_J (kHz)	1.078 ± 0.012		1.280 ± 0.017		1.168 ± 0.026	
D_{JK} (kHz)	-8.735 ± 0.075		-46.831 ± 0.112		-28.714 ± 0.631	
σ^f	0.077		0.085		0.075	
κ^g	−0.9804		−0.9803		−0.9802	
Δ^h	−0.702		−1.212		−1.861	
N^i	43		28		20	

[f] Standard deviation of the overall fit.
[g] Asymmetry parameter.
[h] Inertia defect $\Delta = I_c - I_b - I_a$.
[i] Number of transitions used in the fit.

Table 4: Comparison of the observed inertia defect ($\Delta/u\text{Å}^2$) values for the ground and excited torsional state (v) of *trans*-propenal with some other molecules.

Molecules	Inertia defect ($\Delta/u\text{Å}^2$) values			
	$v = 0$	$v = 1$	$v = 2$	$v = 3$
Trans-propenal[j] (CH_2CHCHO)	−0.018	−0.702	−1.212	−1.861
o-cis 3-fluorobenzaldehyde[k] (C_6H_5FCOH)	−0.078	−0.988	−1.876	−2.726
Nitrobenzene[l] ($C_6H_5NO_2$)	−0.481	−1.863	−3.186	−4.470
Benzoyl fluoride[m] (C_6H_5COF)	−0.325	−1.528	−2.765	−3.963
2-Fluorpstyrene[n] ($C_6H_4FC_2H_3$)	−1.215	−2.689	−3.341	−4.380

[j] This work, [k] [17], [l] [18], [m] [19], [n] [20].

calculation was also carried out to evaluate the spectroscopic constants, dipole moment, and the structural parameters of the *trans* conformer of propenal. Finally, the experimentally determined rotational and CD constants were compared with the best set of values obtained after a series of DFT calculations.

2. Experimental Details

The spectrometer used in the present work is basically a 50 kHz source-modulated system combined with a free space glass discharge cell of 1.5 m in length and 10 cm in diameter. The cell is fitted with two Teflon lenses at each end. A high voltage DC regulated power supply (6 kV, 1300 mA) procured from Glassman, Japan was used to apply a DC voltage through a flow of low pressure precursor gases. The cell is connected with a high vacuum pump at one end and to the sample holder section through a glass port on the other.

Klystrons and Gunn diodes followed by frequency doubler (Millitech model MUD-15-H23F0 and MUD-10-LF000) have been used as radiation sources. Millimeter wave radiation was fed into the cell by a waveguide horn and Teflon lens. A similar horn and lens arrangement was used to focus the millimeter-wave power onto the detector after propagating through the cell. The output frequency of the millimeter wave radiation was frequency modulated by a bidirectional square-wave of 50 kHz [24] and the signal from the detector (Millitech model DBT-15-RP000 and DXP-10-RPFW0) was amplified by a 100 kHz tuned preamplifier and detected by a phase-sensitive lock in amplifier in the 2f mode. The output of the lock in amplifier was connected to an oscilloscope or a chart recorder for signal display. The spectrometer was calibrated by measuring standard OCS signals in the entire frequency range. After calibration, the uncertainty in frequency measurement has been estimated to be ±0.10 MHz. A block diagram of the spectrometer is shown in Figure 1. Details of the spectrometer used have been described elsewhere [25, 26].

Propenal (CH_2CHCOH) was produced inside the absorption cell by applying a DC glow discharge through a low pressure (~5–10 mTorr) flow of allyl alcohol (CH_2CHCH_2OH) vapor. The discharge current was maintained at around 5 mA with an applied voltage of 1.0 kV. A mechanical on/off type discharge was found to be suitable to observe good signals of propenal. Signals could be observed at room temperature. However, a controlled flow of liquid nitrogen vapor through the cell helps in improving the signal intensity. The observed signals of propenal appeared as sharp lines immediately after the DC discharge was applied but started losing intensity with time.

3. Computational Method

Quantum chemical computations were performed using GAUSSIAN 09W package [27]. Density functional methods with various functionals were used to calculate the structural parameters, dipole moment, total energy (sum of electronic

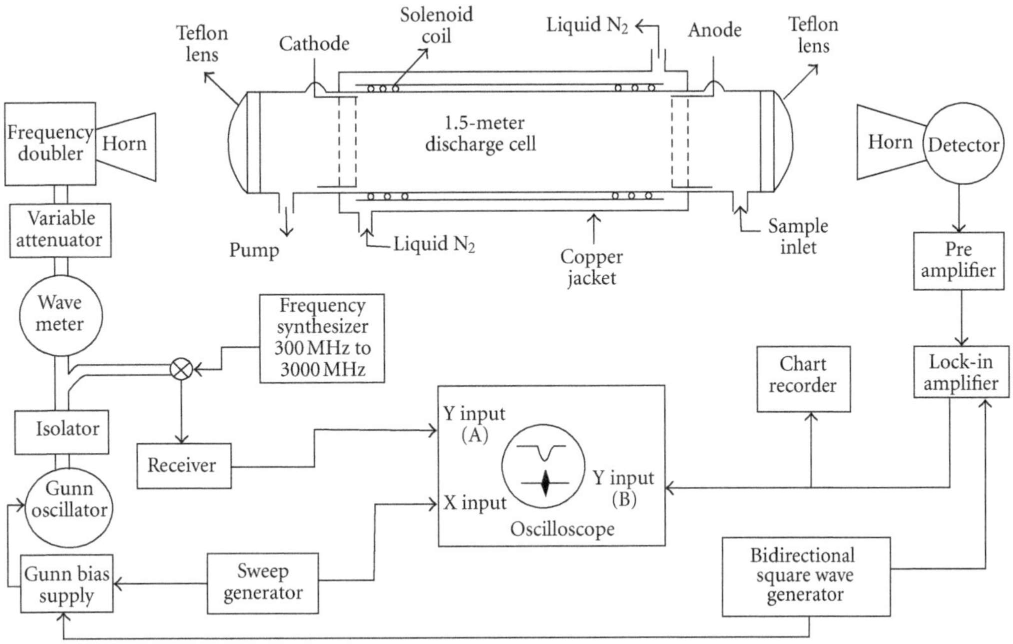

FIGURE 1: Block diagram of source-modulated millimeter wave spectrometer with DC discharge facility.

and zero point energy) as well as the rotational and centrifugal distortion constants of *trans*-propenal. The geometry optimization was carried out using different functionals like Becke 3-term correlation functional(B3LYP) with basis sets 6-31+g(d, p) to 6-311++g(d, 2p), Becke three-Parameter hybrid functional and Perdew/Wang 91 nonlocal correlation functional (B3PW91) method with different basis sets from 6-31 g to 6-311++g(d, 2p), modified Perdew-Wang one-parameter hybrid model taking basis sets from 6–31 g to 6-311++g(d, 2p) and Perdew, Burke, and Ernzerhof functional (PBEPBE) with the basis sets 6-311 g to 6-311++g(d, 2p). The frequency calculation along with its anharmonicity was done on optimized geometry. The objective of this DFT calculation is to compare the structural parameters and rotational constants of *trans*-propenal with the experimentally observed values in its ground state. The molecular drawing is done by using GAUSSVIEW5.0 [28].

4. Rotational Spectrum and Analysis

4.1. Ground State. The ground state rotational spectrum of the *trans* conformer of propenal was predicted in the frequency range 60.0-99.0 GHz using the rotational and centrifugal distortion constants reported earlier [14]. $J = 7 \leftarrow 6$ to $J = 11 \leftarrow 10$ series of transitions along with their different $K_{-1}K_{+1}$ components falls within this frequency range. Different components in each $J+1 \leftarrow J$ series were measured. The observed lines were found very close to their predicted values. Finally, 224 a- and b-type R- and Q-branch transitions consisting of all previous microwave [1, 3], millimeter-wave [14], and present data were used to perform a kind of global fit to the semirigid rotor Watson's S-reduction Hamiltonian (I^r-representation) [29] to determine a set of three rotational, five quartic, and two sextic centrifugal distortion constants. The shifts in frequency of the absorption lines from their rigid rotor positions due to centrifugal distortion effect were found to be less than that of propynal [23]. The observed and measured transition frequencies by us corresponding to $J = 7 \leftarrow 6$ to $11 \leftarrow 10$ series are listed in Table 1. The ground state spectroscopic constants obtained for *trans*-propenal using the global fit are listed in Table 2. The small negative value of the inertia defect ($\Delta = -0.018\, u\text{Å}^2$) demonstrates that the equilibrium configuration of *trans*-propenal is planar. The agreement between the derived set of spectroscopic constants and those obtained earlier [1, 3, 14] with commercial samples indicates that the newly assigned transition frequencies of Table 1 definitely belong to *trans*-acrolein (*trans*-propenal), a discharge product of allyl alcohol vapor. Figure 2 shows the observed trace of the $K_{-1} = 3$ doublet of $J = 9 \leftarrow 8$ transition immediately after the DC discharge was applied. The trace remained visible for a couple of minutes on the oscilloscope screen with gradually diminishing intensity.

4.2. Excited Torsional States. From an analysis of the ultraviolet [5] and far infrared spectrum [4] of acrolein vapor the first four excited torsional levels were found to lie around $157\,\text{cm}^{-1}$ ($v = 1$), $312\,\text{cm}^{-1}$ ($v = 2$), $468\,\text{cm}^{-1}$ ($v = 3$) and $623\,\text{cm}^{-1}$ ($v = 4$), respectively. Wagner et al. [3] have reported a few low J transitions of trans- acrolein in the 18.0–36.0 GHz for the first three ($v = 1$, $v = 2$ and $v = 3$) excited torsional states and determined only the rotational constants B and C for each of these excited states. In this work, we have extended the analysis of rotational transitions in each of the above three excited states up to 99.0 GHz which has resulted in the determination of three rotational and two quartic centrifugal distortion constants for all the three torsional

Table 5: Calculated ground state rotational constants of *trans*-propenal (CH_2CHCOH) with various models and basis sets.

Model	Basis set	A_0 (MHz)	B_0 (MHz)	C_0 (MHz)
DFT				
B3LYP	6-31+g(d, p)	47421.368	4599.276	4189.265
B3LYP	6-31++g(d, 2p)	47462.833	4596.304	4190.461
B3LYP	6-311++g(d, 2p)	47713.427	4621.512	4213.560
B3PW91	6-31g	47562.896	4564.835	4164.742
B3PW91	6-31++g(d, 2p)	47445.030	4617.117	4207.640
B3PW91	6-311++g(d, 2p)	47640.454	4642.978	4230.737
MPW1PW91	6-31g	47680.556	4583.545	4181.241
MPW1PW91	6-31++g(d, 2p)	47532.149	4635.391	4223.524
MPW1PW91	6-311++g(d, 2p)	47719.894	4660.612	4246.017
PBEPBE	6-311g	46829.581	4528.525	4119.901
PBEPBE	6-311++g(d, 2p)	46855.368	4594.974	4175.330
PBEPBE	6-31++g(d, 2p)	46605.708	4567.810	4150.825
Expt.[o]		47353.729	4659.4894	4242.7034

[o]This work.

Table 6: Comparison of the molecular bond lengths, dipole moment, and total energy of *trans*-propenal calculated by various methods and basis sets with the experimental values.

Models	Basis sets	Bond lengths between							Dipole Moment (D)	Energy (eV)
		1C–2H	1C–3H	1C–4C	4C–5H	4C–6C	6C–7H	6C–8O		
B3LYP	6-31+g(d, p)	1.088	1.085	1.340	1.087	1.474	1.112	1.218	3.515	−5220.511
	6-31++g(d, 2p)	1.087	1.084	1.340	1.086	1.474	1.111	1.218	3.502	−5220.755
	6-311++g(d, 2p)	1.084	1.081	1.335	1.083	1.474	1.109	1.211	3.464	−5221.979
B3PW91	6-31g	1.088	1.084	1.341	1.086	1.465	1.106	1.239	3.542	−5216.647
	6-31++g(d, 2p)	1.085	1.082	1.334	1.084	1.471	1.111	1.208	3.491	−5218.334
	6-311+g(d, 2p)	1.085	1.082	1.334	1.084	1.417	1.111	1.208	3.434	−5219.694
MPW1PW91	6-31g	1.086	1.083	1.339	1.085	1.463	1.104	1.236	3.564	−5217.218
	6-31++g(d, 2p)	1.086	1.083	1.336	1.085	1.470	1.110	1.213	3.481	−5219.204
	6-311++g(d, 2p)	1.084	1.081	1.332	1.083	1.469	1.109	1.205	3.443	−5220.238
PBEPBE	6-311g	1.093	1.090	1.348	1.092	1.467	1.115	1.250	3.392	−5213.844
	6-31++g(d, 2p)	1.096	1.092	1.349	1.095	1.475	1.123	1.228	3.394	−5214.252
	6-311++g(d, 2p)	1.093	1.089	1.344	1.092	1.474	1.121	1.221	3.367	−5215.531
Expt.[p]		1.089	1.081	1.341	1.084	1.468	1.113	1.215	3.117	

[p]Ref. [13].

excited states. The new assigned transitions along with those reported earlier [3] are also shown in Table 1 along with the ground state transitions. The excited state data were also used to fit to the same semirigid rotor Watson's S-reduction Hamiltonian (I^r-representation) [29]. Three rotational and two quartic (D_J and D_{JK}) CD constants were used to fit the data. The contribution of other CD parameters was found to be negligible while fitting the excited state data. The derived spectroscopic constants and inertia defect values for the three torsional excited states are shown in Table 3. The more negative inertia defect values for successive torsional excited states indicate that the excited state lines arise from an out-of-plane vibration, in this case, COH group torsion about C–C single bond. The observed inertia defect values for the ground and torsional excited states of *trans*-propenal and some other related molecules are compared in Table 4. The plots of inertia defect values with torsional quantum numbers for *trans*-propenal along with other molecules are shown in Figure 3 for comparison.

4.3. Computational Results. Propenal is a slightly asymmetric prolate top molecule ($\kappa = -0.9806$). The optimization of geometry for the *trans* conformer of propenal was tested by employing various levels of theory and basis sets. However, the computed rotational and centrifugal distortion constants and the structural parameters obtained with model MPW1PW91 model with 6-31++g(d, 2p) basis set were found to be in good agreement with the observed values. Calculated values of ground state rotational constants of trans-propenal obtained with various models and basis sets are shown in Table 5. Results obtained with DFT MPW1PW91/6-31++g(d, 2p) have been compared with the corresponding experimental values in Table 2. For optimized geometry of *trans*-propenal the calculated energy

TABLE 7: Comparison of the molecular bond angles of *trans*-propenal calculated by various methods and basis sets with the experimental values.

Models	Basis sets	Bond angles between								
		H2–C1–H3	H2–C1–C4	H3–C1–C4	C1–C4–H5	C1–C4–H6	H5–C4–C6	C4–C6–H7	C4–C6–O8	H7–C6–O8
B3LYP	6-31+g(d,p)	116.799	121.067	122.133	122.342	121.089	116.569	115.154	124.162	120.684
	6-31++g(d,2p)	116.834	121.043	122.123	122.314	121.153	116.533	115.123	124.165	120.712
	6-311++g(d,2p)	116.849	120.973	122.177	122.347	121.069	116.583	114.830	124.348	120.822
B3PW91	6-31g	116.488	121.219	122.22	122.091	121.314	116.596	115.381	123.994	120.626
	6-31++g(d,2p)	116.699	121.104	122.196	122.363	120.926	116.711	115.126	124.248	120.625
	6-311+g(d,2p)	116.922	120.852	122.226	122.414	120.821	116.766	114.689	124.414	120.897
MPW1PW91	6-31g	116.516	121.201	122.284	122.146	121.259	116.594	115.443	123.945	120.612
	6-31++g(d,2p)	116.938	120.930	122.131	122.419	120.864	116.717	115.023	124.176	120.800
	6-311++g(d,2p)	116.964	120.807	122.227	122.491	120.714	116.794	114.691	124.394	120.914
PBEPBE	6-311g	116.573	121.132	122.295	122.136	121.724	116.139	115.404	123.451	120.604
	6-31++g(d,2p)	116.976	120.764	122.260	122.329	120.985	116.685	114.796	124.334	120.870
	6-311++g(d,2p)	116.998	120.664	122.337	122.394	120.846	116.762	114.486	124.549	120.964
Expt.[q]		118.0	119.8	122.2	122.4	120.3	117.3	114.7	123.9	121.3

[q]Ref. [13].

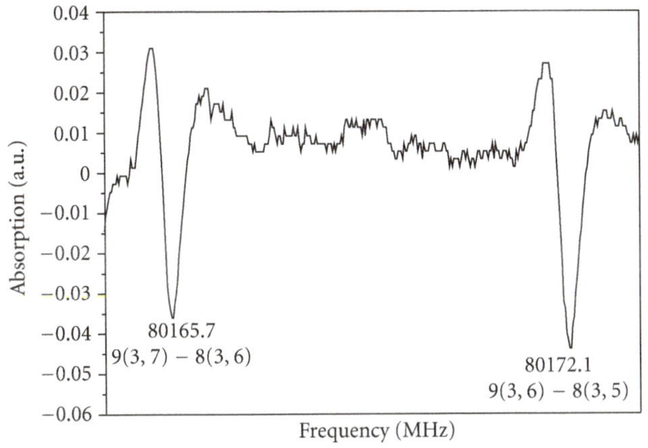

FIGURE 2: Observed trace of the $K_{-1} = 3$ doublet of $J = 9 \leftarrow 8$ transition of *trans*-propenal produced by DC discharge.

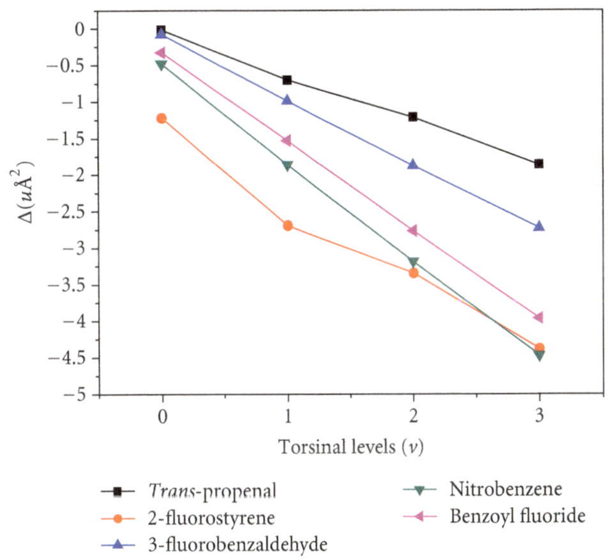

FIGURE 3: Plot of the variation of inertia defect values with torsional state quantum number for *trans*-propenal and some other molecules.

is −5219.204 eV and the dipole moment is 3.481 D. The number and labeling of atoms in propenal molecule as shown in Figure 4. Bond lengths and angles have been computed using different models and basis sets and are shown in Tables 6 and 7, respectively.

5. Conclusion

An efficient method of generating *trans*-propenal (*trans*-acrolein) in the gas phase by applying a DC glow discharge through a low pressure vapor of allyl alcohol inside the absorption cell has been presented. The gas phase rotational spectra of the *trans* conformer of propenal produced in this way has been recorded and analyzed in the frequency range 60.0–99.0 GHz for the ground as well as three torsional excited states ($v = 1$, 2 and 3). The asymmetric top K_{-1} K_{+1}-components of different transitions having J values 6 to 10 have been measured. The observed transition frequencies along with the previously reported data [1, 3, 14] were fitted to a standard asymmetric-top Watson's S-reduction Hamiltonian (I^r-representation) to determine ground state rotational and centrifugal distortion constants. Analysis of the rotational transitions for the three excited torsional states has been extended up to 99.0 GHz which has enabled us to determine the three rotational and two centrifugal distortion constants. The small negative value of the inertia defect ($\Delta = -0.018\,u\text{Å}^2$) in the ground vibrational state ($v = 0$) and the linear variation of the inertia defect values with torsional quantum number ($v = 1, 2, 3$) demonstrate that the equilibrium configuration of *trans*-propenal is planar as noticed in case of 3-fluorobenzaldehyde, benzoyl fluoride,

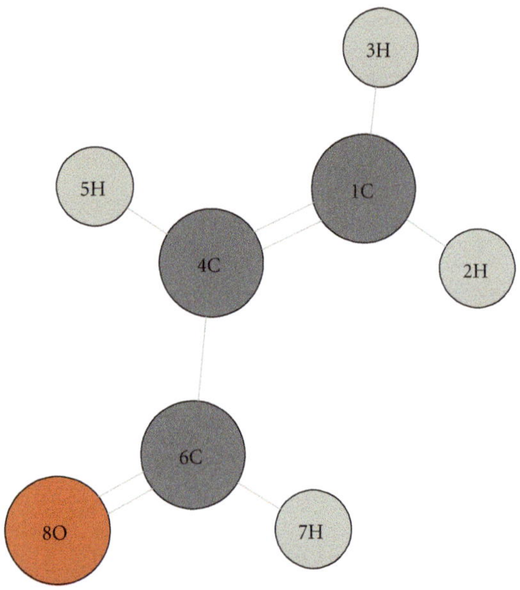

FIGURE 4: Optimized geometry of *trans*-propenal molecule and the numbering of atoms

and nitrobenzene (Figure 4). The existence of a slightly bent or twisted–COH group would have resulted in a zig-zag behavior in the variation of inertia defect values with torsional quantum number as observed in the case of 2-fluorostyrene (Figure 4). To compare the experimental results with theory, DFT calculations were performed using various models and basis sets. However, it was found that MPW1PW91 model with 6-31++g (d, 2p) basis set produced the best values of rotational and quartric centrifugal distortion constants which are close to the experimental values.

Acknowledgment

The authors would like to thank Mr. A. K. Bhattacharya for his technical assistance during the course of this work.

References

[1] E. A. Cherniak and C. C. Costain, "Microwave spectrum and molecular structure of trans-acrolein," *The Journal of Chemical Physics*, vol. 45, no. 1, pp. 104–110, 1966.

[2] J. M. Hollis, P. R. Jewell, F. J. Lovas, A. Remijan, and H. Møllendal, "Green bank telescope detection of new interstellar aldehydes: propenal and propanal," *Astrophysical Journal*, vol. 610, no. 1, pp. L21–L24, 2004.

[3] R. Wagner, J. Fine, J. W. Simmons, and J. H. Goldstein, "Microwave spectrum, structure, and dipole moment of s-trans acrolein," *The Journal of Chemical Physics*, vol. 26, no. 3, pp. 634–637, 1957.

[4] R. K. Harris, "Vibrational assignments for glyoxal, acrolein and butadiene," *Spectrochimica Acta*, vol. 20, no. 7, pp. 1129–1141, 1964.

[5] J. C. D. Brand and D. G. Williamson, "Near-ultra-violet spectrum of propenal," *Discussions of the Faraday Society*, vol. 35, pp. 184–191, 1963.

[6] J. M. Hollas, "The electronic absorption spectrum of acrolein vapour," *Spectrochimica Acta*, vol. 19, no. 9, pp. 1425–1426, E1–E2, 1427–1441, 1963.

[7] A. C. P. Alves, J. Christoffersen, and J. M. Hollas, "Near ultra-violet spectra of the s-trans and a second rotamer of acrolein vapour," *Molecular Physics*, vol. 20, no. 4, pp. 625–644, 1971.

[8] A. C. P. Alves, J. Christoffersen, and J. M. Hollas, "Erratum to "Near ultra-violet spectra of the s-trans and a second rotamer of acrolein vapour"," *Molecular Physics*, vol. 21, no. 2, p. 384, 1971.

[9] A. Krantz, T. D. Goldfarb, and C. Y. Lin, "A simple method for assigning vibrational frequencies to rapidly equilibrating rotational isomers," *Journal of the American Chemical Society*, vol. 94, no. 11, pp. 4022–4024, 1972.

[10] C. E. Blom, R. P. Miller, and H. Günthard, "S-trans and S-cis acrolein: trapping from thermal molecular beams and uv-induced isomerization in argon matrices," *Chemical Physics Letters*, vol. 73, pp. 483–486, 1980.

[11] L. A. Carreira, "Raman spectrum and torsional potential function of acrolein," *Journal of Physical Chemistry*, vol. 80, no. 11, pp. 1149–1152, 1976.

[12] C. E. Blom and A. Bauder, "Microwave spectrum, rotational constants and dipole moment of s-cis acrolein," *Chemical Physics Letters*, vol. 88, no. 1, pp. 55–58, 1982.

[13] C. E. Blom, G. Grassi, and A. Bauder, "Molecular structure of s-cis- and s-trans-acrolein determined by microwave spectroscopy," *Journal of the American Chemical Society*, vol. 106, no. 24, pp. 7427–7431, 1984.

[14] M. Winnewisser, G. Winnewisser, T. Honda, and E. Hirita, "Ground state centrifugal distortion constants of trans-acrolein, CH_2=CH-CHO from the microwave and millimeter wave rotational spectra," *Zitschrift, Naturforsch*, vol. 30, pp. 1001–1014, 1975.

[15] A. R. W. McKellar, D. W. Tokaryk, and D. R. T. Appadoo, "The far-infrared spectrum of acrolein, CH_2CHCHO: the ν_{18} fundamental and $(\nu_{17} + \nu_{18})$—ν_{18} hot bands," *Journal of Molecular Spectroscopy*, vol. 244, no. 2, pp. 146–152, 2007.

[16] L.-H. Xu, X. Jiang, H. Shi et al., "10 μm High-resolution spectrum of trans-acrolein: rotational analysis of the ν_{11}, ν_{16}, ν_{14} and $\nu_{16} + \nu_{18}$—ν_{18} bands," *Journal of Molecular Spectroscopy*, vol. 268, no. 1-2, pp. 136–146, 2011.

[17] J. L. Alonso and R. M. Villamañán, "Rotational isomerism in monofluorobenzaldehydes," *Journal of the Chemical Society, Faraday Transactions*, vol. 85, no. 2, pp. 137–149, 1989.

[18] J. H. Høg, L. Nygaard, and G. Ole Sørensen, "Microwave spectrum and planarity of nitrobenzene," *Journal of Molecular Structure*, vol. 7, no. 1-2, pp. 111–121, 1971.

[19] R. K. Kakar, "Microwave spectrum of benzoyl fluoride," *The Journal of Chemical Physics*, vol. 56, no. 3, pp. 1246–1252, 1972.

[20] R. M. Villamañán, J. C. López, and J. L. Alonso, "On the planarity of 2-fluorostyrene," *Journal of the American Chemical Society*, vol. 111, no. 17, pp. 6487–6491, 1989.

[21] S. Saito, "Laboratory microwave spectroscopy of interstellar molecules," *Applied Spectroscopy Reviews*, vol. 25, pp. 261–296, 1989.

[22] A. I. Jaman, P. Hemant Kumar, and P. R. Bangal, "Rotational spectrum of propyne observed in a DC glow discharge and DFT calculation," *Asian Journal of Spectroscopy, Special Issue*, pp. 43–48, 2010.

[23] A. I. Jaman, R. Bhattacharya, D. Mondal, and A. Kumar Das, "Millimeterwave spectral studies of propynal (HCCCHO) produced by DC discharge and ab initio DFT calculation,"

Journal of Atomic, Molecular, and Optical Physics, vol. 2011, Article ID 439019, 8 pages, 2011.

[24] J. F. Verdieck and C. D. Cornwell, "Radio-frequency spectrometer with bidirectional square wave frequency modulation," *Review of Scientific Instruments*, vol. 32, no. 12, pp. 1383–1386, 1961.

[25] A. I. Jaman, "Millimeterwave spectroscopy of transient molecules produced in a DC discharge," *Pramana*, vol. 61, no. 1, pp. 85–91, 2003.

[26] A. I. Jaman, "Millimeter wave spectrum of ICN, a transient molecule of chemical and astrophysical interest," *Journal of Physics*, vol. 80, no. 1, Article ID 012006, 2007.

[27] M. J. Frisch, G. W. Trucks, and H. B. Schlegel, *Gaussian 09, Revision A. 1*, Gaussian, Wallingford, Conn, USA, 2009.

[28] GaussView 5. 0.

[29] J. K. G. Watson, "Aspects of quartic and sextic centrifugal effects on rotational energy levels," in *Vibrational Spectra and Structure*, J. R. Durig, Ed., vol. 6, pp. 1–89, Mercel Dekker, New York, NY, USA, 1977.

12

The Effect of Nanoparticle Size on Cellular Binding Probability

Vital Peretz,[1] Menachem Motiei,[2] Chaim N. Sukenik,[1] and Rachela Popovtzer[2]

[1] *The Department of Chemistry, The Institute of Nanotechnology and Advanced Materials, Bar-Ilan University, Ramat Gan 52900, Israel*
[2] *Faculty of Engineering, The Institute of Nanotechnology and Advanced Materials, Bar-Ilan University, Ramat Gan 52900, Israel*

Correspondence should be addressed to Rachela Popovtzer, rachela.popovtzer@biu.ac.il

Academic Editor: Zeev Zalevsky

Nanoparticle-based contrast agents are expected to play a major role in the future of molecular imaging due to their many advantages over the conventional contrast agents. These advantages include prolonged blood circulation time, controlled biological clearance pathways, and specific molecular targeting capabilities. Recent studies have provided strong evidence that molecularly targeted nanoparticles can home selectively onto tumors and thereby increase the local accumulation of nanoparticles in tumor sites. However, there are almost no reports regarding the number of nanoparticles that bind per cell, which is a key factor that determines the diagnostic efficiency and sensitivity of the overall molecular imaging techniques. Hence, in this research we have quantitatively investigated the effect of the size of the nanoparticle on its binding probability and on the total amount of material that can selectively target tumors, at a single cell level. We found that 90 nm GNPs is the optimal size for cell targeting in terms of maximal Au mass and surface area per single cancer cell. This finding should accelerate the development of general design principles for the optimal nanoparticle to be used as a targeted imaging contrast agent.

1. Introduction

Imaging plays a critical role in overall cancer management; in diagnostics, staging, radiation planning, and evaluation of treatment efficiency. Conventional imaging technologies for cancer detection such as CT, MRI, and ultrasound can be categorized as structural imaging modalities. They are able to identify anatomical patterns and to provide basic information regarding tumor location, size, and spread based on endogenous contrast. However, these imaging modalities are not efficient in detecting tumors and metastases that are smaller than 0.5 cm [1], and they can barely distinguish between benign and cancerous tumors. Molecular imaging is an emerging field that integrates molecular biology, chemistry, physics, and medicine in order to gain understanding regarding biological processes and to identify diseases based on molecular markers, which appear before the clinical presentation of the disease.

Recently, much research has focused on the development of targeted nanoparticles for use as contrast agents for molecular imaging. These include superparamagnetic nanoparticles for MRI [2–6], quantum dots for optical imaging [7–9], and gold nanoparticles (GNPs) for optical imaging [10, 11] and CT [12–14].

GNPs are a class of contrast agents with unique optical properties. They are well known for their strong interactions with visible light through the resonant excitations of the collective oscillations of the conduction electrons within the particles [15]. As a result, local electromagnetic fields near the particle can be many orders of magnitude higher than the incident fields, and the incident light around the resonant-peak wavelength is scattered very strongly. The resonance condition is determined from absorption and scattering spectroscopy and is found to depend on the shape, size, and dielectric constants of both the metal and the surrounding material. This localized surface plasmon resonance (LSPR) has led to the development of a wide range of biochemical detection assays [16] and various nanoprobes for optical imaging of cancer [17, 18].

In order to study whether incubation of the cancer cells with different sizes of GNPs can improve the coverage of the cells' surface, 15, 70, and 150 nm GNPs were incubated with head and neck cancer cells (A431) in different incubation orders, as illustrated in Table 2. We hypothesized that small

GNPs, when introduced to the cells in a second cycle (after larger GNPs were incubated), will fill the gaps between the larger GNPs on the cells' surfaces.

GNPs are also an ideal CT contrast agents. The ability of CT to distinguish between different tissues is based on the fact that different tissues provide different degrees of X-ray attenuation, where the attenuation coefficient is determined by the atomic number and electron density of the tissue; the higher the atomic number and electron density, the higher the attenuation coefficient. The atomic number and electron density of gold (79 and 19.32 g/cm^3, resp.) are much higher than those of the currently used iodine (53 and 4.9 g/cm^3) and, therefore, gold induces a strong X-ray attenuation. When the gold particles are linked to specific-targeting ligands, such as monoclonal antibodies or peptides, these nanoparticles can selectively tag a wide range of medically important targets, for example, specific cancer cells, with high affinity and specificity. In addition, gold nanoparticles proved to be nontoxic and biocompatible *in vivo* [19, 20].

Recently, Hainfeld et al. [21] showed that GNPs can enhance the visibility of millimeter-sized human breast tumors in mice, and that active tumor targeting (with anti-Her2 antibodies) is more efficient than passive targeting. They also showed that the specific uptake of the targeted GNPs in the tumor's periphery was 22-fold higher than in surrounding muscle. Another recent study demonstrated enhanced CT attenuation of bombesin-functionalized GNPs that selectively targeted cancer receptor sites that are over-expressed in prostate, breast, and small-cell lung carcinoma [22]. In our own research, [23], we recently demonstrated that a small tumor, which is currently undetectable through anatomical CT, is enhanced and becomes clearly visible by the molecularly-targeted GNPs. We further showed that the CT number of molecularly targeted head and neck tumor is over five times higher than the corresponding CT number of an identical but untargeted tumor, and that active tumor targeting is more efficient and specific than passive targeting.

These studies have provided strong evidence that nanoparticles accumulate *in vivo* on the tumor. However, there are almost no reports regarding the number of nanoparticles that bind per cell, which is a key factor that determines the diagnostic efficiency and sensitivity of the overall molecular imaging techniques. Hence, in this research, we have quantitatively investigated the effect of the size of the nanoparticle on its binding probability and on the total amount of material that can selectively target tumors, on a single cell level. We have further investigated the ability to increase the amount of contrast material that binds per cell by simultaneously targeting nanoparticles in different sizes or in consequent cycles.

2. Methods

2.1. Gold Nanospheres Synthesis, Conjugation, and Characterization. Gold nanospheres (10, 15, and 30 nm) were synthesized by citrate reduction according to the method described by Turkevitch et al. [24]. Larger GNPs (70, 90, and 150 nm) were synthesized using the seed mediated growth method [25]. Briefly, gold seeds were synthesized using an aqueous $HAuCl_4$ solution (0.25 mL of 0.05 M solution) and adding it to 50 mL H_2O and boiling. After boiling, 1.75 mL of 1% citrate (1.75 mL of a solution that was 114 mg in 10 mL H_2O) were added to the solution and stirring was continued for 20 min. The solution was cooled to room temperature and used directly for further experiments. This method produced gold nanospheres with a diameter of 15 nm [26]. In order to enlarge the nanoparticles, 170 mL DD water along with 0.44 mL of 1.4 M $HAuCl_4$ and 26 mL seed solution to make 90 nm GNPs or 31 mL seed solution to make 70 nm GNPs was added to a 400 mL Erlenmeyer flask. Then, 3.72 mL of 0.1 M 2-mercaptosuccinic acid (MSA) was added as the reducing agent. The solution was stirred for 1 min and left overnight. According to this method 70 and 90 nm gold nanospheres were synthesized.

In order to prevent aggregation and to stabilize the particles in physiological solutions, a layer of polyethylene glycol (mPEG-H) was absorbed onto the GNPs. This layer also provides the chemical groups that are required for antibody conjugation (SH-PEG-COOH). The SH-PEG layer consisted of a mixture of 15% SH-PEG-COOH (Mw. 3400) and 85% SH-PEG-methyl (Mw. 5000), both obtained from Creative PEGWorks, Winston Salem, NC. The PEG solution was added to the GNPs solution and stirred for 3 hours. It was then centrifuged in order to get rid of excess PEG citrate and MSA. The ratio of PEG molecules to GNPs was calculated based on a footprint area of 0.35 nm^2 [27].

2.1.1. Conjugation of Antibodies to GNPs. In order to specifically bind to the SCC cancer cells, anti-EGFR (epidermal growth factor receptor, Erbitux, Merck KGaA), antibodies, which bind exclusively to the EGF receptor, were conjugated to the outer coating of the nanoparticles. The interaction between the GNP and the antibody are based on electrostatic attraction between the negatively charged heterofunction PEG (SH-PEG-COOH) and the positive segment of the antibody in pH 7.4.

EGFR conjugation to 70 nm GNPs: 1.31 mg of SH-PEG-COOH and 11 mg of SH-PEG were dissolved in 2 mL DD water and added to the GNPs solution with stirring for 3 hours. The solution was centrifuged and 15 mL of 5 mg/mL of anti-EGFR were added to the solution with stirring for 1 hour in order to get rid of excess anti-EGFR. The solution was kept at 4°C. The conjugation of anti-EGFR to all other GNPs sizes was achieved using the same method with different quantities.

2.2. In Vitro Cell Targeting Study Using GNPs. In order to study the effect of nanoparticle size on its binding probability, 15, 70, and 150 nm GNPs were incubated with head and neck cancer cells (A431); (Group A with 15 nm GNPs, group B with 70 nm GNPs, and group C with 150 nm GNPs.) A431 cells (1.5×10^6) in 5 mL DMEM medium containing 5% FCS, 0.5% penicillin, and 0.5% glutamine were incubated for a quantitative cell binding study (each experimental group was run in triplicate). Each group was incubated 3 times with access amount of anti-EGFR-coated GNPs for 30 minutes at 37°C. After incubation, the medium

TABLE 1: The effect of each additional incubation cycle on the amount of contrast material that binds per cell. Each cell sample contained 1.5×10^6 cells. The cells were incubated with the different sized GNPs in consequent cycles for 30 min at 37°C each cycle. Each experimental group was run in triplicate. The GNPs were added in excess.

The effect of each additional incubation cycle	Cell type	GNPs size (nm)	Comment
A	A431	90	The cells were incubated with the GNPs in consequent cycles
B	A431	90, 90	
C	A431	90, 30	
D	A431	90, 30, 10	

TABLE 2: The effect of incubation of different sized GNPs in consequent cycles experiment. Each cell sample contained 1.5×10^6 cells. The cells were incubated with the different sized GNPs in consequent cycles for 30 min at 37°C each cycle. Each experimental group was run in triplicate. The GNPs were added in excess amount.

The effect of combination of different sized GNPs in consequent cycles	Cell type	GNPs size (nm)	Comment
A	A431	15, 70, 150	The cells were incubated with the different size of GNPs in consequent cycles
B	A431	150, 70, 15	
C	A431	150, 150, 150	

was washed twice with PBS followed by addition of 1 mL of aqua-regia. After evaporation of the acid, the sediment was dissolved in 5 mL of 0.05 M HCl. The gold concentrations of the samples were quantified by Flame Atomic Absorption Spectroscopy (SpectrAA 140, Agilent Technologies).

We have further investigated whether we are able to increase the amount of contrast material that binds per cell by subsequent cycles of binding, and how many GNPs can bind to a single cancer cell after each cycle of incubation. Therefore, 90, 30, and 10 nm GNPs were incubated in consequent cycles with the A431 cells (each cycle for 30 min at 37°C) as illustrated in Table 1. After each incubation cycle, the amount of gold (Au mass/cell) was measured using FAAS. Then, the number of GNPs that were bound in each cycle could be calculated.

3. Results and Discussion

3.1. Gold Nanospheres Synthesis, Conjugation and Characterization. We have successfully synthesized GNPs in various sizes, ranging from 10 nm up to 150 nm. Figure 1 shows TEM images of 10, 30, 70, and 150 nm GNPs. As can be seen, the small GNPs (~10 nm) have a relatively large size distribution (25%, Figure 1(a)), while particles that are larger than 30 nm are more homogeneous with a very narrow size distribution (Figures 1(b), 1(c), and 1(d)). The surface plasmon resonances of the various size GNPs are illustrated in Figure 2. It can be seen that when the nanospheres are enlarged there is a red shift in the surface plasmon resonance of the particles (from 525 nm to 580 nm).

GNPs were successfully coated with PEG and anti-EGFR antibody. The antibody conjugated GNPs were stable for up to three months, confirmed by their maintenance of the same plasmon resonance.

3.2. Quantitative Investigation of the Size Effect of the Nanoparticle on Its Binding Probability. Before studying the effect of nanoparticle size on its binding probability, we have evaluated the specificity of the interaction between the antibody-coated GNPs and the A431 SCC cancer cells (which highly express the EGF receptor). Two types of GNPs (50 μL of 25 mg/mL) were introduced to the SCC head and neck cancer cells (2.5×10^6 cells). The first was specifically coated with anti-EGFR antibody, while the second, which was used as a negative control, was coated with a nonspecific antibody (anti-Rabbit IgG). Atomic absorption spectroscopy measurements quantitatively demonstrated that the active tumor targeting (anti-EGFR coated GNPs) was significantly more specific than the control experiment (anti-Rabbit IgG coated GNPs). The A431 cells took up $26.3 \pm 2.3\,\mu$g of targeted GNPs (3.9×10^4 GNPs per A431 cell); while parallel cells in the negative control experiment absorbed only $0.2 \pm 0.01\,\mu$g of GNPs (3.4×10^3 GNPs per cell). Our results correlate well with previously published studies, which report that head and neck SCC express from 2×10^4 to 2×10^6 EGFRs/cell [28, 29].

In order to quantitatively investigate the effect of nanoparticle size on its binding probability (on a single cell level), head and neck cancer cells were incubated with different size GNPs (15, 70, and 150 nm) for 30 min. For maximal binding, the particles were incubated three times with the cancer cells. Figure 3 shows the total amount of gold (Au mass/cell) that binds per cell for nanoparticles of different sizes (15, 70, and 150 nm). The results clearly demonstrate that larger particles produce larger amounts of gold per cancer cell. For the 15 nm GNPs, only 0.0018 ng of gold was bound to a single cancer cell (A431) while for the largest particles, 150 nm, 0.145 ng of gold was bound to a single cancer cell.

Once we have quantitatively measured (using FAAS) the total amount of gold that was bound to a single cancer cell, the exact number of nanoparticles and the GNPs' surface area per cancer cell could then be calculated. Table 3 shows the total Au mass, the number of GNPs of different sizes,

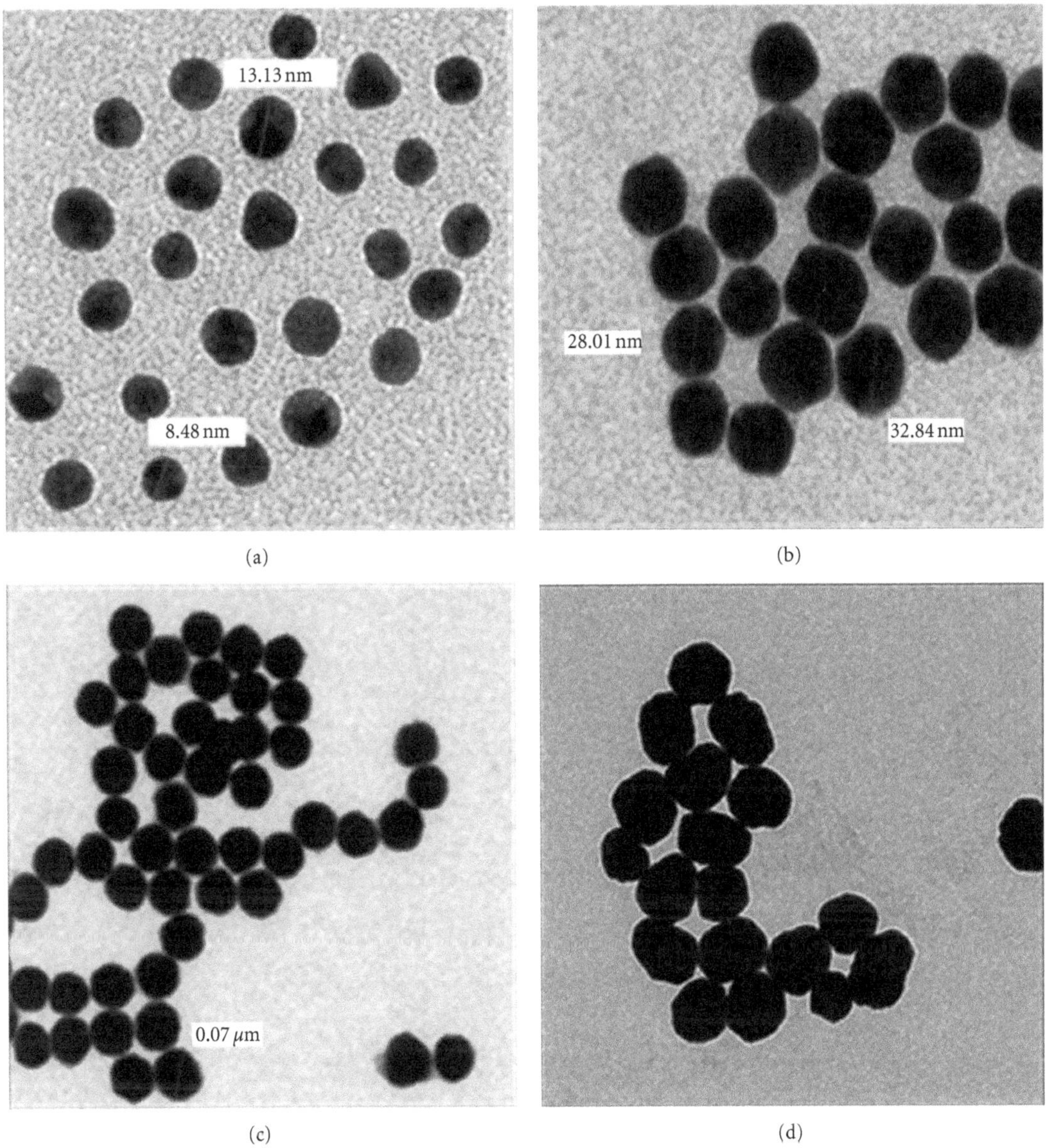

Figure 1: TEM images of different sized GNPs (a) 10 nm, (b) 30 nm, (c) 70 nm, and (d) 150 nm.

Table 3: Quantitative analysis: number of GNPs of different sizes that are bound to a single cancer cell, total Au mass and the GNPs' surface area per single cell.

GNP size (nm)	Number of GNP per cell	Au/cell (ng)	Total surface area (m^2)
15	54000	0.00186	3.8×10^{-11}
30	39000	0.01	1.1×10^{-10}
90	12000	0.124	3.05×10^{-10}
150	4200	0.145	2.9×10^{-10}

and the surface area of the GNPs that are bound to a single cancer cell.

These results clearly demonstrate that smaller particles have a higher probability to bind to cancer cells (via antibody-antigen interaction) than larger particles (Figure 4(a)). The probability of 15 nm GNPs to bind to cancer cells is about 13 times more than the probability of 150 nm GNPs. However, larger particles produce larger amounts of Au mass per cell, as well as larger surface area, as illustrated in Figures 4(b) and 4(c). Nevertheless, particles larger than 90 nm only slightly increased the Au mass/cell and the surface area/cell.

We have further investigated whether incubation of the cancer cells with different sizes of GNPs can improve the coverage of the cells' surface. We have hypothesized that small GNPs, when introduced to the cells in a second cycle (after larger GNPs were incubated), will fill the gaps between the larger GNPs on the cells' surface. However, as can be seen in Figure 5, maximum coverage (or max. Au mass/cell) was obtained for the largest GNPs (column C, 0.14 ngr Au/cell). It has also been demonstrated that the order of the incubation (between the cells and the GNPs) is critical. When 15 nm GNPs were introduced first to the cells, overall, a much smaller amount of gold was bound (column A).

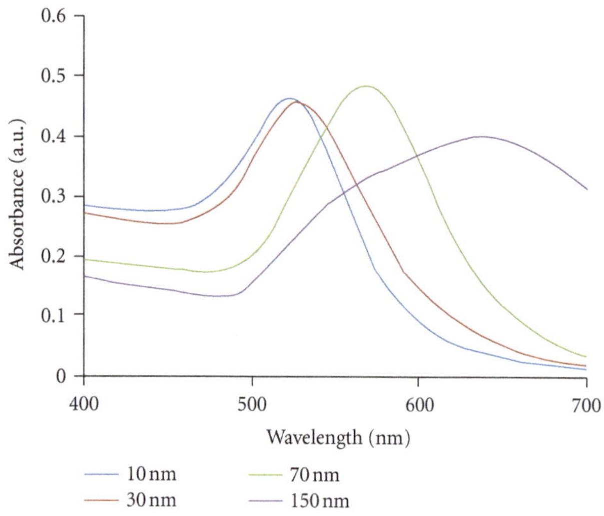

FIGURE 2: UV-Vis spectroscopy of 10, 30, 70, and 150 nm gold nanospheres.

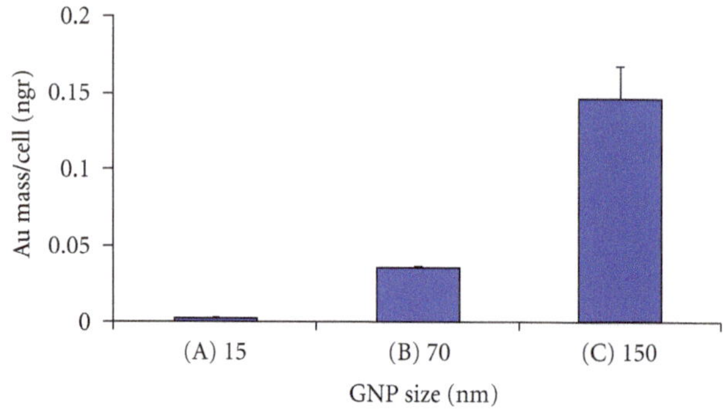

FIGURE 3: Quantitative measurements using FAAS of Au mass/cell for different sizes of GNPs. Each cell sample contained 1.5×10^6 cells and was incubated 3 times with the GNPs. The GNPs were added in excess. The error bars represent the standard deviation of three samples.

We have further investigated whether we are able to increase the amount of contrast material that binds per cell by consequent cycles of binding, and how many GNPs were bound to a single cancer cell after each cycle of incubation. As seen in Figure 6, the first incubation is the most critical. After one incubation with 90 nm GNPs, 0.1048 ngr of gold (=14258 GNPs) was bound to a single cancer cell. In the second cycle of incubation with 90 nm GNPs (Figure 6, column B), only a relatively small number of GNPs were bound (2742 90 nm GNPs (20%)). Adding smaller GNPs (30 and 15 nm, Figure 6 columns C and D) barely influenced the amount of gold per cell (0.0015 ngr (5772 30 nm GNPs (1%) for 30 nm GNPs, and 0 ngr for the 15 nm GNPs). It has been also demonstrated that the first antibody antigen interaction (first incubation between the cells and the GNPs) is the most effective (Figure 6, column A).

4. Summery and Conclusions

In order to develop general design principles for nanoparticles to be used as *in vivo* imaging contrast agents, we

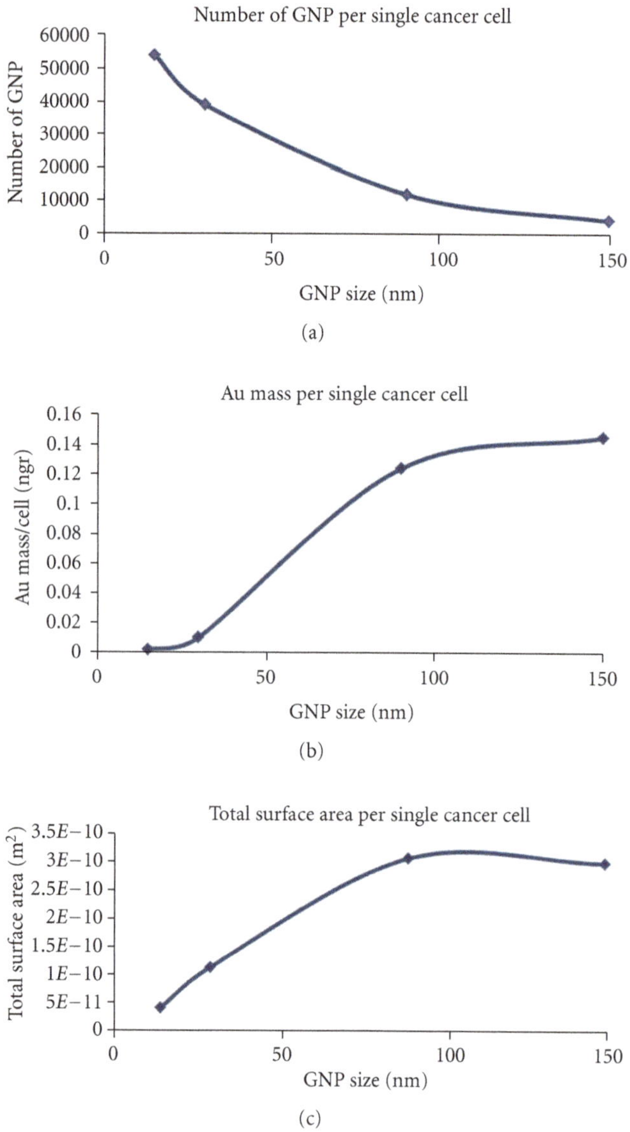

FIGURE 4: correlation between GNPs' sizes, number, mass and surface area for a single cancer cell: (a) number of GNP per cell, (b) Au mass/cell, (c) surface area/cell.

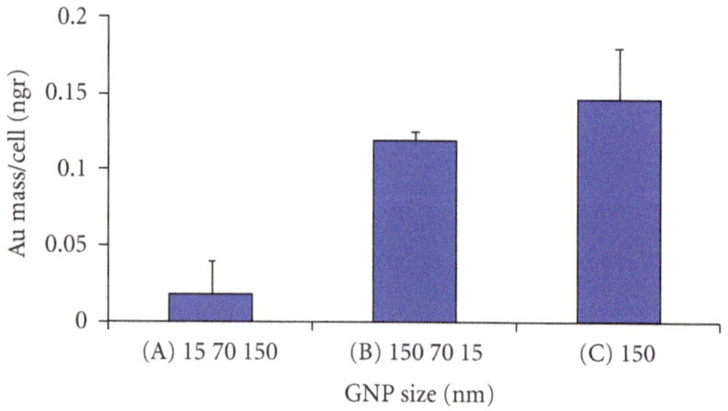

FIGURE 5: Atomic absorption measurement of gold per cell. Each column shows 3 incubations differing in GNP size and order of application.

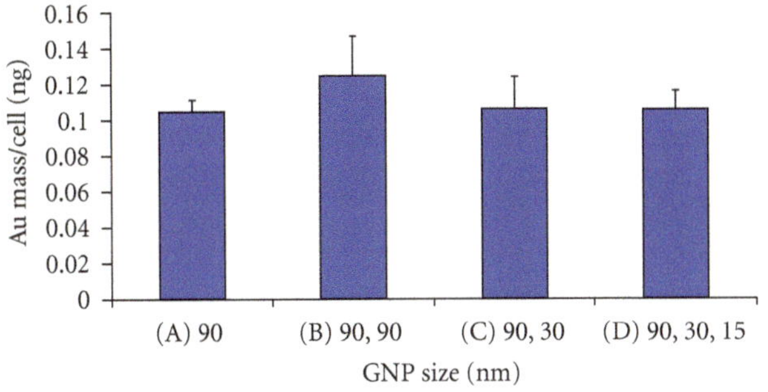

FIGURE 6: Atomic absorption measurement of gold concentration per cell for various cycles and different sizes of GNPs.

have quantitatively investigated the effect of the size of the nanoparticle on its binding probability and on the total amount of material that can selectively target tumors, on a single cell level. We found that 90 nm GNPs are the optimal size for cell targeting both in terms of maximal Au mass and surface area per single cell. For *in vivo* applications, 90 nm is in the right size range since the particles should be larger than ~15 nm to avoid rapid clearance by the kidneys or uptake in the liver, and smaller than ~150 nm to avoid filtration in the spleen [30]. It has been demonstrated that smaller particles, despite having higher binding probability, produce a smaller amount of Au mass per cell as well as a smaller surface area. Particles that are larger than 90 nm only slightly increased the Au mass/cell and decreased the surface area/cell. It has been also demonstrated that the first incubation is the most critical. However, subsequent incubation can increase the amount of contrast material by about 20%. The results of this study should accelerate the development of general design principles for the optimal nanoparticle to be used as a targeted imaging contrast agent.

References

[1] H. Rusinek, D. P. Naidich, G. McGuinness et al., "Pulmonary nodule detection: low-dose versus conventional CT," *Radiology*, vol. 209, no. 1, pp. 243–249, 1998.

[2] R. Lamerichs, "MRI-based molecular imaging using nanoparticles," *Cellular Oncology*, vol. 30, no. 2, p. 100, 2008.

[3] C. Sun, O. Veiseh, J. Gunn et al., "In vivo MRI detection of gliomas by chlorotoxin-conjugated superparamagnetic nanoprobes," *Small*, vol. 4, no. 3, pp. 372–379, 2008.

[4] R. Kopelman, Y. E. Lee Koo, M. Philbert et al., "Multifunctional nanoparticle platforms for in vivo MRI enhancement and photodynamic therapy of a rat brain cancer," *Journal of Magnetism and Magnetic Materials*, vol. 293, no. 1, pp. 404–410, 2005.

[5] Y. E. L. Koo, G. R. Reddy, M. Bhojani et al., "Brain cancer diagnosis and therapy with nanoplatforms," *Advanced Drug Delivery Reviews*, vol. 58, no. 14, pp. 1556–1577, 2006.

[6] A. M. Neubauer, H. Sim, P. M. Winter et al., "Nanoparticle pharmacokinetic profiling in vivo using magnetic resonance imaging," *Magnetic Resonance in Medicine*, vol. 60, no. 6, pp. 1353–1361, 2008.

[7] X. H. Gao and S. M. Nie, "Long-circulating QD probes for in-vivo tumor imaging," *Nanosensing: Materials and Devices*, vol. 5593, pp. 292–299, 2004.

[8] P. Diagaradjane, J. M. Orenstein-Cardona, N. E. Colón-Casasnovas et al., "Imaging epidermal growth factor receptor expression in vivo: pharmacokinetic and biodistribution characterization of a bioconjugated quantum dot nanoprobe," *Clinical Cancer Research*, vol. 14, no. 3, pp. 731–741, 2008.

[9] Y. Guo, D. Shi, J. Lian et al., "Quantum dot conjugated hydroxylapatite nanoparticles for in vivo imaging," *Nanotechnology*, vol. 19, no. 17, Article ID 175102, 2008.

[10] C. Loo, A. Lowery, N. Halas, J. West, and R. Drezek, "Immunotargeted nanoshells for integrated cancer imaging and therapy," *Nano Letters*, vol. 5, no. 4, pp. 709–711, 2005.

[11] A. M. Gobin, M. H. Lee, N. J. Halas, W. D. James, R. A. Drezek, and J. L. West, "Near-infrared resonant nanoshells for combined optical imaging and photothermal cancer therapy," *Nano Letters*, vol. 7, no. 7, pp. 1929–1934, 2007.

[12] R. Popovtzer, A. Agrawal, N. A. Kotov et al., "Targeted gold nanoparticles enable molecular CT imaging of cancer," *Nano Letters*, vol. 8, no. 12, pp. 4593–4596, 2008.

[13] D. Kim, S. Park, H. L. Jae, Y. J. Yong, and S. Jon, "Antibiofouling polymer-coated gold nanoparticles as a contrast agent for in vivo X-ray computed tomography imaging," *Journal of the American Chemical Society*, vol. 129, no. 24, pp. 7661–7665, 2007.

[14] J. F. Hainfeld, D. N. Slatkin, T. M. Focella, and H. M. Smilowitz, "Gold nanoparticles: a new X-ray contrast agent," *British Journal of Radiology*, vol. 79, no. 939, pp. 248–253, 2006.

[15] K. H. Su, Q. H. Wei, X. Zhang, J. J. Mock, D. R. Smith, and S. Schultz, "Interparticle coupling effects on plasmon resonances of nanogold particles," *Nano Letters*, vol. 3, no. 8, pp. 1087–1090, 2003.

[16] C. T. Campbell and G. Kim, "SPR microscopy and its applications to high-throughput analyses of biomolecular binding events and their kinetics," *Biomaterials*, vol. 28, no. 15, pp. 2380–2392, 2007.

[17] P. K. Jain, I. H. ElSayed, and M. A. El-Sayed, "Au nanoparticles target cancer," *Nano Today*, vol. 2, no. 1, pp. 18–29, 2007.

[18] I. H. El-Sayed, X. Huang, and M. A. El-Sayed, "Surface plasmon resonance scattering and absorption of anti-EGFR antibody conjugated gold nanoparticles in cancer diagnostics: applications in oral cancer," *Nano Letters*, vol. 5, no. 5, pp. 829–834, 2005.

[19] E. E. Connor, J. Mwamuka, A. Gole, C. J. Murphy, and M. D. Wyatt, "Gold nanoparticles are taken up by human cells but do not cause acute cytotoxicity," *Small*, vol. 1, no. 3, pp. 325–327, 2005.

[20] T. S. Hauck, A. A. Ghazani, and W. C. W. Chan, "Assessing the effect of surface chemistry on gold nanorod uptake, toxicity, and gene expression in mammalian cells," *Small*, vol. 4, no. 1, pp. 153–159, 2008.

[21] J. F. Hainfeld, M. J. O'Connor, F. A. Dilmanian, D. N. Slatkin, D. J. Adams, and H. M. Smilowitz, "Micro-CT enables microlocalisation and quantification of Her2-targeted gold nanoparticles within tumour regions," *British Journal of Radiology*, vol. 84, no. 1002, pp. 526–533, 2011.

[22] N. Chanda, V. Kattumuri, R. Shukla et al., "Bombesin functionalized gold nanoparticles show in vitro and in vivo cancer receptor specificity," *Proceedings of the National Academy of Sciences of the United States of America*, vol. 107, no. 19, pp. 8760–8765, 2010.

[23] T. Reuveni, M. Motiei, Z. Romman, A. Popovtzer, and R. Popovtzer, "Targeted gold nanoparticles enable molecular CT imaging of cancer: an in vivo study," *International Journal of Nanomedicine*, vol. 6, pp. 2859–2864, 2011.

[24] J. Turkevich, P. C. Stevenson, and J. Hillier, "A study of the nucleation and growth processes in the synthesis of colloidal gold," *Discussions of the Faraday Society*, vol. 11, pp. 55–75, 1951.
[25] J. Niu, T. Zhu, and Z. Liu, "One-step seed-mediated growth of 30-150 nm quasispherical gold nanoparticles with 2-mercaptosuccinic acid as a new reducing agent," *Nanotechnology*, vol. 18, no. 32, Article ID 325607, 2007.
[26] G. Frens, "Controlled nucleation for regulation of particle-size in monodisperse gold suspensions," *Nature-Physical Science*, vol. 241, no. 105, pp. 20–22, 1973.
[27] W. P. Wuelfing, S. M. Gross, D. T. Miles, and R. W. Murray, "Nanometer gold clusters protected by surface-bound monolayers of thiolated poly(ethylene glycol) polymer electrolyte," *Journal of the American Chemical Society*, vol. 120, no. 48, pp. 12696–12697, 1998.
[28] R. Todd and D. T. W. Wong, "Epidermal growth factor receptor (EGFR) biology and human oral cancer," *Histology and Histopathology*, vol. 14, no. 2, pp. 491–500, 1999.
[29] P. Stanton, S. Richards, J. Reeves et al., "Epidermal growth factor receptor expression by human squamous cell carcinomas of the head and neck, cell lines and xenografts," *British Journal of Cancer*, vol. 70, no. 3, pp. 427–433, 1994.
[30] F. Hallouard, N. Anton, P. Choquet, A. Constantinesco, and T. Vandamme, "Iodinated blood pool contrast media for preclinical X-ray imaging applications—a review," *Biomaterials*, vol. 31, no. 24, pp. 6249–6268, 2010.

Electron-Pair Densities with Time-Dependent Quantum Monte Carlo

Ivan P. Christov

Physics Department, Sofia University, 1164 Sofia, Bulgaria

Correspondence should be addressed to Ivan P. Christov; ipc@phys.uni-sofia.bg

Academic Editor: Keli Han

We use sets of de Broglie-Bohm trajectories to describe the quantum correlation effects which take place between the electrons in helium atom due to exchange and Coulomb interactions. A short-range screening of the Coulomb potential is used to modify the repulsion between the same spin electrons in physical space in order to comply with Pauli's exclusion principle. By calculating the electron-pair density for orthohelium, we found that the shape of the exchange hole can be controlled uniquely by a simple screening parameter. For parahelium the interelectronic distance, hence the Coulomb hole, results from the combined action of the Coulomb repulsion and the nonlocal quantum correlations. In this way, a robust and self-interaction-free approach is presented to find both the ground state and the time evolution of nonrelativistic quantum systems.

1. Introduction

The electronic many-body problem is of key importance for the theoretical treatments of physics and chemistry. A typical manifestation of the quantum many-body effects is the electron correlation which results from the Coulomb and exchange interactions between the electrons combined with the underlying quantum nonlocality. Since in general the electron correlation reshapes the probability density in configuration space, it is difficult to elucidate this effect for higher dimensions. Therefore, to better understand the effects of electron correlation in atoms and molecules, one needs, besides one-particle quantities such as the electron density function, to consider also extensions which explicitly incorporate many-body effects. Such an appropriate quantity is the electronic pair-density function which represents the probability density of finding two electrons at distance $\mathbf{u}$ from each other [1, 2]:

$$I(\mathbf{u},t) = \left\langle \Psi(\mathbf{R},t) \left| \sum_{i<j} \delta\left[\left(\mathbf{r}_i - \mathbf{r}_j\right) - \mathbf{u}\right] \right| \Psi(\mathbf{R},t) \right\rangle, \quad (1)$$

where $\mathbf{r}_i$ is the position of the ith electron and the many-body wave function $\Psi(\mathbf{R},t)$ resides in configuration space with arguments being the instantaneous coordinates of all electrons $\mathbf{R} = (\mathbf{r}_1, \mathbf{r}_2, \ldots, \mathbf{r}_N)$.

The importance of the electron-pair density, also known as electron position intracule, comes from the fact that it can be associated with experimental data obtained from X-ray scattering, and it can also be used to visualize the notion of exchange and correlation holes which surround the quantum particles. However, the calculation of the many-body wave function in (1) is hampered by the computational cost which scales exponentially with system dimensionality. Therefore, different approximations have been employed in order to calculate the electronic pair densities. These include Hartree-Fock (HF) approximation as well as Hylleraas type explicitly correlated wave functions represented as product of HF function and pair-correlation factors [3–6]. Other (e.g., quantum Monte Carlo [7]) approaches use appropriate Slater-Jastrow-type many-body wave functions which involve number of parameters, which after optimization can be used to calculate the average in (1).

Here we calculate the electron-pair densities for helium atom in $2\,^1$S and $2\,^3$S states using the recently proposed time-dependent quantum Monte Carlo (TDQMC) method which employs sets of particles and quantum waves to describe the ground state and the time evolution of many-electron systems [8–13]. In TDQMC each electron is described statistically as an ensemble of walkers which represent different replicas of that electron in position space, where each walker is guided by a separate time-dependent de Broglie-Bohm pilot wave. The correlated guiding waves obey a set of coupled time-dependent Schrödinger equations (TDSE), where the electron-electron interactions are accounted for using explicit nonlocal Coulomb potentials. In the TDQMC algorithm the preparation of the ground state of the quantum system involves a few steps which include initialization of the Monte Carlo (MC) ensembles of walkers and guide waves, followed by their concurrent propagation in complex time toward steady state in the presence of random component in walker's motion to account for the processes of quantum drift and diffusion. Once the ground state is established, the real-time quantum dynamics can be studied, for example, the interaction of atoms and molecules with external electromagnetic fields. The large speedup of the calculations when using TDQMC comes from the fact that walker's distribution reproduces the amplitude (or modulus square) of the many-body wave function, while its phase is being disregarded as it is not needed for most applications. Also, the TDQMC method can be implemented very efficiently on parallel computers where tens of thousands of coupled Schrödinger equations can be solved concurrently for affordable time.

2. General Theory

The TDQMC is an *ab initio* method with respect to the electron correlation in that it does not involve explicit pair-correlation factors which may become too complex when used for larger systems. For a system of N electrons, the many-body wave function obeys the Schrödinger equation:

$$i\hbar\frac{\partial}{\partial t}\Psi(\mathbf{R},t) = -\frac{\hbar^2}{2m}\nabla^2\Psi(\mathbf{R},t) + V(\mathbf{R})\Psi(\mathbf{R},t), \quad (2)$$

where $\nabla = (\nabla_1, \nabla_2, \ldots, \nabla_N)$. The potential $V(\mathbf{R})$ in (2) is a sum of electron-nuclear, electron-electron, and external potentials:

$$\begin{aligned} V(\mathbf{r}_1,\ldots,\mathbf{r}_N) &= V_{e-n}(\mathbf{r}_1,\ldots,\mathbf{r}_N) + V_{e-e}(\mathbf{r}_1,\ldots,\mathbf{r}_N) \\ &+ V_{\text{ext}}(\mathbf{r}_1,\ldots,\mathbf{r}_N,t) = \sum_{k=1}^{N} V_{e-n}(\mathbf{r}_k) \\ &+ \sum_{k>l}^{N} V_{e-e}(\mathbf{r}_k - \mathbf{r}_l) + V_{\text{ext}}(\mathbf{r}_1,\ldots,\mathbf{r}_N,t). \end{aligned} \quad (3)$$

For Hamiltonians with no explicit spin variables the exchange effects can be accounted for efficiently using screened Coulomb potentials as described in [10]. The simple idea behind this approach is that the short-range screened Coulomb potential ensures full-scale Coulomb interaction between only electron replicas (MC walkers) which are not too close to each other, in accordance with Pauli's exclusion principle. The use of screened Coulomb potentials is beneficial in that it eliminates the need of using antisymmetrized products of guiding waves in the Broglie-Bohm guiding equation for the velocity of the walkers. Instead, the many-body wave function is replaced by a simple product:

$$\Psi^k(\mathbf{r}_1,\mathbf{r}_2,\ldots,\mathbf{r}_N,t) = \prod_{i=1}^{N}\varphi_i^k(\mathbf{r}_i,t), \quad (4)$$

where $\varphi_i^k(\mathbf{r}_i,t)$ denote the individual time-dependent guide waves with indexes i and k for the electrons and the walkers, respectively. Then, the guiding equations for the Monte Carlo walkers read

$$\mathbf{v}\left(\mathbf{r}_i^k\right) = \frac{\hbar}{m}\operatorname{Im}\left[\frac{1}{\varphi_i^k(\mathbf{r}_i,t)}\nabla_i\varphi_i^k(\mathbf{r}_i,t)\right]_{\mathbf{r}_i=\mathbf{r}_i^k(t)}. \quad (5)$$

On the other side, the guide waves obey a set of coupled TDSE:

$$\begin{aligned} i\hbar\frac{\partial}{\partial t}\varphi_i^k(\mathbf{r}_i,t) = \Bigg[& -\frac{\hbar^2}{2m}\nabla_i^2 + V_{e-n}(\mathbf{r}_i) \\ & + \sum_{j\neq i}^{N} V_{e-e}^{\text{eff}}\left[\mathbf{r}_i - \mathbf{r}_j^k(t)\right] \\ & + V_{\text{ext}}(\mathbf{r}_i,t)\Bigg]\varphi_i^k(\mathbf{r}_i,t), \end{aligned} \quad (6)$$

where the effective electron-electron potential $V_{e-e}^{\text{eff}}[\mathbf{r}_i - \mathbf{r}_j^k(t)]$ is expressed as a Monte Carlo sum over the smoothed walker distribution [9]:

$$V_{e-e}^{\text{eff}}\left[\mathbf{r}_i - \mathbf{r}_j^k(t)\right] = \frac{1}{Z_j^k}\sum_{l=1}^{M} V_{e-e}^{\text{scr}}\left[\mathbf{r}_i - \mathbf{r}_j^l(t)\right] K\left(\frac{\left|\mathbf{r}_j^l(t) - \mathbf{r}_j^k(t)\right|}{\sigma_j^k(\mathbf{r}_j,t)}\right), \quad (7)$$

where

$$Z_j^k = \sum_{l=1}^{M} K\left(\frac{\left|\mathbf{r}_j^l(t) - \mathbf{r}_j^k(t)\right|}{\sigma_j^k(\mathbf{r}_j,t)}\right), \quad (8)$$

where K is a smoothing kernel and Z_j^k is the weighting factor. The width $\sigma_j^k(\mathbf{r}_j^k,t)$ of the kernel in (7) is a measure for the characteristic length of nonlocal quantum correlations within the ensemble of walkers which represent the jth electron. In practice, the parameter $\sigma_j^k(\mathbf{r}_j^k,t)$ is determined by variationally minimizing the ground state energy of the quantum system [13].

In our calculation a Coulomb potential screened by an error function is used [10]:

$$V_{e-e}^{\text{scr}}\left[\mathbf{r}_i - \mathbf{r}_j^l(t)\right] = V_{e-e}\left[\mathbf{r}_i - \mathbf{r}_j^l(t)\right]\operatorname{erf}\left[\frac{\left|\mathbf{r}_i - \mathbf{r}_j^l(t)\right|}{r_j^s\delta_{si,sj}}\right], \quad (9)$$

where the Kronecker symbol $\delta_{si,sj}$ restricts the screening effect to the repulsion between only the same-spin walkers, while the value of screening parameter r_i^s is estimated from the Hartree-Fock approximation.

In the approach outlined previously, a self-interaction-free dynamics in physical space is achieved, where the separate walkers do not share guiding waves which represent different distributions. In order to calculate the many-body probability distribution in configuration space, a separate auxiliary set of walkers with primed coordinates $\mathbf{r}_i'^k$ is introduced which is guided by an antisymmetric wave function:

$$\mathbf{v}'\left(\mathbf{r}_i'^k\right) = \frac{\hbar}{m}\operatorname{Im}\left[\frac{1}{\Psi'^k\left(\mathbf{r}_1',\ldots,\mathbf{r}_N',t\right)}\nabla_i\Psi'^k\left(\mathbf{r}_1',\ldots,\mathbf{r}_N',t\right)\right]_{\mathbf{r}_j'=\mathbf{r}_j'^k(t)}, \quad (10)$$

where $\Psi'^k(\mathbf{r}_1',\ldots,\mathbf{r}_N',t)$ is an antisymmetrized product (Slater determinant or a sum of Slater determinants) of the time-dependent guide waves $\varphi_i^k(\mathbf{r}_i,t)$ of (6):

$$\Psi'^k\left(\mathbf{r}_1',\mathbf{r}_2',\ldots,\mathbf{r}_N',t\right) = A\prod_{i=1}^{N}\varphi_i^k\left(\mathbf{r}_i',t\right). \quad (11)$$

From (10) and (11) one can see that each walker with primed coordinates samples the many-body wave function, and thus it belongs to all guide waves (i.e., it represents an indistinguishable electron). The distribution of these walkers can be used to directly estimate the average in (1) by reducing it to (for states with spherical symmetry)

$$I(u,t) \propto \sum_i K_i\left[\frac{\left|r_{12}'^i(t)-u\right|}{\sigma_{12}^i}\right], \quad (12)$$

where $r_{12}'^i(t) = |\mathbf{r}_1'^i(t) - \mathbf{r}_2'^i(t)|$. In other words, the pair-density function can be simplified to a smoothed histogram (or a kernel density estimation with kernel K_l and bandwidth σ_{12}^i [14]) over the ensemble of the distances between the primed walkers.

3. Exchange and Coulomb Correlations in Helium

The two major sources of electron-electron correlation are due to the symmetry of the quantum state and due to the Coulomb repulsion. Here we consider first the effect of the exchange correlation on the pair-density function of helium atom. Although the electron-pair densities for helium have been analyzed by different techniques, they have never, to the author's knowledge, been studied using time-dependent methods.

In order to examine the electron correlation which is due to the exchange interaction, we consider the spin-triplet ground state of helium (orthohelium). The preparation of the ground state is described elsewhere [11, 12]. In the calculation here we use up to 100 000 Monte Carlo walkers and the same number of guiding waves, which are propagated over 2000 complex time steps (see (5) through (10)) in the presence of random component in walker's motion such that each walker samples the distribution given by its own guiding wave. In order to determine the screening parameter r_i^s of (9), we invoke the Hartree-Fock approximation, where for $\sigma_j^k(\mathbf{r}_j^k,t) \rightarrow \infty$ the Coulomb potential in (7) reduces to a simple (unweighted) sum of the Coulomb potentials due to all walkers. Because of the spherical symmetry of the 2 ^{3}S state r_i^s is being varied until minimizing the mean integrated squared error of the walker's distribution against the probability distribution obtained from an independent Hartree-Fock solution (e.g., in [15]). Figure 1 shows the probability distributions obtained from TDQMC for the optimizing value of $r_i^s = r^s = 1.13$ a.u. in (9). The blue and the green lines show the densities of the walkers guided in physical space (see (5) through (9)), respectively, while the red line represents the radial distribution of the walkers guided in configuration space (see (10)). In these calculations a new accurate algorithm for kernel density estimation was used [16]. Notice that all probability distributions throughout this paper are normalized to unity.

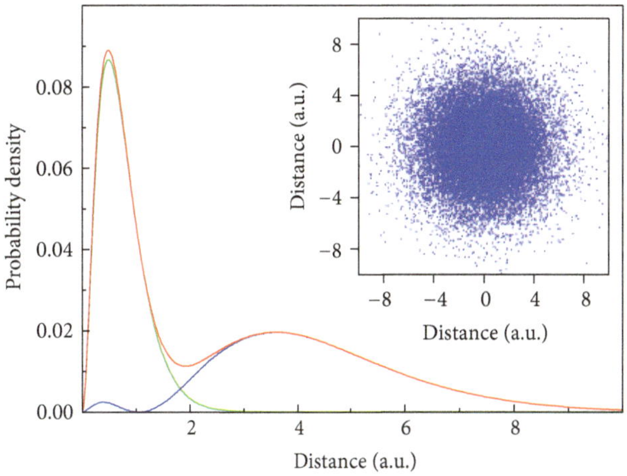

Figure 1: Radial electron density for the ground state of orthohelium, for MC walkers guided in physical space (blue and green lines), and for MC walkers guided in configuration space (red line). The inset shows the projection of the coordinates of the MC walkers in the x-y plane.

The electron-pair density for the ground state was calculated very efficiently by simply performing kernel density estimation over the ensemble of distances between the primed walkers. The result is shown in Figure 2(a) where the blue and the red lines present the cases with and without exchange interaction, respectively. The lack of exchange ($r_i^s \rightarrow 0$ in (9)) leads to a full (unscreened) Coulomb repulsion, which in the limit of infinite nonlocal correlation length ($\sigma_j^k(\mathbf{r}_j^k,t) \rightarrow \infty$) becomes equivalent to the Hartree approximation. Figure 2(b) shows the difference between the two curves in Figure 2(a), which in fact depicts the shape of the exchange hole for the 2 ^{3}S state of helium (see also e.g., [5]). Note that the exchange hole in our calculation may differ from other results because the distribution of the Monte Carlo walkers varies in radial direction as $r^2R^2(r)$

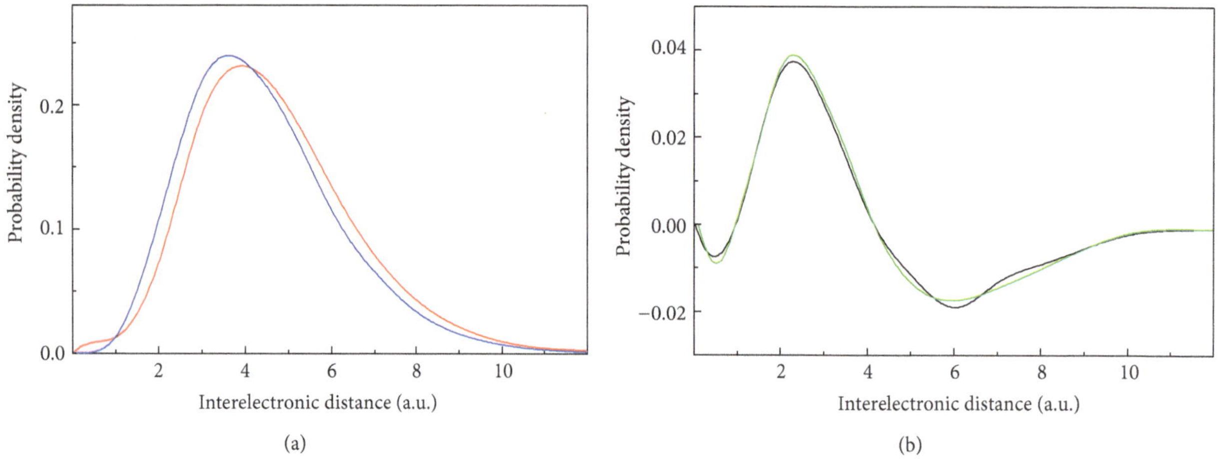

FIGURE 2: Electron-pair density as function of the interelectronic distance, for the ground state of orthohelium. (a) Red line—no screening (no exchange), blue line—short-range screened Coulomb potentials. Exchange hole (b) for screened Coulomb potentials (black) and for Hartree-Fock exchange (green).

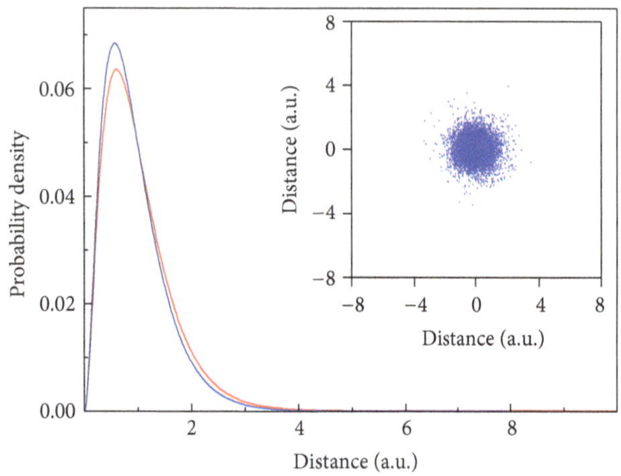

FIGURE 3: Radial electron density for the ground state of parahelium, for MC walkers guided in physical space (red line), and from the Hartree-Fock approximation (blue line). The inset shows the projection of the coordinates of the MC walkers in the x-y plane.

instead of as $R^2(r)$, where $R(r)$ is the radial wave function. The green line in Figure 2(b) shows the exchange hole obtained from an independent Hartree-Fock calculation with no potential screening. It is seen that the two curves are close where the deviations for larger interelectronic distances are mainly due to the fast decrease of the walker's density away from the core. As the screening parameter r_i^s tends to zero both the height and the width of the exchange hole decrease until the two curves in Figure 2(b) become very close, with the only remaining difference being a result of purely Coulomb correlations.

For the ground state of the 2 ^{1}S (para)helium, the quantity of interest is the Coulomb hole which occurs due to the repulsion of the closely spaced walkers. Figure 3 shows the probability distribution of the ground state walkers as compared to the Hartree-Fock calculation, while Figure 4(a) depicts the corresponding interelectronic distances for the two cases. The Coulomb hole calculated as the difference between the two curves is presented in Figure 4(b) which is close to previous results by other methods [3]. As the nonlocal correlation length $\sigma_j^k(\mathbf{r}_j^k, t)$ tends to infinity, both the height and the width of the Coulomb hole decrease until the two curves in Figure 4(b) coincide. Thus, in our approach where the exchange and the Coulomb correlations are accounted for by solely modifying the potential of electron-electron interaction in physical space, the two parameters r_i^s and $\sigma_j^k(\mathbf{r}_j^k, t)$ may ensure a smooth transition between the Hartree, the Hartree-Fock, and the fully correlated approximations to the electron-electron interaction. It is important to point out that in the $\hbar/m \rightarrow 0$ limit the quantum drift in (6) vanishes and so does the width of the quantum wave packet. Therefore, for an isolated atom the quantum correlation length $\sigma_j^k(\mathbf{r}_j^k, t)$ tends to zero in this limit, and if there are no exchange effects ($r_i^s \rightarrow 0$), the ensemble of quantum particles governed by (5) and (6) transforms to an ensemble of classical particles with the only force being due to the standard Coulomb repulsion between these particles.

4. Conclusions

In this paper, it has been shown that for charged particles, the quantum correlation effects which occur due to the exchange and Coulomb correlations can adequately be described by sets of de Broglie-Bohm walkers within the time-dependent quantum Monte Carlo framework. A short-range screening of the Coulomb potential ensures that each replica of a given electron interacts with only those replicas of the rest of the same spin electrons which are sufficiently apart to respect Pauli's exclusion principle in space. On the other hand, the electron-electron interaction is modified by the quantum nonlocality which demands that each replica of a given electron interacts with the replicas of the other electrons which are within the range of the nonlocal quantum correlation length. This concept allows one to build a robust, self-consistent, and self-interaction-free approach to find

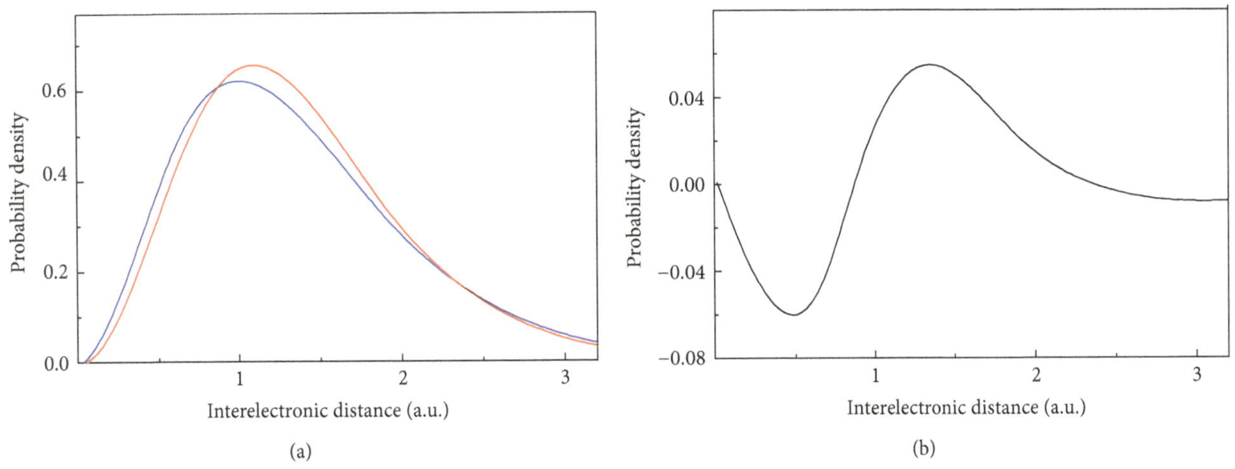

FIGURE 4: Electron-pair density as function of the interelectronic distance for the ground state of parahelium. (a) Red line—correlated result, blue line—Hartree-Fock approximation. The Coulomb hole (b).

both the ground state and the time evolution of quantum systems. It is demonstrated here that the otherwise awkward procedure for calculating the pair distribution functions of para- and orthohelium atom can be simplified to the level of finding the ground state probability distributions of the corresponding Monte Carlo walkers.

Besides the relative ease of its implementation, another advantage of using TDQMC is the affordable time scaling it offers which is almost linear with the system dimensionality. This is especially valid when using multicore parallel computers where little communication overhead between the different processes can be achieved, thus utilizing the inherent parallelism of the Monte Carlo methods. This nears the TDQMC to other efficient procedures for treating many-body quantum dynamics such as the time-dependent density functional approximation which, however, suffers systematic self-interaction problems due to the semiempirical character of the exchange-correlation potentials.

Acknowledgments

The author gratefully acknowledges support from the National Science Fund of Bulgaria under Grant DCVP 02/1 (SuperCA++). Computational resources from the National Supercomputer Center (Sofia) are gratefully appreciated.

References

[1] A. J. Coleman, "Density matrices in the quantum theory of matter: energy, intracules and extracules," *International Journal of Quantum Chemistry*, vol. 1, supplement 1, pp. 457–464, 1967.

[2] A. J. Thakkar, "Extracules, intracules, correlation holes, potentials, coefficients and all that," in *Density Matrices and Density Functionals*, R. Erdahl and V. H. Smith Jr, Eds., pp. 553–581, Reidel, New York, NY, USA, 1987.

[3] C. A. Coulson and A. H. Neilson, "Electron correlation in the ground state of helium," *Proceedings of the Physical Society*, vol. 78, no. 5, p. 831, 1961.

[4] R. J. Boyd and C. A. Coulson, "The Fermi hole in atoms," *Journal of Physics B*, vol. 7, no. 14, pp. 1805–1816, 1974.

[5] N. Moiseyev, J. Katriel, and R. J. Boyd, "On the Fermi hole in atoms," *Journal of Physics B*, vol. 8, no. 8, pp. L130–L133, 1975.

[6] P. M. W. Gill, D. O'Neill, and N. A. Besley, "Two-electron distribution functions and intracules," *Theoretical Chemistry Accounts*, vol. 109, no. 5, pp. 241–250, 2003.

[7] B. M. Austin, D. Y. Zubarev, and W. A. Lester, "Quantum monte carlo and related approaches," *Chemical Reviews*, vol. 112, no. 1, pp. 263–288, 2012.

[8] I. P. Christov, "Correlated non-perturbative electron dynamics with quantum trajectories," *Optics Express*, vol. 14, no. 15, pp. 6906–6911, 2006.

[9] I. P. Christov, "Dynamic correlations with time-dependent quantum Monte Carlo," *Journal of Chemical Physics*, vol. 128, no. 24, Article ID 244106, 2008.

[10] I. P. Christov, "Polynomial-time-scaling quantum dynamics with time-dependent quantum Monte Carlo," *The Journal of Physical Chemistry A*, vol. 113, pp. 6016–6021, 2009.

[11] I. P. Christov, "Correlated electron dynamics with time-dependent quantum Monte Carlo: three-dimensional helium," *Journal of Chemical Physics*, vol. 135, no. 4, Article ID 044120, 2011.

[12] I. P. Christov, "Erratum "Correlated electron dynamics with time-dependent quantum Monte Carlo: three-dimensional helium"," *Journal of Chemical Physics*, vol. 135, no. 14, Article ID 149902, 2011.

[13] I. P. Christov, "Exploring quantum non-locality with de Broglie-Bohm trajectories," *Journal of Chemical Physics*, vol. 136, no. 3, Article ID 034116, 2012.

[14] B. W. Silverman, *Density Estimation for Statistics and Data Analysis, Monographs on Statistics and Applied Probability*, Chapman and Hall, London, UK, 1986.

[15] S. E. Koonin and D. C. Meredith, *Computational Physics*, Addison-Wesley, 1990.

[16] Z. I. Botev, J. F. Grotowski, and D. P. Kroese, "Kernel density estimation via diffusion," *The Annals of Statistics*, vol. 38, no. 5, pp. 2916–2957, 2010.

14

Multispark Discharge in Water as a Method of Environmental Sustainability Problems Solution

E. M. Barkhudarov,[1] I. A. Kossyi,[1] Yu. N. Kozlov,[2] S. M. Temchin,[1] M. I. Taktakishvili,[1] and Nick Christofi[3]

[1] *A.M. Prokhorov General Physics Institute of RAS (GPI RAS), Vavilov Street 38, Moscow 119991, Russia*
[2] *Semenov Institute of Chemical Physics of RAS, Kosygin Street 4, Moscow 119991, Russia*
[3] *Edinburgh University, Edinburgh EH9 3JF, UK*

Correspondence should be addressed to I. A. Kossyi; kossyi@fpl.gpi.ru

Academic Editor: Elena Tatarova

Multispark discharge excited in water is described, and its useful physical and chemical properties are discussed in the light of some environmental issues. Discharge of such a type generates hot and dense plasmoids producing intense biologically active UV radiation and chemically active radicals, atoms, and molecules. Simultaneously, discharge creates strong hydrodynamic perturbations and cavitation bubbles. Particular attention is given to factors influencing on water purity with special reference to discharge application for effective sterilization of water and its cleaning of harmful chemicals. The gas discharges of this type show considerable promise as a means for solving some actual plasma-chemical problems. The above-mentioned discharge properties have been demonstrated in a series of laboratory experiments, which proved the efficiency of disinfection of potable and waste water, water cleaning of pesticide (herbicide) contaminations, and conversion (recovery) of natural methane.

1. Introduction

High voltage electric discharge in water [1, 2] has been considered as a potential method of water treatment to kill microorganisms and to clean it of harmful contaminations negating the use of chemicals that leads to by-products which may additionally compromise human health [3–5]. Factors favoring their use include the generation of UV radiation, acoustic, shock waves, chemically active substances, cavitation processes, pyrolysis, and hydrolysis. There are also possible synergetic effects following physical and chemical reactions.

Among the different means of in-liquid electric discharge, a novel method involves multielectrode (multispark) slipping (gliding) discharges (SSDs) [6] which may have some advantages over the two-electrode systems generally used at present [1, 7].

The present work describes the construction of a multispark discharger and discusses results of experimental investigation of SSD-based methods of water disinfection and their application in plasma-chemical technology for solving some of environmental problems, such as conversion (recovery) of methane (as well as other natural hydrocarbons), and water cleaning of pesticide (herbicide) contamination.

2. Treatment System

The apparatus used to treat liquids is shown schematically in Figures 1 and 2. The basic components were a chamber filled with water, a multielectrode system for exciting of slipping surface discharge, and high voltage power supply (Figure 1). The multielectrode discharge system (Figure 2) was similar in design to that previously described in [6, 8, 9]. The discharger consisted of a set of annular electrodes mounted on a dielectric tube surrounding a back-current conductor. A gas (air, argon, oxygen, etc.) was injected through a set of holes into water between the electrodes, producing fine gas bubbles. Discharge in each interelectrode gap was produced throughout the system, including the metal electrodes, a dielectric substrate, a gas bubble, and water.

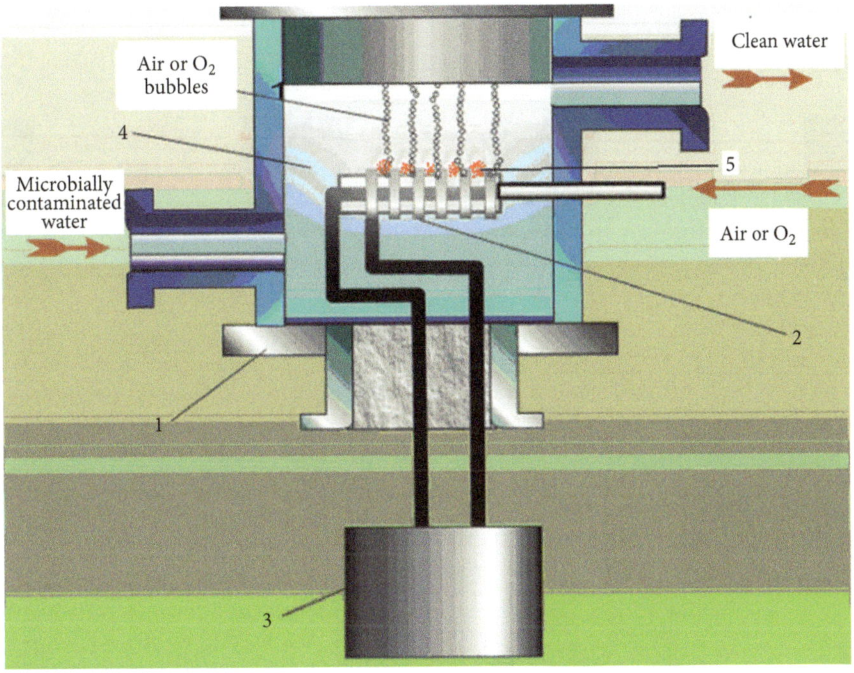

FIGURE 1: Scheme of multispark discharge disinfection of water. (1) Chamber; (2) multispark discharger; (3) generator of high voltage pulses; (4) cleaning water; (5) plasma of gliding discharge.

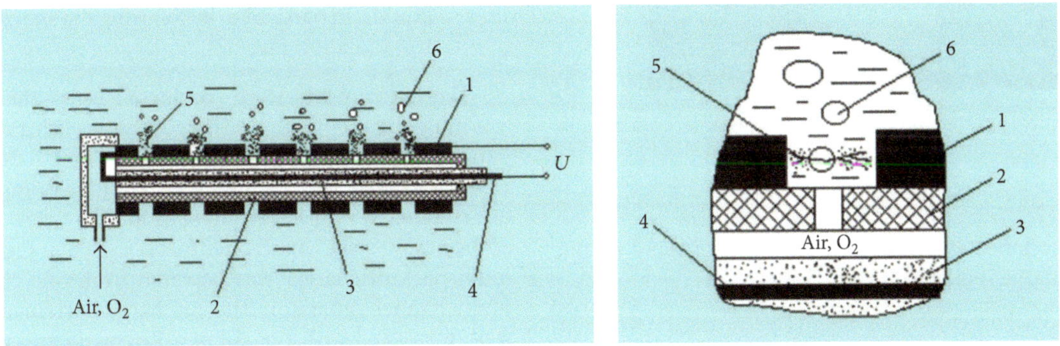

(a) (b)

FIGURE 2: Multielectrode gliding surface discharge facility. (1) Electrodes; (2, 3) dielectric tube; (4) back-current rod; (5) discharge plasma; (6) gas bubbles.

The initial plasma channel may be thought as originating in ordinary gas discharge in a gas bubble if the electric field therein is higher than the gas breakdown threshold [10, 11]. But in actual fact, a large (sometimes dominant) part in the interelectrode plasma formation could be played by a gliding discharge along the dielectric surface with the subsequent interaction of discharge plasma with electrodes and explosive microplasma production on their surface [12] (see Figure 3). There are just these processes that have been considered to be operative in the case when multispark discharger works in the gas medium [13].

When the high voltage pulse is applied to the immersed in the aqueous medium discharger (shown in Figure 2), plasma bunches (plasmoids) appear almost simultaneously between electrodes. Reasoning from their characteristics these plasmoids can be classified (in accordance with the recently adopted terminology) as "microplasma" formations, involved in various applications [14]. According to the results of previously performed experiments, the electron density in plasmoids attains 10^{17} cm^{-3}, and the gas temperature 4000–5000 K [15]. According to [16], explosive metallic plasma is a source of intensive hard UV radiation.

A typical photograph of the operating system is shown in Figure 4.

The principal advantage of the multispark system lies in the following peculiarities of their construction.

(i) The area of the surface of all electrodes contacting water in the multielectrode version can be minimized by introducing insulating dielectric screens ensuring the SSD operation in high-conducting water (up to conductivities of 10^4 μS cm^{-1}) without substantial

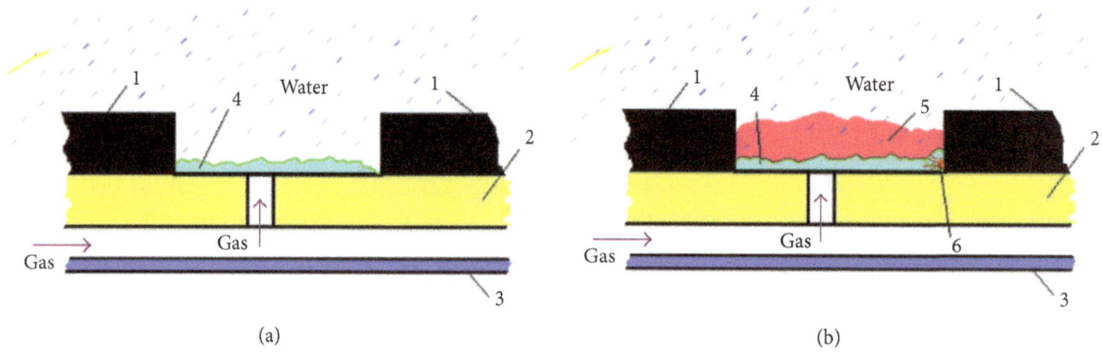

FIGURE 3: Two consecutive phases of plasma production in each interelectrode gap. (1) Electrodes; (2) dielectric tube; (3) back-current rod; (4) gliding surface discharge; (5) metallic plasma; (6) unipolar arc.

FIGURE 4: Typical photograph of multispark discharger operating in water.

reduction of the efficiency of energy supply to the discharge region.

(ii) The discharger has no pointed electrodes; the working surface of the electrodes (unprotected by dielectric screens) is developed and is either a part of cylindrical surface of tubular electrodes or the plane surface at the exit sections of the tube. Thus, the principal advantage of SSD system lies in the decrease in the discharge load of each electrode (thereby enhancing the erosion resistance on the system as a whole), which ultimately substantially increases the lifetime of the system.

(iii) The dischargers can affect the aqueous (liquid) medium through several simultaneously acting mechanisms, among them the direct influence of discharge plasma, the action of UV radiation generated by microscopic discharges, the chemical action of chemically active radicals, atoms, and molecules produced in discharges, and the hydrodynamic action through microscopic cavitation bubbles.

(iv) Cleansing action and bactericidal effect of a multispark discharge in the water medium unessentially depend on electrode material. Nevertheless among the tested metals (Fe, Mo, Cu, Ti, etc.), just stainless steel and titanium have been selected as materials exhibiting the most promise for working as a detail of multispark discharger. Just these two metals have been used in electrodischarge systems applied in the General Physics Institute (GPI RAS) for solution of water purification problems.

(v) The discharge gaps could be distributed in such a way as to increase the efficiency of the discharge action on liquids, in particular, by focusing the shock waves and UV radiation flux [17].

The experiments were conducted using the high voltage multichannel (5 channels) generator with the following parameters: high voltage amplitude, $U \leq 20\,\text{kV}$; pulse repetition frequency, $f \leq 100\,\text{Hz}$; capacitive storage energy of one channel, $W \leq 2\,\text{J}$, and pulse duration, $\tau \approx 5\,\mu\text{s}$. The circuit of the output stage of each channel is shown in Figure 5. Each multispark discharger was powered from one channel of a multichannel generator. The discharge current and voltage were measured with the aid of a Rogowski coil and voltage divider. The signals shown on the Figure 6 were recorded with an oscilloscope (TDS 3012). These measurements allowed the determination of the energy density (J cm^{-3}) released in liquid.

3. Multispark Electric Discharge in Water as a Source of UV Radiation, Ozone, and Hydrogen Peroxide

Figure 7 shows a schematic of the experiment intended to investigate a multispark SSD in water as a source of UV radiation, ozone, and hydrogen peroxide. Multielectrode discharger (2) is positioned in a cell (1) with water. A high voltage pulse produces a plasma channel between the electrodes. The gas leaving the reactor (as a working gas air or oxygen has been applied) flows into a quartz cell (3) intended for determining the ozone content by the method of absorption spectroscopy. In the course of the experiments, the production of H_2O_2 was also measured. UV radiation was measured in the presence and absence of water in the reactor chamber.

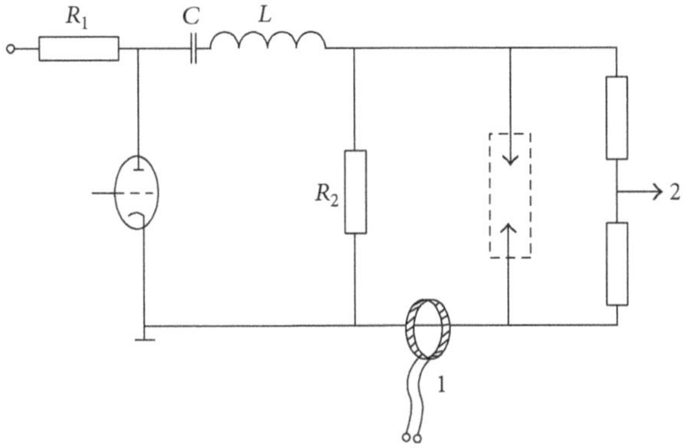

Figure 5: Output stage of one channel of the high voltage pulses power supply. (1) Rogowski coil; (2) voltage divider. R_1, R_2-resistors; C-capacitor; L-inductor.

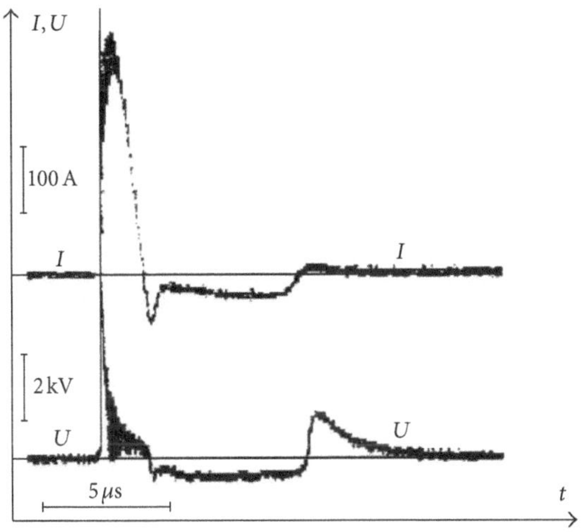

Figure 6: Typical oscillograph trace of SSD current and voltage.

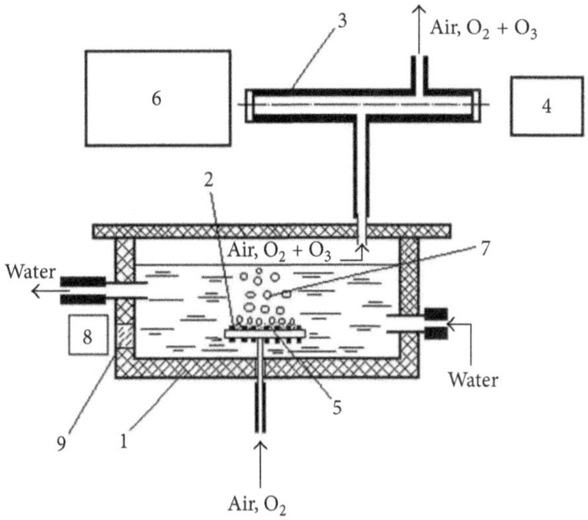

Figure 7: Experimental layout. (1) Vessel filled with water; (2) multispark discharger; (3) diagnostic quartz cell; (4) deuterium lamp; (5) discharge plasma; (6) MDR-3 monochromator; (7) gas bubbles; (8) MUM-1 monochromator; and (9) quartz window.

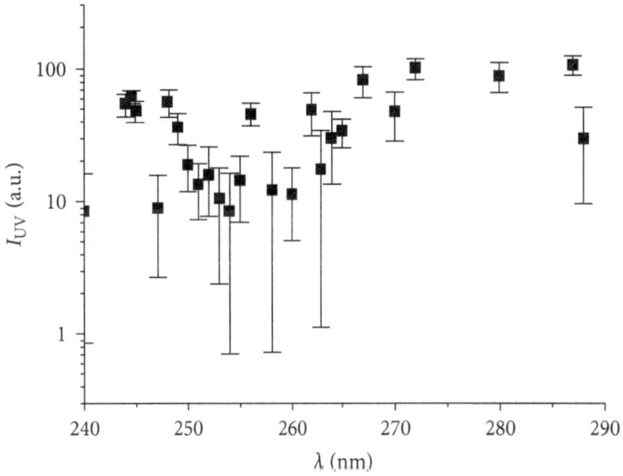

Figure 8: Spectrum of soft UV radiation from multispark discharge in the water.

The discharge emission spectrum in the region $230 < \lambda_d < 300$ nm was measured with the help of an MUM-1 monochromator ((8), Figure 7) and with an FEU-142 photomultiplier. Typical spectra of UV emission from the discharge are shown in Figure 8.

Chemical (actinometric) measurements have been used as well. In this case, the UV intensity was deduced from photolysis of an irradiated $K_3Fe(C_2O_4)_3$ solution with a phenanthroline admixture.

This technique was described in [18] and successfully used in [13] to study the multispark discharge in gaseous (Ar) medium.

To measure the O_3 content in the gas flowing from the reactor, we used both spectroscopic and chemical methods. The scheme of measurements of the O_3 content in O_2 is shown in Figure 7. From attenuation of the UV radiation passing through the cell, the O_3 density in the gas was determined by the absorption method. The spectral interval used to determine the ozone content corresponded to the Hartley absorption band with the maximum near $\lambda_d \cong 255.5$ nm.

In the case of application of air as working gas, the O_3 content was determined by the chemical method from the reaction between O_3 and potassium iodide in the water solution [19].

Figure 9 shows the ozone density in the diagnostic cell as a function of the repetition frequency of high voltage pulses (f) for a discharge in water (for various oxygen flow rates). Restriction of f values by amounts of the order of 100 Hz is not critical and appears explicable only on the basis of improper technical equipment of laboratory.

In the experiments when the oxygen flow rate through the interelectrode gaps and the water-filled reactor was $w \cong 15$ L/min, the ozone density in the oxygen flow was equal to $n_{O_3} \cong (1\text{-}2)\,10^{15}$ cm^{-3}.

The H_2O_2 content in water treated by the electric discharge was measured by the iodide-molybdate method

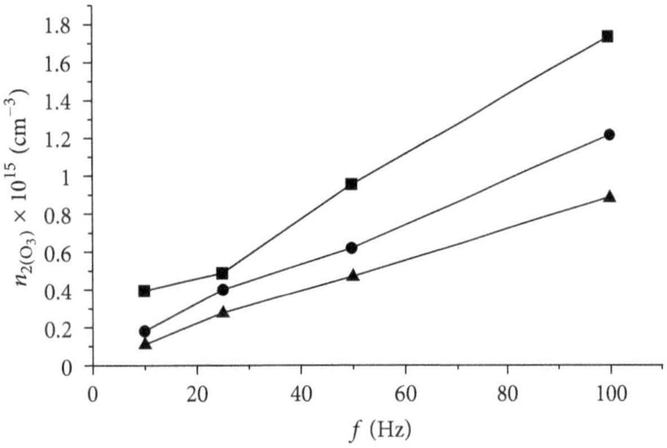

FIGURE 9: Ozone density in the diagnostic cell as a function of the repetition frequency of multispark discharge in the tap water for various flow rates of O_2: ■-w = 10 L min^{-1}; ●-15 L min^{-1}; ▲-20 L min^{-1}.

described in [20] and used in [13] to determine the intensity of hard UV radiation of the gliding surface discharge in argon.

The measurements of hydrogen peroxide production that were carried out in a discharge in water with injected argon showed that a series of discharges for 6-7 minutes in 250 cm^3 of water produced H_2O_2 with a mean density of $n_{H_2O_2} \cong 2 \cdot 10^{-3}$ mol L^{-1} $\cong 1.2 \cdot 10^{18}$ cm^{-3}. The energy cost of production of one H_2O_2 molecule in this case is $h_{H_2O_2} \leq 1.5 \cdot 10^2$ eV/mol.

The performed experiments demonstrated that for the SSD in the water-gas mixture, at least two factors are realized among the factors that are usually invoked to explain the sterilization effect of electric discharges. These are the generation of UV radiation and the production of biologically active ozone and hydrogen peroxide.

It is possible to estimate, using the results of measurements, the effectiveness of these two factors in the degradation of microorganisms during operation of the electric-discharge systems under study.

Examining the UV radiation from the discharge, we have to take into consideration that according to [21] the strongest bactericidal effect is produced by ultraviolet rays with wavelengths from 295 to 220 nm (the "bactericidal" spectral region).

Measurements performed in our work (see [9]) showed that the radiation spectrum of the multispark discharge in water contains the biologically active component, and the intensity of this component increases substantially as the pulse energy increases.

Based on the results of absolute measurements of UV radiation by the actinometric method, we estimate the intensity of the flux of bactericidal rays per pulse discharge as $P_{UV(i)} \approx 3 \cdot 10^6$ μW/cm^2 [9].

Given this intensity, in turn, the effectiveness of the action of radiation on *E. coli* bacteria can be estimated from the known relation [21]

$$n_b \cong n_{b0} \exp\left(\frac{-P_{UV} t_a}{k_b}\right), \quad (1)$$

where n_b is the number of bacteria in a unit volume that remain living after bactericidal irradiation (cm^{-3}), n_{b0} is the initial number of bacteria in a unit volume (cm^{-3}), P_{UV} is the mean intensity of the flux of bactericidal rays (μW cm^{-2}), t_a is the irradiation time (s), and k_b = 2500 is the bacterial tolerance factor.

For the case of repetitive discharge, expression (1) can be rewritten in the form

$$n_b \cong n_{b0} \exp\left(\frac{-P_{UV(i)} \tau f t_a}{k_b}\right), \quad (2)$$

where τ is pulse duration (s) and f is the repetition frequency of high voltage pulses (Hz).

It is easy to see that for $P_{UV(i)} \sim 3 \cdot 10^6$ μW cm^{-2}, $\tau = 5$ μs and f = 100 Hz, the exposure time equal to a few seconds is sufficient to decrease the number of bacteria in water by a factor of ten. This means that the energy cost of treating water by bactericidal UV rays is of the order of $\xi_{UV} \approx$ (1-2) 10^{-4} kW h L^{-1}.

Under the experimental arrangement shown in Figure 7, ozone generated in the discharge has no time to dissolve in water and is almost completely removed by the air (oxygen) flow into the space over the water reactor. In principle, it is possible to construct a reactor such that the produced ozone will be completely "entrapped" in the water being treated. Let us estimate how effective the role of ozone in the sterilization action of discharge may be in this case.

As follows from the data presented in [19], the effect of ozone dissolved in water on microorganisms becomes significantly stronger when the O_3 content reaches the threshold level $[n_{O_3}]_{th} \cong 8 \cdot 10^{16}$ cm^{-3}. Over $[n_{O_3}]_{th}$, the *E. coli* bacteria content decreases by more than four orders of magnitude.

It is easy to see that the bactericidal treatment capacity of ozone can be as high as

$$w_{O_3} \cong \frac{n_{O_3} w_b}{[n_{O_3}]_{th}} \cong 25 \text{ L h}^{-1}, \quad (3)$$

where w_{O_3} is the water-treatment rate (L h^{-1}) and w_b is the air flow rate through the discharge facility (L h^{-1}). Then, the energy cost of water treatment by ozone generated in the discharge (assuming that it is completely dissolved in water) can reach $\xi_{O_3} \cong 3 \cdot 10^{-4}$ kW h L^{-1} which is comparable with the energy cost of sterilization by UV radiation.

Finally, we estimate the effectiveness of a possible bactericidal action of the multispark discharge in water due to the production of hydrogen peroxide. Special microbiological studies carried out by us showed that an addition of hydrogen peroxide as a level of $n_{H_2O_2} \sim 10^{17}$ cm^{-3} to tap water allows the number of *E. coli* bacteria to be reduced by one order of magnitude. This means that the experimentally measured rate of H_2O_2 production ensures the energy cost of water sterilization at the level $\xi_{H_2O_2} \sim 10^{-4}$ kW h L^{-1}, which is close to the energy cost of sterilization by ozone production in discharge.

Hence, the performed direct measurements of UV radiation and chemically active products evidence that described below multispark slipping surface discharge (SSD) in water with air as an working gas is promising for water sterilization since two effects only, examined in our work, can ensure the energy cost as low as $\xi \cong 10^{-4}$ kW h L^{-1} for reducing the *E. coli* bacteria content by one order of magnitude (i.e., with a generator with a mean power of 1 kW, it is possible to reach a water treatment rate of the order of 10 m^3 h^{-1}).

It should be pointed out that possibility to apply for multispark discharger excitation of practically every gas or gaseous mixtures offers great opportunities for action on a microbiological component through the different chemically active atoms and radicals. However, in this work authors have restricted for water sterilization by the application only of air or oxygen taking into account that based on application of these gases discharger will be simplex and cheapest.

4. Multispark Electric Discharge Disinfection of Microbially Contaminated Liquids

As a step of our activity experimental investigation of effectiveness of disinfection action of multispark discharge on the water containing *Escherichia coli* and its viruses (coliphages) has been carried out [22].

The apparatus used to treat liquids is the same as shown schematically in Figure 1. The discharge device—multispark discharger—was situated in the treatment chamber through which water contaminated with microorganisms was pumped. Water contaminated with *E. coli* or viruses (somatic coliphages) can be used to test the killing efficiencies of the discharge system. Samples of water for microbiological analyses were taken via a sampling port; triplicate samples in 10 mL sterile bottles being removed for analysis.

Escherichia coli (NCIMB 86; ATCC 4157) was grown overnight in nutrient broth (oxoid) at 37°C. The cultures were diluted to population densities of approximately 10^6 cfu mL^{-1} with tap water and placed in treatment chamber containing the multispark discharger.

Water samples treated by the electric discharges were removed from the system at varying time intervals and bacterial killing assessed using spread plate counting methodology. *Escherichia coli* was determined by spreading 100 μL aliquots of diluted samples onto nutrient agar plates. Occasionally, MacConcey agar (HMSO 1994) and a spiral platter were utilized. Replicate plates were incubated at 37°C for 24 h. Coliphages were estimated by a plaque assay utilizing *E. coli C* (ATCC 13706) as the host bacterium. Dilutions of treated samples were spread onto lawns of *E. coli C*, sensitive to a broad spectrum of coliphages, and the number of plaques formed after 24 h incubation counted.

Figure 10 shows the effect of multispark discharges on microorganisms in the water. The fraction of surviving bacteria and viruses (N/N_0) is plotted versus the energy density (J cm^{-3}) released in water. Each point in the plot presents the mean of three measurements. Deviation from the mean did not exceed 15%. Numerous experiments were carried out using *E. coli,* and all showed a similar killing efficiency of the multispark discharge system. Data of microbial killing in liquids containing tap water-microbe combinations and a conductivity of 100 μS cm^{-1} are presented. It is evident from Figure 10 that the viruses were killed using a lower energy input to the liquid. *Escherichia coli* required an energy input

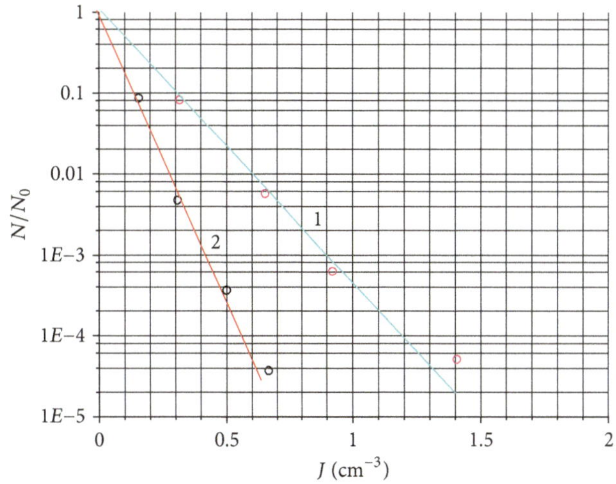

FIGURE 10: Changes in populations of *Escherichia coli* and viruses (N) in treated water relative to the initial populations (N_0) as a function of specific energy release (J cm^{-3}) during the treatment. Potable water with a conductivity σ = 100 μS cm^{-1} was used. f = 10 Hz. The initial (N_0) concentration of *E. coli* was $\approx 10^6$ colony-forming units mL^{-1} and that of coliphages $\approx 10^7$ plaque-forming units mL^{-1}. (1) *E. coli;* (2) coliphages.

of 0.3 J cm^{-3} (approx. 10^{-4} kW h L^{-1}) to reduce the population by a factor 10 (1 log reduction) while coliphages required an energy input of 0.15 J cm^{-3} for the same result.

The used multispark discharger regimes are identical with the regimes previously investigated [9] where an examination was made of the generation of biologically active UV radiation, ozone, hydrogen peroxide, and other active species (see preceding section of this paper). Measurements carried out during the present study allowed calculation of energy costs of the disinfection action using multispark electric dischargers, and these were as low as 10^{-4} kW h L^{-1} for bacteria. These values verified the bacterial action of discharges in the water predicted in the preceding section and confirmed that the main factors affecting microbial destruction in the water were UV radiation and the production of biologically active chemicals. The latter are not involved in treatment systems utilizing UV lamps which would be unable to generate highly reactive chemical species. Acoustic and shock waves generated by multispark discharge also played a part in microbial disinfection but, in addition, they facilitated the mixing of treated water, delivering reactive chemical species to all parts of the treatment system.

The possibility that disinfection using electric discharges might lead to the production of toxic by-products was tested by the input of energy as high as ~1 J cm^{-3} into water. Water samples were analyzed for a range of substances and physical appearance by the Certification Control-Analytical Center (Moscow State University, Russia). The water was tested for color, turbidity, pH, ammonium, Fe, Pb, Cr, fluorite, chlorite, nitrate, and sulphate. The quality of the treated water fulfilled the necessary standards of the European Union (Council Directives on the quality of water intended for human consumption 80/778/EEC and the new drinking water Directive 98/83/EC adopted by the Council on 3 November 1998). The results for Fe were particularly important as the electrodes used in the study were manufactured from stainless steel. Erosion of multispark discharger is small and does not affect overall concentrations in water. In addition, incubations of multispark discharge treated water with microorganisms were carried out to test whether the killing action continued. This could be due to the persistence of oxidizing species produced by the discharge but these were rapidly quenched within the system following treatment. There were no increased effects on *E. coli* added to system containing plasma treated compared with nonplasma-treated tap water. This is contrary to results obtained with two-electrode discharges [1] and could be explained by quite low level of operated multispark discharger electrodes sputtering and as a result extremely low level (in comparison with the two-electrode system) of content of metallic clusters responsible (according to [1]) for prolonged action of discharge on a microbial population. It is of interest to note that a multispark discharge treatment of short duration could sterilize tap water containing *E. coli* and coliphage. The duration was short enough for the cost-effective treatment of water supplies (<5 min); contact time being in the region of minutes rather than the 30 mins used in chlorination.

This study concentrated on verifying the predictions of microbial killing made originally in [9] and utilized *E. coli* and coliphage as representative organisms. No attempt has been made at this stage to examine the effect of multispark discharge plasma on the other bacteria (Gram-positive or -negative types), viruses, or spores (bacterial or fungal). Preliminary experiments have been performed to determine only the effect of multispark plasma on the oocysts of *Cryptosporidium* (a protozoan parasite causing gastrointestinal disorders), which are resistant to chlorination. The microscopic examination of cysts after treatment showed cell wall degradation and an inability to induce excystation in the organism.

It is of interest to investigate the possibility of using the multispark system described to treat industrial and domestic wastewater. The first attempt at such an application has been taken in [8, 23]. Water treatment was carried out using wastewater directly abstracted from final effluent stream at the Livingston Wastewater Treatment Plant in West Lothian, Scotland, UK. The scheme of system for wastewater treatment is shown in Figure 11. Results of SSD action on a final effluent stream are presented in Figure 12. It was shown that a specific energy of 1.25–1.5 J cm^{-3} was required to achieve 1 log reduction in bacterial (faecal coliforms/total aerobic heterotrophs) content. This study has demonstrated the effectiveness of the multispark dischargers in microbial disinfection of wastewater. The system can be engineered to eradicate microbial populations to levels governed by legislation by increasing treatment time or energy input.

5. Plasma-Chemical Converter of Methane on the Basis of Multielectrode Discharger

One from the currently important ecological problem consists in utilization of gases accompanying oil recovery. Yearly more than billion cubic meters of associated gases are burning down worldwide. Russian oil producing companies for compensation of an ecological harm are paying near 500 rubles for each 1000 m^3 of burning petroleum gas.

Presented work objective is the investigation of possibility of natural hydrocarbons (namely, CH_4) recovery in plasma-chemical reactor based on the SSD. Traditional for GPI research multispark dischargers have been used with only one key distinctive feature of their construction: as a discharge formative gas methane (or any other utilizable natural hydrocarbons) has been applied.

The diagram of the experiment is shown schematically in Figure 13. A multielectrode discharger is introduced into the reaction chamber in the form of an organic glass vessel filled with water (volume $V \sim 0.25$ L). When a high-voltage pulse is applied to the discharger, a system of plasma formations (plasmoids) in which the decomposition of hydrocarbons takes place is formed in bubbles of methane or methane-oxygen mixture in the gaps between the electrodes. The source of high voltage pulses was a generator producing single pulses or operating in the pulse-periodic regime. The pulse-repetition rate was $f \leq 50$ Hz, the pulse duration was $\tau_p \approx 1\,\mu$s, and the pulse amplitude was $U_p \approx 40$ kV.

We analyzed samples of the gas taken at the outlet of the reaction volume. Analysis of the gas passing through the discharger was carried out using the following techniques:

(i) special ITT IK/VP test tubes (OOO Impul's) used for determining the contents of acetylene (C_2H_2), carbon dioxide (CO_2), and carbon monoxide (CO);

(ii) SPECORD IR spectrograph used for determining the acetylene content;

(iii) gas chromatograph used for determining the concentration of methane (CH_4) and hydrogen (H_2).

Figure 14 shows the characteristic spectrograms obtained on the SPECORD IR spectrograph. The main absorption lines of CH_4, C_2H_2, and CO can be distinguished (in subsequent analysis of the experimental results, CO was disregarded). The lines of the nearest unsaturated hydrocarbon ethylene C_2H_4 are also very weak (at the noise level). In analysis of the efficiency of the plasma-chemical conversion of methane, it is expedient (see [24]) to use such parameters as the degree of conversion α expressed in fractions (in other words, the fraction of methane fed to the reactor and converted into a certain product at the output) and the energy value ε of conversion (i.e., the energy value of transformation of methane molecules in eV/molecule). If we disregard for simplicity the small amounts of ethylene formed as a result of methane treatment, we can assume that mainly two reactions occur in the plasma-chemical reactor: pyrolysis reaction,

$$CH_4 \longrightarrow C + 2H_2 \quad (4)$$

and the reaction of transformation of methane into acetylene,

$$2CH_4 \longrightarrow C_2H_2 + 3H_2 \quad (5)$$

It can be seen from simplified reaction formulas (4) and (5) that the volume of the reaction products exceeds the

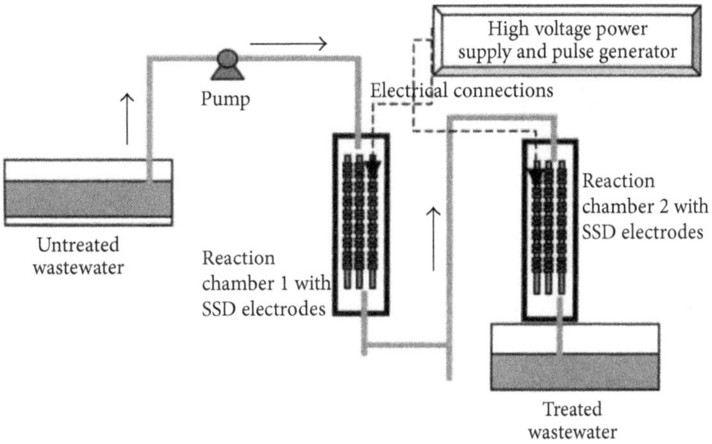

FIGURE 11: Diagrammatic representation of continuous wastewater treatment using system of multispark dischargers.

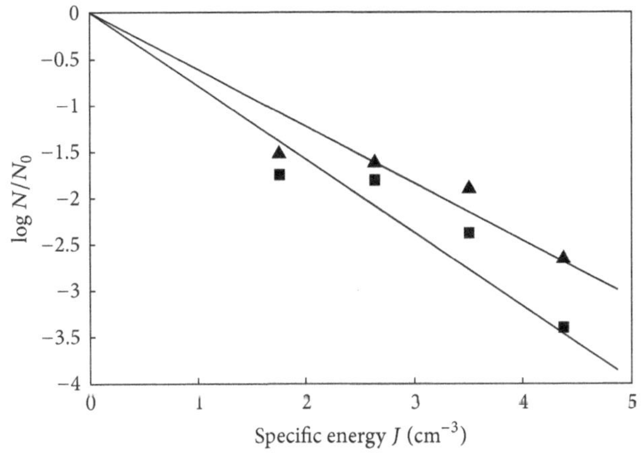

FIGURE 12: Log bacterial population (N/N_0) changes versus specific energy released in water during the multispark discharger operation. ▲-Total aerobic heterotrophic bacteria (22°C); ■-faecal coliforms (37°C).

volume of the primary mixture. For this reason, the measurements of concentration of methane and decomposition products at the reactor outlet cannot be directly used for estimating the degree of conversion.

It can easily be shown [24, 25] that the degree of conversion α_1 of methane into carbon and hydrogen according to reaction (4) and the degree of conversion α_2 of methane into acetylene according to reaction (5) are connected with experimentally determined concentrations C_{CH_4}, $C_{C_2H_2}$, and C_{H_2} by the relations

$$\alpha_1 = \frac{4C_{H_2} - 3(1 - C_{CH4})}{1 + C_{CH_4}},$$
$$\alpha_2 = \frac{4(1 - C_{H_2} - C_{CH_4})}{1 + C_{CH_4}}, \quad (6)$$
$$\alpha_0 = \frac{1 - C_{CH_4}}{1 + C_{CH_4}},$$

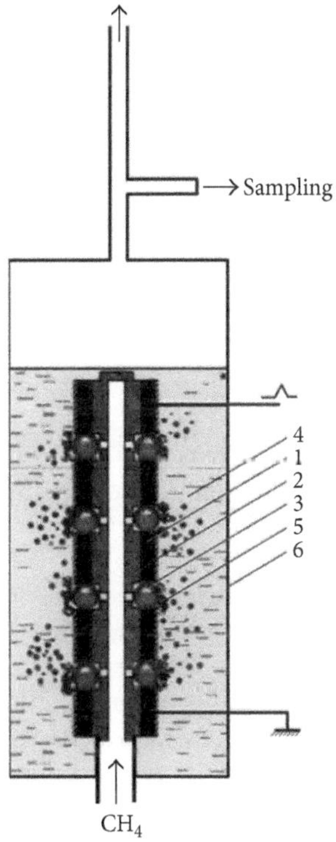

FIGURE 13: Schematic of the experiment. (1) Dielectric tube; (2) annular electrodes; (3) working gas (CH_4) bubbles; (4) water; (5) plasma in the interelectrode gaps; (6) reaction chamber.

where $\alpha_0 = \alpha_1 + \alpha_2$ is the total degree of conversion of methane over channels (4) and (5), which is determined in the given experiment.

The energy value of the reaction of decomposition of a methane molecule (in other words, the value of formation of products) is defined by the relation

$$\varepsilon_n = \frac{\overline{P}}{\alpha_n q_{CH_4}}, \quad (7)$$

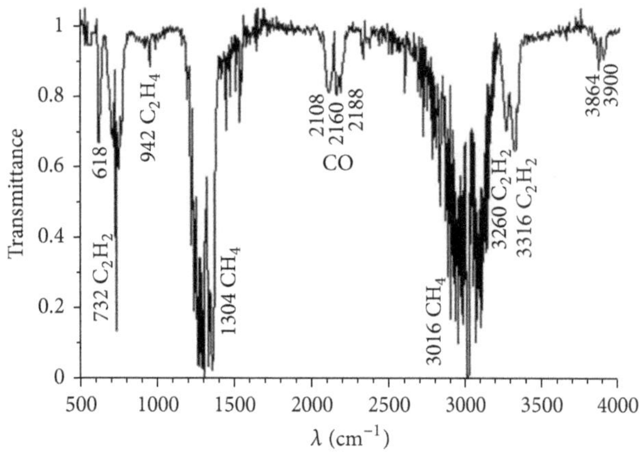

FIGURE 14: Characteristic adsorption IR spectrum of a working gas sample taken at the reactor outlet.

where $n = 0, 1, 2$ is the power supplied to the reactor, q_{CH_4} is the methane flow rate, and $\overline{P}$ is the average microwave power.

The dependences of flow rate q_{CH_4} of methane and of the energy value on its decomposition and the formation of products on the degree of conversion of methane are shown in Figures 15 and 16.

The dependence of the degree of conversion of methane on its flow rate shown in Figure 15 closely fits to the inverse proportionality function

$$\alpha_0 = \frac{A}{q_{CH_4}}. \quad (8)$$

Using iterations, we find that $A = 0.02809$ L/min. The fact that experimental points fit well to functional dependence (8) suggests that this dependence is preserved in a certain interval of q_{CH_4} beyond the range of the values studied experimentally. This in turn raises hopes that if we could implement a regime with the methane flow rate on the order of 0.1 L/min, the degree of conversion would increase to ~ 28%. The same results could be obtained by increasing the repetition rate of discharge pulses to 1 kHz for a methane flow rate of 1 L/min. By increasing the pulse repetition rate to 3 kHz for the same methane flow rate, we could reach a degree of conversion as high as 84%. The implementation of basically attainable degrees of conversion involves modernization of the generator of high voltage pulses and the design of the discharger, which will form the basis of subsequent experimental investigation. It is evident that without special justification these increased degrees of conversion are looking rather as a wishful thinking.

It can be seen from Figure 16 that the energy value of the conversion is almost independent of the methane flow rate and amounts to approximately 5 eV/molecule. Such energy value is close to record-low values for the atmospheric pressure (see, e.g., [25]).

The fact that the energy value of conversion is almost independent of the methane flow rate in the entire range of its variation in the experiment is an additional argument in favor of the possibility of a substantial increase in the degree of conversion due to passage to small values of q_{CH_4}.

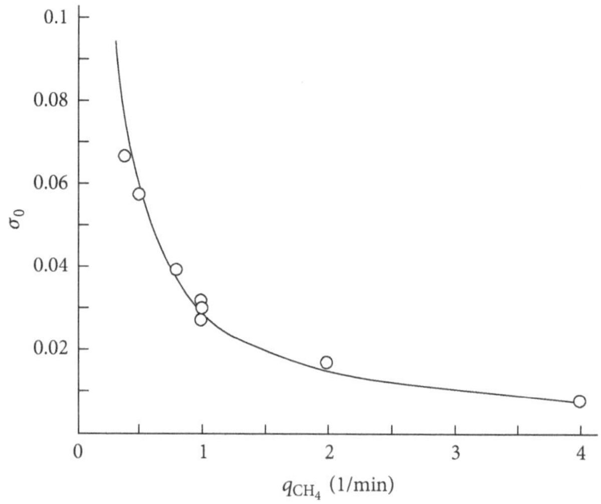

FIGURE 15: Dependence of the total degree of conversion of methane on its flow rate.

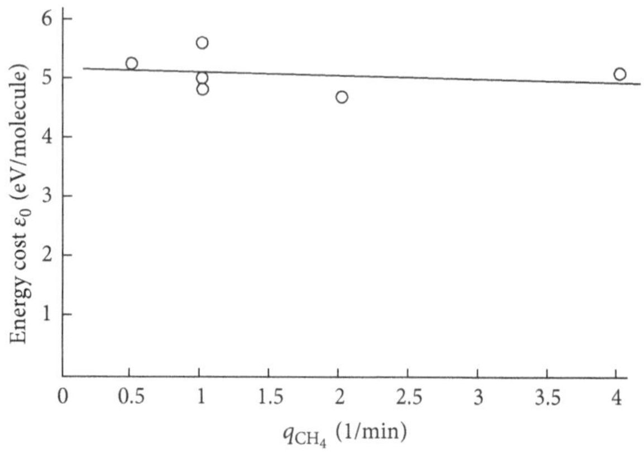

FIGURE 16: Dependence of the energy value of conversion of methane on its flow rate.

Analysis made in [26] shows that a high efficiency of methane conversion processes characteristic of the described technology is due to peculiarities contained just in discharges localized in an interelectrode gaps. Fast heating (up to 4000–5000 K) of the gas propagating between the electrodes through the area occupied by microplasma leads to the effective decomposition of hydrocarbon. At the same time, fast cooling of the gas penetrating into the surrounding water is followed by quenching phenomena, and the level of the parent-gas decomposition does not change.

The low energy value of methane decomposition and the possibility of elevating the degree of conversion justify the application of the method of plasma-chemical action for solving the topical problem of recovery of natural blowouts of hydrocarbons. In this connection, the role of pyrolysis in the methane decomposition is of interest in its own right. If the contribution from reaction (4) is significant, it is expedient to determine the form and efficiency of the production of carbon accompanying the decomposition of methane.

The experiments performed in accordance with the diagram in Figure 13 have shown that if the multispark discharge is initiated in water using CH_4 as the bubble-forming gas, most carbon particles appearing in water as a result of plasma-chemical decomposition of methane precipitate.

Analysis of the precipitate shows that its main part is nanosize carbon. Figure 17 shows the characteristic size distribution of carbon particles as a function of the time of electric-discharge treatment of water, which was determined using Fotokor dynamic scattering spectrometer. The typical photograph of nanocarbon produced in the course of methane recovery by multispark discharge in the water is shown on the Figure 18.

The rate of production of nanosized particles in the discharge, which was determined by evaporation of the SSD-processed liquid and weighing the precipitate, was about 35 mg/h. This means that the energy value of production of nanocarbon upon decomposition of methane in the SSD is 0.3 kW h/g. The measured value is close to that obtained for arc discharges with carbon electrodes in water, in which carbon is formed in the liquid as a result of destruction of the electrodes [27, 28].

The structure of the precipitate was determined using a LAB RAM HR 800 Raman spectrometer from the Raman shift. Fractions of disordered graphite and carbon were detected.

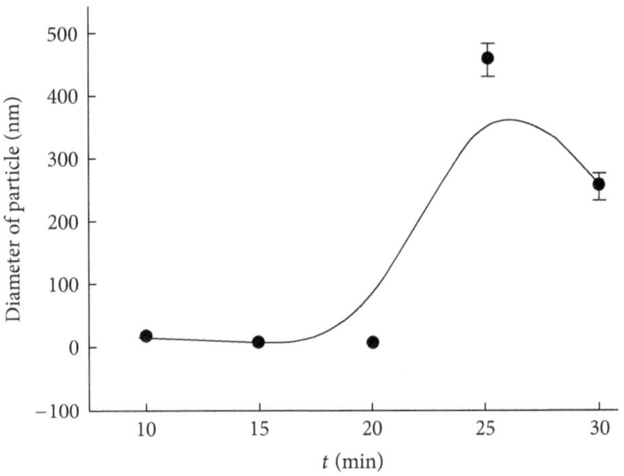

FIGURE 17: Dependence of the average size of carbon particles produced in the reactor on the time of electric discharge processing of methane.

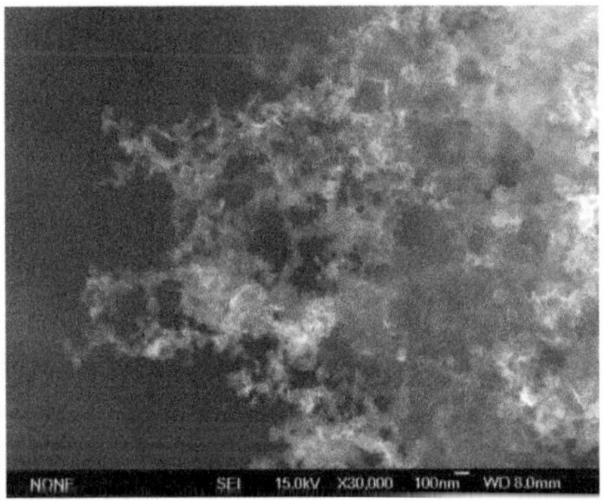

FIGURE 18: The typical photograph of nanocarbon produced in the course of methane recovery by multispark discharge in the water.

6. Water Cleaning of 2,4-Dichlorophenoxyacetic Additive

Polychlorinated biphenyls (PCBs), among man-made pollutions, deserve particular attention. These compounds were synthesized in 1920 s, and with their advent new materials with unique thermophysical and electrical insulating properties became available.

However, in spite of the presence of a number of unique properties, these compounds were withdrawn from industrial processes already in 1970 s. This is due to the fact that PCBs were implicated in a number of incidents in different countries by causing mass intoxication and exerting a detrimental effect on the health of humans on a large scale.

The PCBs are no longer manufactured but remain in the environment, so that the search for ways of their destruction is one of the urgent problems of the day. At the General Physics Institute of RAS, experiments were carried out to examine possibility of electric discharge (SSD) in water as an efficient and inexpensive method for cleaning the manufacturing water of PCBs. Instead of a toxic PCB in our experiments, we used a 2,4-D dichlorophenoxyacetic acid (2,4-D). This material was chosen for plasmachemical decomposition, because the configuration of the 2,4-D molecule somewhat resembles PCB. More exactly, the 2,4-D molecule, like the PCB congeners, contains a doubly chlorine-substituted benzene ring with attached acetic acid.

The experimental procedure was as follows. Two solid particles of 2,4-D (97%) of weight 40 mg were preliminarily dissolved in 10 mL of alcohol. The solution was poured into a polyethylene container with 5 L of distilled water. The acid concentration in the container was estimated at ~8 $\mu g\, cm^{-3}$, that is, about 300 times larger than the maximum allowable ("permissible") concentration.

Decomposition of the acidic additive was accomplished using a multispark discharger mounted in a plexiglass reactor chamber of volume $V = 15 \times 6 \times 4.5\, cm^3$. The multispark discharger, which was placed on inside of the reactor cover, produced a discharge in water. The working gas passed through the discharger was oxygen.

Analysis of the SSD-processed solutions was conducted in the Laboratory of Analytic Environmental Toxicology at the Severtsov Institute of Ecology and Evolution of the Russian Academy of Sciences.

GC/MS (Gas chromatography/Mass spectrometry) analysis of solutions was performed by using a Finnigan TRACE GC Ultra gas chromatograph coupled with a Finnigan Polaris Q mass spectrometer (ion trap). This GC/MS system possessing ultra-high sensitivity allows detection of 2,4-D compound and its possible organic products of fragmentation with sensitivity ~$10^{-9}\, g\, cm^{-3}$.

All our experiments were conducted at fixed values of the initial 2,4-D concentration $N = 8\, \mu g\, cm^{-3}$ and solution volume $V = 250\, cm^3$. In all experiments, a sample of solution was taken from the reactor before processing in order that the initial 2,4-D concentration will be accurately known.

Data of GC/MS measurements ensure complete decomposition of 2,4-D (at the level of sensitivity 10^{-9} g cm^{-3}) in all of the experiments when the processing time was longer than 150 s and the mean power of the high-voltage generator was ~20 W. These experiments give a conservative estimate of the efficiency of plasmachemical decomposition of the organic 2,4-D compound by the use of a multielectrode system excited electric discharge (SSD) in water. A characteristic dependence of the 2,4-D concentration on the duration of SSD processing is presented in Figure 19.

Almost complete (~100%) decomposition of 2,4-D a high-concentrated solution shows that the SSD processing will outperform the traditional reactors. From the experiments, it might be inferred that SSD working in the water containing about 300 maximum allowable concentration of 2,4-D provides almost complete decomposition of liquid solution with expenditure of energy as low as ~ $2 \cdot 10^{-3}$ kW h/L. Accordingly, with a power source ~1 kW it is possible to clean more than 0.5 m^3 of water per hour.

We do not have a clear notion of what mechanism is dominant in the technological process of water cleaning of the 2,4-D additive. Special experiments have yet to be performed to construct a physicochemical model for electric-discharge destruction of the acid (and its decomposition fragments). However, we have good reason to believe that a leading part in destruction is played by plasma-chemical reactions occurring in SSD with the resulting formation of chemically active radicals and molecules.

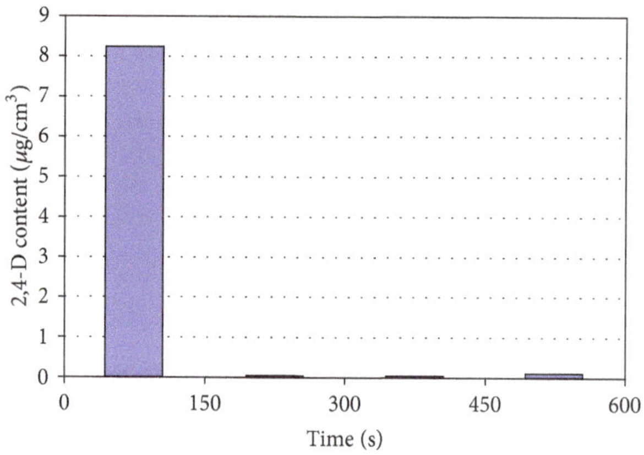

FIGURE 19: 2,4-D content as a function of time of water treatment by means of multispark discharge.

7. Conclusion

A new electric-discharge system, which has been developed and tested at the GPI RAS, has a multitude of potential uses. Examples can be found in the present paper. A plasma-chemical reactor of simple design using a multielectrode (multispark) discharger operating in aqueous medium may serve for efficient disinfection of microbially contaminated potable and waste water, conversion (recovery) of methane, destruction of acidic 2,4-D pollutant.

The SSD-based electrode system is capable of producing multiple microplasma formations in liquid medium at relatively low electrode voltages. Physical and chemical properties peculiar to this type of discharges have been studied experimentally. It is shown that these properties are controlled by the following four factors simultaneously acting upon the liquid (aqueous) medium:

(i) direct influence of electric-discharge plasma possessing a high electron density and relatively high temperatures of the gas and electron component,

(ii) exposure to intense UV radiation emitted by microplasma formations,

(iii) chemical action of chemically active radicals, atoms, and molecules produced in discharges and penetrated the water,

(iv) hydrodynamic action through cavitation bubbles.

For each concrete application, the electric-discharge system may be modified in design so as to increase one or the other of these factors.

The experiments demonstrated high efficiency of multispark discharge in water for solving diversified environmental problems listed above. Note that the dominant mechanism in sterilization of potable and waste water was the biologically active UV radiation and generation of chemically active molecules (ozone, hydrogen peroxide). The achievement of encouraging results in conversion of natural hydrocarbons is credited to the immediate action of microplasma formation on the gas being treated. The success in the accomplishment of water cleaning of 2,4-D is attributed to plasmochemical mechanism of generating chemically active substances.

In conclusion, the multispark discharge in water is being used more and more. Thus, the action of SSD on the organic pollutions has been investigated in [29]. Decomposition of dissolved pentachlorophenol and parachlorophenol under multispark discharge action has been measured. Efficiency of reforming these phenols was as good as 1-2 kJ/mg.

References

[1] V. L. Goryachev, F. G. Rutberg, and V. N. Fedyukovich, "Electric-discharge method of water treatment. Status of the problem and prospects," *Applied Energy*, vol. 36, pp. 35–49, 1998.

[2] L. A. Yutkin, *Electrohydroulic Effect and Industrial Application*, Mashinostroenie, Leningrad, Russia, 1986.

[3] J. Sketchell, H.-G. Peterson, and N. Christofi, "Disinfection by-product formation after biologically assisted GAC treatment of water supplies with different bromide and DOC content," *Water Research*, vol. 29, no. 12, pp. 2635–2642, 1995.

[4] F. X. R. Van Leeuwen, "Safe drinking water: the toxicologist's approach," *Food and Chemical Toxicology*, vol. 38, pp. 851–858, 2000.

[5] U. Von Gunten, A. Driedger, H. Gallard, and E. Salhi, "By-products formation during drinking water disinfection: a tool to assess disinfection efficiency?" *Water Research*, vol. 35, no. 8, pp. 2095–2099, 2001.

[6] PCT, Treatment of Liquid International Patent Application no PCT/GB99/00755, 1999.

[7] L. A. Kul'skii, O. S. Savchuk, and E. Yu. Deinega, *Influence of Electron Field on Process of Water Sterilization*, Nauk. Dumka, Kiev, Ukraine, 1980.

[8] E. M. Barkhudarov, I. A. Kossyi, M. I. Taktakishvili, N. Christofi, and V. Zadiraka Yu, "Multispark generation of plasma in liquids and its utilization in waste water treatment," in *Proceedings of the 13th International Conference on Gas Discharges and their Applications*, vol. 2, pp. 680–683, Strathclyde University, Glasgow, UK, 2000.

[9] A. M. Anpilov, E. M. Barkhudarov, Y. B. Bark et al., "Electric discharge in water as a source of UV radiation, ozone and hydrogen peroxide," *Journal of Physics D*, vol. 34, no. 6, pp. 993–999, 2001.

[10] S. M. Korobeinikov and E. V. Yashin, "Bubble model for breakdown in water at pulsed voltage. Electric discharge in liquid and its industrial application, part 1, Nikolaev, Russia," 1988.

[11] V. L. Goryachev, A. A. Ufimtsev, and A. M. Khodakovskii, "Mechanism of electrode erosion in pulsed discharges in water with a pulse energy of~1 J," *Technical Physics Letters*, vol. 23, no. 5, pp. 386–387, 1997.

[12] A. M. Anpilov, E. M. Barkhudarov, N. K. Berezhetskaya et al., "Source of a dense metal plasma," *Plasma Sources Science and Technology*, vol. 7, no. 2, pp. 141–148, 1998.

[13] Y. B. Bark, E. M. Barkhudarov, Y. N. Kozlov et al., "Slipping surface discharge as a source of hard UV radiation," *Journal of Physics D*, vol. 33, no. 7, pp. 859–863, 2000.

[14] K. H. Becker, K. H. Schoenbach, and J. G. Eden, "Microplasmas and applications," *Journal of Physics D*, vol. 39, no. 3, pp. R55–R70, 2006.

[15] A. M. Anpilov, N. K. Berezhetskaya, V. A. Kop'ev et al., "Explosive-emissive source of a carbon plasma," *Plasma Physics Reports*, vol. 23, no. 5, pp. 422–428, 1997.

[16] N. K. Berezhetskaya, V. A. Kop'ev, I. A. Kossyi, I. I. Kutuzov, and B. M. Tiit, "Explosive emission phenomena on a metal-hot plasma interface," *Zhurnal Tekhnicheskoi Fizikiv*, vol. 61, no. 2, pp. 179–184, 1991 (Russian).

[17] E. M. Barkhudarov, I. A. Kossyi, and M. I. Taktakishvili, "Distributed plasma generation in liquids," in *Proceedings of 13th International Conference on Gas Discharges and their Applications*, vol. 2, pp. 340–342, Strathclyde University, Glasgow, UK, 2000.

[18] C. G. Hatchard and C. A. Parker, "A new sensitive chemical actinometer. II. Potassium ferrioxalate as a standard chemical actinometer," *Proceedings of the Royal Society A*, vol. 235, no. 1203, pp. 518–536, 1956.

[19] V. V. Lunin, M. P. Popovich, and S. N. Tkachenko, *Physical Chemistry of Ozone*, Moscow State University Press, Moscow, Russia, 1998.

[20] J. H. Baxeudale, "The flash photolysis of water and aqueous solutions," *Radiation Research*, vol. 17, no. 3, pp. 312–326, 1962.

[21] B. N. Frog and A. P. Levchenko, *Preparation of Water*, Moscow State University Press, Moscow, Russia, 1996.

[22] A. M. Anpilov, E. M. Barkhudarov, N. Christofi et al., "Pulsed high voltage electric discharge disinfection of microbially contaminated liquids," *Letters in Applied Microbiology*, vol. 35, no. 1, pp. 90–94, 2002.

[23] A. M. Anpilov, E. M. Barkhudarov, N. Christofi et al., "The effectiveness of a multi-spark electric discharge system in the destruction of microorganisms in domestic and industrial wastewaters," *Journal of Water and Health*, vol. 2, no. 4, pp. 267–277, 2004.

[24] A. I. Babaritskii, S. A. Demkin, V. K. Zhivotov et al., *Plasmachemistry-91 (INKhS AN SSSRv)*, vol. 2, pp. 286–303, 1991.

[25] S. I. Gritsinin, P. A. Gushchin, A. M. Davydov, E. V. Ivanov, I. A. Kossyi, and M. A. Misakyan, "Conversion of methane in a coaxial microwave torch," *Plasma Physics Reports*, vol. 35, no. 11, pp. 933–940, 2009.

[26] A. M. Anpilov, E. M. Barkhudarov, N. K. Berezhetskaya et al., "Methane conversion in a multielectrode slipping surface discharge in the two-phase water-gas medium," *Technical Physics*, vol. 56, no. 11, pp. 1588–1592, 2011.

[27] N. Parkansky, O. Goldstein, B. Alterkop et al., "Features of micro and nano-particles produced by pulsed arc submerged in ethanol," *Powder Technology*, vol. 161, no. 3, pp. 215–219, 2006.

[28] N. Sano, "Low-cost synthesis of single-walled carbon nanohorns using the arc in water method with gas injection," *Journal of Physics D*, vol. 37, no. 8, p. L17, 2004.

[29] V. M. Shmelev, N. V. Evtyukhin, Y. N. Kozlov, and E. M. Barkhudarov, "Action of pulsed surface discharge on organic contaminants in water," *Khimicheskaya Fizika*, vol. 23, no. 9, pp. 77–85, 2004 (Russian).

The Advantages of Not Entangling Macroscopic Diamonds at Room Temperature

Mark E. Brezinski[1, 2, 3]

[1] *Center for Optical Coherence Tomography and Modern Physics, Department of Orthopedic Surgery, Brigham and Women's Hospital, 75 Francis Street, MRB-114, Boston, MA 02115, USA*
[2] *Center for Optical Coherence Tomography and Modern Physics, Department of Orthopedic Surgery, Harvard Medical School, 25 Shattuck Street, Boston, MA 02115, USA*
[3] *Department of Electrical Engineering and Computer Science, Massachusetts Institute of Technology, Room 36-360, 77 Massachusetts Avenue, Cambridge, MA 02139, USA*

Correspondence should be addressed to Mark E. Brezinski, mebrezin@mit.edu

Academic Editor: Alan Migdall

The recent paper entitled by K. C. Lee et al. (2011) establishes nonlocal macroscopic quantum correlations, which they term "entanglement", under ambient conditions. Photon(s)-phonon entanglements are established within each interferometer arm. However, our analysis demonstrates, the phonon fields between arms become correlated as a result of single-photon wavepacket path indistinguishability, not true nonlocal entanglement. We also note that a coherence expansion (as opposed to decoherence) resulted from local entanglement which was not recognized. It occurred from nearly identical Raman scattering in each arm (importantly not meeting the Born and Markovian approximations). The ability to establish nonlocal macroscopic quantum correlations through path indistinguishability rather than entanglement offers the opportunity to greatly expand quantum macroscopic theory and application, even though it was not true nonlocal entanglement.

1. Introduction

The ability to observe and control nonlocal macroscopic quantum coherence/correlations, under ambient conditions, would likely have a powerful influence across a wide range of fields. This was achieved recently by Lee et al., in *Science*, establishing phonon field quantum correlations in two spatially separated diamonds [1, 2]. The paper was entitled entitled "Entangling Macroscopic Diamonds at Room Temperature." Two other studies nonlocally correlating reflectors (by our group) and a cesium gas respectfully support the results [3, 4]. However, we will demonstrate on several grounds that while quantum correlations are established between the diamonds, they are not true entanglement.

The work in the Lee et al. paper is essentially a two-arm extension of the DLCZ (Duan, Lukin, Cirac, and Zoller) experiments [5–9]. Figure 1 is a schematic of the key components of the Lee experiment, but a more detailed schematic can be found in Figure 1 of the original paper. An ultrashort pulsed source is used whose outputs can be represented by a collection of single photon wavepackets (each wave packet can only interfere with itself), as they are neither entangled photons nor significant biphoton wavepackets. An MZI interferometer is used where diamonds are present in each arm which contain nearly identical Raman scatterers. The diamonds are 15 cm apart making any interaction between them macroscopic. The optical phonon modes of the diamond allow relatively low decoherence at room temperature because they have very high oscillatory frequencies (40 THz) so are not readily disturbed by thermal energies. A pump pulse is sent through the interferometer of sufficient intensity to entangle with and stimulate the Raman scatterers. A Stokes photon is then emitted, with the diamond and Stokes photon entangled until detection. The extra energy remaining in the diamond (lost from the photon) is in the form of increased phonon field energy levels. If the detector registers one Stokes photon, it could have come from either of the diamond crystals in which one phonon was excited. This will be discussed in more detail below, but because the paths are indistinguishable, the system behaves as if the photon at the beam splitter came from both arms. Paraphrasing Dirac, a single photon wavepacket can only interfere with itself. Therefore, prior to the pump photon being detected, both phonon fields are stimulated. To confirm these results, a probe photon is introduced into the interferometer that interacts with the diamonds producing the anti-Stokes photons. The probe photon must interact with the diamond prior to the Stokes photon being detected. The nature of the detection scheme for the anti-Stokes photon allows determination if one or both phonon fields are stimulated. If we were only looking at one arm, prior to the Stokes photon detection, there is an entanglement between the Stokes photon, phonon field, and anti-Stokes photon. This is somewhat analogous to the nonlocal entanglements in the well-known studies performed by Brune et al. described below which we will use to support our conclusions about the Lee paper [10–12]. The key point of the Lee et al. paper made below is that the two phonon fields are quantum correlated, but not truly entangled as stated in the original paper.

Our analysis is that Lee's explanation, in the *Science* paper, for the quantum correlations generated between diamonds (resulting from the pump photons) is unlikely representative of the actual situation. They postulated a nonlocal entanglement between the diamonds. While we agree that quantum correlations are established, we do not believe that the data or analysis of the experimental design supports true entanglement. The essential points will be made here but the remainder of the paper will expand on these points. First, our examination supports that these nonlocal quantum correlations occur from a combination of paths indistinguishability (for a single photon wavepacket) plus nearly identical local entanglements (Raman scatterers) in each path [13–19]. The source is coherent so building the pulses up from single photon wavepackets (a photon can only interfere with itself) is a useful approach for illustrating the physics. The correlations between diamond phonons do not fit definitions of entanglement laid out, for example, by von Neumann, EPR-B, or GHZ [20–23].

Second, the pump photon/diamond interactions do not (and must not) meet the Born (system-environment coupling weak) or Markovian (memory effects of the environment are negligible) approximations of decoherence theory [15, 17]. This occurs largely from the high frequency of the optical phonons and the strong coupling associated with the Raman scatterers. The results then of the pump photon/diamond interactions are more analogous to single photon wavepacket decoherence theory than nonlocal entanglement (point 1). Environmental interactions are occurring with indistinguishable paths, but in the case of the *Science* paper, coherence is expanded rather than lost (point 2) [15–17]. This demonstrates perhaps the most important point of the paper, that the diamonds can lead to either decoherence (distinct local entanglements and meeting Born-Markov approximations) or coherence expansion (nearly identical local entanglements and not meeting Born-Markov approximations) depending on the setup.

In the next several paragraphs, the topics addressed will be as follows. First, nonlocal correlations will be examined, which can be represented by entangled states or states generated by indistinguishable paths. Second, we review the general definition of entanglement demonstrating why the nonlocal phonon field correlations in the Lee study are not accurately described as being entangled. Third, we discuss that path indistinguishability and the quantum correlations that can be generated. This and the previous paragraphs draw heavily from the work by pioneers that include von Neumman, Mandel, and Shih, as well as insights from recent decoherence theory by Zurcek and Zeh [15–17]. Decoherence theory is particularly useful in illustrating the point of this paper as indistinguishable paths lead to coherence while typical environmental entanglements generally lead to decoherence (with this paper representing an exception). Finally, we will also discuss how the authors represented visibility, concurrence, density operators, and statistical significance (particularly the correlation coefficient), and how these are completely consistent with nonlocal correlations from either indistinguishable paths or entanglement. We do not believe there is a basis to employ a two-mode squeeze state as discussed by Julsgaard et al., for the nonlocal correlations [4]. The appendix will speculate on the role misunderstanding type II SPDC sources and Dirac notation play in the misuse of the term "entanglement."

2. Nonseparable States (Unfactorizable) and Quantum Correlations

In order to discuss quantum correlations, including entangled states and those from path indistinguishability, density operators and their nonseparability will be discussed. The density operator is a Hermitian operator acting on Hilbert space with nonnegative eigenvalues whose sum is 1 (it is not a classical statistical operator). It should not be confused with a classical statistical matrix and it has its greatest value in calculating expectation values of physical properties [24]. A density operator does not specify a unique microscopic configuration, which is not surprising based on its definition and contains the information about superpositions between subsystems. Quantum correlations imply unfactorizable density operators between multiple entities, with quantum entanglement being one type. They demonstrate correlations that exceed those describable by classical mechanics. They can be local or nonlocal, with the latter used extensively in decoherence theory. For simplicity, in this paper, we will approximately describe the coherent portion of the system as the principal and everything else as the environment. We describe the principal as being represented by a pure state density operator, a single vector in Hilbert space (there is no loss of generality as a mixed state can be modeled using purification) [15]. In the Lee system, for clarity, the phonon fields are part of the principal and can be viewed as pure. But as the phonon fields are part of the diamond, the diamond itself is of low purity as the principal only makes up a small portion of the diamond.

Described more formally below, a state describing a pair of nonlocal quantum correlated entities (photons or

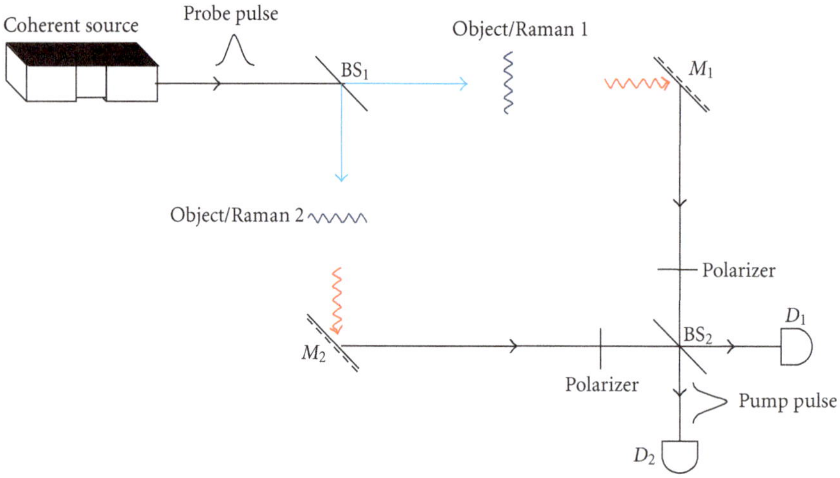

FIGURE 1: This diagram is a simplified version of the interferometer used in the Lee et al. experiments. Components have been removed which are needed for practical application but not for understanding the physical principles.

phonons) has an unfactorizable density operator for the pair that progresses forward in time via linear unitary operators. But in performing the trace operation to obtain the subsystems (e.g., a given diamond phonon field), these subsystems are represented by reduced density operators that move forward in time, unlike the true principal, via nonlinear unitary operators (i.e., the trace gives information on the subsystem statistical averages but is not the complete description of the subsystems) [15, 25]. So for the Lee system, the principal contains both phonon fields that have inseparable density operators.

3. Entanglement

Entanglement, a type of quantum correlation, is a function of superposition and the linearity of Schrödinger's equation, but not generally path indistinguishability (which will be dealt with in a subsequent section) [22]. Here we will limit the discussion to complete entanglement and partial entanglement can be extrapolated from the discussion. Demonstrating interference with entangled photons though does require path indistinguishable (see the Appendix). The entanglement process is described by (as per von Neumman) [26] as follows:

$$|\psi\rangle|a_r\rangle = \left(\sum_i c_i|s_i\rangle\right)|a_r\rangle \longrightarrow |\Psi\rangle = \sum_i c_i|s_i\rangle|a_i\rangle. \quad (1)$$

This is a form which would be used to describe decoherence (or a one-arm Lee experiment) where the principal is given by the wavefunction (ψ) [15, 16]. For two-particle entanglement, the wavefunction is simply replaced by a particle symbol. The arrow describes the unitary transform. The principal is represented in terms of the basis s_i while the basis for the environment is given by a_r. Entanglement represents pairing of the eigenstates. It can be stated in an equivalent form that their conjugate pairs (e.g., position momentum) are completely correlated. So we have two points: (1) with two entangled particles, the two base states s_i and a_i develop a constant relationship; this is the core to entanglement. Measuring one of an entanglement pairs establishes the eigenvalue of both exactly from the superposition. (2) This point will be more clear from the path indistinguishability discussion, but without further interactions, entangled particles continue to have inseparable density operators. This is not true for quantum correlations from path indistinguishability where the inseparability is dependent on such factors as detector time and wavepacket width. We will use the phrase "conditionally inseparable." (3) The initial entanglement generally requires local interaction between atomic/subatomic particles, but can become nonlocal with entanglement swapping, for which we use for illustration the well-known-Brune studies described below [10, 12]. This local-to-nonlocal entanglement can be found both in the Lee and Brune papers.

Equation (1) in the Lee paper (which is a DCLZ equation or one arm of the Lee interferometer) presents the initial local type of entanglement, in the annihilation operator form. This form was introduced by Dirac and expanded upon by Glauber for the quantum theory of light [27, 28]. The equation is

$$|\Psi_S\rangle \approx [1 + \varepsilon_S s^+(l_S)b^+(l_S)]|\text{vac}\rangle. \quad (2)$$

The equation is described in detail in the Lee paper. The essential point is that, for the potential, annihilation operators for the Stokes and phonon modes are in an inseparable product form. It will be seen that this is in contrast to (2) from that paper which is a superposition (below).

As noted, in addition to the Lee study being an extension of the DLCZ experiments, it is analogous to the pioneering experiments by Brune, entangling atoms with fields (and then a second atom) [10–12]. These studies are more useful than the Lee study for understanding the physics of entanglement and entanglement swapping because of the complex design of the Lee study. Its analogy is to a single arm of the

Lee experiment. Rubidium atoms in a Rydberg state were passed through an EM field in the large Q cavity. The atom and field become and remained entangled even after the Rubidium atom exited the system (i.e., until a measurement is made at the output of the device). One can then only speak of the combined Rubidium atom-Q cavity field system as a pure state, which is non-local (this is analogous to the pump-phonon entanglement in the Lee experiment). The nonlocality can be extended even further by sending a second atom after the first (analogous to the probe photon in the Lee study). Here, the second atom becomes non-locally entangled with the first atom (which had already passed through) with perfect correlation (inseparable biparticle wave packet). The second atom non-local entanglement represents entanglement swapping with the field, which is no longer entangled. This demonstrates true nonlocal entanglement of the two atoms as the eigenstates of each, even though passing through the cavity at different times, exactly correlate. The two atoms of course are analogous to the Stokes and anti-Stokes photons in the Lee study and the EM field to the phonon field, except only one arm is used.

For a more formal description of entanglement and its subsystems, we will provide the mathematical framework for one EPR-B particle state. There are two observers of these particles, A and B, separated by a large distance. One of these two entangled qubits is directed at each observer. The specific paths of each are inconsequential as long as no measurement has occurred. Neither does the order of detection nor the times between detection (as opposed to correlations from indistinguishable paths) for these entangled states. The Bell state used here is given by (let them be spin 1/2 particles, with two states, 0 and 1)

$$|\Phi^+\rangle = |\Psi\rangle = \frac{1}{\sqrt{2}}(|0\rangle_A \otimes |0\rangle_B + |1\rangle_A \otimes |1\rangle_B) \quad (3)$$

(Analogous Bell states with entangled energy and spin generated by a SPDC source type II, and the limitations, are discussed in the appendix.) Equation (3) is a true entangled state (spin superposition) in that the result of one observer exactly correlates with the results obtained with the second observer (irrespective of what spin axis is measured); the information of the system is complete. The density operator is given by

$$\begin{aligned} \hat{\rho}_T &= |\Psi\rangle\langle\Psi| = \hat{\rho}_A \otimes \hat{\rho}_B \\ &= \frac{|00\rangle\langle 00| + |11\rangle\langle 00| + |00\rangle\langle 11| + |11\rangle\langle 11|}{2}. \end{aligned} \quad (4)$$

The density operator product is nonfactorizable. If we examine a subsystem, it is an inseparable state as the trace operation of each observer (here, observer B) yields less information than the whole

$$\begin{aligned} \rho^A &= \mathrm{Tr}(\rho) \\ &= \frac{\mathrm{Tr}_B(|00\rangle\langle 00|) + \mathrm{Tr}_B(|11\rangle\langle 00|)}{2} \\ &\quad + \frac{\mathrm{Tr}_B(|00\rangle\langle 11|) + \mathrm{Tr}_B(|11\rangle\langle 11|)}{2} \\ &= \frac{|0\rangle\langle 0|\langle 0 \mid 0\rangle + |1\rangle\langle 0|\langle 1 \mid 0\rangle}{2} \\ &\quad + \frac{|0\rangle\langle 1|\langle 0 \mid 1\rangle + |1\rangle\langle 1|\langle 1 \mid 1\rangle}{2} \\ &= \frac{|0\rangle\langle 0| + |1\rangle\langle 1|}{2} = \frac{I}{2}. \end{aligned} \quad (5)$$

A reduced density operator is generated by the trace operation representing an improper mixed state, losing information about coherences. It is an expectation value. To paraphrase Schrödinger, the best possible knowledge of a whole does not include the best possible knowledge of its parts (if that knowledge is even available) [22]. In other words, the principal is inseparable as any description of the subsystem is incomplete as demonstrated by (5). We will contrast this true entanglement with correlations from indistinguishable paths where they are inseparable within certain experimental limits (e.g., path lengths and detector integration time).

4. Path Distinguishability and First-Order Correlations

Path indistinguishability can lead to nonlocal macroscopic correlations but generally not entanglement. A more complete discussion of coherence and indistinguishability can be found in the pioneering work of Mandel [20], reviewed by Shih (for both single- and two-photon (boson) correlations) [13, 19]. The topic will be addressed here briefly. It should also be noted that our group in a previous paper also established nonlocal macroscopic correlations. Correlations were produced between two reflector arms with path indistinguishability using a thermal source under ambient conditions [3].

We begin looking at path indistinguishability for a single photon entering a beam splitter with the two arms as exit ports (essentially equivalent to the pump photon in the Lee paper). All first-order interference is a single-photon wavepacket interference (as per Dirac), no matter what the intensity, along indistinguishable paths. Second-order correlations are generally the interference of biphoton wavepackets and are reviewed elsewhere [3, 13, 19]. First-order coherence (single-photon wavepacket interference) has a wavefunction given by

$$|\psi\rangle = \alpha|1\rangle_1|0\rangle_2 + \beta|0\rangle_1|1\rangle_2. \quad (6)$$

Here the subscripts 1 and 2 are the two paths and the value in the ket represents occupation number. The alpha and beta terms take into account beam splitter ratios. Note that this is the form of (2) of the Lee paper and is not an entangled state. Equation (2) in the Lee paper was

$$E|\Psi_S\rangle = \left[b_L^{+}(l_S) + e^{-i\varphi_x} b_R^{+}(l_S)\right]|\text{vac}_{\text{vib}}\rangle. \quad (7)$$

Again, the specifics can be found in the original paper but unlike (7), the annihilation operators of the potential are now in a summation form rather than a product form.

Returning to (6), the density operator (in its expanded form) is given by

$$\begin{aligned}\hat{\rho} = {} & |\alpha|^2 |1\rangle_1 |0\rangle_{22} \langle 0|_1 \langle 1| + |\beta|^2 |0\rangle_1 |1\rangle_{22} \langle 1|_1 \langle 0| \\ & + \left[\alpha\beta * |1\rangle_1 |0\rangle_{22} \langle 1|_1 \langle 0| + \text{h.c.}\right].\end{aligned} \quad (8)$$

The first two terms, the diagonal terms, are the DC terms that reduce fringe visibility to a maximum of 50% unless they can be removed (for true entanglement, there are no DC terms and maximum visibility is 100%). When paths are distinguishable, these are the only nonzero terms. The third and fourth terms represent indistinguishable paths and generate interference (h.c. is the Hermitian conjugate or adjoint) (see Figure 3 in the Lee paper, as off-diagonal elements are not exclusive to entanglement as suggested). These off-diagonal elements are complex. It is important to note that the density operator is inseparable only within the constraints of path indistinguishable (e.g., wavepacket width, detector time, path lengths, etc.). Coherence time is an example. For an optical pulse, delay times must be within the coherence time. In contrast, for most entangled states, coherence time is not an issue except when demonstrating interference.

Young's interferometer is useful for illustrating the concepts of path indistinguishability. We will use diamonds similar to the Lee experiment before each slit in the Young's interferometer. Examining the Young's interferometer (Figure 2), if one or the other slit is blocked, the photons are registered on the screen with no interference pattern (NI). If both slits are open, classically it is easy to appreciate when waves pass through the apparatus, and an interference pattern will develop on the screen (I). The sinusoidal peaks in the Young's design are position-dependent interference on the screen (I) due to varying phase relationships. The Young's experiment results hold for a high intensity photon beam, but the interference is still single-photon wavepacket interference. Even when only one photon (or other particle) is coming from the source at a time, a first order interference pattern develops on the detection screen, which is predicted naturally from quantum mechanics but is unexplainable by classical mechanics (which would predict the NI pattern) [13, 14]. This is because quantum mechanics is predicting the interference of potentials (along indistinguishable paths), and not intensities, as long as no measurement is made prior to the screen. There is no measurement of the pump photons in the *Science* paper until after the second beam splitter, so paths are still indistinguishable (in spite of the frequency shifts from the Raman scattering). So two-pump "beams" do not actually interfere as in the classical description of

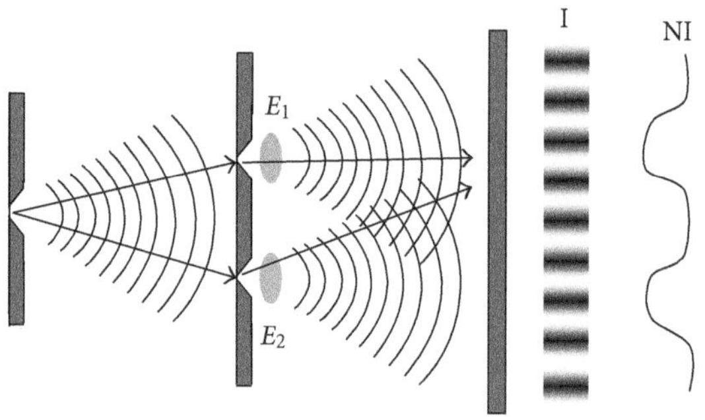

FIGURE 2: Illustration path indistinguishability and the influence environmental entanglements (diamonds) with Young's interferometer. The I is an interference pattern and the NI is no interference pattern. E_1 and E_2 are the diamonds.

interference after the second beam splitter; it is a single-photon interference. Interference of indistinguishable paths potentials (of single-photon wave packets) leads to the interference. Interference is possible when these single photon potential paths are identical with respect to the diamond interactions, as is more formally described in the next paragraph. Quantum correlations are established between the diamonds because they are part of each indistinguishable path that led to the single photon interference.

Now we extend (8) beyond one photon (increase intensity) and include interactions with the environment, E (diamonds) in the form of an inner product. This is a relatively common procedure for describing basic decoherence [15, 17], where the relevance to the diamond experiment will become apparent (though coherence is expanded rather than reduced). The interference pattern at the screen (of the Young's interferometer) is described by the cross terms (off-diagonal) in the density operator (it is in the expanded matrix form) as

$$\begin{aligned}\hat{\rho} = \frac{1}{2}\Big\{ & |\psi_1\rangle\langle\psi_1| + |\psi\rangle_2\langle\psi|_2 \\ & + |\psi_1\rangle\langle\psi_2|\langle E_2 \mid E_1\rangle + |\psi_2\rangle\langle\psi_1|\langle E_2 \mid E_1\rangle\Big\},\end{aligned} \quad (9)$$

where

$$\begin{aligned}|\psi_1\rangle\langle\psi_1| &= \rho_{11}, & |\psi_2\rangle\langle\psi_2| &= \rho_{22}, \\ |\psi\rangle_1\langle\psi|_2 &= \rho_{12}, & |\psi\rangle_2\langle\psi|_1 &= \rho_{21}.\end{aligned} \quad (10)$$

The first two terms are again DC terms and the second two represent interference terms. The wavefunction (in the bras-kets) incorporates all properties of the photons (polarization, bandwidth, photon numbers, etc.) now and not just occupation number. As can be seen from the density operator, the interference pattern is independent of whether the photons come individually or at high intensity (if one of the wavefunctions was zero, interference would still occur). In the density operator equation, 1 and 2 correspond to the two potential paths the photon can take. The density operator contains an inner product (E) in the last two terms that represents the diamonds which can be identical or distinct. The event that occurs at the screen is analogous to detection at D_3 in the *Science* paper.

To illustrate the counter-intuitive interaction of the photons and phonons leading to indistinguishable paths and coherence, Young's experiment will be examined by varying the Raman scattering. As a basic rule of quantum mechanics, which can be found in any introductory quantum mechanics textbook, until a measurement is made potentials are added then squared but once a measurement occurs, intensities (squared potentials) are added. If we initially ignore the E terms (environmental entanglements/diamonds), the pattern on the screen demonstrates interference that comes from the last two terms (off-diagonal) of the density operator (again, even if one photon is coming through at a time). Now, if E_1 and E_2 are substantially different terms (inner product near zero), such as when the Stokes photons are of different frequencies, the third and fourth terms disappear as the paths become distinguishable. Interference is lost in this simple example of environmentally induced decoherence by Raman scattering [15–17]. The similarity of the Raman scattering in each arm affects the degree to which coherence (and interference) is lost (fringe visibility). If E_1 and E_2 are similar (inner product 1) such that Stokes photons are identical from the prospective of detection, the paths are indistinguishable even though the interaction with the diamonds occurred (and changed the frequency), and the interference pattern is maintained. The key point is that indistinguishablity is needed at the *time of measurement* (the detector).

But another critical point is that the Born and Markovian approximations are not met hence decoherence will result. The Born approximation is that the diamond-principal interaction is sufficiently weak and environment (diamond) large such that the principal does not significantly change the diamond. Obviously the coupling is strong (Raman scattering) and the diamond changes significantly (change in phonon frequency). The Markovian approximation, having no memory effects, means that self-correlations within the diamond/environment decay for all practical purposes instantly into the environment. If these two are not met (along with the diamond interactions being identical), then the diamonds become part of the coherent system rather than a source of decoherence. Together, the indistinguishable paths of single-photon coherence, near identical nature of Raman scattering, and not meeting the Born/Markovian approximations resulted in expansion of the coherence (the two diamonds become part of the principal, resulting in quantum correlations). This describes why the two phonon fields become correlated and why it does not require (or include) an explanation of true non-local entanglement between arms.

We suggest that confusion over the distinction between quantum correlations due to entanglement versus path indistinguishability has arisen, at least in part, over a misunderstanding of the type II spontaneous parametric down-conversion (SPDC) source and overextending interpretations of Dirac notation, which is presented in the Appendix. This speculative topic is addressed in the Appendix.

5. General Results of the Lee et al. Paper

So to summarize, in the Lee paper the state, when using a single arm/diamond, is initially a Stokes-phonon(s) entanglement then Stokes-phonon(s)-anti-Stokes entanglement, arising from and remaining consistent with (1). It is an entanglement in the von Neumann sense as measurement of one subsystem exactly determines the state of the other subsystems. When two paths are used, the photon(s) and phonons are then entangled within a given path, but not entanglement of phonons between paths. However, the two diamonds are quantum correlated through path indistinguishability. The use of a coherent pulsed source allows the argument to be built up from single photon wavepacket interference.

As pointed out, the coherence expansion that results requires very specific conditions with respect to the diamonds. First, the high phonon frequency minimizes thermal decoherence. Second, the generated Stokes photons must be essentially identical with respect to detection. Third, the Born and Markovian approximations must not be met. Together, along with the path indistinguishability, this results in quantum correlations between the diamond phonons.

6. Notes on the Probe Photons

Just briefly discussing the probe photons, what is being measured is second-order correlation between detectors D_{a+} and D_{a-}, generated from phonon fields in the two arms, in a superposition. In general, we agree with the author's interpretations of the physical principles of the probe photons, which will not be reviewed here because of space limitations [13, 18, 25]. A quantitative description of these second-order correlations from both entangled photons and indistinguishable paths is best described in terms of the correlation functions, electric field operators, and annihilation operators. These are discussed elsewhere in detail for those interested [13, 19, 27].

7. Notes on the Quantitative Results

The four-quantitative/qualitative results for discussion from the *Science* paper are the density operators presented, concurrence, confidence intervals, and visibility. (1) The density operators in Figures 3 and 4 of the Lee paper describe a coherent state as demonstrated by the off-diagonal coherences, which is not unique to entanglement. (2) There was some confusion in editorials/commentaries on the article that there was 98% concurrence. There was actually a 98% confidence interval that the concordance was positive (which as the reader is aware could mean it was 98% confidence the concurrence was extremely small but positive). The concurrence was positive and somewhere below 35%, values consistent with quantum correlations that are not exclusive to an entangled state [29]. (3) The visibility graph (Figure 2 of the *Science* paper) demonstrates two main points. (A) The second order correlations are phase sensitive with opposite signs due to the beam splitter, which is known for second-order correlations. (B) The correlations between the pump and probe can exceed coincidence rates of classical correlations. These results demonstrate quantum correlations, but are not sufficient for specifically demonstrating quantum entanglement. This statement is also consistent with the experimental design analysis described above.

True entanglement between the phonon fields neither needs to be elicited as an explanation for the results nor leads to be proven in the paper. Though the phrase "entanglement of diamonds" attracts considerable attention, we believe that the establishment of quantum correlations/coherence between two macroscopic objects using path indistinguishability without nonlocal entanglements is far more important to the field. We point out that we have also achieved this with two macroscopic distant reflectors [3]. Path indistinguishability, under the local entanglement conditions described above, leads to quantum correlations. This approach required that no quantum source could be done under ambient conditions and potentially opens the door to a much larger number of applications than straight entanglement.

8. Conclusion

The recent paper in *Science* entitled "Entangling Macroscopic Diamonds at Room Temperature" by C. Lee et al. establishes nonlocal macroscopic quantum correlations between two diamonds. However, while the authors claim the correlations between diamonds represent entanglement, we present why a different underlying mechanism exists which explain the results. The quantum correlations are generated by path indistinguishability of first order correlations (single-photon wavepackets) in combination with essentially identical local entanglement in each arm. Irrespective, the results are of considerable importance. They offer a mechanism for generating macroscopic nonlocal quantum correlations under ambient conditions, which could represent a substantial advance to a wide range of applications.

Appendix

Unfortunately, many examples exist in the literature that treat quantum correlations from path indistinguishability and entanglement as essentially identical, an obstruction to the field and in part likely due to misunderstanding of the widely used SPDC II source (spontaneous parametric downconversion) and misuse of Dirac's notation. Two prominent examples are a 2008 *Nature* review on entanglement and the recent study claiming entanglement between two diamonds in *Science* [1, 30]. A brief review of the SPDC may illustrate the point.

SPDC sources generally use a CW pumped nonlinear crystal to produce two energy entangled photon pairs (including entanglement of uncertainty) [31]. They were initially pursed to test EPR-B. Due to energy conservation, photon pairs' angular frequency and wave number are entangled. According to the standard theory of parametric downconversion, the two-photon state can be written as

$$|\Psi\rangle = \int d\omega_P A(\omega_P) \times \int d\omega_1 d\omega_2 \delta(\omega_1 + \omega_2 - \omega_P) a^+(\omega_1) a^+(\omega_2)|0\rangle, \tag{A.1}$$

where ω represents the angular frequency of the signal (1), idler (2), and pump (p) of the downconversion. The a^+ represents the respective annihilation operators. The delta function represents perfect frequency phase matching of the downconversion (i.e., entanglement). $A(\omega)$ is related to the wavepacket extent and is not critical to the discussion here (but it is when interfering entangled photons). This is a type I SPDC source (no fixed polarization relationship); note that the equation does not require path indistinguishability. With a type II SPDC source, the signal and the idler have orthogonal polarization states (i.e. the energy entangled photons are associated with perpendicular polarizations). The state is given by [32]

$$|\Psi\rangle = \int d\omega_P A(\omega_P) \times \int d\omega_1 d\omega_2 \delta(\omega_1 + \omega_2 - \omega_P) a_o^+(\omega_1) a_e^+(\omega_2)|0\rangle. \tag{A.2}$$

The subscripts on the signal and idler represent different polarization states associated with the entangled energy states (o and e). Again, the energy states are entangled (and thereby the polarization states) without any use of indistinguishable paths.

Now, using a SPDC II source with an interferometer (Figure 3) illustrates both entanglement and path indistinguishability. In this setup, prior to the beam splitter, the photons are both entangled by energy and polarization. After the beam splitter, indistinguishable paths are used to generate interference. Under the correct setup of the polarizers (P_1 and P_2) in each arm, Bell states can be generated which can be used to test, for example, EPR-B. The path indistinguishability after the beam splitter does not cause the entanglement, but rather it is used to generate Bell states from the already entangled states. Authors often abbreviate the wave function for these Bell states, (entangled photons grouped by indistinguishable paths) for example, as $(1/2)(|HV\rangle + |VH\rangle)$. This representation, as seen in the *Nature* review, can be misleading, because it drops the energy/polarization entanglement that exists without the beam splitter, as well as the wavepacket for the biphoton (basically just using the e and o from (A.2) and giving the impression that they are being entangled by the beam splitter) [37]. Interfering light from the SPDC II source in the *Nature* paper, a common yet incorrect statement in Figure 2 of that paper is made, "However, in the regions where the two cones overlap, the state of the photons will be $|HV\rangle + |VH\rangle$. It is around these points that entangled photons are generated." This abbreviated representation of

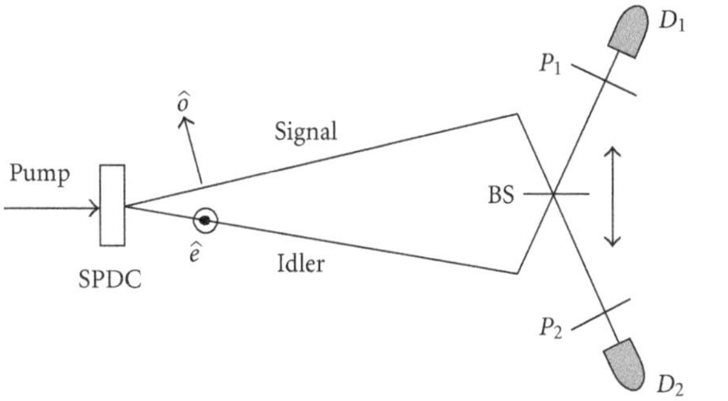

Figure 3: An SPDC type II source using a beam splitter used to generate Bell states.

the state ignores the already entangled energy/polarization in areas outside the overlap (9), as well as the space-time probability density. This leads to a misunderstanding of the physics. Entanglement exists in the areas outside the overlap, not just Bell states. Similarly, path indistinguishability did not lead to entanglement of diamonds in the Lee experiment.

The example also illustrates the misuse of Dirac notation, which seems particularly common in the quantum communication and computer fields. Dirac notation is a powerful shorthand technique for describing quantum information flow. But it is frequently treated as representing the state of a system, which it generally does not do. If we represent a vacuum and photon by $|01\rangle + |10\rangle$, this neither tells us, for example, about the state of the vacuum nor the bandwidth of the photon. But this is how it is often interpreted leading to erroneous conclusions.

Acknowledgments

This paper is sponsored by the National Institutes of Health, Contracts R01-AR44812, R01-EB000419, R01 AR46996, R01- HL55686, R21 EB015851-01, and R01-EB002638.

References

[1] K. C. Lee, M. R. Sprague, B. J. Sussman et al., "Entangling macroscopic diamonds at room temperature," *Science*, vol. 334, no. 6060, pp. 1253–1256, 2011.

[2] L. M. Duan, "Quantum correlation between distant diamonds," *Science*, vol. 334, no. 6060, pp. 1213–1214, 2011.

[3] M. E. Brezinski and B. Liu, "Nonlocal quantum macroscopic superposition in a high-thermal low-purity state," *Physical Review A*, vol. 78, no. 6, Article ID 063824, 13 pages, 2008.

[4] B. Julsgaard, A. Kozhekin, and E. S. Polzik, "Experimental long-lived entanglement of two macroscopic objects," *Nature*, vol. 413, no. 6854, pp. 400–403, 2001.

[5] L. M. Duan, M. D. Lukin, J. I. Cirac, and P. Zoller, "Long-distance quantum communication with atomic ensembles and linear optics," *Nature*, vol. 414, no. 6862, pp. 413–418, 2001.

[6] D. N. Matsukevich and A. Kuzmich, "Quantum state transfer between matter and light," *Science*, vol. 306, no. 5696, pp. 663–666, 2004.

[7] K. S. Choi, H. Deng, J. Laurat, and H. J. Kimble, "Mapping photonic entanglement into and out of a quantum memory," *Nature*, vol. 452, no. 7183, pp. 67–71, 2008.

[8] C. W. Chou, H. De Riedmatten, D. Felinto, S. V. Polyakov, S. J. Van Enk, and H. J. Kimble, "Measurement-induced entanglement for excitation stored in remote atomic ensembles," *Nature*, vol. 438, no. 7069, pp. 828–832, 2005.

[9] T. Chanelière, D. N. Matsukevich, S. D. Jenkins, S. Y. Lan, T. A. B. Kennedy, and A. Kuzmich, "Storage and retrieval of single photons transmitted between remote quantum memories," *Nature*, vol. 438, no. 7069, pp. 833–836, 2005.

[10] M. Brune, E. Hagley, J. Dreyer et al., "Observing the progressive decoherence of the "meter" in a quantum measurement," *Physical Review Letters*, vol. 77, no. 24, pp. 4887–4890, 1996.

[11] M. Brune, S. Haroche, J. M. Raimond, L. Davidovich, and N. Zagury, "Manipulation of photons in a cavity by dispersive atom-field coupling: quantum-nondemolition measurements and generation of Schrödinger cat states," *Physical Review A*, vol. 45, no. 7, pp. 5193–5214, 1992.

[12] T. Meunier, S. Gleyzes, P. Maioli et al., "Rabi oscillations revival induced by time reversal: a test of mesoscopic quantum coherence," *Physical Review Letters*, vol. 94, no. 1, Article ID 010401, 4 pages, 2005.

[13] Y. Shin, *An Introduction to Quantum Optics: Photon and Biphoton Physics*, CRC Press, New York, NY, USA, 2011.

[14] L. Mandel, "Coherence and indistinguishability," *Optics Letters*, vol. 16, no. 23, pp. 1882–1883, 1991.

[15] M. Schlosshuaser, *Decoherence and the Quantum to Classical Transition*, Springer, Melbourne, Australia, 2007.

[16] W. H. Zurek, "Decoherence and the transition from quantum to classical," *Physics Today*, vol. 44, no. 10, pp. 36–44, 1991.

[17] M. A. Neilson and I. L. Chuang, *Quantum Computer and Quantum Information*, Cambridge University Press, Cambridge, UK, 2007.

[18] L. Mandel, "Quantum effects in one-photon and two-photon interference," *Reviews of Modern Physics*, vol. 71, no. 2, pp. S274–S282, 1999.

[19] H. Chen, T. Peng, S. Karmakar, Z. Xie, and Y. Shih, "Observations of anti-correlations in incoherent thermal light fields," *Physical Review A*, vol. 84, Article ID 033835, 2011.

[20] A. Einstein, B. Podolsky, and N. Rosen, "Can quantum-mechanical description of physical reality be considered complete?" *Physical Review*, vol. 47, no. 10, pp. 777–780, 1935.

[21] D. Greenberger, M. Horne, A. Shimony, and A. Zeilinger, "Bell's theorem without inequalities," *American Journal of Physics*, vol. 58, p. 1131, 1990.

[22] E. Schrödinger, "Die gegenwärtige situation in der quantenmechanik," *Naturwissenschaften*, vol. 23, no. 807, pp. 823–844, 1935.

[23] G. Jarger, *Entanglement, Information, and the Interpretation of Quantum Mechanics*, Springer, New York, NY, USA, 2009.

[24] K. Blum, *Density Matrix Theory and Applications*, Plenum Press, New York, NY, USA, 1996.

[25] C. K. Hong, Z. Y. Ou, and L. Mandel, "Measurement of subpicosecond time intervals between two photons by interference," *Physical Review Letters*, vol. 59, pp. 2044–2046, 1987.

[26] J. von Neumann, *Mathematical Foundations of Quantum Mechanics*, chapter 4, Princeton University Press, Princeton, NJ, USA, 1955.

[27] R. J. Glauber, "The quantum theory of optical coherence," *Physical Review*, vol. 130, no. 6, pp. 2529–2539, 1963.

[28] P. A. M. Dirac, "The question theory of the emission and absorption of radiation," *Proceedings of the Royal Society of London A*, vol. 114, no. 767, pp. 243–265, 1927.

[29] R. Hildebrand, "Concurrence revisted," *Journal of Mathematical Physics*, vol. 48, no. 10, Article ID 102108, 23 pages, 2007.

[30] V. Vedral, "Quantifying entanglement in macroscopic systems," *Nature*, vol. 453, no. 7198, pp. 1004–1007, 2008.
[31] P. G. Kwiat, K. Mattle, H. Weinfurter, A. Zeilinger, A. V. Sergienko, and Y. Shih, "New high-intensity source of polarization-entangled photon pairs," *Physical Review Letters*, vol. 75, no. 24, pp. 4337–4341, 1995.
[32] Y. Shih, "Entangled Photons," *IEEE Journal on Selected Topics in Quantum Electronics*, vol. 9, no. 6, pp. 1455–1467, 2003.

16

Energies, Fine Structures, and Hyperfine Structures of the $1s^2 2snp\ ^3P$ ($n = 2$–4) States for the Beryllium Atom

Chao Chen

School of Physics, Beijing Institute of Technology, Beijing 100081, China

Correspondence should be addressed to Chao Chen, chen_chao@tsinghua.org.cn

Academic Editor: Derrick S. F. Crothers

Energies and wave functions of the $1s^2 2snp\ ^3P$ ($n = 2$–4) states for the beryllium atom are calculated with the full-core plus correlation wave functions. Fine structures and hyperfine structures are calculated with the first-order perturbation theory. For the $1s^2 2s2p\ ^3P$ state, the calculated energies, fine structure, and hyperfine structure parameters are in good agreement with the latest theoretical and experimental data in the literature; it is shown that atomic parameters of the low-lying excited states for the beryllium atom can be calculated accurately using this theoretical method. For the $1s^2 2snp\ ^3P$ ($n = 3, 4$) states, the present calculations may provide valuable reference data for future theoretical calculations and experimental measurements.

1. Introduction

In recent years, studies of energies, fine structures, and hyperfine structures of the low-lying excited states for the beryllium atom [1–10] have been of great interest to spectroscopists because there are many strong optical transitions suitable for spectral and hyperfine structure measurements. On the other hand, studies of the low-lying excited states for the beryllium atom play an important role in developing the excited state theory of multielectron atoms and better understanding the complicated correlation effects between electrons. The fine structure comes from the spin-orbit, spin-other-orbit, and spin-spin interactions. The hyperfine structure of atomic energy levels is caused by the interaction between the electrons and the electromagnetic multipole moments of the nucleus. The leading terms of this interaction are the magnetic dipole and electric-quadrupole moments. The fine and hyperfine structure is sensitive to the correlation effects among electrons. Experimentally, some properties of the atomic nucleus can be obtained by investigating the hyperfine structure of the atomic energy levels. The nuclear electric-quadrupole moment, which is difficult to measure directly with nuclear physics techniques, can be determined using the measured hyperfine structure and the accurate theoretical results.

The $1s^2 2s2p\ ^3P$ state of the beryllium atom is of interest since it is the lowest excited state in which hyperfine effects can occur, and the ground state has no hyperfine splitting because it is $J = 0$. It is generally a very demanding task to calculate hyperfine structure accurately. Polarization of the closed shells in the $1s^2$ core, due to the Coulomb interaction with open shells, can have a large effect on the hyperfine structure. Up till now, the most sophisticated theoretical calculations of the hyperfine structure parameters for the $1s^2 2s2p\ ^3P$ state of the Be atom have been carried out using linked-cluster many-body perturbation (LC MBPT) theory [5, 6], Hartree-Fock and CI allowing all SD excitations to correlation orbitals of Slater type by Beck and Nicolaides [7], as well as multiconfiguration Hartree-Fock (MCHF) method [8, 9]. Experimentally, the magnetic dipole and electric-quadrupole hyperfine constants have been determined very accurately with the atomic-beam magnetic-resonance technique [10] for the $1s^2 2s2p\ ^3P$ state in beryllium. To the best of our knowledge, few results on energies, fine structures, and hyperfine structures have been investigated for the $1s^2 2snp\ ^3P$ ($n \geq 3$) states of the beryllium atom due to the restriction of resolution from experiments and the numerical unsteadiness in theoretical calculations.

An elegant and complete variation approach, namely, the full core plus correlation (FCPC) method, has been developed by Chung [11, 12]. This method has been successfully applied to three- and four-electron systems, with the $1s^2$ core. Many elaborate calculations, especially for the dipole polarizabilities [13], quadrupole and octupole polarizabilities [14], and total atomic scattering factors [15], show that FCPC wave functions have a reasonable behavior over the whole configuration space for three-electron systems. This method has also been used to calculate the hyperfine structure of the $1s^2ns\,^2S$ and $1s^2np\,^2P$ states (n = 2–5) for the lithium isoelectronic sequence; the results are in good agreement with the Hylleraas calculations and with the experiment data [16]. As is well known, theoretical calculations of the hyperfine structure parameters depend sensitively on the behavior of the wave function in the proximity of the nucleus. In addition, core polarization effects for the low l states need to be included in the nonrelativistic wave function. It would be interesting to find out whether the FCPC wave function can also be successful for calculating hyperfine structure parameters of low-lying excited states for the beryllium atom. In this work, the FCPC wave functions are carried out on the $1s^22snp\,^3P$ (n = 2–4) states of the beryllium atom. The energies, fine structures, and hyperfine structures are calculated and compared with the data available in the literature. The purpose of this work is to explore the capacity of the FCPC wave function to calculate the atomic parameters of the low-lying excited states for the beryllium atom and provide more reliable theoretical data to stimulate further experimental measurements.

2. Theory

According to the FCPC method [11, 12], the wave function for the four-electron $1s^22snp\,^3P$ state can be written as

$$\Psi(1,2,3,4) = A\left[\Phi_{1s1s}(1,2)\Phi_{2snp}(3,4) + \sum_i C_i\Phi_{n(i),l(i)}(1,2,3,4)\right], \tag{1}$$

where A is an antisymmetrization operator. Φ_{1s1s} is a predetermined $1s^2$-core wave function which is represented by a CI basis set,

$$\Phi_{1s1s}(1,2) = A\sum_{k,n,l} C_{knl}r_1^kr_2^n\exp(-\beta_lr_1-\rho_lr_2)Y_l(1,2)\chi(1,2), \tag{2}$$

the angular part is

$$Y_l(1,2) = \sum_m\langle l,m,l,-m\mid 0,0\rangle Y_{lm}(\theta_1,\varphi_1)Y_{l-m}(\theta_2,\varphi_2). \tag{3}$$

χ (1,2) is a two-electron singlet spin function. The linear and nonlinear parameters in (2) are determined by optimizing the energy of the two-electron core. The factor $\Phi_{2snp}(3,4)$ represents the wave function of the two outer electrons which is given by

$$\Phi_{2snp}(3,4) = A\sum_{k,n,l} d_{knl}r_3^kr_4^n\exp(-\lambda_lr_3-\eta_lr_4)Y_l(3,4)\chi(3,4), \tag{4}$$

the angular part is

$$Y_l(3,4) = \sum_m\langle l,m,l+1,-m\mid 0,0\rangle Y_{lm}(\theta_3,\varphi_3)Y_{l+1-m}(\theta_4,\varphi_4). \tag{5}$$

The latter wave function of (1) describes the core relaxation and the intrashell electron correlation in the four-electron system. It is given by

$$\Phi_{n(i),l(i)}(1,2,3,4) = \varphi_{n(i),l(i)}(R)Y_{l(i)}^{\mathrm{LM}}(\Omega)\chi_{\mathrm{SS}_z}, \tag{6}$$

where

$$\varphi_{n(i),l(i)}(R) = \prod_{j=1}^4 r_j^{n_j}\exp\left(-\alpha_jr_j\right). \tag{7}$$

A different set of α_j is used for each $l(i)$. The angular part is

$$\begin{aligned} Y_{l(i)}^{\mathrm{LM}}\left(\hat{R}\right) = &\sum_{m_j}\langle l_1l_2m_1m_2\mid l_{12}m_{12}\rangle \\ &\times\langle l_{12}l_3m_{12}m_3\mid l_{123}m_{123}\rangle \\ &\times\langle l_{123}l_4m_{123}m_4\mid \mathrm{LM}\rangle\prod_{j=1}^4 Y_{l_jm_j}\left(\Omega_j\right). \end{aligned} \tag{8}$$

To simplify notation, this angular function is simply denoted as

$$l(i) = [(l_1,l_2)l_{12},l_3]l_{123},l_4, \tag{9}$$

with the understanding that l_{123} and l_4 couple into L, the total orbital angular momentum. There are three possible spin functions for the $1s^22snp\,^3P$ state, namely,

$$\begin{aligned} \chi_1 &= [(s_1,s_2)0,s_3]\frac{1}{2},s_4, \\ \chi_2 &= [(s_1,s_2)1,s_3]\frac{1}{2},s_4, \\ \chi_3 &= [(s_1,s_2)1,s_3]\frac{3}{2},s_4. \end{aligned} \tag{10}$$

For the radial basis functions of each angular-spin component, a set of linear and nonlinear parameters is chosen. These parameters are determined in the energy optimization process. For each set of l_1, l_2, l_3, and l_4, we try all possible $l(i)$ and χ and keep the ones which make significant contribution to the energy in (1).

The fine structure perturbation operators [1, 2] are given by

$$\widehat{H}_{\mathrm{FS}} = \widehat{H}_{\mathrm{so}}+\widehat{H}_{\mathrm{soo}}+\widehat{H}_{\mathrm{ss}}, \tag{11}$$

where the spin-orbit, spin-other-orbit, and spin-spin operators are

$$\hat{H}_{SO} = \frac{Z}{2c^2}\sum_{i=1}^{4}\frac{\vec{l}_i \cdot \vec{s}_i}{r_i^3},$$

$$\hat{H}_{SOO} = -\frac{1}{2c^2}\sum_{\substack{i,j=1\\ i\neq j}}^{4}\left[\frac{1}{r_{ij}^3}\left(\vec{r}_i - \vec{r}_j\right)\times\vec{p}_i\right]\cdot\left[\vec{s}_i + 2\vec{s}_j\right], \quad (12)$$

$$\hat{H}_{SS} = \frac{1}{c^2}\sum_{\substack{i,j=1\\ i<j}}^{4}\frac{1}{r_{ij}^3}\left[\vec{s}_i\cdot\vec{s}_j - \frac{3\left(\vec{s}_i\cdot\vec{r}_{ij}\right)\left(\vec{s}_j\cdot\vec{r}_{ij}\right)}{r_{ij}^2}\right].$$

To calculate the fine structure splitting, the LSJ coupling scheme is used:

$$\Psi_{LSJJ_Z} = \sum_{M,S_Z}\langle LSMS_Z \mid JJ_Z\rangle\Phi_{LSMS_Z}. \quad (13)$$

The fine structure energy levels are calculated by first-order perturbation theory

$$(\Delta E_{FS})_J = \left\langle\Phi_{LSJJ_Z}\left|\hat{H}_{so} + \hat{H}_{soo} + \hat{H}_{ss}\right|\Phi_{LSJJ_Z}\right\rangle. \quad (14)$$

For an N-electron system, the hyperfine interaction Hamiltonian can be represented as follows [17, 18]:

$$\hat{H}_{hfs} = \sum_{k=1} T^{(k)}\cdot M^{(k)}, \quad (15)$$

where $T^{(k)}$ and $M^{(k)}$ are spherical tensor operators of rank k in the electronic and nuclear spaces, respectively. The $k = 1$ term represents the magnetic-dipole interaction between the magnetic field generated by the electrons and nuclear magnetic dipole moments, and the $k = 2$ term the electric quadrupole interaction between the electric-field gradient from the electrons and the nonspherical charge distribution of the nucleus. The contributions from higher-order terms are much smaller and can often be neglected.

In the nonrelativistic framework, the electronic tensor operators, in atomic units, can be written as

$$T^{(1)} = \frac{\alpha^2}{2}\sum_{i=1}^{4}\left[2g_l r_i^{-3}l_i^{(1)} - \sqrt{10}g_s\left\{s_i^{(1)}C_i^{(2)}\right\}^{(1)}r_i^{-3} + \frac{8\pi}{3}g_s s_i^{(1)}\delta(\mathbf{r}_i)\right], \quad (16)$$

$$T^{(2)} = -\sum_{i=1}^{4} r_i^{-3}C_i^{(2)},$$

where $g_l = (1 - m_e/M)$ is the orbital electron g factor, and $g_s = 2.0023193$ is the electron spin g factor. M is the nuclear mass. The tensor $C_i^{(2)}$ is connected to the spherical harmonics $Y_{lm}(i)$ by

$$C_m^{(l)} = \sqrt{\frac{4\pi}{2l} + 1}\mathbf{Y}_{lm}. \quad (17)$$

The hyperfine interaction couples the electronic angular momenta $\mathbf{J}$ and the nuclear angular momenta $\mathbf{I}$ to a total angular momentum $\mathbf{F} = \mathbf{I} + \mathbf{J}$. The uncoupling and coupling hyperfine constants are defined in atomic units as [17, 18]:

$$a_C = \left\langle\gamma LSM_LM_S\left|\sum_{i=1}^{N}8\pi\delta^3(r_i)s_0(i)\right|\gamma LSM_LM_S\right\rangle$$

(Fermi contact),

$$a_{SD} = \left\langle\gamma LSM_LM_S\left|\sum_{i=1}^{N}2C_0^{(2)}(i)s_0(i)r_i^{-3}\right|\gamma LSM_LM_S\right\rangle$$

(Spin dipolar),

$$a_l = \left\langle\gamma LSM_LM_S\left|\sum_{i=1}^{N}l_0(i)r_i^{-3}\right|\gamma LSM_LM_S\right\rangle$$

(orbital),

$$b_q = \left\langle\gamma LSM_LM_S\left|\sum_{i=1}^{N}2C_0^{(2)}(i)r_i^{-3}\right|\gamma LSM_LM_S\right\rangle$$

(electric quadrupole),

$$A_J = \frac{\mu_I}{I}\frac{1}{[J(J+1)(2J+1)]^{1/2}}\left\langle\gamma J\left\|T^{(1)}\right\|\gamma J\right\rangle,$$

$$A_{J-1,J} = \frac{\mu_I}{I}\frac{1}{[J(2J-1)(2J+1)]^{1/2}}\left\langle\gamma J-1\left\|T^{(1)}\right\|\gamma J\right\rangle,$$

$$B_J = 2Q\left[\frac{2J(2J-1)}{(2J+1)(2J+2)(2J+3)}\right]^{1/2}\left\langle\gamma J\left\|T^{(2)}\right\|\gamma J\right\rangle, \quad (18)$$

where $M_L = L$ and $M_S = S$. In these expressions, μ_I is the nuclear magnetic moment and Q is the nuclear electric quadrupole moment. I is the nuclear spin, and J is the atomic electronic angular moment.

3. Results and Discussions

In order to achieve accurate calculation results for various properties of the low-lying excited states for the beryllium atom, the choice of basis function with sufficiently high quality is critical and it is our major concern. The seven l components (0,0), (1,1), (2,2), (3,3), (4,4), (5,5), (6,6) altogether 159 terms are used for the $1s^2$ core. The Φ_{2snp} in (1) has four angular components, l is summed from 0 to 3 with the angular components (0,1), (1,2), (2,3), and (3,4), and the number of terms in Φ_{2snp} ranges from 36 to 15. Most of the other correlation effects are included in (6), which accounts for the intershell as well as the intrashell correlations. Many relevant angular and spin couplings are important for the energy, these basis functions are tried to include in $\Phi_{n(i),l(i)}$ (1, 2, 3, 4) with significant energy contribution. For each set of orbital angular momenta l_1, l_2, l_3, and l_4, there could be several ways to couple this set into the desired total orbital angular momentum. In this work,

TABLE 1: Nonrelativistic energies of the $1s^2 2snp\,^3P$ (n = 2–4) states for the beryllium atom (in a.u.).

	This work	Hibbert[a]	Weiss[b]
$1s^2 2s2p\,^3P$	−14.56637	−14.5184	−14.51844
$1s^2 2s3p\,^3P$	−14.39839	−14.3510	−14.35106
$1s^2 2s4p\,^3P$	−14.36248		−14.31530

[a]Reference [1].
[b]Reference [2].

TABLE 2: Fine structure splittings ν_{J-J} of the $1s^2 2snp\,^3P_J$ (n = 2–4) states for the beryllium atom (in cm^{-1}).

	$1s^2 2s2p\,^3P_J$	$1s^2 2s3p\,^3P_J$	$1s^2 2s4p\,^3P_J$
ν_{2-1}	2.36	0.35	0.13
Experiment[a]	2.35 (.2)		
Other theory[b]	2.53		
ν_{1-0}	0.64	0.092	0.034
Experiment[a]	0.64 (.1)		
Other theory[b]	0.71		

[a]Reference [3].
[b]Reference [4].

for $1s^2 2snp\,^3P$ states, the important angular series (l_1, l_2, l_3, l_4) are $(0, 0, l, (l+1))$, $(0, 1, l, l)$, $(l, l, 0, 1)$, and so forth. In both cases, the value of l is from 0 to 6, as the energy contribution from set with $l > 6$ is small and negligible. In order to get the high-quality wave function, the number of angular-spin components in the $\Phi_{n(i),l(i)}$ wave functions ranges from 15 to 66, and the number of terms in the $\Phi_{n(i),l(i)}$ of (6) is about 790. The linear and nonlinear parameters are individually optimized in the energy minimization process. Using the Rayleigh-Ritz variational method, the basic wave function Ψ and the corresponding eigenvalue E are determined.

Nonrelativistic energies of the $1s^2 2snp\,^3P$ (n = 2–4) states for the beryllium atom are given in Table 1. As Table 1 shows, for the $1s^2 2snp$ ($n = 2, 3$) ^3P states, the nonrelativistic energies in this work are lower and better than those of Hibbert and Weiss [1, 2], the improvement ranging from 0.0479 a.u. to 0.0473 a.u. Hibbert and Weiss reported a set of large-scale configuration interaction (CI) calculations for the $1s^2 2snp$ ($n = 2, 3$) ^3P states, which can give an accurate approximation for each state, but it may tend to obscure the global picture of the spectrum which is so transparent in the other approach. The work of Hibbert and of Weiss did not include any intrashell correlation in the $1s$ shell, as the calculations were of transitions in the outer subshells. The correlation energy of the $1s$ shell is almost independent of the nuclear charge and also of the number of additional electrons outside the $1s$ shell. For Be, it is about 0.0457 a.u. and this accounts for the main difference between earlier work and the present; more accurate results are presented in Table 1. Of course, for the calculation of hyperfine parameters, correlation within the $1s$ shell is crucial in obtaining accurate hyperfine parameters, and this has been achieved in the present work. For the $1s^2 2s4p\,^3P$ state, the present calculation from the FCPC method is also lower than the result of Weiss [2].

If including the effects of the spin-orbit, spin-other-orbit, and spin-spin interactions, the energies of the fine structure resolved J levels are obtained. In this work, the fine structure splittings of the triplet states are calculated with the H_{so}, H_{soo}, and H_{ss} operators using the first-order perturbation theory. Table 2 gives the fine-structure splittings of the $1s^2 2snp\,^3P_J$ (n = 2–4) states for the beryllium atom. The experimental Be $2s2p\,^3P_J$ splittings are 2.35 ($J = 2 \rightarrow 1$) and 0.64 ($J = 1 \rightarrow 0$) cm^{-1} [3]. They agree with our prediction 2.36 and 0.64 cm^{-1}. Although many theoretical studies have been done on the BeI excited systems, the published theoretical fine structure results are scarce. One exception is Laughlin, Constantinides, and Victor [4]. They use a model potential calculation and predict the splittings to be 2.53 and 0.71 cm^{-1} for the $1s^2 2s2p\,^3P_J$ state, which should be considered as quite good in view of the simplicity in their computation and fall in experimental uncertainties. Present calculations for this state are more accurate due to correlation effect well described in this method. For the experiment, the splitting of $1s^2 2s3p\,^3P_J$ ($J = 1, 0$) is not resolved. But the splitting from the $J = 2$ state to the $J = 1, 0$ is determined to be 0.37 cm^{-1}. In this work, the calculated splittings are 0.35 ($J = 2 \rightarrow 1$) and 0.092 ($J = 1 \rightarrow 0$) cm^{-1}. This implies that the predicted splitting from $J = 2$ to the center of gravity of $J = 1$ and 0 should be 0.373 cm^{-1}. It agrees with the experiment. The good agreement with experiment could be used as the indication of the accuracy of the wave function constructed here. For the $1s^2 2s4p\,^3P_J$ state, our calculated splittings are hoped to offer reference for further experimental measurements.

The hyperfine structure parameters of the $1s^2 2snp\,^3P$(n = 2–4) states for the beryllium atom are calculated in this work: Fermi contact a_c, the spin dipolar a_{SD}, the orbital a_l, and the electric quadrupole b_q. In the present calculation, $Q = 0.0530b$, $\mu_I = -1.177492$ nm, $I = 1.5$ for Be are taken from [19]. The hyperfine interaction in the $1s^2 2s2p\,^3P$ state for the beryllium atom is of interest since it is the lowest excited state in which hyperfine effects can occur, which has

TABLE 3: The hyperfine structure parameters (in a.u.) and coupling constants (in MHz) of the $1s^2 2s2p\ ^3P$ state for the beryllium atom.

Method	a_c	a_{sd}	a_l	b_q	A_2	A_1	Reference
LC MBPT[a]	9.2319	−0.06490	0.30478	−0.1156	−124.21		[5, 6]
HF + SDCI[b]	9.2738	−0.06656	0.30014	−0.1097	−124.76	−139.77	[7]
FE MCHF[c]	9.2349	−0.06564	0.30261	−0.1150			[8]
MCHF	9.2416	−0.06587	0.30329	−0.11570	−124.50	−139.35	[9]
This work	9.2436	−0.06523	0.30201	−0.11588	−124.51	−139.36	
Experiment					−124.5368	−139.373	[10]

[a] Linked-cluster many-body perturbation theory.
[b] Hartree-Fock and CI allowing all SD excitations to correlation orbitals of Slater type.
[c] Finite-element multiconfiguration Hartree-Fock.

TABLE 4: The hyperfine structure parameters (in a.u.) and coupling constants (in MHz) of the $1s^2 2snp\ ^3P\,(n = 3, 4)$ states for the beryllium atom.

	a_c	a_{sd}	a_l	b_q	A_2	A_1
$1s^2 2s3p\ ^3P$	12.029	−0.00898	0.04276	−0.01788	−151.59	−153.61
$1s^2 2s4p\ ^3P$	12.118	−0.00338	0.01610	−0.00686	−151.92	−152.68

been studied over the past four decades [5–10]. Table 3 gives the hyperfine structure parameters of the $1s^2 2s2p\ ^3P$ state for the beryllium atom through the FCPC wave function to compare with data in the literature. As can be seen from Table 3, the present results for hyperfine structure parameters are better than the earlier theoretical results [5–7] in whole. The present calculations also agree with the results by FE MCHF (finite-element multiconfiguration Hartree-Fock) method [8] to two significant figures. The calculated Fermi contact term a_c in this work differs from the results from the latest calculation through MCHF method [9] by only 0.07%, and the differences for the other terms are on the order of a few parts in a thousand. This means that the wave function used in the present work is reasonable and accurate in the full configuration space. The hyperfine coupling constants A_J are also listed in Table 3 to compare with results from other calculations and experiments. Our calculated hyperfine coupling constants agree perfectly with the experimental value [10] to four significant figures. That is also true for the MCHF calculation of [9]. It is shown that hyperfine structure parameters of the low-lying excited states for the beryllium atom can be calculated accurately using the present FCPC wave function. For the $1s^2 2snp\ ^3P\ (n = 3, 4)$ states, to the best of our knowledge, there is no report on hyperfine structure parameters in the literature. The present predictions for the hyperfine structure parameters and coupling constants are listed in Table 4, which may provide valuable reference data for other theoretical calculations and experimental measurements in future.

4. Summary

In this work, energies, fine-structure splittings, and hyperfine structure parameters of the $1s^2 2snp\ ^3P\ (n = 2–4)$ states for the beryllium atom are calculated with the FCPC wave functions. The obtained nonrelativistic energies are much lower than the previous published theoretical values. The calculated fine structure splittings are in good agreement with experiment. For the $1s^2 2s2p\ ^3P$ state, the calculated hyperfine structure parameters are in good agreement with the latest theoretical and experimental data in the literature; it is shown that hyperfine constants of the low-lying excited states for the beryllium atom can be calculated accurately using this kind of wave function. For other states, the present predicted hyperfine structure parameters may provide valuable reference data for future theoretical calculations and experimental measurements.

Acknowledgments

The author is grateful to Dr. Kwong T. Chung for his computer code. The work is supported by National Natural Science Foundation of China and the Basic Research Foundation of Beijing Institute of Technology.

References

[1] A. Hibbert, "Oscillator strengths of transitions involving $2s3l\,^3L$ states in the beryllium sequence," *Journal of Physics B*, vol. 9, no. 16, pp. 2805–2811, 1976.
[2] A. W. Weiss, "Calculations of the $2sns\,^1S$ and $2p3p\,^{3,1}P$ Levels of Be I," *Physical Review A*, vol. 6, no. 4, pp. 1261–1266, 1972.
[3] L. Johansson, "The spectrum of the neutral beryllium atom," *Arkiv For Fysik*, vol. 23, pp. 119–128, 1962.
[4] C. Laughlin, E. R. Constantinides, and G. A. Victor, "Two-valence-electron model-potential studies of the Be I isoelectronic sequence," *Journal of Physics B*, vol. 11, no. 13, pp. 2243–2250, 1978.
[5] S. N. Ray, T. Lee, and T. P. Das, "Many-body theory of the magnetic hyperfine interaction in the excited state ($1s^2 2s2p\ ^3P$) of the beryllium atom," *Physical Review A*, vol. 7, no. 5, pp. 1469–1479, 1973.
[6] S. N. Ray, T. Lee, and T. P. Das, "Study of the nuclear quadrupole interaction in the excited ($2\,^3P$) state of the beryllium atom by many-body perturbation theory," *Physical Review A*, vol. 8, no. 4, pp. 1748–1752, 1973.
[7] D. R. Beck and C. A. Nicolaides, "Fine and hyperfine structure of the two lowest bound states of Be- and their first

two ionization thresholds," *International Journal of Quantum Chemistry*, vol. 26, supplement 18, pp. 467–481, 1984.

[8] D. Sundholm and J. Olsen, "Large MCHF calculations on the hyperfine structure of Be($^3P^O$): the nuclear quadrupole moment of ^{9}Be," *Chemical Physics Letters*, vol. 177, no. 1, pp. 91–97, 1991.

[9] P. Jönsson and C. F. Fischer, "Large-scale multiconfiguration Hartree-Fock calculations of hyperfine-interaction constants for low-lying states in beryllium, boron, and carbon," *Physical Review A*, vol. 48, no. 6, pp. 4113–4123, 1993.

[10] A. G. Blachman and A. Lurio, "Hyperfine structure of the metastable $(1s^2 2s2p)\,^3P$ states of $_4Be^9$ and the nuclear electric quadrupole moment," *Physical Review*, vol. 153, no. 1, pp. 164–176, 1967.

[11] K. T. Chung, "Ionization potential of the lithiumlike $1s^2 2s$ states from lithium to neon," *Physical Review A*, vol. 44, no. 9, pp. 5421–5433, 1991.

[12] K. T. Chung, X. W. Zhu, and Z. W. Wang, "Ionization potential for ground states of berylliumlike systems," *Physical Review A*, vol. 47, no. 3, pp. 1740–1751, 1993.

[13] Z. W. Wang and K. T. Chung, "Dipole polarizabilities for the ground states of lithium-like systems from $Z = 3$ to 50," *Journal of Physics B*, vol. 27, no. 5, pp. 855–864, 1994.

[14] C. Chen and Z. W. Wang, "Quadrupole and octupole polarizabilities for the ground states of lithiumlike systems from $Z = 3$ to 20," *The Journal of Chemical Physics*, vol. 121, no. 9, pp. 4171–4174, 2004.

[15] C. Chen and Z. W. Wang, "Total atomic scattering factors for the ground states of the lithium isoelectronic sequence from Na^{8+} to Ca^{17+}," *The Journal of Chemical Physics*, vol. 122, no. 2, Article ID 024305, 5 pages, 2005.

[16] X. X. Guan and Z. W. Wang, "The hyperfine structure of the $1s^2ns\,^2S$ and $1s^2np\,^2P$ states ($n = 2, 3, 4$, and 5) for the lithium isoelectronic sequence," *The European Physical Journal D*, vol. 2, no. 1, pp. 21–27, 1998.

[17] J. Carlsson, P. Jönsson, and C. Froese Fischer, "Large multiconfigurational Hartree-Fock calculations on the hyperfine-structure constants of the ^{7}Li $2s\,^2S$ and $2p\,^2P$ states," *Physical Review A*, vol. 46, no. 5, pp. 2420–2425, 1992.

[18] A. Hibbert, "Developments in atomic structure calculations," *Reports on Progress in Physics*, vol. 38, no. 11, pp. 1217–1338, 1975.

[19] P. Raghavan, "Table of nuclear moments," *Atomic Data and Nuclear Data Tables*, vol. 42, no. 2, pp. 189–291, 1989.

17

Statistical Complexity of Low- and High-Dimensional Systems

Vladimir Ryabov[1] and Dmitry Nerukh[2]

[1] *Department of Complex System, School of Systems Information Science, Future University Hakodate, 116-2 Kamedanakano-Cho, Hakodate-Shi, Hakodate, Hokkaido 041-8655, Japan*
[2] *Non-Linearity and Complexity Research Group, Aston University, Birmingham B4 7ET, UK*

Correspondence should be addressed to Dmitry Nerukh, d.nerukh@aston.ac.uk

Academic Editor: Keli Han

We suggest a new method for the analysis of experimental time series that can distinguish high-dimensional dynamics from stochastic motion. It is based on the idea of statistical complexity, that is, the Shannon entropy of the so-called ϵ-machine (a Markov-type model of the observed time series). This approach has been recently demonstrated to be efficient for making a distinction between a molecular trajectory in water and noise. In this paper, we analyse the difference between chaos and noise using the Chirikov-Taylor standard map as an example in order to elucidate the basic mechanism that makes the value of complexity in deterministic systems high. In particular, we show that the value of statistical complexity is high for the case of chaos and attains zero value for the case of stochastic noise. We further study the Markov property of the data generated by the standard map to clarify the role of long-time memory in differentiating the cases of deterministic systems and stochastic motion.

1. Introduction

Statistical complexity is a measure that had been introduced by Crutchfield and Young in 1989 [1]. It has been proven useful for describing various complex systems, including those with hundreds of degrees of freedom [2]. According to our earlier paper [3], the statistical complexity of high-dimensional trajectories generated by the dynamics of an ensemble of water molecules grows up to the time scale of 1 microsecond, that is, an extremely long-time interval for a typical molecular dynamics simulation. Moreover, this property is much less pronounced for so-called surrogate time series that have exactly the same power spectrum and, hence, autocorrelation function as the original time series.

For example, in Figure 1 we plot the dependence of statistical complexity on the length of the time series for the symbolic data obtained from a Poincaré the section of 3D velocities describing the motion of a hydrogen atom in an ensemble of 392 water molecules [3]. The details of computing the atomic trajectories as well as the method used for partitioning the phase space and obtaining a symbolic string from the initially floating point data can be found in [4]. In the same figure, we draw the curves calculated for so-called phase-shuffled surrogate time series [5], the data having identical autocorrelation function, and hence power spectrum as the original velocity trajectories. One can notice significant differences between the statistical complexity of the physical and the artificially generated data.

We then put forward a hypothesis that this property, that is a high value of statistical complexity, can be used for distinguishing between deterministic and stochastic systems (see also [6]). The phenomenon of the complexity growth with the length of time series that ensures the difference between the cases of deterministic and stochastic behaviour remains still unexplored. In order to elucidate the mechanism that makes the value of complexity high, we performed numerical experiments with the standard map (known also as the Chirikov-Taylor map) [7], one of the most studied paradigmatic models in nonlinear dynamics. We observed that statistical complexity was high indeed in the case of the standard map, and it had much lower value for the surrogate time series, being close to zero for the case of noncorrelated noise from a random number generator.

For the purpose of estimating statistical complexity for a symbolic time series, we utilize the CSSR algorithm [8] that

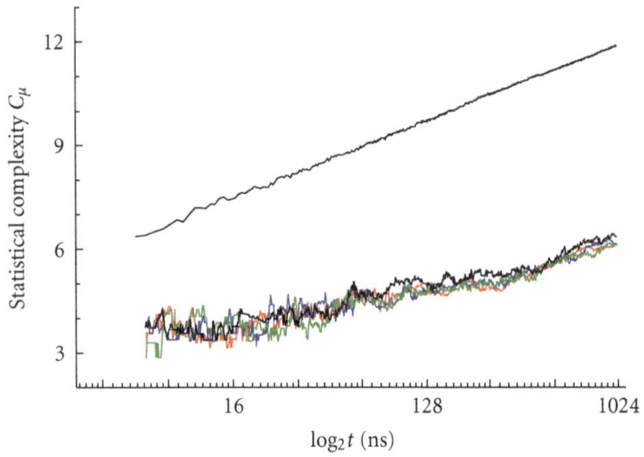

FIGURE 1: Statistical complexity versus the (log of) length of the analysis interval for the hydrogen velocity time series (top curve) and four surrogate time series (bottom curves): three independent realisations of the phase-shuffling algorithm (red, green, and blue), and single time series of a white noise passed through a low-pass linear filter (black) (from [3]). Note that the value of statistical complexity for the data obtained from simple random number generator is close to zero and does not depend on the length of time series for large enough value of the latter (not shown).

had been reported as an efficient, reliable, and easy to use software. The algorithm constructs an ϵ-machine, a Markov-chain with l-step memory, which constitutes a probabilistic model for the analysed data series. Statistical complexity measures an information content of the ϵ-machine via its Shannon entropy. Our analysis shows, however, that although the CSSR algorithm always converges well and produces a finite value of complexity, in some cases the approximation of data with a Markov-chain-type model is inadequate, making the complexity value dependent on the length of the analysed data. Finally, we came to the conclusion that, at least in the case of standard map, the main reason for the growth of complexity is the property of stickiness of periodic islands in the chaotic sea, a generic phenomenon in Hamiltonian systems [9]. It has been noted in [10] that due to the sticking property of the regular component in a subcritical domain ($K < 0.9716$), the dynamics of the standard map is subdiffusive that can be well approximated with a continuous time random walk model. Anomalous properties of the temporal behaviour of nonextensive entropy, a generalization of the usual Boltzmann-Gibbs entropy, have been also analysed in [11].

In the present work, we mainly study the domain of $K \gg 1$ where the area occupied by periodic islands is small, and the chaotic motion can be expected to be strongly mixing and ergodic. Nevertheless, as our results show, the presence of stickiness is still an important factor defining the long-term statistical measures. In terms of the CSSR algorithm, the property of stickiness breaks the independence of the data points separated by a history long-time interval, thus making the Markov-chain approximation invalid.

Finally, we discuss a conjecture that the property of the non-Markovianity of the ϵ-machine and growth of statistical complexity can be used in a constructive way for distinguishing deterministic and stochastic behaviours. The problem of detecting determinism in a noise looking chaotic time series is a long standing one. An extensive review of the issues related to the difference between chaos and noise and to inherent difficulties encountered in the high-dimensional cases can be found in [12]. We suggest that there is a significant difference between the statistical complexities of Hamiltonian chaos and coloured noise with identical power spectrum, the main reason for which consists in the presence of the long-time memory in time series obtained from Hamiltonian systems. This property originates from the stickiness of periodic islands that are abundant in the chaotic sea due to multiple resonances that occur in the phase space.

We would also like to note that since the phase space of Hamiltonian systems has a complicated structure of chaotic areas intermingled with periodic islands, this leads in some cases to the necessity of distinguishing between chaos and complex quasiperiodic motion. A measure called *orbital complexity* had been introduced for this purpose in the context of analysing the orbital motions of planets [13–15]. This measure, although being based on the calculation of the Shannon entropy (but in the spectral domain), has quite different meaning, purpose, and scope of applicability compared to statistical complexity.

2. Systems and Method of Analysis

The standard map is defined as

$$\begin{aligned} p_{n+1} &= p_n + K \sin \theta_n \bmod 2\pi, \\ \theta_{n+1} &= \theta_n + p_{n+1} \bmod 2\pi, \end{aligned} \tag{1}$$

where K is a single parameter defining the dynamics of this system. In all the calculations below, the value of the parameter K has been chosen at $K = 6.908745$.

First, at the step called "symbolization," the original real-valued time series is transformed to a symbolic sequence by introducing a suitable partitioning of the phase space (Figure 2(a)).

At the next stage, the sequence of symbols is transformed to the sequence of histories, the l-symbol strings representing a refinement of the partitioning in the phase space [3]. ϵ-machine reconstruction requires a grouping of histories to "causal states," based on the analysis of the predictive properties of each history by one step forward in time. Finally, the statistical complexity is calculated from the ϵ-machine as Shannon entropy of the probability distribution of the causal states:

$$C_\mu = -\sum_{i=1}^{N_c} p_i \log p_i, \tag{2}$$

here p_i are probabilities of the causal states in the ϵ-machine, and N_c is the total number of the causal states.

3. Numerical Experiments

We have calculated the statistical complexity using the algorithm called CSSR [8] for the standard map and plotted

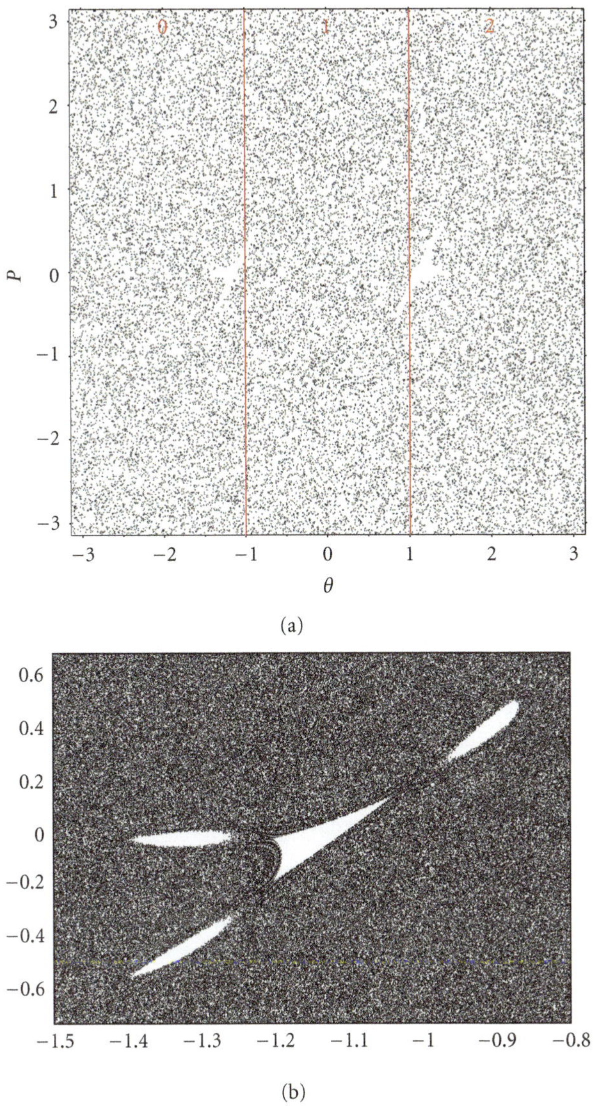

FIGURE 2: The standard map. Symbolization with a three-symbol alphabet (a). Two periodic islands are embedded into the chaotic sea. One of them zoomed is shown in (b).

the graphs of complexity versus the amount of data (the length of symbolic sequence). We have also studied how the calculated values depend on the method of partitioning the map, initial conditions, and the parameter K of the system.

A typical plot of statistical complexity C_μ and the number of causal states for the history length $l = 2 \cdots 9$ are shown in Figure 3.

The results for the surrogate data generated using the same trajectory of the standard map are shown in Figure 4. Changing the initial conditions, the type of partitioning the phase space at the stage of symbolization, and/or value of the parameter K brings qualitatively the same results, that is, there is a significant difference between the complexity values calculated for the data obtained from the map and those for the surrogate time series. It should be noted though that the complexity value moderately increases with parameter K as shown in Figure 5. This behaviour is similar to that of other

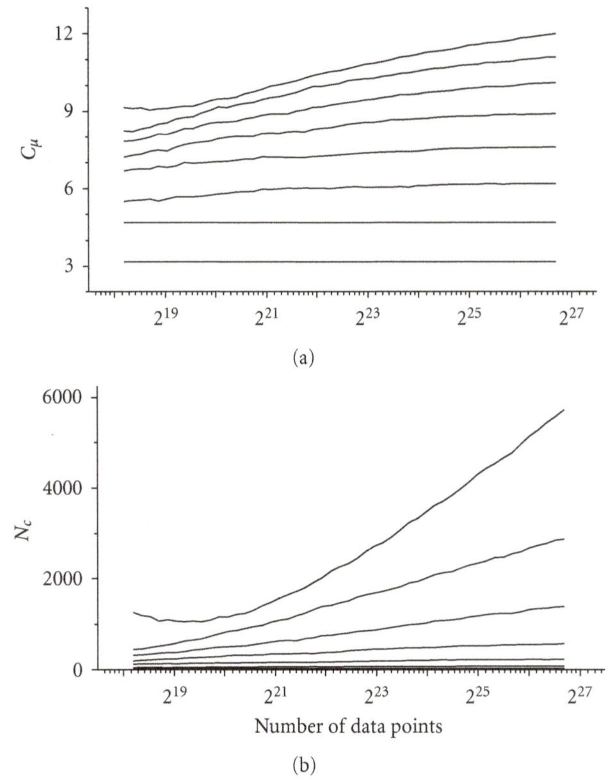

FIGURE 3: The values of C_μ and the number of causal states for various history lengths (from bottom to top $l = 2 \cdots 9$) for the standard map trajectory as a function of the data length.

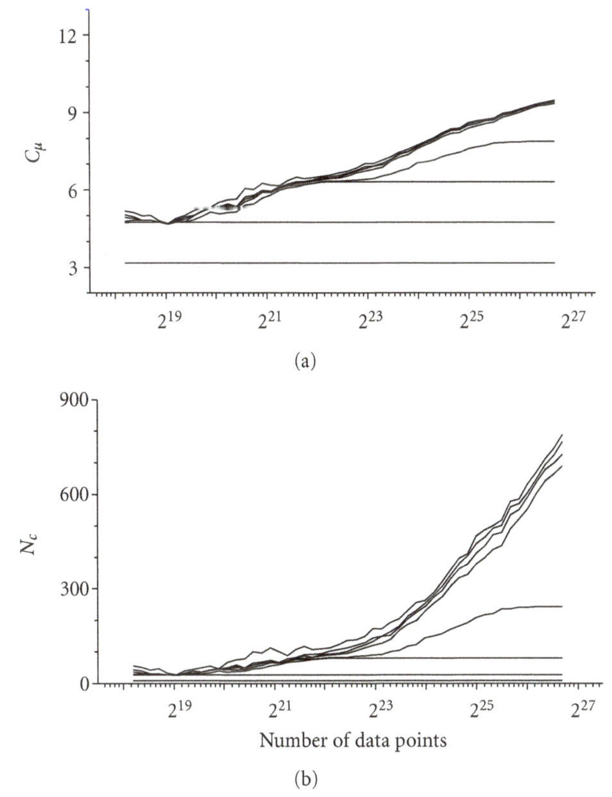

FIGURE 4: Same as in Figure 3 but for the random surrogate.

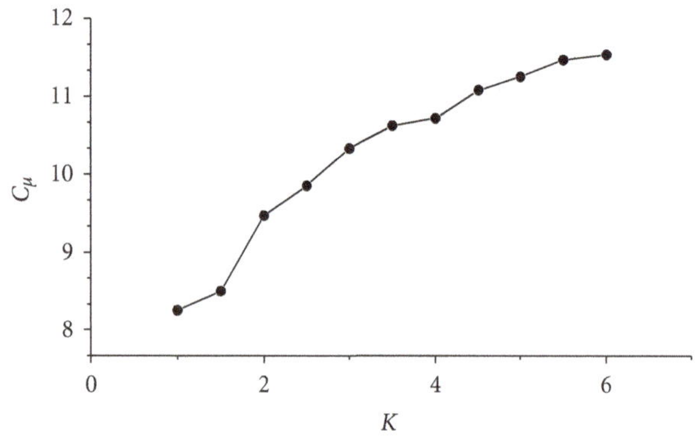

FIGURE 5: Dependence of statistical complexity on the parameter K (the number of data points is 10^8).

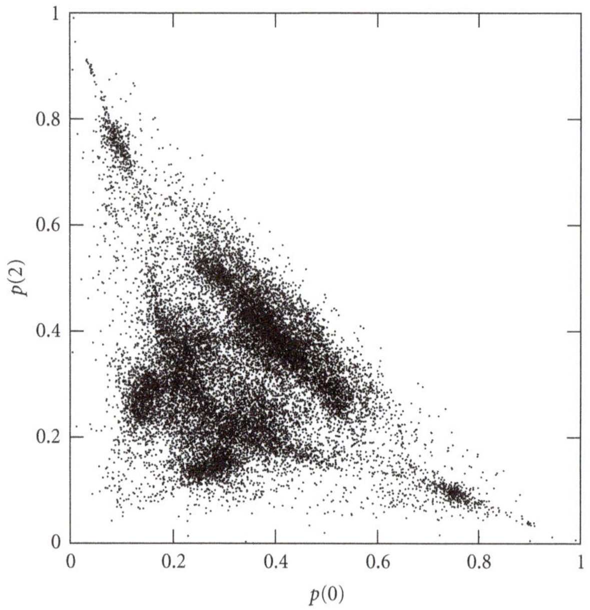

FIGURE 6: Conditional distributions of the next symbol for all histories at history length $l = 8$. Every point represents a history. The total number of points (histories) is 3^8.

characteristics used in nonlinear dynamics, like Lyapunov exponents or measure of the chaotic area reported in [16].

4. A Hypothesis on Markov Property

In this section, we would like to demonstrate that the large complexity values observed in the case of the standard map are caused by the presence of certain segments in the chaotic trajectory (which become histories after symbolization) that do not possess a property necessary for building a Markov-chain from the data. Consider the stage when the symbolic string has been converted to the sequence of histories, that is, symbolic words of length l. The Markov-chain (ϵ-machine) can be built from such a sequence, if the conditional probability distribution of the next symbol in the symbolic sequence depends only on the l-symbol string preceding the symbol, and it is independent on the previous symbol, that

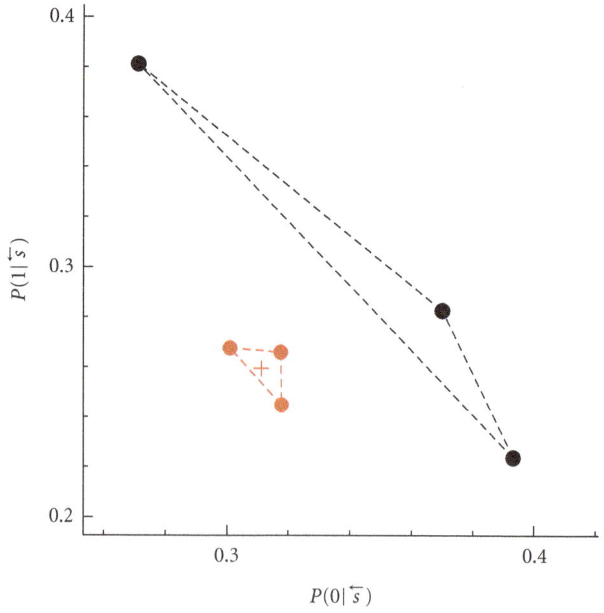

FIGURE 7: Conditional probability distribution for the history with the largest deviation from the Markov property (black triangle). Adding a symbol to the history changes drastically the position of the point in the diagram. Three circles correspond to adding "0," "1," or "2" to the history $l = 8$ (crosses). The red triangle represents a randomly chosen history with clear Markov property, that is, the distribution of probabilities does not depend on the added symbol.

is, the symbol that occurred $l + 1$ time steps before. In other words, if we consider the conditional probability distribution for a given history, it should not change (in statistical sense), if we increase the length of a history by one symbol to the past.

In Figure 6, we present a scatter diagram that demonstrates the distribution of the conditional probabilities for each history at the history length $l = 8$. Every point in the diagram corresponds to a single history. The large spread around the point with coordinates $(1/3, 1/3)$ evidences significant difference compared to the case of uniform distribution. The change from $l = 8$ to $l = 9$ does not change the overall pattern of point distribution shown in Figure 6. However, the analysis of the conditional probabilities for individual histories reveals huge changes in the position of points depending on the extra symbol added at the beginning of the history. In Figure 7, we depict the conditional distribution for the next symbol for two histories: one that shows strong deviation from Markov property and a "normal" one, that is, a randomly chosen history. The large deviation in the distribution of conditional probabilities can be concluded from a big distance between the vertexes of the upper triangle (distribution of the conditional probabilities at $l = 9$) and the cross corresponding to the conditional probability at $l = 8$. The probabilities at $l = 9$ are computed by adding one of the three symbols (012) at the beginning of the history of $l = 8$.

Finally, we would like to show that the segments of chaotic trajectories that correspond to the history with large deviation from the Markov property are located in the areas

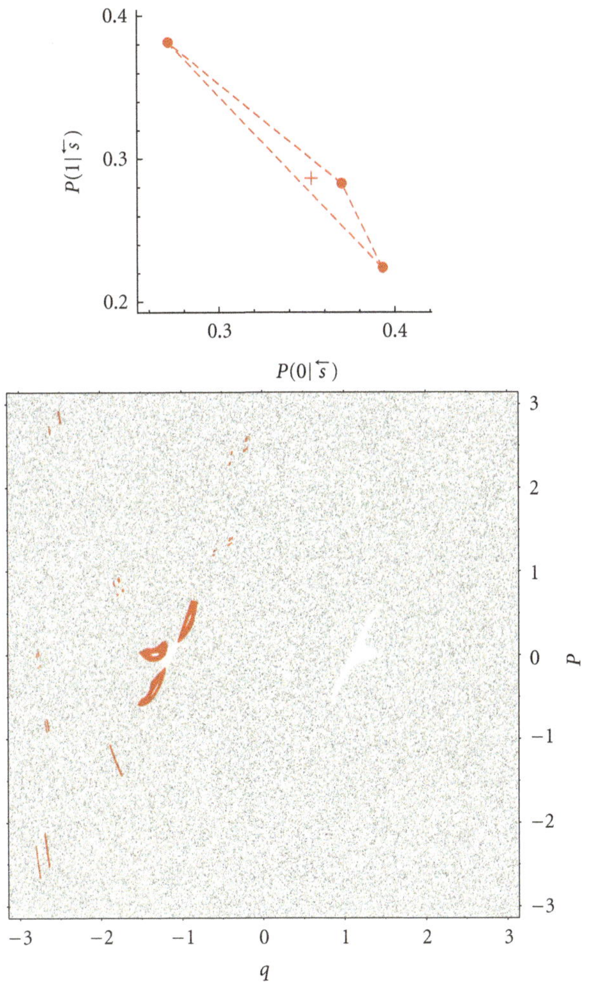

FIGURE 8: Parts of chaotic trajectory corresponding to the history with large deviation from the Markov-chain property. Only points corresponding to the central three symbols in the history are shown. Apparently, the history includes one of the two periodic islands shown in Figure 2.

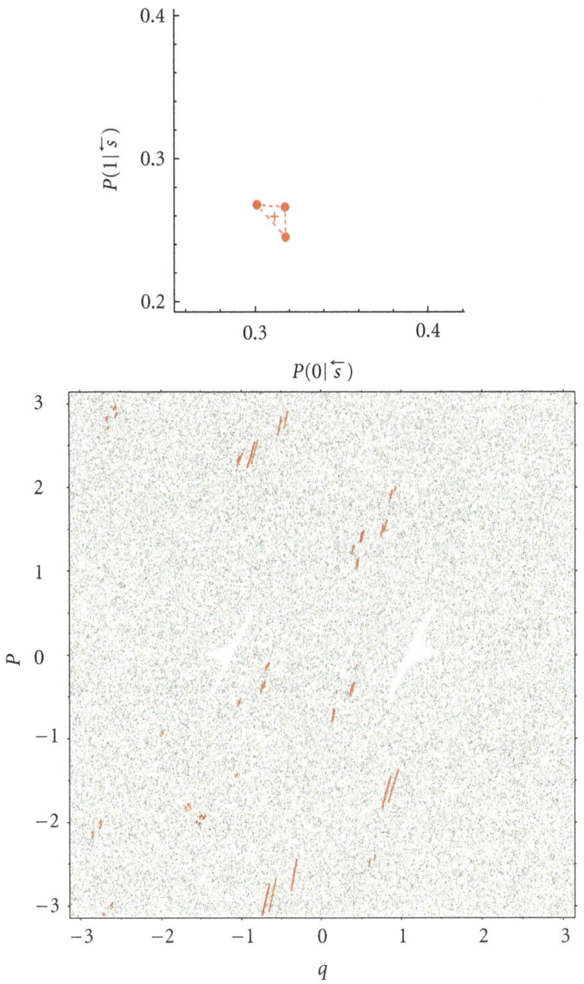

FIGURE 9: Parts of chaotic trajectory corresponding to the history with no deviation from the Markov-chain property. Only points corresponding to the central three symbols in the history are shown. Apparently, the history does not include any of the two periodic islands shown in Figure 2.

of the phase space close to periodic islands. For this purpose, we plotted in Figure 8 only the points that correspond to the history with large deviation from the Markov-chain property. A comparison to Figure 2 suggests that the history with large deviation in the distributions is located close to the periodic islands. Therefore, we suppose that the breaking of Markovianity can be interpreted as a manifestation of the well-known phenomenon of "stickiness" [9] of trajectories in the areas close to periodic islands. Prolonged wandering of a trajectory around the island is equivalent to existing of long-time memory in the corresponding segments of the chaotic time series. Figure 8 should be also compared to Figure 9, which presents the segments of the chaotic trajectory corresponding to a history possessing the Markov property. Apparently, it has no relation to periodic islands. Such histories represent a vast majority in the ensemble of 3^l histories, the non-Markovian histories constituting only a fraction of percent.

5. Discussion

It has been demonstrated in this paper that statistical complexity appears to be a useful measure for distinguishing Hamiltonian chaos in low- and high-dimensional systems from correlated noise with identical autocorrelation function. Its value for the symbolic time series calculated from the dynamics of Hamiltonian systems is substantially larger than that for a white noise time series or the time series obtained from the phase shuffling surrogate algorithm. Our explanation of the origin of this phenomenon in terms of Markov-chain theory consists in breaking down the Markov property by the symbolic sequences obtained from Hamiltonian systems.

We believe that the large value of complexity observed in our numerical experiments is defined by the presence of periodic islands with sticky borders in the phase space of Hamiltonian systems. The stickiness of certain areas in the phase space leads to long-time memory effects that are responsible for breaking down the statistical independence of the future states from the past ones. This, in turn, makes the procedure of grouping the histories into causal states constituting the core of CSSR algorithm unstable. As a result, the algorithm finds more and more causal states necessary for building the ϵ-machine as a Markov chain, and the value of complexity grows with the number of causal states.

References

[1] J. P. Crutchfield and K. Young, "Inferring statistical complexity," *Physical Review Letters*, vol. 63, no. 2, pp. 105–108, 1989.

[2] D. P. Feldman, C. S. McTague, J. P. Crutch-field et al., "The organization of intrinsic computation: complexity-entropy diagrams and the diversity of natural information processing," *Chaos*, vol. 18, no. 4, Article ID 043106, 15 pages, 2008.

[3] D. Nerukh and V. Ryabov, "Computational mechanics of molecular systems," in *Computational Mechanics Research Trends, Computer Science, Technology and Applications*, H. P. Berger, Ed., Nova Science, 2010.

[4] D. Nerukh, V. Ryabov, and R. C. Glen, "Complex temporal patterns in molecular dynamics: a direct measure of the phase-space exploration by the trajectory at macroscopic time scales," *Physical Review E*, vol. 77, no. 3, Article ID 036225, 2008.

[5] J. Theiler, S. Eubank, A. Longtin, B. Galdrikian, and J. Doyne Farmer, "Testing for nonlinearity in time series: the method of surrogate data," *Physica D*, vol. 58, no. 1-4, pp. 77–94, 1992.

[6] J. M. Amigó, S. Zambrano, and M. A. F. Sanjuán, "Combinatorial detection of determinism in noisy time series," *EPL*, vol. 83, no. 6, Article ID 60005, 2008.

[7] B. V. Chirikov, "A universal instability of many-dimensional oscillator systems," *Physics Reports*, vol. 52, no. 5, pp. 263–379, 1979.

[8] C. R. Shalizi and K. L. Shalizi, "Blind construction of optimal nonlinear recursive predictors for discrete sequences," in *Proceedings of the Uncertainty in Artificial Intelligence, 20th Conference*, M. Chickering and J. Halpern, Eds., pp. 504–511, AUAI Press, 2004.

[9] G. M. Zaslavsky, "Chaos, fractional kinetics, and anomalous transport," *Physics Report*, vol. 371, no. 6, pp. 461–580, 2002.

[10] J. H. Misguich, J.-D. Reuss, Y. Elskens, and R. Balescu, "Motion in a stochastic layer described by symbolic dynamics," *Chaos*, vol. 8, pp. 248–256, 1998.

[11] F. Baldovin, C. Tsallis, and B. Schulze, "Nonstandard entropy production in the standard map," *Physica A*, vol. 320, pp. 184–192, 2003.

[12] G. Boffetta, M. Cencini, M. Falcioni, and A. Vulpiani, "Predictability: a way to characterize complexity," *Physics Reports*, vol. 356, no. 6, pp. 367–474, 2002.

[13] N. T. Faber, C. M. Boily, and S. Portegies Zwart, "On time-dependent orbital complexity in gravitational N-body simulations," *Monthly Notices of the Royal Astronomical Society*, vol. 386, no. 1, pp. 425–439, 2008.

[14] H. E. Kandrup, B. L. Eckstein, and B. O. Bradley, "Chaos, complexity, and short time Lyapunov exponents: two alternative characterisations of chaotic orbit segments," *Astronomy and Astrophysics*, vol. 320, no. 1, pp. 65–73, 1997.

[15] I. V. Sideris and H. E. Kandrup, "Chaos and the continuum limit in the gravitational N-body problem. II. Nonintegrable potentials," *Physical Review E*, vol. 65, no. 6, Article ID 066203, 2002.

[16] I. I. Shevchenko, "Isentropic perturbations of a chaotic domain," *Physics Letters A*, vol. 333, no. 5-6, pp. 408–414, 2004.

18

A First-Principles-Based Potential for the Description of Alkaline Earth Metals

Johannes M. Dieterich, Sebastian Gerke, and Ricardo A. Mata

Institut für Physikalische Chemie, Universität Göttingen, Tammannstrasse 6, 37077 Göttingen, Germany

Correspondence should be addressed to Johannes M. Dieterich, jdieter@gwdg.de

Academic Editor: David Wales

We present a set of Gupta potentials fitted against highest-level *ab initio* data for interactions of the alkaline earth metals: beryllium, magnesium, and calcium. Reference potential energy curves have been computed for both pure and mixed dimers with the coupled cluster method including corrections for basic set incompleteness and relativistic effects. To demonstrate their usability for the efficient description of high-dimensional complex energy landscapes, the obtained potentials have been used for the global optimization of 38- and 42-atom clusters. Both pure and mixed compositions (binary and ternary clusters) were investigated. Distinctive trends in the structure of the latter are discussed.

1. Introduction

Metallic clusters have become over the years a subject of intense study, both theoretical as well as experimental [1]. Interest stems from the distinct properties they reveal when compared to the bulk phase and how these may change as a function of the cluster size. Different compositions (in binary, ternary, and higher mixtures) can also lead to new chemical and physical phenomena. Nanoalloys are a prime example of how both factors can be combined for material design and application in catalysis [2, 3]. The computational study of their structures is a challenging task for two interlacing reasons. On the one hand, the number of local minima is considered to scale exponentially with the cluster size, making the search for the global minimum *NP-hard* [4]. This property reflects back on all algorithms designed to explore the energy landscape of such systems. On the other hand, a suitable theoretical description of the interactions in play is required. It needs to be accurate enough to properly describe the energy landscape for a wide range of bonding patterns. It should also be amenable to computation, meaning that the computation of several hundred many-body interactions can be carried out in a sensible time frame. This is even more important since multiple thousands of these computations are required for a proper sampling of the energy landscape.

One of the most successful approaches to the study of metallic clusters has been the combination of fitted potentials with global optimization algorithms [5–8]. The former are usually obtained by fitting experimental data or electronic structure results to an analytic expression. The brute force use of quantum mechanical methods is impractical due to the computational cost, particularly linked to its scaling relative to the system size. Even semiempirical methods may be too costly as the prefactors are high enough to hinder a proper sampling of conformational space.

In this work, we have made use of correlated wave function methods to calculate the two-body interaction potential of alkaline earth metals (Be, Mg, and Ca). Emphasis has been placed on obtaining converged energy profiles relative to basic set, relativistic, and electronic correlation effects. The high-level reference data thus obtained was mapped to a two-body Gupta-type potential, [9] which in turn could be used to explore the structure of pure, binary, and ternary clusters. A few comments should be made about this choice of approach. First of all, it follows a bottom-to-top rationale that no information about nano- or macroscopic materials is used in the fit. It is purely based on first principles results that no empirical information (aside from the form of the chosen potential) has been included. This can certainly be seen as an advantage, since it allows us to improve the description in a systematic way. However, since the reference data has been computed with computationally demanding methods, it is not possible to benchmark the fit by repeating calculations for a selected test set of clusters. In fact, some of the terms included in the energy expression would be hard to obtain even for a 3-atom system. The advantages and disadvantages of our choice are later discussed in the text.

2. Methods and Techniques

Both for the cluster structure optimization as well as the potential fit, the OGOLEM framework for global optimization was used. Its features have been introduced in a series of publications [10–12]. Therefore, we will restrict ourselves to a brief discussion of the relevant features. The OGOLEM framework is loosely based on genetic algorithms as described in [13] replacing the generation-based scheme with the more efficient genetic pool scheme. While standard generation-based schemes feature serial bottlenecks at the end of every generation, a pool-based scheme removes this constraint through constant updates of a genetic population, allowing for a more efficient parallelization of the algorithm. As a side effect, elitism is a built-in feature of any genetic pool scheme, therefore removing the need to define additional criteria for it. Since the genetic pool contains all current solution candidates, parent individuals are chosen from it (father based on ranked fitness, mother randomly) and subject to the usual genetic operations: crossover and mutation. The crossover operator used for the global potential fit is a one-point genotype operator accompanied by a genotype mutation (probability 5%). For the cluster structure optimization, our implementation of a phenotype operator [11] is used, again accompanied by a genotype mutation (probability 5%). It should be noted here that no explicit exchange mutation (as, e.g., proposed in [6] and applied in [11]) was used. The phenotype implementation already includes some internal exchange which proved effective enough for lightly mixed clusters as targeted in this study.

In the case of cluster structure optimization, the solution candidates are then subject to a graph-based collision and dissociation detection. Should a candidate structure show either, it will be rejected and does not enter the subsequent local optimization. In the case of the potential fits, no such restriction is applied. Finally, it is attempted to add the fitter of the two locally optimized individuals to the genetic pool. This operation is only successful if it does not violate the fitness-based diversity criterion. After a defined number of these iterations, a converged solution pool is obtained, containing the global minimum candidate. In the case of cluster structure optimizations, such candidate is only accepted if four independent runs yield the same individual.

3. Global Fit of Potentials

All two-body interactions of beryllium, magnesium, and calcium have been fitted against highest-level *ab initio* data. The numerical data will be published elsewhere [27]. To obtain the highest possible accuracy at a still affordable computational footprint, different levels of theory based on wave function methods are combined as follows:

$$E_{\text{inter}} = E_{\text{HF}}^{\infty} + \Delta E_{\text{CCSD(T)}}^{\infty} + \Delta E_{\text{rel}} + \Delta E_{\text{Q}}, \quad (1)$$

where E_{HF}^{∞} is the CBS[3, 4, 5]-extrapolated HF/aug-cc-pCVXZ [14–16] energy as proposed by Feller [17], $\Delta E_{\text{CCSD(T)}}^{\infty}$ is the CBS[4, 5]-extrapolated correlation energy using the CCSD(T)/aug-cc-pCVXZ ($X = Q, 5$) level of theory with the $X - 3$ formula, ΔE_{rel} is a relativistic correction using a Douglas-Kroll Hamiltonian at the CCSD(T)/aug-cc-pCVTZ-DK [15] level of theory, and ΔE_{Q} is the quadruples contribution to the correlation energy obtained with CCSDT(Q)/aug-cc-pVTZ with the frozen core approximation in place. All calculations were performed with the Molpro2010.1 program package [18]. The CCSDT(Q) runs were carried out by the MRCC program [19, 20] interfaced to the latter.

The quality of this data set is high enough to reproduce the experimental dissociation energy of 11.1 kJ mol $^{-1}$ and equilibrium distance of 2.45 Å for the beryllium dimer [21] and can be expected to be of similar quality for the other interactions. Additionally, it provides a consistent data set for all pairs. For the latter property, the inclusion of all electrons in the $\Delta E_{\text{CCSD(T)}}^{\infty}$ term calculation and the inclusion of relativistic effects are of particular importance.

The Gupta potential [9] used is of the regular form

$$E(a,b) = A(a,b) \cdot \exp\left[-p(a,b)\left(\frac{r_{ab}}{r_0(a,b)} - 1\right)\right] - \sqrt{\chi(a,b)^2 \cdot \exp\left[-2 \cdot q(a,b)\left(\frac{r_{ab}}{r_0(a,b)} - 1\right)\right]}, \quad (2)$$

where r_{ab} is the distance between atoms a and b, and $A(a,b)$, $p(a,b)$, $r_0(a,b)$, $\chi(a,b)$, and $q(a,b)$ are the parameters to be fitted against the reference.

Due to the rigid nature of the Gupta potential, a weighting of data points was necessary to guarantee a good fit. This weighting followed the rationale that an exact reproduction of the depth and position of the minimum is most important. A good reproduction of the attractive part of the potential was the second target, and less focus was placed on reproducing the repulsive part. We consider these to be reasonable design principles, reflecting the standard demands on potentials. Used weighting factors are tabulated in the supplementary information (see Supplementary Table in the Supplementary Material available online at doi: 10.1155/2012/648386).

The derived potentials are depicted in Figures 1 and 2 with the numerical values of the parameters to four digits precision available in Table 2. Perhaps one of the most striking features, upon inspection of the figures, is the difficulty in describing the weak-bonding regime. Some of the potential curves show a close to linear profile on approaching the minimum. This is the case for the Be-Be interaction and, less drastically, for the Be-Mg interactions. In the former case, a clear platteau is visible. Under the constraints of the potential

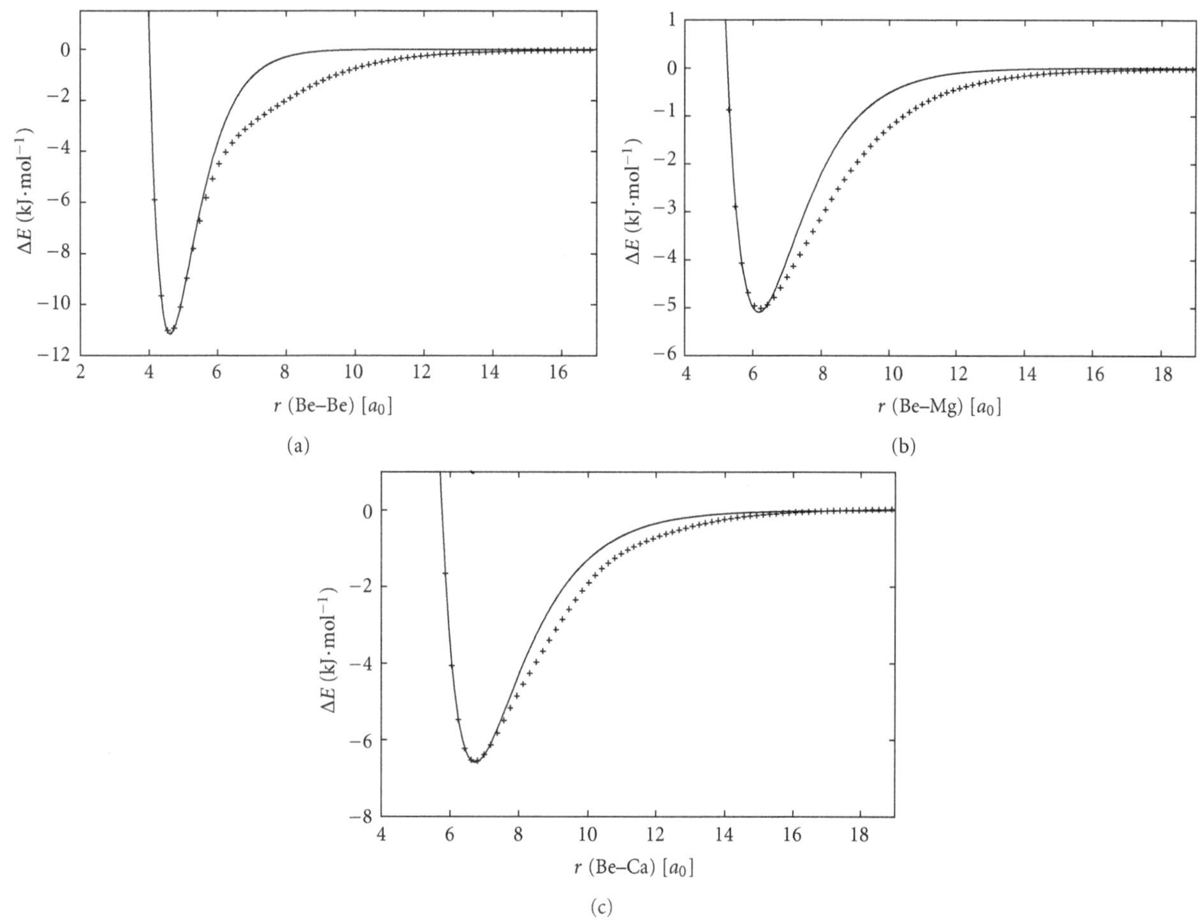

FIGURE 1: Derived Gupta potentials for Be-Be, Be-Mg, and Be-Ca interactions.

TABLE 1: Quality of the fit for all alkaline earth interactions. Distance regime r_{AB} in Å and absolute and average deviation in kJ/mol. Absolute deviation includes weights; average deviation is weight-free.

Pair	r_{AB}	N_{ref}	Abs. dev.	Avg. dev.
Be-Be	2.0 → 10.0	81	47.43	0.59
Be-Mg	2.5 → 15.0	126	25.71	0.20
Be-Ca	2.7 → 15.0	124	25.80	0.21
Mg-Mg	3.0 → 15.0	121	13.79	0.11
Mg-Ca	3.2 → 15.0	119	11.42	0.10
Ca-Ca	3.5 → 15.0	116	16.60	0.14

form chosen, it is not possible to correctly reproduce this behavior without significantly affecting the description of the minimum. Nevertheless, all fitted potentials accurately describe the position and depths of the minimum correctly and are in overall good agreement with the reference. The Mg-Ca and Ca-Ca fits reproduce extremely well the reference data. Numerical information on the fitting quality can be obtained from Table 1. It should be noted though that the depth of the potential needs to be taken into account. The average deviation of 0.14 kJ·mol $^{-1}$ for the Ca-Ca interaction (minimum depth approximately 11 kJ·mol^{-1}) is less severe than the average deviation of 0.11 kJ·mol $^{-1}$ for the Mg-Mg interaction (minimum depth approximately 5 kJ·mol $^{-1}$). Further enhancements in the description would ultimately require another potential type, either another rigid potential more suitable for these interactions or a more flexible potential form. Both Morse potentials and LJ-type potentials were found to be unsuitable to overcome this principle problem. In a recent study by Li et al. [22], the Tang-Toennies potential model was used to fit experimental data of homogeneous alkaline earth dimers. The attractive part of the Be-Be interaction could not be perfectly described in this case either.

TABLE 2: Numerical values to four digits precision for the fitted Gupta potentials. All values in atomic units.

Parameter	Be-Be	Be-Mg	Be-Ca	Mg-Mg	Mg-Ca	Ca-Ca
A	1.7943	2.1964	1.3955	2.7232	2.5646	0.8815
p	4.2656	2.0473	3.5205	2.8024	2.0231	3.3835
r_0	2.0323	1.4388	2.3759	1.8582	1.7852	3.9567
χ	−0.4088	0.2066	0.0913	−0.0221	−0.1453	−0.5749
q	2.7536	1.1548	1.6065	0.7203	0.9789	2.8282

Further enhancement to the potential would also be possible through parametrization of three-body terms. These

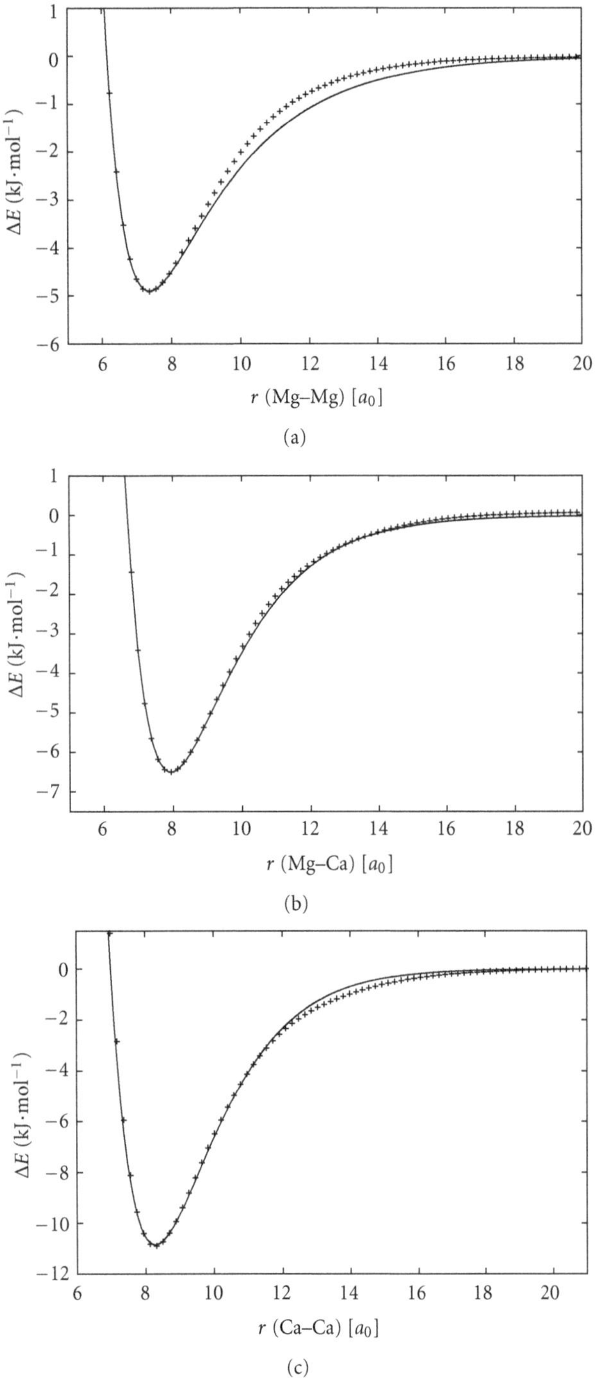

FIGURE 2: Derived Gupta potentials for Mg-Mg, Mg-Ca and Ca-Ca interactions.

would have to be computed at a lower level of theory due to the large number of points needed and the size increase in the system. The computation of quadruple excitations is particularly costly and would be hard to perform in systems other than dimers. A possible approach would be to add an effective 3-body term in agreement with experimental structural data or by using simulation results at a lower level. Caution should be taken in computing such a term

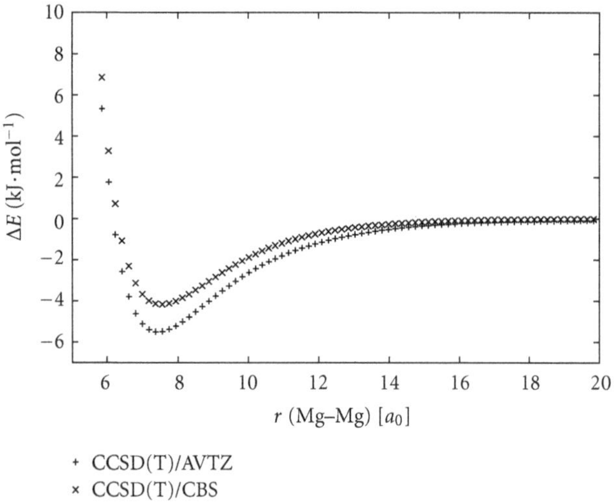

FIGURE 3: Mg-Mg interaction energy curves obtained at the CCSD(T)/CBS and CCSD(T)/AVTZ levels of theory.

from three-atom systems for two reasons. First of all, it is expected that basic set superposition effects (BSSE/BSIE) can contaminate the potential. Most importantly, we note that many-body stabilization is overestimated when considering only 3-body interactions [25]. To illustrate the BSSE/BSIE problem, we compare in Figure 3 the energy profile for the Mg dimer computed at the CCSD(T)/aug-cc-pCVTZ level (CCSD(T)/AVTZ) and the energy obtained from the first two terms in (1) (CCSD(T)/CBS). The difference between the two sets of data is exclusively due to differences in the basic set. The use of a triple-zeta quality basic set leads to a clear overestimation of the well depth. The CCSD(T)/AVTZ level of theory predicts the equilibrium distance at 7.4 a_0 with a dissociation energy of 5.1 kJ·mol $^{-1}$ in contrast to the CCSD(T)/CBS prediction of 7.6 a_0 and 4.0 kJ·mol $^{-1}$, respectively. This amounts to an error of approximately 20% in the dissociation energy. If one were to estimate three-body terms with the triple-zeta basis, an overestimation will be expected. The basic functions of a third atom can contribute to the basic space of the neighboring dimer, resulting in a biased potential. Only close-to-CBS values could be used for correctly estimating 3-body contributions.

In general, we expect that the inclusion of 3-body terms should amount to an overall compression of the structure which would, in turn, induce local structural changes [25]. This could, however, be balanced by even higher-order terms in the many-body expansion. Work in this direction is underway.

4. Cluster Structure Optimization

To demonstrate the real-world applicability of the derived potentials, they have been used in the global optimization of medium-sized alkaline earth clusters. We focused on clusters of 38 alkaline earth atoms since this size typically exhibits the most interesting structural behaviour in the medium size regime [8]. To check whether the observed structural trends

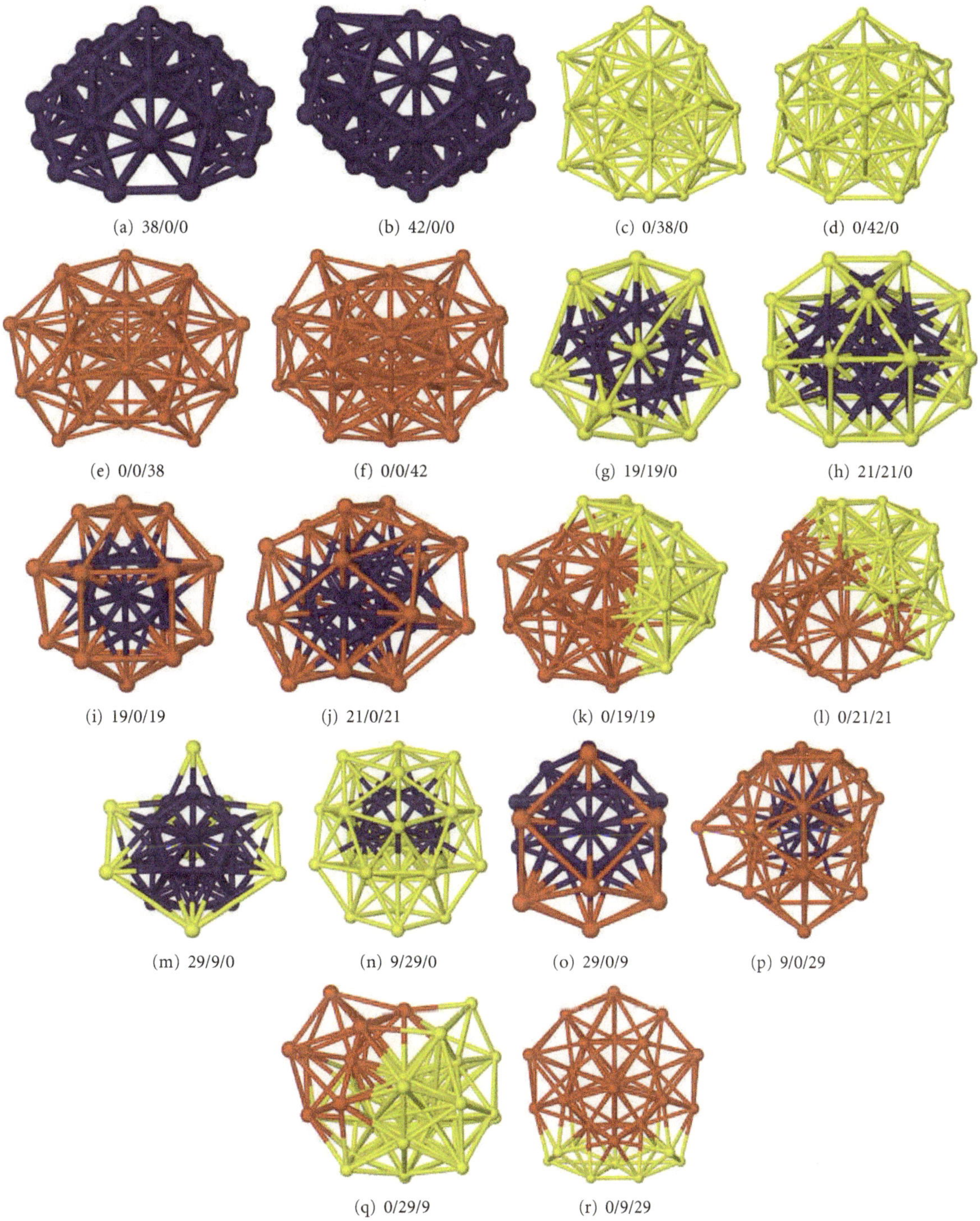

FIGURE 4: Global minimum candidate structures of homogenous and binary clusters of the alkaline earth metals beryllium (indigo), magnesium (yellow), and calcium (red). The caption $X/Y/Z$ denotes the number of beryllium atoms X, of magnesium atoms Y, and calcium atoms Z. All graphics are obtained with Jmol [23] and POV-ray [24].

are specific to this cluster size, similar compositions in 42 atom clusters have been optimized. The structural data will be available from the Cambridge Cluster Database [26] after publication. All global minimum candidate structures are depicted in Figures 4 and 5.

The homogeneous clusters show icosahedral structural motifs. Depending on the atom in play, the structure varies slightly. While Be_{38}, Ca_{38}, and Ca_{42} possess mirror plane symmetry and seem to be magic numbers, Be_{42}, Mg_{38}, and Mg_{42} lack a number of atoms in defined positions, which is

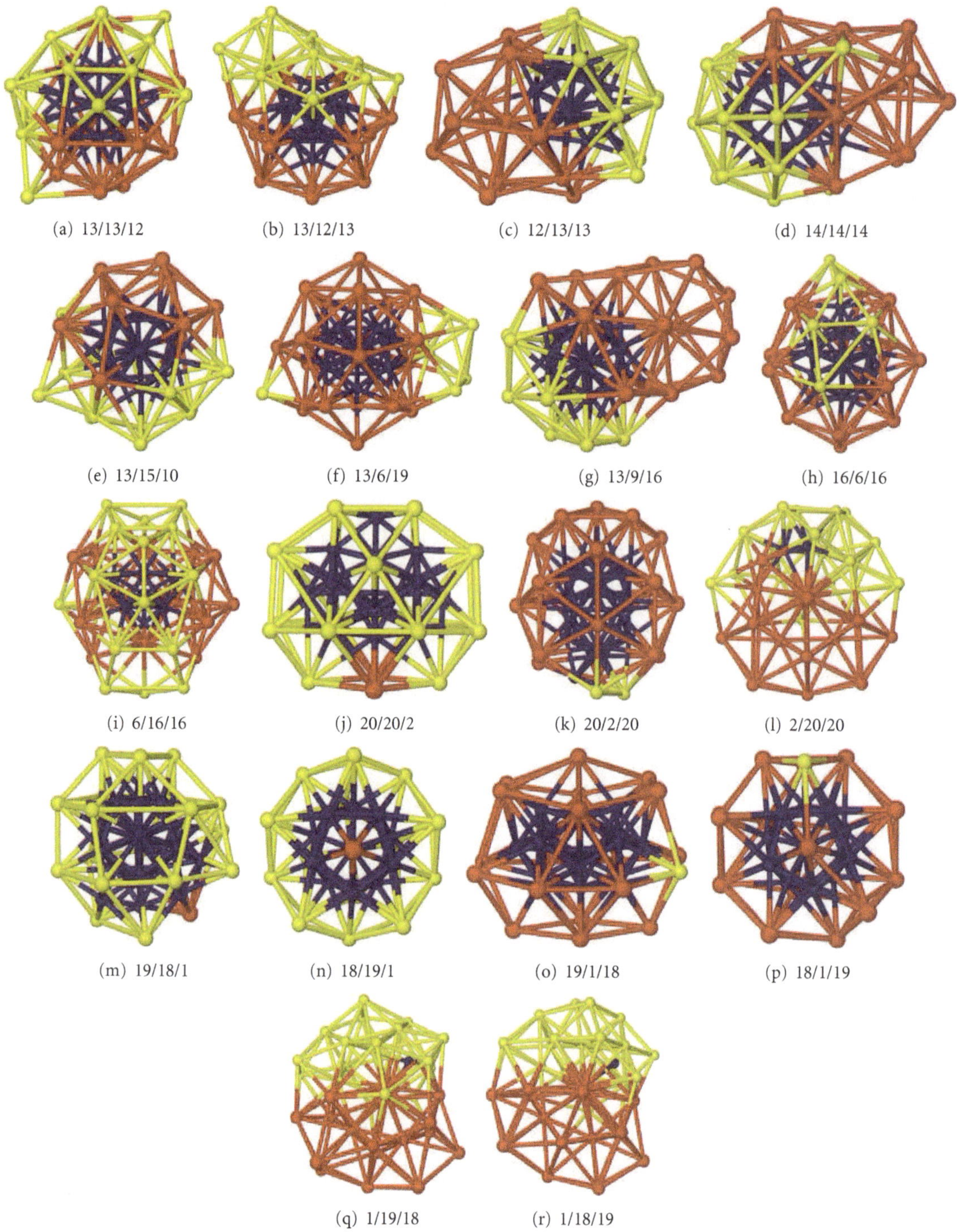

FIGURE 5: Global minimum candidate structures of ternary clusters of the alkaline earth metals: beryllium (indigo), magnesium (yellow), and calcium (red). The caption $X/Y/Z$ denotes the number of beryllium atoms X, of magnesium atoms Y, and calcium atoms Z. All graphics are obtained with Jmol [23] and POV-ray [24].

clear through visual inspection. It should be noted that no stable *fcc* structure could be located for any of the alkaline earth metals.

The same principle motifs hold true for the binary compositions. Common features are icosahedral substructures and real or pseudo mirror plane symmetry. Additional structural motifs can be observed for all binary clusters. First of all, a segregation of atom types can be observed in the form of the well-known *core-shell* structures [8] for all clusters containing beryllium. Beryllium forms an icosahedral core which can be easily explained with the potential profiles. The Be-Be interaction exhibits a deep and narrow minimum at a short distance. In contrary, the Mg-Mg and Ca-Ca interactions are both either not as deep (magnesium) or not as narrow (both magnesium and calcium). The formation of *core-shell* structures is also supported by the shape of the Be-Mg and Be-Ca potentials. In both cases, the minimum is located at longer distances than the Be-Be equilibrium distance and is not as deep as the Be-Be one. Obviously, the system must maximize the number of Be-Be contacts for an energetically low structure, which is only the case for a small icosahedral beryllium core.

A segregation of atom types can also be observed for the Mg/Ca binary compositions albeit not in the form of *core-shell* structures. Again, the potentials provide evidence for this behaviour. The Ca-Ca interactions possess a deeper minimum than the Mg-Ca interaction which in turn is slightly deeper than the Mg-Mg interaction. The system must therefore maximize the number of Ca-Ca contacts, followed by the number of Mg-Ca contacts. Since the equilibrium distance of the Ca-Ca is longer than the Mg-Ca and Mg-Mg one, a *core-shell* structure would require a very high Mg/Ca ratio. As can be seen from Figure 4(q), even a 29 : 9 ratio is not sufficiently high for such behaviour. In any other ratio, calcium forms the icosahedral backbone of the structure, with the magnesium atoms literally melting on that backbone, as can be seen, for example, in Figures 4(l) and 4(r). The resulting structures may probably be best described as *Janus* particles [8], possessing both magnesium and calcium character on the surface. Closely related is the *ball-and-cup* structure found, for example, in Figure 4(l).

The same design principles hold true when moving to ternary compositions. In the most simple case, when substituting single atoms, the binary cluster structure is slightly distorted but remains overall unchanged. This can be, for example, clearly seen in the transformation from the binary Be_{21} Mg_{21} (Figure 4(h)) to the ternary Be_{20} Mg_{20} Ca_2 (Figure 5(j)) cluster. Once the composition contains more atoms of the third species, the cluster structure is again subject to the principle rules that have been formulated earlier. Beryllium forms a small icosahedral core, with magnesium and calcium segregating around it. This behaviour is most pronounced in the Be_{13} Mg_{15} Ca_{10} (Figure 5(e)) and Be_6 Mg_{16} Ca_{16} (Figure 5(i)) cluster structures. In the earlier cluster, the beryllium core is large enough in comparison to the number of magnesium and calcium atoms to allow forming two half-shells around the core. In the latter, the core is small enough so that the calcium atoms form the shell and magnesium atoms remain at the surface. This ordering is due to the dissociation energy of the Be-Ca interaction being higher than the one of the Be-Mg interaction.

It is possible to conclude that alkaline earth clusters in the studied size regime seem to obey well-defined and rational building rules when using the Gupta model. A possible fault, and one which will be addressed in later work [27], is the problematic description of the beryllium atom. It is unclear how the deviations in the fit can influence the cluster structures. This, however, requires a more flexible functional form than the Gupta potential.

5. Conclusions

Gupta potentials for all bimetallic interactions involving beryllium, magnesium, and calcium are derived from highest-level *ab initio* data using global optimization techniques. All potentials reproduce the position and depths of the minimum correctly. The potentials have been subsequently used for the global optimization of medium-sized cluster structures, namely, up to ternary 42 atom clusters.

The structures obtained reveal several systematic trends. Clusters containing beryllium will form beryllium cores surrounded by a shell of the other atoms in play. Magnesium and calcium segregate, forming a calcium backbone with magnesium on the surface.

Acknowledgment

The authors gratefully acknowledge the financial support from the German Excellence Initiative through the Free Floater Research Group program of the University of Göttingen.

References

[1] R. Ferrando, J. Jellinek, and R. L. Johnston, "Nanoalloys: from theory to applications of alloy clusters and nanoparticles," *Chemical Reviews*, vol. 108, no. 3, pp. 845–910, 2008.

[2] B. F. G. Johnson, "From clusters to nanoparticles and catalysis," *Coordination Chemistry Reviews*, vol. 190–192, pp. 1269–1285, 1999.

[3] C. L. Bracey, P. R. Ellis, and G. J. Hutchings, "Application of copper-gold alloys in catalysis: current status and future perspectives," *Chemical Society Reviews*, vol. 38, no. 8, pp. 2231–2243, 2009.

[4] L. T. Wille and J. Vennik, "Computational complexity of the ground-state determination of atomic clusters," *Journal of Physics A*, vol. 18, no. 8, pp. L419–L422, 1985.

[5] B. Hartke, "Global optimization," *WIREs Computational Molecular Science*, vol. 1, no. 6, pp. 879–887, 2011.

[6] R. L. Johnston, "Evolving better nanoparticles: genetic algorithms for optimising cluster geometries," *Dalton Transactions*, no. 22, pp. 4193–4207, 2003.

[7] L. O. Paz-Borbón, T. V. Mortimer-Jones, R. L. Johnston, A. Posada-Amarillas, G. Barcaro, and A. Fortunelli, "Structures and energetics of 98 atom Pd-Pt nanoalloys: potential stability of the Leary tetrahedron for bimetallic nanoparticles," *Physical Chemistry Chemical Physics*, vol. 9, no. 38, pp. 5202–5208, 2007.

[8] L. O. Paz-Borbón, R. L. Johnston, G. Barcaro, and A. Fortunelli, "Structural motifs, mixing, and segregation effects in 38-atom binary clusters," *Journal of Chemical Physics*, vol. 128, no. 13, Article ID 134517, 2008.

[9] R. P. Gupta, "Lattice relaxation at a metal surface," *Physical Review B*, vol. 23, no. 12, pp. 6265–6270, 1981.

[10] J. M. Dieterich and B. Hartke, "OGOLEM: global cluster structure optimisation for arbitrary mixtures of flexible molecules. A multiscaling, object-oriented approach," *Molecular Physics*, vol. 108, no. 3-4, pp. 279–291, 2010.

[11] J. M. Dieterich and B. Hartke, "Composition-induced structural transitions in mixed Lennard-Jones clusters: global reparametrization and optimization," *Journal of Computational Chemistry*, vol. 32, no. 7, pp. 1377–1385, 2011.

[12] N. Carstensen, J. M. Dieterich, and B. Hartke, "Design of optimally switchable molecules by genetic algorithms," *Physical Chemistry Chemical Physics*, vol. 13, no. 7, pp. 2903–2910, 2011.
[13] D. E. Goldberg, *Genetic Algorithms in Search, Optimization and Machine Learning*, Kluwer Academic Publishers, 1989.
[14] D. E. Woon and T. H. Dunning, "Gaussian basis sets for use in correlated molecular calculations. V. Core-valence basis sets for boron through neon," *Journal of Chemical Physics*, vol. 103, no. 11, pp. 4572–4585, 1995.
[15] B. P. Prascher, D. E. Woon, K. A. Peterson, T. H. Dunning, and A. K. Wilson, "Gaussian basis sets for use in correlated molecular calculations. VII. Valence, core-valence, and scalar relativistic basis sets for Li, Be, Na, and Mg," *Theoretical Chemistry Accounts*, vol. 128, no. 1, pp. 69–82, 2011.
[16] J. Koput and K. A. Peterson, "Ab initio potential energy surface and vibrational-rotational energy levels of X," *Journal of Physical Chemistry A*, vol. 106, no. 41, pp. 9595–9599, 2002.
[17] D. Feller, "Application of systematic sequences of wave functions to the water dimer," *Journal of Chemical Physics*, vol. 96, no. 8, pp. 6104–6114, 1992.
[18] H.-J. Werner, P. J. Knowles, R. Lindh et al., Molpro, version 2010.1, a package of ab initio programs, 2010, http://www.molpro.net.
[19] Mrcc, a string-based quantum chemical program suite written by M. Kállay, see also M. Kállay, P. R. Surján, Journal of Chemical Physics , vol.115, pp. 2945, 2001, http://www.mrcc.hu.
[20] M. Kállay and J. Gauss, "Approximate treatment of higher excitations in coupled-cluster theory," *Journal of Chemical Physics*, vol. 123, no. 21, Article ID 214105, 2005.
[21] J. M. Merritt, V. E. Bondybey, and M. C. Heaven, "Beryllium dimer-caught in the act of bonding," *Science*, vol. 324, no. 5934, pp. 1548–1551, 2009.
[22] P. Li, J. Ren, N. Niu, and K. T. Tang, "Corresponding states principle for the alkaline earth dimers and the van der waals potential of Ba2," *Journal of Physical Chemistry A*, vol. 115, no. 25, pp. 6927–6935, 2011.
[23] Jmol: an open-source java viewer for chemical structures in 3d, http://www.jmol.org/.
[24] Pov-ray—the persistence of vision raytracer, http://www.povray.org/.
[25] E. Blaisten-Barojas and S. N. Khanna, "Development of a first-principles many-body potential for beryllium," *Physical Review Letters*, vol. 61, no. 13, pp. 1477–1480, 1988.
[26] The cambridge cluster database, http://www-wales.ch.cam.ac.uk/CCD.html.
[27] J. M. Dieterich, S. Fischmann, and R. A. Mata, In preparation.

19

Uniformly Immobilizing Gold Nanorods on a Glass Substrate

Hadas Weinrib, Amihai Meiri, Hamootal Duadi, and Dror Fixler

Faculty of Engineering and The Institute of Nanotechnology and Advanced Materials, Bar-Ilan University, Ramat Gan 52900, Israel

Correspondence should be addressed to Dror Fixler, dror.fixler@biu.ac.il

Academic Editor: Rachela Popovtzer

The goal of this paper is to immobilize gold nanoparticles uniformly on a glass substrate. In order to attach gold-nanorods (GNRs) to an area of a few squared microns surface of glass substrate without preliminary coating of the GNR, 3-(Mercaptopropyl)trimethoxysilane molecules were used as linker while using different methods. These methods included placing the glass slide and the GNR (1) inside a tube without any motion; (2) inside a shaker; (3) in a fan setup. The fan setup included a tube containing the GNR solution and the glass slide at a vertical position, when the fan blows above the tube, producing turbulations in the liquid. Each method was evaluated according to the density and the homogeneousness of the GNR monolayer on the surface. The uniformity of the monolayer was demonstrated using AFM images of different areas on the slides, and the effectiveness of the protocol was demonstrated by calculating the average density of the GNR on the surface using image processing and analysis software. It was found that while both the shaker and the fan setups improved the monolayer density, the fan setup improved the density by a factor of more than two than the density found using the shaker.

1. Introduction

Nanoparticles play a significant role in an increasing number of researches and variety of applications. Recently, gold nanoparticles (GNPs) have gained popularity, and they serve as promising agents due to their favorable optical properties, such as an enhanced absorption cross-section [1] and scattering properties [2], biocompatibility [3], and well-developed bioconjugation protocols [4]. The increment in the number of applications using GNP has led to an ongoing demand for developing new techniques for immobilization of GNP to the substrate surface. Immobilization of GNP to substrates is required due to their applicability in various fields, among them: optical sensing using effects like surface plasmon resonance (SPR) and surface-enhanced Raman scattering (SERS), using nanoparticles as immobilizers for biomolecules, electron transfer enhancement, and labeling of biomolecules [5, 6].

Immobilization of the particles on the surface can be achieved by using various methods, which may be divided into two groups: physical attachment [7, 8] which is mainly used to obtain a multilayer of particles or a thin film, and chemical attachment [9, 10], which demands plating the substrate with linker molecules and is used to obtain a high-quality monolayer or structure of organized layers. While some of the chemical attachment methods take advantage of the ionic nature of the gold as well as its affinity toward thiol, others require the use of linker molecules [5].

Most of the common methods for chemical attachment between GNP and a substrate use self-assembled monolayers that contain organic groups, especially amine and thiol, while the use of charged polymer as a linker medium serves as an alternative [11].

The principle of using linker molecules is based on the ability of these molecules to self-arrange to what is called a self-assembled monolayer. The linker molecules are in fact bifunctional molecules where one end binds to the substrate, and the other is ready to bind to gold nanoparticles.

Previous studies [12–15] have shown that gold colloids can be self-assembled from solution onto a functionalized glass surface. The self-assembled monolayer is stabilized by attractive electrostatic interactions. Aminopropyltrimethoxy silane (APTMS or APS) has been commonly used to obtain amine-functionalised nonmetal substrate surface for

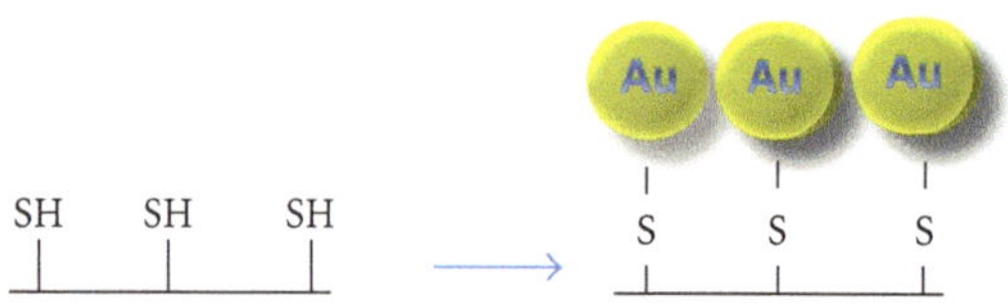

FIGURE 1: Chemical attachment between S–H and gold (Au): on the left the original thiol moleculse and on the right the gold particles chemicaly bounded to the thiol molecules.

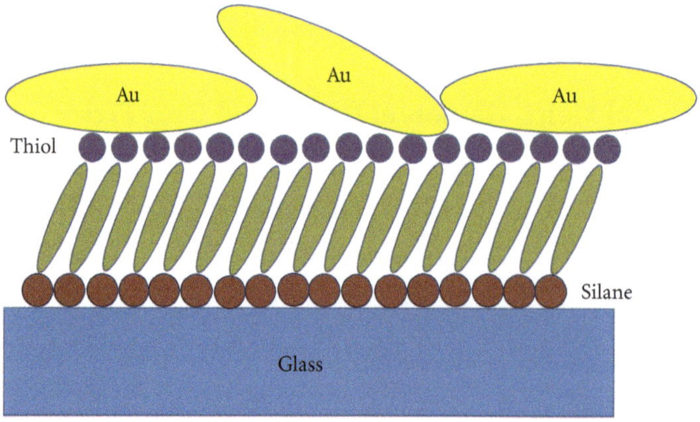

FIGURE 2: Scheme of the attachment between the GNR and the glass substrate by silane head and thiol tail.

the attachment of gold colloidal nanoparticles [10, 12, 14, 15]. This utilization has been used for a variance of purposes, mainly for application using SPR [12, 14, 15] and SERS effects [10, 13].

Usually, utilization of this method would be successful with gold nanospheres (GNSs) rather than gold nanorods (GNRs) because GNSs are usually produced in a negatively charged citrate medium, while GNR are usually produced using a positively charged CTAB medium [16].

The positive charge of the top amine group of APTMS is a more suitable instrument in attracting the negatively charged GNS. Although one could stripe the solvent and achieve natural GNP [17], other options like negatively charged coupling agents are preferable due to the difficulty in the total removal of all the remnants of the medium. Among these agents, we can state negatively charged polymer [11] or thiol groups.

Using thiol molecules to link GNP to substrates has been extensively studied not only because of their unique physical and chemical properties but also because of their easy preparation and good performance.

S–H head groups are used on noble metal substrates due to the strong affinity of sulfur in these metals (Figure 1).

The sulfur gold interaction is semicovalent and has a strength of approximately 45 kcal/mol. In addition, gold is an inert and biocompatible material that can withstand harsh chemical cleaning treatments. These traits make thiol molecules attractive candidates for a wide range of applications [18–20]. Among these applications are technologies for the biosensor industry [21] and building of nanostructures for fabrication of nanodevices [9].

In most cases, thiol molecules are used as linkers for metal-metal binding, by using dithiol [22–25], or as a GNP coating in order to prevent self-aggregation [24]. However, thiol molecules could be useful for binding metal to other substrates such as glass. Glass-metal binding is accomplished by using molecules with thiol tail groups and head groups, which are substrate-specific. In case of a glass substrate, an optional functional group is silane [26].

Upon working with GNR, a common problem of aggregation arise as GNR tend to self-aggregate in the solution or on the surface causing a disorder in the layer on the surface. The prevalent solution is an additional step of preliminary coating the GNR by dithiols.

To our knowledge, there are no simple routine methods for coating uniformly a relatively wide area with high-quality, densely packed monolayer of GNR without preliminary coating the GNR. In this work, we used thiosilanes as bifunctional molecules in order to attach the GNR to a glass substrate when silane head groups are attached to the glass substrate and the GNR are chemically bound to the thiol tail groups as shown in Figure 2.

In order to simplify the process and avoid preliminary coating of the GNR, we used a shaker or a central processing unit (CPU) fan to prevent self aggregation in the solution and on the substrate and to increase the probability of a single GNR to chemically bound to linker molecules on the glass, what improves the quality of the monolayer.

2. Materials and Methods

GNRs were synthesized using the seed-mediated growth method [21]. Their size, shape, and uniformity were characterized using transmission electron microscopy (TEM) and the resultant shape was 25 nm × 65 nm, with a narrow size distribution (10%) [27] (Figure 3).

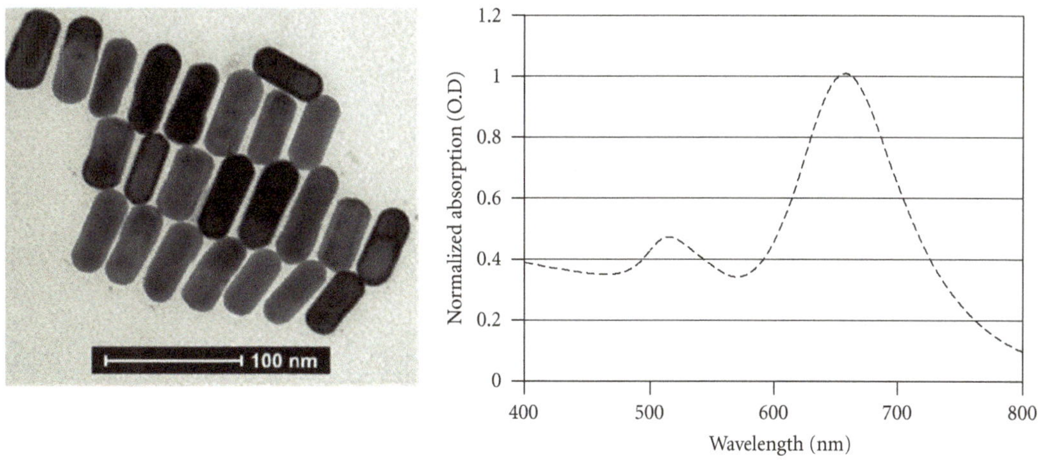

(a) (b)

FIGURE 3: TEM image of gold nanorods and corresponding absorption spectrum.

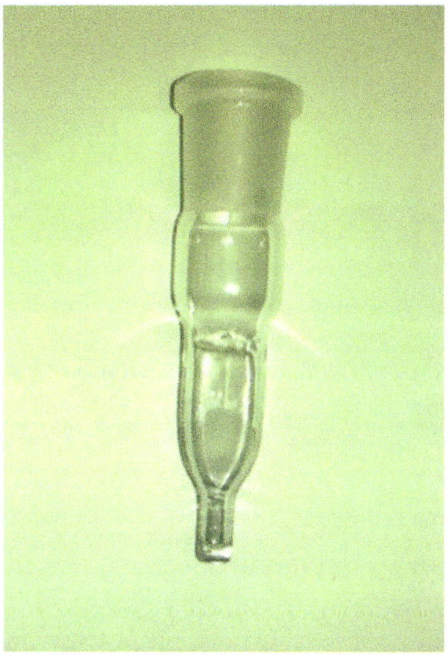

FIGURE 4: Digital image of the custom-made tube.

FIGURE 5: The setup of the system including a CPU fan of top of the tube contains the glass slide inside the GNR solution.

FIGURE 6: Scheme of dispersion of the GNR in the fan setup.

3-(Mercaptopropyl)trimethoxysilane was purchased from Sigma-Aldrich (St. Louis, MO, USA).

Cover glass slides (2.2 cm × 1.2 cm) with a diameter of 0.13–0.17 mm were cleaned in a piranha solution (3 : 1 (v/v) H_2SO_4/H_2O_2) that causes vigorous oxidation for 90 min.

A special glass tube was custom produced by us (Figure 4). The tube was composed of a flat part that enables vertical positioning of the slide and a cylindrical part for possible future use of a magnetic stirrer. Vertical positioning of the glass slide is important for obtaining homogeneous binding of the GNR to the slide.

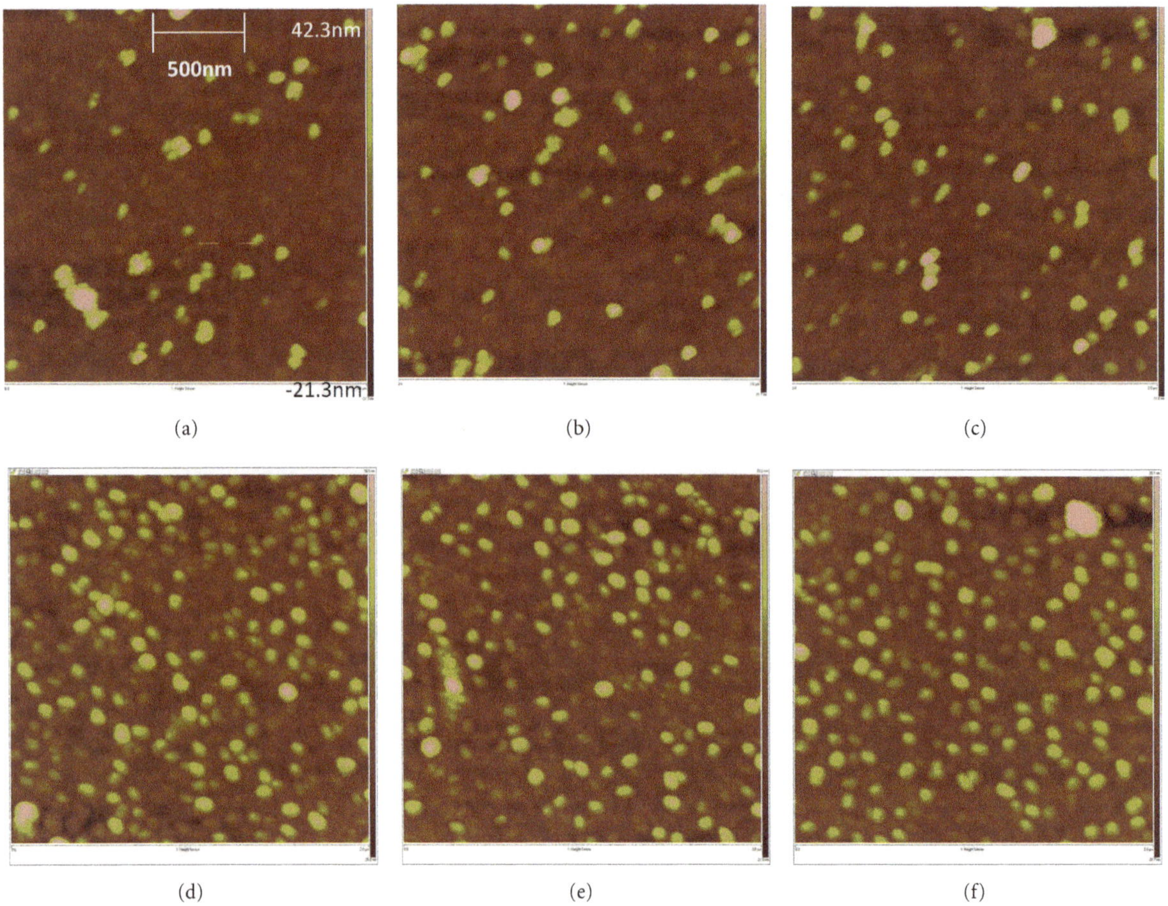

Figure 7: AFM images of the GNR on the glass slides. (a)–(c) are images of different areas of a slide that was coated using a shaker; (d)–(f) are images of different areas of a slide that was coated using a CPU fan.

After cleaning, the piranha was removed, and the slides were rinsed repeatedly a few times with water. Subsequently, the slides were rinsed repeatedly with ethanol and then incubated in thiosilane solution—an MPTMS 1% ethanol solution for 2.5 hours. The slides were double-rinsed with ethanol and sonicated for 10 min after each rinse. Prior to further modification, the slides were dried at 70°C for 5 minutes. AFM measurements and imaging were carried out using a ICON scanning probe microscope (Bruker AXS, Santa Barbara, CA, USA). All images were obtained using the tapping mode with a single TESP silicon probe (force constant of 20–80 N/m, Bruker, Camarillo, CA, USA). The resonance frequency of this cantilever was approximately 307–375 kHz. The scan angle was maintained at 0 degrees, and the images were captured in the retrace direction with a scan rate of 1.5 Hz (resp. for the scan size was 3000 × 3000 nm). The aspect ratio was 1 : 1, and image resolution was 1024 samples/line. Analysis of the image was done using the NanoScope software.

The data was processed using ImageJ software. ImageJ (image processing and analysis) is a public domain, Java-based image processing program developed at the National Institutes of Health [28].

3. Results and Discussion

Our first goal was to create a monolayer of the GNR.

When rinsing the glass vertically in the GNR solution without any intervention produced no attachment to the glass was seen. In order to improve the results, different manipulations were tried. First, the slides were induced in the GNR colloid for 2 hours in a shaker in order to have maximal contact of the GNR with the surface and try to avoid self-aggregation. Afterwards, the slides were rinsed in ethanol, sonicated for 5 min, and dried at 70°C for 5 minutes. The results can be seen in Figures 7(a)–7(c).

Placing the tube in a shaker improved the attachment of the GNR to the glass, but further improvement was required.

Next, we sought to improve the density, uniformity, and homogeneousness of the monolayer without adding steps of preliminary process such us prior coating of the GNR.

In order to improve the quality of the monolayer, the shaker was replaced by a CPU fan. The fan was placed on top of the tube that contained the slides in the GNR solution. The intensity of the blow was controlled by the power supplier as shown in Figure 5.

Both methods, shaker and fan, despite their difference, are based on the assumption that the motion of the GNR solution increases the probability of a single GNR to chemically bound to linker molecules on the glass. A scheme of movement of the GNR in the fan setup is shown in Figure 6.

In order to compare the density of the monolayer obtained by using the different methods, AFM images were taken. To confirm the uniformity of the monolayer, images from different areas of the slide were taken for every sample.

While the use of a shaker produced a relatively uniform monolayer (see Figures 7(a)–7(c)), the introduction of the fan to the protocol resulted in a denser layers can be seen in Figures 7(d)–7(f).

In order to quantify the improvement in density between the methods we used a Java image processing program called ImageJ. A particle analysis facility was used on the images, which occupied an area of $2\,\mu m \times 2\,\mu m$ on the glass slide. We converted the images into 8-bit images and adjusted the lower threshold to 90, while the upper threshold was kept at 255. This choice of threshold ensured the exclusion of the surface and the binding layer from the count.

The improvement in the quality of the monolayer is clearly demonstrated where the use of the fan increased the particle density significantly to 180.16 ± 39 particles per frame ($2\,\mu m \times 2\,\mu m$) in comparison to 80.2 ± 32 particles per frame for the shaker.

4. Conclusions

The results reported in this work indicate that using thiosilane molecules combined with utilization of a fan in the setup applies a uniform high-quality monolayer of GNR on a few micron squared area of glass substrate, which is a relatively wide field.

In order to improve the density and the homogeneousness of GNR monolayer and yet avoid preliminary coating of the GNR, two methods were tested. In the first method, the slides were rinsed in the GNR solution and deposited in a shaker. In the second method, the slides were placed in the setup blown with a fan from above. Both methods improved the attachment of the GNR to the glass, but the latter produced significantly better results by increasing the quality of the monolayer. The AFM images clearly demonstrate that the density of the monolayer using the fan is higher than using a shaker. In addition, the monolayer that was obtained was denser than the one that was achieved by Niidome et al. [11] using a negatively charged polymer. Comparison between AFM images of random different areas of the slide indicates the uniformity of the monolayer rather than one image of smaller area.

As was demonstrated, using the fan setup improved the quality of the GNR monolayer on the glass. This suggested that the method enables simplifying the process of attaching GNR to glass substrate. Such a process may serve as a first step towards the development of a novel super resolution method based on GNR attached to an observed object.

References

[1] M. A. El-Sayed, "Some interesting properties of metals confined in time and nanometer space of different shapes," *Accounts of Chemical Research*, vol. 34, no. 4, pp. 257–264, 2001.

[2] P. K. Jain, K. S. Lee, I. H. El-Sayed, and M. A. El-Sayed, "Calculated absorption and scattering properties of gold nanoparticles of different size, shape, and composition: applications in biological imaging and biomedicine," *Journal of Physical Chemistry B*, vol. 110, no. 14, pp. 7238–7248, 2006.

[3] E. E. Connor, J. Mwamuka, A. Gole, C. J. Murphy, and M. D. Wyatt, "Gold nanoparticles are taken up by human cells but do not cause acute cytotoxicity," *Small*, vol. 1, no. 3, pp. 325–327, 2005.

[4] S. Kumar, J. Aaron, and K. Sokolov, "Directional conjugation of antibodies to nanoparticles for synthesis of multiplexed optical contrast agents with both delivery and targeting moieties," *Nature Protocols*, vol. 3, no. 2, pp. 314–320, 2008.

[5] X. Luo, A. Morrin, A. J. Killard, and M. R. Smyth, "Application of nanoparticles in electrochemical sensors and biosensors," *Electroanalysis*, vol. 18, no. 4, pp. 319–326, 2006.

[6] R. Ankri, A. Meiri, S. I. Lau, M. Motiei, R. Popovtzer, and D. Fixler, "Surface plasmonresonance coupling and diffusion reflection measurements for real-time cancer detection," *Journal of Biophotonics*. In press.

[7] X. Xu, M. Stevens, and M. B. Cortie, "In situ precipitation of gold nanoparticles onto glass for potential architectural applications," *Chemistry of Materials*, vol. 16, no. 11, pp. 2259–2266, 2004.

[8] L. Wang, W. Mao, D. Ni, J. Di, Y. Wu, and Y. Tu, "Direct electrodeposition of gold nanoparticles onto indium/tin oxide film coated glass and its application for electrochemical biosensor," *Electrochemistry Communications*, vol. 10, no. 4, pp. 673–676, 2008.

[9] H. X. He, H. Zhang, Q. G. Li, T. Zhu, S. F. Y. Li, and Z. F. Liu, "Fabrication of designed architectures of Au nanoparticles on solid substrate with printed self-assembled monolayers as templates," *Langmuir*, vol. 16, no. 8, pp. 3846–3851, 2000.

[10] O. Seitz, M. M. Chehimi, E. Cabet-Deliry et al., "Preparation and characterisation of gold nanoparticle assemblies on silanised glass plates," *Colloids and Surfaces A*, vol. 218, no. 1–3, pp. 225–239, 2003.

[11] Y. Niidome, H. Takahashi, S. Urakawa, K. Nishioka, and S. Yamada, "Immobilization of gold nanorods on the glass substrate by the electrostatic interactions for localized plasmon sensing," *Chemistry Letters*, vol. 33, no. 4, pp. 454–455, 2004.

[12] N. Nath and A. Chilkoti, "Label-free biosensing by surface plasmon resonance of nanoparticles on glass: optimization of nanoparticle size," *Analytical Chemistry*, vol. 76, no. 18, pp. 5370–5378, 2004.

[13] E. J. Bjerneld, F. Svedberg, and M. Käll, "Laser induced growth and deposition of noble-metal nanoparticles for surface-enhanced Raman scattering," *Nano Letters*, vol. 3, no. 5, pp. 593–596, 2003.

[14] N. Nath and A. Chilkoti, "A colorimetric gold nanoparticle sensor to interrogate biomolecular interactions in real time on a surface," *Analytical Chemistry*, vol. 74, no. 3, pp. 504–509, 2002.

[15] T. Okamoto, I. Yamaguchi, and T. Kobayashi, "Local plasmon sensor with gold colloid monolayers deposited upon glass substrates," *Optics Letters*, vol. 25, no. 6, pp. 372–374, 2000.

[16] J. Pérez-Juste, I. Pastoriza-Santos, L. M. Liz-Marzán, and P. Mulvaney, "Gold nanorods: synthesis, characterization and applications," *Coordination Chemistry Reviews*, vol. 249, no. 17-18, pp. 1870–1901, 2005.

[17] X. Xu, T. H. Gibbons, and M. B. Cortie, "Spectrally-selective gold nanorod coatings for window glass," *Gold Bulletin*, vol. 39, no. 4, pp. 156–165, 2006.

[18] G. Schmid, S. Peschel, and T. Sawitowski, "Two-dimensional arrangements of gold clusters and gold colloids on various surfaces," *Zeitschrift fur Anorganische und Allgemeine Chemie*, vol. 623, no. 5, pp. 719–723, 1997.

[19] T. Ohgi, H. Y. Sheng, and H. Nejoh, "Au particle deposition onto self-assembled monolayers of thiol and dithiol molecules," *Applied Surface Science*, vol. 130-132, pp. 919–924, 1998.

[20] A. Doron, E. Joselevich, A. Schlittner, and I. Willner, "AFM characterization of the structure of Au-colloid monolayers and their chemical etching," *Thin Solid Films*, vol. 340, no. 1, pp. 183–188, 1999.

[21] B. Nikoobakht and M. A. El-Sayed, "Preparation and growth mechanism of gold nanorods (NRs) using seed-mediated growth method," *Chemistry of Materials*, vol. 15, no. 10, pp. 1957–1962, 2003.

[22] E. Hutter, S. Cha, J. F. Liu et al., "Role of substrate metal in gold nanoparticle enhanced surface plasmon resonance imaging," *Journal of Physical Chemistry B*, vol. 105, no. 1, pp. 8–12, 2000.

[23] M. D. Musick, C. D. Keating, L. A. Lyon et al., "Metal films prepared by stepwise assembly. 2. Construction and characterization of colloidal Au and Ag multilayers," *Chemistry of Materials*, vol. 12, no. 10, pp. 2869–2881, 2000.

[24] C. N. R. Rao, G. U. Kulkarni, P. J. Thomas, and P. P. Edwards, "Metal nanoparticles and their assemblies," *Chemical Society Reviews*, vol. 29, no. 1, pp. 27–35, 2000.

[25] Z. M. Qi, I. Honma, M. Ichihara, and H. Zhou, "Layer-by-layer fabrication and characterization of gold-nanoparticle/ myoglobin nanocomposite films," *Advanced Functional Materials*, vol. 16, no. 3, pp. 377–386, 2006.

[26] Y. Wang, L. Q. Chen, Y. F. Li, X. J. Zhao, L. Peng, and C. Z. Huang, "A one-pot strategy for biomimetic synthesis and self-assembly of gold nanoparticles," *Nanotechnology*, vol. 21, no. 30, Article ID 305601, 2010.

[27] R. Ankri, V. Peretz, M. Motiei, R. Popovtzer, and D. Fixler, "A new method for cancer detection based on diffusion reflection measurements of targeted gold nanorods," *International Journal of Nanomedicine*, vol. 7, pp. 449–455, 2012.

[28] T. J. Collins, "ImageJ for microscopy," *BioTechniques*, vol. 43, no. 1, pp. 25–30, 2007.

Permissions

The contributors of this book come from diverse backgrounds, making this book a truly international effort. This book will bring forth new frontiers with its revolutionizing research information and detailed analysis of the nascent developments around the world.

We would like to thank all the contributing authors for lending their expertise to make the book truly unique. They have played a crucial role in the development of this book. Without their invaluable contributions this book wouldn't have been possible. They have made vital efforts to compile up to date information on the varied aspects of this subject to make this book a valuable addition to the collection of many professionals and students.

This book was conceptualized with the vision of imparting up-to-date information and advanced data in this field. To ensure the same, a matchless editorial board was set up. Every individual on the board went through rigorous rounds of assessment to prove their worth. After which they invested a large part of their time researching and compiling the most relevant data for our readers. Conferences and sessions were held from time to time between the editorial board and the contributing authors to present the data in the most comprehensible form. The editorial team has worked tirelessly to provide valuable and valid information to help people across the globe.

Every chapter published in this book has been scrutinized by our experts. Their significance has been extensively debated. The topics covered herein carry significant findings which will fuel the growth of the discipline. They may even be implemented as practical applications or may be referred to as a beginning point for another development. Chapters in this book were first published by Hindawi Publishing Corporation; hereby published with permission under the Creative Commons Attribution License or equivalent.

The editorial board has been involved in producing this book since its inception. They have spent rigorous hours researching and exploring the diverse topics which have resulted in the successful publishing of this book. They have passed on their knowledge of decades through this book. To expedite this challenging task, the publisher supported the team at every step. A small team of assistant editors was also appointed to further simplify the editing procedure and attain best results for the readers.

Our editorial team has been hand-picked from every corner of the world. Their multi-ethnicity adds dynamic inputs to the discussions which result in innovative outcomes. These outcomes are then further discussed with the researchers and contributors who give their valuable feedback and opinion regarding the same. The feedback is then collaborated with the researches and they are edited in a comprehensive manner to aid the understanding of the subject.

Apart from the editorial board, the designing team has also invested a significant amount of their time in understanding the subject and creating the most relevant covers. They scrutinized every image to scout for the most suitable representation of the subject and create an appropriate cover for the book.

The publishing team has been involved in this book since its early stages. They were actively engaged in every process, be it collecting the data, connecting with the contributors or procuring relevant information. The team has been an ardent support to the editorial, designing and production team. Their endless efforts to recruit the best for this project, has resulted in the accomplishment of this book. They are a veteran in the field of academics and their pool of knowledge is as vast as their experience in printing. Their expertise and guidance has proved useful at every step. Their uncompromising quality standards have made this book an exceptional effort. Their encouragement from time to time has been an inspiration for everyone.

The publisher and the editorial board hope that this book will prove to be a valuable piece of knowledge for researchers, students, practitioners and scholars across the globe.

List of Contributors

Yao Xu, Ramachandran Gnanasekaran and David M. Leitner
Department of Chemistry and Chemical Physics Program, University of Nevada, Reno, NV 89557, USA

Henryk T. Flakus and Anna Jarczyk-Jędryka
Institute of Chemistry, University of Silesia, 9 Szkolna Street, 40-006 Katowice, Poland

Brenda Dana
Department of Electrical Engineering, Faculty of Engineering, Tel-Aviv University, 69978 Tel-Aviv, Israel

Israel Gannot
Department of Biomedical Engineering, Faculty of Engineering, Tel-Aviv University, 69978 Tel-Aviv, Israel

E. A. McCoy and G. S. McDonald
Joule Physics Laboratory, School of Computing, Science and Engineering, Materials & Physics Research Centre, University of Salford, Greater Manchester M5 4WT, UK

J. Sánchez-Curto and P. Chamorro-Posada
Departamento de Teor´ıa de la Se˜nal y Comunicaciones e Ingenier´ıa Telem´atica, Universidad de Valladolid, ETSI Telecomunicaci´on, Campus Miguel Delibes Paseo Bel´en 15, E-47011 Valladolid, Spain

J.M. Christian
Joule Physics Laboratory, School of Computing, Science and Engineering, Materials & Physics Research Centre, University of Salford, Greater Manchester M5 4WT, UK
Departamento de Teor´ıa de la Se˜nal y Comunicaciones e Ingenier´ıa Telem´atica, Universidad de Valladolid, ETSI Telecomunicaci´on, Campus Miguel Delibes Paseo Bel´en 15, E-47011 Valladolid, Spain

Mark T. Oakley and Roy L. Johnston
School of Chemistry, University of Birmingham, Edgbaston, Birmingham B15 2TT, UK

David J. Wales
University Chemical Laboratories, Lensfield Road, Cambridge CB2 1EW, UK

L. Sobczyk, B. Czarnik-Matusewicz, M. Rospenk and M. Obrzud
Faculty of Chemistry, University of Wrocław, Joliot-Curie 14, 50-383 Wrocław, Poland

Alexandr Gorski, Sylwester Gawinkowski, Roman Luboradzki, Marek Tkacz and Jacek Waluk
Institute of Physical Chemistry, Polish Academy of Sciences, Kasprzaka 44/52, 01-224 Warsaw, Poland

Randolph P. Thummel
Department of Chemistry, University of Houston, Houston, TX 77204-5003, USA

N. T. Zinner
Department of Physics, Harvard University, Cambridge, MA 02138, USA
Department of Physics and Astronomy, University of Aarhus, 8000 Aarhus, Denmark

Betül Karaçoban and Leyla Özdemir
Department of Physics, Sakarya University, 54187 Sakarya, Turkey

Ossama Kullie
Laboratoire de Chimie Quantique, Institute de Chimie de Strasbourg, CNRS et Universit´e de Strasbourg, 4 rue Blaise Pascal, 67070 Strasbourg, France

A. I. Jaman and Rangana Bhattacharya
Experimental Condensed Matter Physics Division, Saha Institute of Nuclear Physics, Sector 1, Block AF, Bidhannagar, Kolkata 700 064, India

Vital Peretz and Chaim N. Sukenik
The Department of Chemistry, The Institute of Nanotechnology and Advanced Materials, Bar-Ilan University, Ramat Gan 52900, Israel

Menachem Motiei and Rachela Popovtzer
Faculty of Engineering, The Institute of Nanotechnology and Advanced Materials, Bar-Ilan University, Ramat Gan 52900, Israel

Ivan P. Christov
Physics Department, Sofia University, 1164 Sofia, Bulgaria

E. M. Barkhudarov, I. A. Kossyi, S. M. Temchin and M. I. Taktakishvili
A.M. Prokhorov General Physics Institute of RAS (GPI RAS), Vavilov Street 38, Moscow 119991, Russia

Yu. N. Kozlov
Semenov Institute of Chemical Physics of RAS, Kosygin Street 4, Moscow 119991, Russia

Nick Christofi
Edinburgh University, Edinburgh EH9 3JF, UK

Mark E. Brezinski
Center for Optical Coherence Tomography and Modern Physics, Department of Orthopedic Surgery, Brigham and Women's Hospital, 75 Francis Street, MRB-114, Boston, MA 02115, USA
Center for Optical Coherence Tomography and Modern Physics, Department of Orthopedic Surgery, Harvard Medical School, 25 Shattuck Street, Boston, MA 02115, USA
Department of Electrical Engineering and Computer Science, Massachusetts Institute of Technology, Room 36-360, 77 Massachusetts Avenue, Cambridge, MA 02139, USA

Chao Chen
School of Physics, Beijing Institute of Technology, Beijing 100081, China

Vladimir Ryabov
Department of Complex System, School of Systems Information Science, Future University Hakodate, 116-2 Kamedanakano-Cho, Hakodate-Shi, Hakodate, Hokkaido 041-8655, Japan

Dmitry Nerukh
Non-Linearity and Complexity Research Group, Aston University, Birmingham B4 7ET, UK

Johannes M. Dieterich, Sebastian Gerke and Ricardo A. Mata
Institut f¨ur Physikalische Chemie, Universit¨at G¨ottingen, Tammannstrasse 6, 37077 G¨ottingen, Germany

Hadas Weinrib, Amihai Meiri, Hamootal Duadi and Dror Fixler
Faculty of Engineering and The Institute of Nanotechnology and Advanced Materials, Bar-Ilan University, Ramat Gan 52900, Israel